Jürgen-Hinrich Fuhrhop, Guangtao Li

Organic Synthesis

Further Titles of Interest

C. Bittner et al.

Organic Synthesis Workbook II

2001, XI, 291 pp
ISBN 3-527-30415-0

H.-G. Schmalz, T. Wirth (Eds.)

Organic Synthesis Highlights V

2003, approx. 320 pp
ISBN 3-527-30611-0

M. M. Green, H. Wittcoff

Organic Chemistry Principles and Industrial Practice

2003, approx. 280 pp
ISBN 3-527-30289-1

H. F. Ebel, C. Bliefert, W. E. Russey

The Art of Scientific Writing

2003, approx. 500 pp
ISBN 3-527-29829-0

Jürgen-Hinrich Fuhrhop, Guangtao Li

Organic Synthesis

Concepts and Methods

WILEY-VCH

WILEY-VCH GmbH & Co. KGaA

Authors

Prof. Dr. J.-H. Fuhrhop
Department of Chemistry/Organic Chemistry
Free University of Berlin
Takustraße 3
14195 Berlin
Germany

Dr. G. Li
Department of Chemistry/Organic Chemistry
Free University of Berlin
Takustraße 3
14195 Berlin
Germany

Cover
The cover shows the model of a catenane mono-layer in form of a molecular wire on polysilicon. Titanium is then vapor condensed through a sha-dow mask as a second layer of wires on top and aligned perpendicular with respect to the polysili-con-catenane wire. A change of potential will mo-ve the catenane rings against each other. The cur-rents which flow through the wires will be chan-ged. A molecular-based device on the one-nano-meter scale is thus realized (see page 408 ff. The original drawing was kindly provided by Profes-sor J. F. Stoddart). This picture gives a valid exam-ple of the modern "do it all at once" approach.

Library of Congress Card No.: applied for

British Library Cataloguing-in-Publication Data
A catalogue record for this book is available from the British Library.

Bibliographic information published by Die Deutsche Bibliothek
Die Deutsche Bibliothek lists this publication in the Deutsche Nationalbibliografie; detailed bibliographic data is available in the Internet at <http://dnb.ddb.de>

© 2003 WILEY-VCH Verlag GmbH & Co. KGaA, Weinheim

Printed in the Federal Republic of Germany
Printed on acid-free paper

Typesetting K+V Fotosatz GmbH, Beerfelden
Printing Strauss Offsetdruck GmbH, Mörlenbach
Bookbinding Litges & Dopf Buchbinderei GmbH, Heppenheim

(Softcover) ISBN 3-527-30273-5
(Hardcover) ISBN 3-527-30272-7

Dedicated to the Free University of Berlin in gratitude for the
friendly and efficient support given to us over many years

Foreword

The subject of synthetic organic chemistry has undergone a vigorous and rich evolution for over a century. Nonetheless, it continues to advance in power and scope and to develop in new directions at a remarkable pace. The enormous expansion of the information base of organic synthesis, especially over the past fifty years, is both a source of satisfaction and a challenge to those in the field. One challenge is to find new and better ways of organizing, presenting, and teaching this ever more complex and fascinating body of knowledge. New and excellent text books (and reviews) on organic synthesis must be a critical aspect of the transmission of the major facts and principles of synthesis to successive generations of chemists. Fuhrhop and Li's *"Organic Synthesis"* renders a valuable service in this regard. It discusses a broad range of synthetic topics and presents concisely much important and interesting modern chemistry. Students of synthetic chemistry from intermediate to advanced levels will gain a better command of organic synthesis from this text.

E. J. Corey

Contents

Preface to the Third Edition

This book was written for the advanced chemistry student and for the research chemist. Its purpose is to convey knowledge about concepts, methods and target molecules of modern organic synthesis. At the beginning of each chapter important concepts are summarized including:

- systematic evaluation of the arrangement of functionality in carbon chains, carbocycles and heterocycles
- methods for achieving regio- and stereoselectivity in carbon–carbon bond formation and the conversion of functional groups
- special reagents, intramolecular reactions and reaction conditions, which enforce the formation of thermodynamically unfavored products in high yield
- solid-state syntheses of biopolymers and multicomponent mixtures
- the "do it all at one approach", in which synthesis, supramolecular architecture and functionality are optimized together.

The synthetic methods described were selected under the aspects of applicability, simplicity, didactic value and future aspects. Applicability implies that the methods have been used repeatedly in complex syntheses. Simplicity means that the method is not too time-consuming and does not inevitably rely on costly equipment. Didactic value means that we wanted to demonstrate principles of selectivity control. Future aspects allowed us to include a few syntheses of compounds which are needed for most promising purposes such as nanometer-sized computer elements, charge separating porphyrin assemblies, and the self-optimizing combinatorial reaction systems, which, in the end, might, however, not be successful. The book is framed at the beginning and the end by a short introduction into Corey's synthon approach and a few recent synthetic sketches to be meant as first exercises in retrosynthetic analysis.

We should like to acknowledge the help and advice we received from many students and colleagues at the Free University of Berlin. The VCH's Editorial Department has given us vigorous support throughout the time of writing and correcting. Above all we thank our typist Mrs. Regina Stück, who gave her best to transfer difficult-to-read manuscripts and reference lists to several intermediate and a final typescript with endless patience and great skill.

December 2002

Jürgen-Hinrich Fuhrhop
Guangtao Li

List of Abbreviations

η^3	ligand with a coordination number 3
Acm	acetamidomethyl
Adoc	adamantyloxycarbonyl
6-APA	6-aminopenicillinic acid
7-ACA	7-amino-cephalosporanic acid
9-BBN	9-borabicyclo-3.3.1-nonane
a^2	acceptor-synthon active at C2
Ac	acetyl
ACS	American Chemical Society
AD	asymmetric dihydroxylation
AIBN	a,a'-azoisobutyronitrile
Am	amyl, pentyl
ANA	3-amino-nocardicinic acid
Ar	aryl
BINAP	2,2'-dihydroxy-1,1'-binaphthyl
Boc	*tert*-butoxycarbonyl
BSP	1-benzenesulfinic piperidine
Btea	benzyltriethylammonium
Bu	butyl
Bz	benzoyl
Bn	benzyl
C_{60}	fullerene
CAE	capillary array electrophoresis
CAS	Chemical Abstract Service
CboCl	carbobenzoxy chloride = benzyl chloroformate
Cbz	benzyloxycarbonyl
CC	column chromatography
COD	cyclooctadiene
COT	cyclooctatriene
cp	cyclopentadienyl
Δ	heat
d^2	donor–synthon active at C2
DABCO	1,4-diazabicyclo[2.2.2]-octane
DAST	diethylamino sulfur trifluoride

dba	bisbenzylidene acetone
DBN	diazabicyclo[4.3.0]-non-5-ene
DBU	1,8-diazabicyclo[5.4.0]-undec-7-ene
DCC	N,N'-dicyclohexylcarbodiimide
DCE	1,2-dichloroethane
DCL	dynamic combinatorial library
DCU	dicyclohexylurea
DDQ	2,3-dichloro-5,6-dicyano benzoquinone
DET	diethyl (+)- or (−)-tartrates
DHP	dihydropyrane
DHQ	dihydroquinine
DHQD	dihydroquinidine
DIBAL	diisobutylaluminum hydride
diphos	$Ph_2PCH_2CH_2PPh_2$
DIPT	diisopropyl (+)- or (−)-tartrates
DISIAB	diisoamylborane
DMAP	4-(dimethylamino)pyridine
DME	dimethoxyethane, glyme
DMF	N,N-dimethylformamide
DMPES	isopropyldimethylsilyl
DMPO	5,5-dimethyl-1-pyrroline-N-oxide
DMSO	dimethylsulfoxide
DMT	(+)- or (−)-dimethyltartrate
DMTr	dimethoxytrityl
Dnp	dinitrophenyl
DNPH	dinitrophenylhydrazine
DODAC	dimethyldioctadecylammonium chloride
DOPA	3,4-dioxyphenylalanine
Ec	ethylcarbamoyl
EDA	ethylenediamine
EDCI	1-ethyl-3-[3-(diethylamino)propyl] carbodiimide
EDTA	ethylenediamine tetra-N-acetic acid
e.e.	enantiomeric excess
EG	ethyleneglycol
en	ethylenediamine
Et	ethyl
EWG	electron withdrawing group
FGI	functional group inversion
FMOC	9-Fluorenylmethoxycarbonyl
GC	gas chromatography
GMF	5-(glucosylmethyl)furan
Hal	halide
HPLC	high pressure liquid chromatography
HBT	1-hydroxy-1*H*-benzotriazole
HMPTA	hexamethylphosphoric triamide
IIDQ	1-isobutyloxycarbonyl-2-isobutyloxy-1,2-dihydroquinoline

Ipc$_2$BCl	chlorodiisopinocamphenylboran
Im	imidazole
Ipc	isopinocampheyl
i-Pr	isopropyl
KAPA	potassium 3(aminopropyl)-amide
l	large
L	ligand
LAH	lithium aluminum hydride
LDA	lithium diisopropylamide ·
LNA	locked DNA
L-Selectride	lithium tri-*sec*-butylhydroborate
m	medium
M	metal
MCPBA	*m*-chloroperbenzoic acid
Me	methyl
MEM	2-methoxyethoxy-methyl
MsT	1-mesitylenesulfonyl-1*H*-1,2,4-triazole
MTBE	*tert*-butyl methylether
NBS	N-bromosuccinimide
p-NBSA	*p*-nitrobenzenesulfonylazide
NIS	N-iodosuccinimide
NMO	methylmorpholine N-oxide
NMP	N-methyl-2-pyrrolidinone
Non	nonyl
Nu	nucleophile
Oct	octyl
Oac	acetate
Ox	oxazolinones
pbp	diphenylbutadiene
PCR	polymerase chain reaction
dppf	diphenylphosphinoferrocene
PEG	polyethyleneglycol
PFG	pulse field gradient
PG	prostaglandin, letters correspond to double bond arrangements
Ph	phenyl
PHAL	phthalazine, connected to DHQ (Sharpless osmylation)
Phth	phthaloy
pin	pinanyl
PMB	*p*-methoxybenzoic acid
PMHS	polymethylhydrogen siloxane
PMRI	partially modified retro-inverso peptides
PNA	peptide nucleic acids
p-NBSA	nitrobenzenesulfonylazide
PPY	4-(pyrrolidine-1-yl)pyridine
Pr	propyl

PPA	polyphosphoric acid
prot.	Protected
PS	polystyrene
PxCl	9-chloro-9-phenyl-9*H*-xanthene = pixyl chloride
PYR	pyrimidine, connected to DHQ (Sharpless osmylation)
r	radical reaction site
RCM	ring closing metathesis
ROM	ring opening metathesis
ROMP	ring opening metathetic polymerization
s	small
SAD	Sharpless asymmetric dihydroxylation
SCAL	safety catch amide linker
SFS	sodium formaldehyde sulfoxylate
stien	stilbene diamine
TBDMS, TBS	*tert*-butyl-dimethylsilyl ether
Tbeoc	2,2,2-tribromoethoxycarbonyl
TCA	trichloroacetonitrile
Tceoc	2,2,2-trichloroethoxycarbonyl
TEMPO	1-oxy-2,2,6,6-tetramethylpiperidine
TESCl	triethylchlorosilane
TET	triethyl silyl ethers
TFA	trifluoroacetic acid
Tfac	trifluoroacetyl
Tfmes	trifluormethanesulfonyl
TG	Tentagel®
Th	thienyl = thiophenolaten
Thex	1,1,2-trimethylpropyl
THF	tetrahydrofuran
Thp	tetrahydropyranyl ethers
TLC	thin layer chromatography
TMAD	N',N'',N'''-tetramethylazodicarboxamide
TMEDA	tetramethylethylene diamine
TMOF	trimethylorthoformate
TMS	trimethyl silyl
TMSDEA	N,N-diethyltrimethylsilylamine
Tos	tosylate = *p*-toluene sulfonate
TosMIC	tosylmethyl isocyanide
TPP	*meso*-tetraphenylporphyrin
TpsCl	triisopropylbenzenesulfonyl chloride
Tpte	4-tritylphenylthio-ethyl
Trit	trityl, triphenylmethyl
TTN	thallium(III)trinitrate
und	undecyl
X	halides or pseudo-halides (Tos, N_3)

1
Synthesis of Carbon Chains

1.1
Introduction

The task of organic synthesis is to form carbon–carbon bonds by application of known or conceivable chemical operations. Two molecular or ionic units ("synthons") are thus connected (Corey, 1967 A; Seebach, 1979). Synthesis is accompanied by functional group inversions (FGI), which may make molecules larger (condensation), change heteroatoms (substitution, addition–elimination), or raise or lower the oxidation number of carbon atoms (redox). If you are "cooking", the usual routine in a chemical lab, this is how you will spend most of your time. Your major carbon skeleton will be provided by companies like Aldrich, Sigma or Merck and you will in all probability just combine them by condensation reactions, exchange an OH group in place of NH_2, or introduce isotopes as molecular labels. Nevertheless, it may also happen that you have to put a small carbon chain on 1,4-androstadiene-3,17-dione or that you have to connect two large molecules by a C–C bond. The first chapter tries to teach both the classic reactions by which carbanions and electropositive carbonyl carbons can be connected and also the spontaneous electron flow between two alkenes catalyzed by palladium as it is practiced in the modern "*syn-thesis*" (Gr. together-putting) of large and complicated molecules. A few recent practical procedures will be outlined in order to characterize the surprising simplicity of these methods. In particular, it will be shown that non-polar reactions can often be combined in a one-pot series of reactions, whereas polar reactions require workup and isolation after each step. This is caused by the often extreme conditions of the polar reactions (anhydrous, inert gas, highly acidic or basic), as compared to the much less demanding conditions of their non-polar counterparts.

The most common synthetic reactions are polar: a negatively polarized ("electronegative") carbon atom (electron **d**onor, **d**) of one synthon is combined with a positively polarized ("electropositive") carbon atom (electron **a**cceptor, **a**) of another synthon. Donor synthons are formal carbanions which donate an electron pair to an acceptor synthon, usually an electroneutral carbonyl group. It is the preparation of the donor synthon, which usually dictates the choice of reaction conditions such as reagent, solvent, anhydrous, inert gas, acidic, neutral, or basic conditions, and temperature. The donor also determines the pathways, which reactions take as well as the yield and the structure of the final products. Polar reactions are fast and most suitable for adding small molecules with molecular weights up to about 200 to any substrate. If the educts are both very large, reaction times become long, and unwanted side-reactions (elimination, redox, etc.) often cannot be controlled. Polar addition reactions often produce one or two chiral centers (*) and stereoselectivity is highly desirable, to avoid tedious separation of diastereomers and enantiomers.

Since the mid-1980s, synthetic procedures have been developed, which apply non-polar synthons, in particular the π-electrons of alkenes or aromatic compounds, as donors and organopalladium(II) halides, RPdX, as acceptors. Pd-HX adducts are released as electroneutral leaving groups and, after removal of HX by a base, the Pd(0) complex is formed again and may activate a CX-bond in a catalytic cycle (Beletskaya, 2000; Dedieu, 2000). These reactions are slow, but side-reactions are less important than in polar addition reactions. They yield achiral 1,3-dienes and are most useful to combine large molecules under easy reaction conditions.

16 electrons on Pd

In addition to the palladium-catalyzed reactions there are several other non-polar synthetic reactions. They often occur under mild laboratory conditions and imply a carbon radical and an alkene or alkyne. Radical reactions, however, are mostly restricted to intramolecular reactions. Within a single molecule, a defined radical is formed at one center only, which then attacks neighboring alkene or alkyne groups. Finally the unpaired electron must be removed within the same molecule by an intramolecular chain termination reaction. The unsaturated groups accept a radical at one end and create a new radical at the other end, which then reacts with another double bond or forms a carbocycle. The notation of radical synthons uses closed circles to represent radical precursors and open circles for radical acceptors (Ryu, 1996).

• = donor atom
○ = acceptor atom

A fourth important type of reactions is preferred for the synthesis of carbo- and heterocycles. It involves the covalent linking of two carbon–carbon double bonds by inter- or intramolecular rearrangements (electrocyclic reactions, metathesis, Claisen rearrangement). Six-π-electron systems are particularly useful for one-step syntheses of six-membered rings (Diels–Alder) or the replacement of C–O by C–C bonds (Claisen). Terminal vinyl groups can be combined to form macrocycles in presence of ruthenium complexes (metathesis). Such rearrangements are often triggered by light or heat and acid–base catalysis can often be avoided.

Diels–Alder reaction

EWG = electron withdrawing group

Claisen rearrangement

The cyclization reactions will be covered in the carbocycle and heterocycle Chapters 2 and 4. In this chapter we concentrate on the polar reactions and catalytic reactions characterized in the Schemes of page 2.

1.2
Functional Group Arrangements

Polar reactions usually involve carbonyl compounds as electron acceptors and a functional carbanion bound to a metal (Li, Mg, Sn, Cu-Li) or electropositive non-metal (B, Si, P). The reactive site of a polar synthon may either be on carbon

Tab. 1.1 Examples of types of synthon.

synthon type	example	reagent	functional group
d^0	$CH_3\overset{\ominus}{S}$	CH_3SH	$-\overset{\mid}{\underset{\mid}{C}}-S-$
d^1	$\overset{\ominus}{C}{\equiv}N$	$\overset{\oplus}{K}\,\overset{\ominus}{C}N$	$-C{\equiv}N$
d^2	$\overset{\ominus}{C}H_2-CHO$	CH_3CHO	$-CHO$
d^3	$\overset{\ominus}{C}{\equiv}C-\overset{\mid}{\underset{\mid}{C}}-NH_2$	$\overset{\oplus}{Li}\,\overset{\ominus}{C}{\equiv}C-\overset{\mid}{\underset{\mid}{C}}-NH_2$	$-\overset{\mid}{\underset{\mid}{C}}-NH_2$
alkyl d	$\overset{\ominus}{C}H_3$	$LiCH_3$	—
a^0	$\overset{\oplus}{P}(CH_3)_3$	$(H_3C)_2P-Cl$	$-P(CH_3)_2$
a^1	$HRC{=}CH\,Pdl$	$H_2C{=}\underset{I}{\overset{H}{C}}$	$-\underset{H}{\overset{}{C}}{=}CH_2$
a^1	$H_3C\overset{\overset{OH}{\mid}}{\underset{\oplus}{C}}CH_3$	$H_3C\overset{\overset{O}{\parallel}}{C}CH_3$	$\overset{\diagdown}{\underset{\diagup}{C}}{=}O$
a^2	$H_2\overset{\oplus}{C}\overset{\overset{O}{\parallel}}{C}CH_3$	$Br-H_2C\overset{\overset{O}{\parallel}}{C}CH_3$	$\overset{\diagdown}{\underset{\diagup}{C}}{=}O$
a^3	$H_2\overset{\oplus}{C}\overset{}{\underset{H}{C}}{=}\overset{\overset{O^{\ominus}}{\mid}}{C}OR$	$H_2C{=}\underset{H}{C}\overset{\overset{O}{\parallel}}{C}OR$	$\overset{\overset{O}{\parallel}}{C}OR$
alkyl a	$\overset{\oplus}{C}H_3$	$(CH_3)_3\overset{\oplus}{S}\,\overset{\ominus}{Br}$	—

atom C^1, which is part of the functional group (d^1 or a^1), or on remote carbon atoms C^n (d^n or an; $n{\geq}2$). Synthons are accordingly numbered with respect to the relative positions of a functional group (FG) and the reactive carbon atom. If the carbon atom C-1 of the functional group itself is reacting, one has a d^1- or a^1-synthon. If the carbon atom C-2 next to the functional group (the α-carbon atom) is the reaction center, we call it a d^2- or a^2-synthon. If the β-carbon atom C-3 is the reactive one, we assign d^3 or a^3 to the corresponding synthon, etc. Alkyl donor synthons without functional groups may also form covalent bonds with acceptor synthons. In such cases we speak of d^0- or alkylating synthons. Tab. 1.1 gives examples of typical synthons.

The following obvious rules apply to the arrangement of functionality in the product ("target molecule") of synthesis:

Reacting synthons	Products
alkyl a + alkyl d	non-functional
alkyl a + d^1, alkyl d + a^1	monofunctional
$a^1 + d^1$	1,2-difunctional
$a^1 + d^2$, $a^2 + d^1$	1,3-difunctional
$a^1 + d^3$, $a^2 + d^2$, $a^3 + d^1$	1,4-difunctional

Examples for arrangements of functionality in bisfunctional compounds are given in the scheme below.

If an open-chain organic molecule contains an electron acceptor *and* an electron donor site, the two reactive carbon atoms may be combined intramolecularly. This corresponds to the synthesis of a monocyclic compound. Intramolecular reactions of electron donor and acceptor sites in cyclic starting materials produce spirocyclic, fused, or bridged polycyclic compounds.

1.3
Umpolung

Reagents with carbonyl-type groupings exhibit a^1 or (if a,β-unsaturated) a^3 properties. In the presence of acidic or basic catalysts they may react as enol-type electron donors (d^2 or d^4 reagents). This reactivity pattern is considered as "normal". It allows, for example, syntheses of 1,3- and 1,5-difunctional systems via aldol-type ($a^1 + d^2$) or Michael-type ($a^3 + d^2$) additions.

If hetero atoms are introduced or exchanged, the "normal" reactivity of a carbon atom may be inverted (e.g. $a^1 \rightarrow d^1$, $d^2 \rightarrow a^2$), or a given reactivity may shift from one carbon atom to another (e.g. $a^3 \rightarrow a^4$). It is also possible to change reactivity by adding carbon fragments (e.g. CN^-, $RC\equiv C^-$, carbenoids) to functional groups. All these processes leading to changes of the synthon type have been called "*umpolung*" (German: dipole inversion; Wittig, 1951). In rare cases the polarity of hetero atoms may also be inverted, e.g. $R–S^{(d)}–H– \rightarrow R–S^{(a)}– S^{(a)}–R$.

In retro-synthetic analyses (see Chapter 9) it is often useful to consider an "umpolung" of a given reagent, especially if the target molecule contains 1,2- or 1,4-difunctional systems. Tab. 1.2 summarizes some typical umpolung reactions and some specific synthons with their equivalent reagents (Seebach, 1979).

1.4
Alcohols

Alcohols can be synthesized by the addition of carbanions (alkylating donor synthons) to carbonyl compounds (Still, 1976) or epoxides (a^1-synthons). Both reactions produce new chiral centers if the attached carbon atom carries two different substituents or is "prochiral". Alkylation of carbonyl groups often produces chiral alcohols. The stereoselectivity increases when either the carbonyl substituents or the reacting carbanions are bulky. The Felkin–Anh model explains these observations with the assumption of a reactant-like transition state in a staggered conformation (steric repulsion). Polar substituents such as chlorine are also as far away as possible from the entering carbanion (electronic repulsion). Repulsion effects of the substituents are found to be more important than interactions involving the carbonyl oxygen.

Tab. 1.2 Typical umpolung reactions.

umpolung type	chemical reactions
	exchange of hetero atoms or reduction
$a^1 \longrightarrow d^1$	$\underset{\text{Br}}{\overset{\text{H}}{a\,C}} \quad \xrightarrow[-\text{HBr}]{+\text{Ph}_3\text{P}} \quad d\,C{=}\overset{\oplus}{P}\text{Ph}_3$
$a^1 \longrightarrow d^1$	$C{-}X \quad \xrightarrow[-\text{MX}]{+\text{Mg or}+2\text{M}} \quad \overset{d}{C}{-}\text{MgX},\ C{-}\text{M}$
$a^1 \longrightarrow d^1$	$\underset{\text{H}}{\overset{\text{O}}{a}} \quad \xrightarrow[-\text{H}_2\text{O}]{+\text{SH}\ \ \text{SH}} \quad \xrightarrow{-\overset{\oplus}{\text{H}}} \quad \overset{d}{C} \overset{\text{S}}{\underset{\text{S}}{}}$
$a^1 \longrightarrow d^1$	$\underset{\text{X}}{\overset{\text{O}}{a}} \quad \xrightarrow[-x^{\ominus}]{+\text{Fe(CO)}_4^{\ominus\ominus}} \quad \overset{\text{O}}{\underset{\text{Fe(CO)}_4}{d}}$
$a^1 \longrightarrow a^3$	$\underset{R\ \ \ a\ \ \ R}{\overset{\text{O}}{}} \quad \xrightarrow{\text{MgBr}} \quad \xrightarrow{+\overset{\oplus}{\text{H}}} \quad \underset{R\ R}{\overset{\overset{\oplus}{\text{OH}}_2}{}} a$
$a^1 \longrightarrow d^3$	$\underset{R}{\overset{\text{H}\ \ \text{O}}{a}} \quad \xrightarrow[-\overset{\oplus}{\text{H}}]{+\text{RSH}\ \ +2[\text{O}]} \quad \underset{\text{O}\ \ \text{O}}{\overset{R\ S\ \ \ \ \ d\ \ \ \ \text{O}}{}}R$
$a^3 \longrightarrow a^4$	$\underset{R}{\overset{\text{O}}{a}} \quad \xrightarrow{[:\text{CH}_2]} \quad \underset{a}{\overset{\text{O}}{}}R$
	oxidation of alkenes
$d^{1,2} \longrightarrow a^{1,2}$	$d{=}d \quad \xrightarrow[-\text{RCOOH}]{+\text{RCO(OOH)}} \quad \underset{a\ \ a}{\overset{\text{O}}{}}$
$d^2 \longrightarrow a^2$	$\underset{d}{\overset{\text{H}\ \ \text{O}}{}} \quad \xrightarrow[-\text{HBr}]{+\text{Br}_2} \quad \underset{a}{\overset{\text{Br}\ \ \ \text{O}}{}}$
	addition of carbon fragments
$a^1 \longrightarrow d^1$	$\underset{\text{Ar}\ \ a\ \ \text{H}}{\overset{\text{O}}{}} \quad \xrightarrow{+\text{CN}^{\ominus}} \quad \xrightarrow{\text{OH}^{\ominus}} \quad \underset{\text{Ar}\ \ \ \text{CN}}{\overset{\text{OH}}{\overset{d}{}}}$
$a^1 \longrightarrow d^3$	$\underset{R\ \ a\ \ \text{H}}{\overset{\text{O}}{}} \quad \xrightarrow{+\ \ominus{\equiv}\text{H}} \quad \xrightarrow{-\overset{\oplus}{\text{H}}} \quad \underset{R}{\overset{\text{O}}{}} \overset{d}{\underset{\ominus}{\equiv}}$

S, M, L = small, medium, large substituents

The diastereoselectivity reached in Grignard-type additions to ketones ranges from low to moderate. It improves if sterically hindered Schiff bases are used instead of ketones and if an enantiomerically pure amino ether is added (Inoue, 1993). Chiral substituents on the imine nitrogen also help.

ee 90–98%

In alkyl, alkenyl, and alkynyl anions all of the nonbonding electrons are localized on carbon atoms. These anions are therefore more reactive as enolate donor synthons. The ease of carbanion formation increases in the same order as the s-character of the CH bond: $C–CH < C=CH < C\equiv CH$. sp^3-Alkyl anions are less stable when the carbon atom is highly substituted: tertiary < secondary < primary. Alkyl, aryl, and alkenyl carbanions are usually produced from the corresponding halides by metal–halogen exchange, i.e. by the reduction of a carbon–halogen bond. Halides at sp^3-carbon atoms are more reactive than at sp^2- or sp-carbon atoms. Only alkynes with a terminal CH-group can be directly deprotonated by strong bases.

The classical "hard" counterions for carbanions are lithium and magnesium (Schlosser, 1973). Synthesis of lithiated carbanions occurs either from bromides or iodides and lithium metal or with *n*-butyl lithium (BuLi) and an aryl bromide or a CH-active compound (see Section 1.8, enolates). Lithium organic compounds are more reactive than the magnesium analogues and should only be used under an inert atmosphere (Ar, N_2). BuLi is commercially available in the form of a suspension in high-boiling hydrocarbons and can therefore be applied in hydrocarbon solutions, where it occurs as a mixture of hexamers, tetramers, and ill-defined oligomers. These aggregates are split by complexation of the lithium ion with tetramethylethylene diamine (TMEDA). The butyl anion of the resulting monomer is then reactive enough to deprotonate benzene. Magnesium organic (=Grignard) compounds are made by reduction of carbon halides with magnesium turnings. The latter need to be activated by iodine, which forms basic magnesium iodides on the surface, binds traces of water, and removes protons. Magnesium organyls are soluble in ethers but not in hydrocarbons, and must be prepared freshly.

Grignard reagents RMgX are in equilibrium with R_2Mg and MgX_2. In THF monomers RMgX. $(THF)_2$ predominate, while in diethyl ether cyclic and linear oligomers of RMgX are the principal species. Addition of TMEDA produces only monomers in diethyl ether solutions.

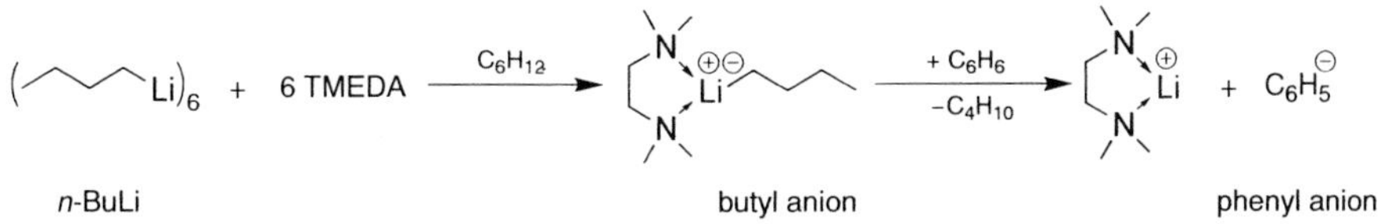

n-BuLi   butyl anion   phenyl anion

Upon addition to carbonyl compounds, Li- or Mg-alkoxides are formed, which are hydrolyzed in workup procedures with slightly acidic (NH_4Cl) aqueous solutions. The organic product is extracted with ether or chloroform and the lithium or magnesium salt separates into the acidic aqueous phase.

Acylations of alkenes and arenes may be carried out with acid chlorides and Friedel–Crafts type catalysis. The following scheme combines two Grignard and Friedel–Craft reactions as they were used in a total synthesis (Faul, 2001). The acylation is regioselective, because the neighboring methyl groups prevent attack of the other carbon atoms of the phenyl ring. The Grignard reagent was made from a phenyl bromide and magnesium turnings in the presence of some activating ethylene dibromide, which formed an inactive magnesium-dialkyl compound. The Grignard reagent reacted with a "Weinreb amide" (see Section 1.6) to give the ketone in almost quantitative yield (Nahm, 1981; de Luca, 2001). After its isolation, a second Grignard reaction led to a tertiary alcohol. The Grignard reaction is satisfactory only when at least one educt molecule is of low molecular weight. Two large educts would react exceedingly slowly, especially at the low temperatures that are necessary with the Weinreb amide. In the present case, the molecules are still mobile enough in solution to react satisfactorily.

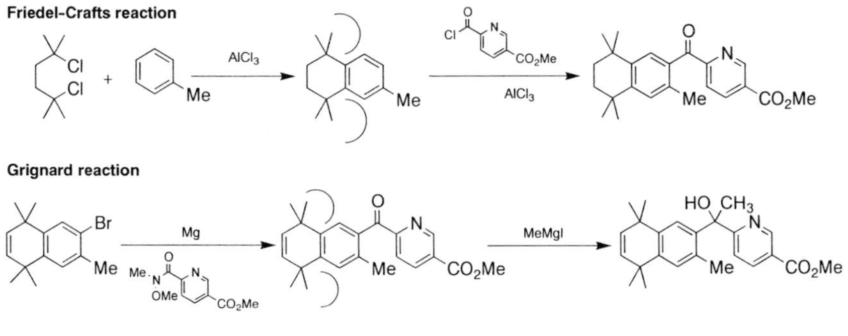

Friedel-Crafts reaction

Grignard reaction

Methyl 6-[3,5,5,8,8-Pentamethyl-2-5,8-dihydronaphthyl)carbonyl]pyridine-3-carboxylate. A suspension of magnesium turnings (0.12 g, 4.85 mmol) in dry THF (0.5 mL) was treated with 3 drops of 1,2-dibromoethane under N_2 followed by a solution of phenyl bromide (0.90 g, 3.22 mmol) in dry THF (4 mL), allowing the reaction to exotherm during the addition. The Grignard solution was stirred at room temperature for 1 h and transferred via cannula to a –78 °C solution of the Weinreb amide (0.36 g, 1.61 mmol) in dry THF (4 mL). The reaction mixture was stirred for 15 min at –78 °C, then warmed to room temperature and stirred for 1 h. The reaction was quenched with 1.0 N HCl (10 mL) and extracted with EtOAc. The organic layer was dried (MgSO₄), and the solvent was removed in vacuo to give a solid alcohol intermediate after chromatography (Faul, 2001).

The conversion of complex aldehydes and ketones to secondary and tertiary alcohols is preferably done with zinc organyls (Reformatsky reaction). These reagents are less reactive and more selective than the corresponding Grignard reagents, from which they are often obtained by reaction with anhydrous zinc chloride. A particularly efficient method for the alkylation of carbonyl compounds uses the alkyl bromide directly. In this process both reagents are poured over a column containing heated zinc granules (Ruppert, 1976). With cyclic ketones the Reformatsky reaction is often highly stereoselective (Mazur, 1960).

Attempts to synthesize transition metal alkyls under Grignard conditions fail. Methyl iron, nickel, titanium, or platinum compounds, for example, could not be isolated at room temperature. Metal–carbon bonds are usually weaker in the case of transition metals than the corresponding bonds in main group metal organyls, and unpaired electrons lead to decomposition in radical chain reactions. The major, often dominating, "side-reaction" is the spontaneous β-elimination of metal hydrides. Only σ-C ligands without β-hydrogen atoms yield long-lived transition metal–

carbon bonds. Nevertheless ionic carbon–copper bonds can be established by metal–metal interchange of copper salts with lithium or magnesium organyls *in situ*. Highly polarizable or "soft" carbanions for Michael reactions are obtained. Lithium organyls can thus be made softer and more nucleophilic by addition of copper(I) iodide. Soluble thermolabile lithiumdiorganocuprates, R_2CuLi, are formed. In alkylation reactions, however, only one of the two alkyl groups is transferred and a large excess of the unstable cuprates is often needed. If the copper iodide is replaced by copper(I) diphenyl phosphites, boiling THF solutions can be used (Martin, 1988). The cross coupling of Grignard reagents and alkyl bromides is particularly effective with Cu(I) reagents at low temperature. The yield of alkane homodimers is then negligibly small. Most important: the nucleophilicity of copper organyls makes reactions fast and large educts can be coupled. The simplest case of a copper organic reaction is the substitution of a halogen at a saturated carbon atom by an alkyl group. Organocopper reagents exhibit strong carbanionic capacity, and attack ester groups only slowly (Bergbreiter, 1975). Ketones, however, should be protected. The relative reactivity of substrates toward lithium diorganocuprates is as follows: acid chlorides >aldehydes > tosylates >epoxides > bromides > ketones > esters > nitriles (Posner, 1975). Dibromides of high molecular weight can be copper methylated in quantitative yield and then react with two equally large iodides to give long-chain compounds which are not accessible with hard organometals. Substitution of a vinylic halide was used in the synthesis of juvenile hormones (Corey, 1968 B). An epoxide behaves electronically like a carbonyl group. Substituted epoxides are attacked by organocopper reagents at the least hindered carbon atom and form alcohols (Johnson, 1974 A). With a,β-unsaturated epoxides *trans*-allylic alcohols are produced selectively by 1,4-addition (Carruthers, 1973; Posner, 1972).

* metal-halogen exchange at > or = 0 °C:

$$R_2\overset{\ominus}{Cu} + R'X \rightleftharpoons RR'\overset{\ominus}{Cu} + Rx \rightleftharpoons \cdots$$

Methylation and ring-opening of epoxides proceeds smoothly with Me$_2$CuCNLi$_2$ with high regioselectivity. The less hindered carbon is methylated and occurs as an *anti* attack of the oxirane oxygen (Dias, 2001). Vinylbromides are connected to epoxides by replacing the bromide first by "hard" lithium then by "very soft" cuprate(I) ions bound to thiophenolate (=thienyl=th) and Li-isocyanide. Reaction conditions are very mild and high yields are obtained with large, polyfunctional educts. Alkenes and hydroxyl groups are tolerated (Smith, 2001). Most enones are reduced to anion radicals by organocuprates. It is likely that this reaction is connected with the alkylation. Neither the formation of anion radicals or of conjugate adducts is observed when the redox potential of the enone becomes too negative. The enones behave here like benzoquinones.

To summarize: Alcohols are made from carbonyl compounds or epoxides and alkyl halides, using a Grignard reaction, often supported by copper, to combine them.

1.5

Synthesis of Alkenes and their Coupling Reactions

Alkenyl anions (West, 1961) constitute $d^{1,2}$-synthons, because the C=C group remains in the products. It may be subject to further synthetic operations. Reductive coupling of carbonyl compounds to yield olefins is achieved with titanium(0), which is freshly prepared by reduction of titanium(III) salts with LiAlH$_4$ or with potassium. The removal of two carbonyl oxygen atoms is driven by TiO$_2$ formation. Yields are often excellent even with sensitive or highly hindered olefins (McMurry, 1974, 1976 A, B).

(85%)

Alkynyl anions are more stable (p$K_a \approx 22$) than the more saturated alkyl or alkenyl anions (p$K_a \approx 40$–45). They may be obtained directly from terminal acetylenes by treatment with strong base, e.g. sodium amide (pK_a of NH$_3 \approx 35$) or BuLi. Magnesium acetylides are made by proton–metal exchange reactions with more reactive Grignard reagents, e.g. CH$_3$MgCl. Copper and mercury acetylides are formed directly from the corresponding metal acetates and acetylenes under neutral conditions (Coates, 1977; Houghton, 1979). Acetylide anions react as good nucleophiles with alkyl bromides (Ames, 1968) or carbonyl compounds. Terminal acetylenes react with copper(II) salts or with copper(I) salts in the presence of oxygen to form bisacetylides, which undergo alkyne coupling (Glaser coupling) in aqueous ammonium solutions or hot pyridine (Eglington, 1963). This reaction has been used for the synthesis of polyenes (Sondheimer, 1963). Unsymmetrical dialkynes can be made if one of the alkynes is submitted to an "umpolung" by bromination and is then allowed to react with a copper(I) acetylide (Cadiot–Chodkiewicz coupling; Eglington, 1963).

Glaser coupling

(90%)

(11%)

Cadiot-Chodkiewicz coupling

$$\text{Ph}-C\equiv C-H \;+\; Br-C\equiv C-COOH \xrightarrow[\text{(NMP), r.t.}]{\text{Cu}_2\text{Cl}_2/\text{C}_4\text{H}_8\text{NH}} \; \text{Ph}-C\equiv C-C\equiv C-COOH$$

(75%)

The most common educts for olefin synthesis are carbonyl compounds and carbanions, which are stabilized by electropositive phosphorus or silicon substituents. Sulfur also stabilizes carbanions, but is difficult to remove from the intermediate addition products.

$$R_3Si-CH_2Cl \xrightarrow[-LiCl]{+\,2\,Li} \left[R_3Si\cdots\overset{\ominus}{C}H_2\right]Li^{\oplus}$$

$$Ph_3P-CH_2-OMe \xrightarrow[-BuH]{+\,BuLi} \left[\begin{array}{c} \overset{\oplus}{Ph_3P}-\overset{\ominus}{C}\overset{H}{\underset{OMe}{<}} \end{array}\right] \longleftrightarrow \left[Ph_3P=C\overset{H}{\underset{OMe}{<}}\right]$$

"Ylide" "Ylene"

$$R-\underset{S}{\overset{S}{<}}\!\!\rangle \xrightarrow[-BuH]{+\,BuLi} \left[R-\underset{S}{\overset{S}{<}}\!\!\overset{\ominus}{\rangle}\right]Li^{\oplus} \qquad \text{1,3-Dithian-2-ide}$$

$$\underset{H_3C}{\overset{O}{\overset{\|}{\underset{CH_3}{S}}}} \xrightarrow[-H_2]{+\,NaH} \left[\underset{H_3C}{\overset{O}{\overset{\|}{\underset{CH_3}{\overset{\ominus}{S}}}}}\right]Na^{\oplus} \qquad \text{"Dimsyl" anion}$$

$$\underset{/\;\backslash}{\overset{O}{\overset{\|}{C}}_a} \;+\; \underset{/\;\backslash}{\overset{Y}{\underset{d}{\overset{\|}{C}}}}{}^{\ominus} \longrightarrow \left[\begin{array}{c} \overset{\ominus}{O}\;\longrightarrow\;Y \\ \overset{}{>}\!\!C-C\!\!<\end{array}\right] \longrightarrow \begin{array}{c}\overset{\ominus}{O}\cdots Y \\ \overset{}{\backslash}C=C\overset{}{/}\end{array}$$

$\cdots Y = \cdots SiR_3$ Peterson olefination

$\cdots Y = \cdots\overset{\oplus}{PR_3}, \;\cdots\overset{O}{\underset{OR}{\overset{\|}{P}-OR}}$ Wittig, Horner, Wadsworth–Emmons olefinations

$\cdots Y = \cdots SR, \;\cdots\overset{O}{\underset{R}{\overset{\|}{S}}}$ not useful, low tendency to $YO^{\ominus}$ elimination

$$\left[\begin{array}{ll} \cdots Y = \cdots\overset{\oplus}{SR_2}, \;\cdots\overset{\oplus}{\underset{R}{\overset{O}{S-R}}} & \text{oxirane formation:} \\[2mm] \cdots Y = \cdots Cl & \overset{\ominus}{O}\!\!>\!\!C-C\overset{Y}{<} \longrightarrow \;>\!\!C\overset{O}{\triangle}C\!\!< \;+\; Y^{\ominus} \end{array}\right]$$

Such carbanions add to the carbonyl group and the resulting oxy anion attacks atom Y intramolecularly. The oxide Y–O⁻ is then eliminated and a new C=C bond is formed. The driving force of the reaction is the formation of a thermodynamically favored YO bond together with an expanded coordination sphere. By far the most important reactions based on this scheme are the Wittig (Wittig, 1980; Maercker, 1965; Cadogan, 1979) and Horner–Wadsworth (Wadsworth, 1977) reactions. An alkylidene-triphenyl-phosphorane $(Y=PPh_3^+)$ or alkylidene phosphonate $(Y=PO(OR)_2)$ is generated in situ by the treatment of a phosphonium salt or a phosphonate with strong base and then reacts with a carbonyl compound to yield an olefin. Alkyl groups adjacent to the phosphonium center lose a proton when treated with base. A zwitterion ("ylide") is formed, which is stabilized by d-p π-bonding or "ylene" formation. The strength of base required depends on the substituents on the carbon atom that is to be deprotonated. The more these substituents are able to stabilize an adjacent negative charge, the more stable and the less reactive will the ylide be. Phosphorus ylides may, depending on the substituents, behave as non-functional (=alkyl) synthons or as d^1, d^2, ...d^n-synthons. Carbanions derived from phosphonic acids are also frequently used in synthesis (Walker, 1972; Cadogan, 1979). The C^--P^+ bond is so polar, that it adds spontaneously to the equally polar carbonyl group of aldehydes and ketones, essentially in the same manner as magnesium and lithium organyls. The carbanion adds to the carbon atom of the carbonyl group forming a C–C bond and the phosphorus to the carbonyl oxygen atom. A four-membered ring is formed. This ring splits open spontaneously and the phosphorus carries the oxygen away from the carbonyl group to release a C=C double bond. Triphenylphosphine oxide is split off and is then supposed to precipitate from organic solution to be filtered off. This last experimental detail often proves difficult. Phosphonate esters, on the other hand, degrade upon hydrolysis and then dissolve in water. The generally good yields, the mild reaction conditions, and the non-interference of ester and olefinic functions, as well as the high degree of control over double bond position and *trans*-stereochemistry, (Schlosser, 1970; Bestmann, 1979) are good reasons for the widespread popularity of this type of olefin synthesis. The most common synthons for alkene synthesis are carbonyl compounds and alkyl halides, and the general method of combining them is the Wittig reaction.

$$P(C_6H_5)_3 + CH_3Br \longrightarrow \left[(C_6H_5)_3\overset{\oplus}{P}\text{-}CH_3\right]\overset{\ominus}{Br} \xrightarrow{\text{NaH or PhLi}} \left[(C_6H_5)_3\overset{\oplus}{P}\text{-}\overset{\ominus}{CH_2}\right] \rightleftharpoons \left[(C_6H_5)_3P{=}CH_2\right]$$

The reactivity of the Wittig reagents, alkylidene-triphenylphosphoranes, is determined by the substituents of the ylide carbon atom. If these do not stabilize the carbanion, the phosphoranes will be markedly nucleophilic and unstable towards water, and will react with carbonyl groups at low temperatures. However, if the alkylidene groups bear electron withdrawing groups, the negative charge of the carbanion will be delocalized, the anion will be more stable to water, and the nucleophilicity of the carbon atom will decrease. The different behaviors of phosphoranes and phosphonates can be explained by the negative charge of the phosphonates in contrast to the electroneutral phosphoranes. If this charge is delocalized in an *a*-acylated phosphonate (a 5-center-6-electron system), the electron density and the nucleophilicity of the central carbon atom are high. In contrast phosphoranes with an electron withdrawing group at the carbanion are not reactive because the negative charge and nucleophilicity of the carbon atom are lowered (4-center-4-electron system).

reactive towards:			
R-CHO	+ +	+	−
RR'CO	+ +	−	−
H₂O, O₂	+ +	±	−

increasing nucleophilicity

Much work has been carried out in order to achieve control of either *cis*- or *trans*-alkene formation (Maercker, 1965; Bergelson, 1964; Schlosser, 1970; Bestmann, 1979). It has become increasingly clear that there is no single dominant Wittig transition state geometry and, therefore, no simple scheme to explain *cis/trans* selectivities. Ylides, e.g. $Ph_3P=CH-CO_2Et$, can be (*E*)- or (*Z*)-selective, depending on the solvent and substrate (Vedejs, 1988 A, B, 1990). The assumption of two limiting geometries for the cyclic transition state proved to be helpful in explaining experimental findings. The *cis*-selective state is thought to contain a puckered 1,2-oxaphosphetane ring and the *trans*-selective state a planar one. The puckered conformer presumably survives better at a low temperature and in solvents of low polarity.

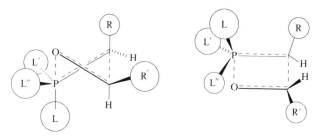

puckered: R,R´-cis   planar: R,R´-trans

Carotenoids, for example, are produced on the 1000 tons scale by Wittig-type re-actions. Actually most of the knowledge of electronic factors in Wittig synthesis originates from industrial experiments on carotene synthesis and was first stated in the patent literature. One of the driving forces of industrial investigations was the need to replace the expensive, air- and moisture-sensitive phenyllithium base by less esoteric bases (Pommer, 1960, 1977). This became feasible when the methylene hydrogen was acidified and the ylide was stabilized by either conju-gated polyenes or carbonyl groups. In these syntheses, mixtures of *cis* and *trans* olefination have been observed, but chemical or photochemical conversion to the natural and more stable *all-trans*-carotenoids was always possible (Reif, 1973). *In situ* generation of the carbonyl component by oxidative cleavage of the phosphor-ane leads to symmetrical olefins (Pommer, 1977). The synthesis of vitamin D3 from a sensitive dienone was another early success of phosphorus ylide synthesis (Inhoffen, 1958 A). This Wittig reaction could be carried out without any isomeri-zation of the diene. An excess of the ylide was needed, presumably because the alkoxides formed from the hydroxy group in the educt removed some of the ylide. Modern variations of the Wittig reaction still use BuLi, but allow a quick olefina-tion of tin organyls at low temperature. The tin organyls have been purified by in-verted-phase (SiO$_2$-C$_{18}$) chromatography and can be further used in palladium-cat-alyzed Stille reactions (see later in this section, Pazos, 2001). The *cis*-double bond was introduced here with commercial (*E*)-3-methylpent-2-en-4-yn-1-ol (Betzer, 1997).

"C$_{15}$ salt"   industrial process

(i) MeOH; –30 °C
(ii) +NaOMe
     15 min; 0–2 °C
(iii) -Ph$_3$PO
     counter-current
     extraction
(iv) [I$_2$], isomerizn.
     of 30% (11-cis)

vitamin A acetate   (98%)

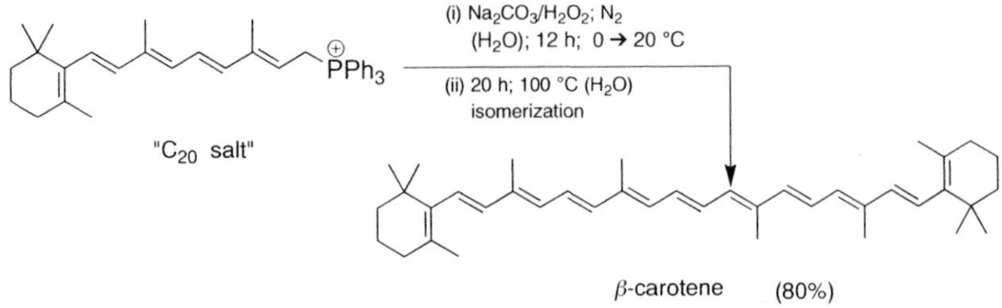

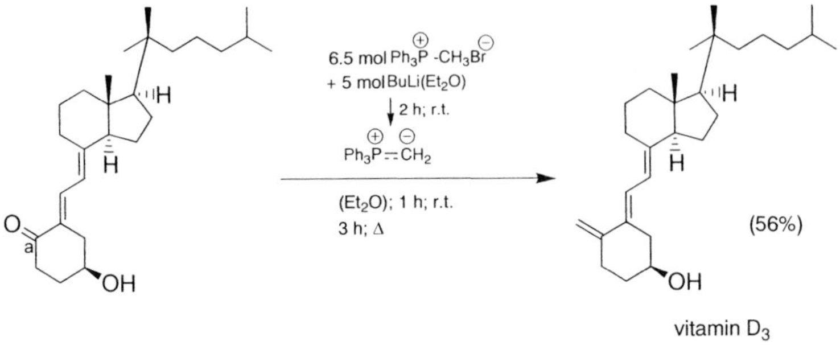

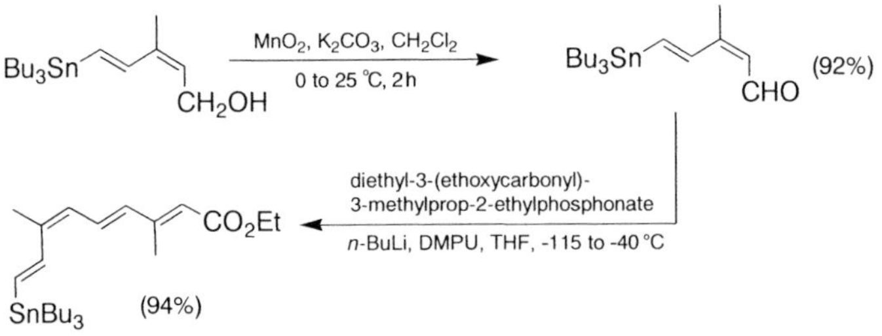

(i) Na$_2$CO$_3$/H$_2$O$_2$; N$_2$
(H$_2$O); 12 h; 0 → 20 °C

(ii) 20 h; 100 °C (H$_2$O)
isomerization

"C$_{20}$ salt"

β-carotene   (80%)

6.5 mol Ph$_3$P -CH$_3$Br
+ 5 mol BuLi(Et$_2$O)

2 h; r.t.

Ph$_3$P=CH$_2$

(Et$_2$O); 1 h; r.t.
3 h; Δ

(56%)

vitamin D$_3$

Bu$_3$Sn ~~~ CH$_2$OH

MnO$_2$, K$_2$CO$_3$, CH$_2$Cl$_2$
0 to 25 °C, 2h

Bu$_3$Sn ~~~ CHO   (92%)

diethyl-3-(ethoxycarbonyl)-
3-methylprop-2-enylphosphonate

n-BuLi, DMPU, THF, -115 to -40 °C

CO$_2$Et

SnBu$_3$ (94%)

Ethyl (2E,4E,6Z,8E)-3,7-Dimethyl-9-(tri-n-butylstannyl)-nona-2,4,6,8-tetraenoate. A cold (0 °C) solution of diethyl 3-(ethoxycarbonyl)-3-methylprop-2-enylphosphonate (1.80 g, 6.80 mmol) in THF (15.0 mL) was treated with DMPU (1.62 mL, 13.38 mmol) and n-BuLi (2.8 mL, 2.35 M in hexane, 6.57 mmol). After stirring at 0 °C for 20 min, the mixture was cooled to –115 °C and a solution of the aldehyde (1.74 g, 4.53 mmol) in THF (15.0 mL) was slowly added. The reaction mixture was stirred at –115 °C for 30 min, and then it was allowed to warm to –40 °C to complete the reaction. A saturated aqueous NH$_4$Cl solution was added, and the reaction mixture was extracted with Et$_2$O. The combined organic layers were washed with water and brine, dried (MgSO$_4$), and evaporated. Purification of the residue by column chromatography (SiO$_2$-C$_{18}$, 60:40 CH$_3$CN/MeOH) afforded 2.10 g of the tetraene (94%) as a yellow oil.

Electron withdrawing groups at the ylide carbon atom, on the other hand, give rise to *trans*-stereoselectivity. The *trans*-olefin is also stereoselectively produced when phosphonate diesters are used, because the elimination of a phosphate ester anion is slow (Wadsworth, 1977). The *trans* selectivity of Wittig and Horner reactions is reliable in synthesis and has frequently been exploited. Stereoselective syntheses of simple *cis*-olefins have been widely explored experimentally. They are best enforced by using low temperatures and alkyl substituents at the carbonyl and phosphane components (Bestmann, 1976). Strongly dissociated base systems like $KN(TMS)_2$/18-crown-6 and/or the use of electrophilic trifluoroethyl-phosphonoesters lead to Z-unsaturated esters (Still, 1983). A modern example of a fluororetinal precursor synthesis (Francesch, 1997) is given below. Again the olefination was fast at low temperature and the new double bond was E with respect to the vinylic fluorine and Z with respect to the carbon chain. *all-Z*-5,8,11,14,17-Eicosapentenoic (EPA) and *all-Z*-4,7,10,13,16,19-docosahexenoic acid (DHA) are the most valuable health components in human food. They have been synthesized by a series of Wittig reaction at –10 °C. Sodium-bis(trimethylsilyl)amide was used as base (Sandri, 1995).

EPA

Ethyl (2E)-4-[(tert-Butyldiphenylsilyl)oxy]-2-fluorobut-2-enoate. *n*-BuLi (10.6 mL, 1.6 M in hexane, 16.9 mmol) was added with a syringe to a cooled (–78 °C) solution of diethyl (fluorocarbethoxymethyl)phosphonate (4.11 g, 16.9 mmol) in THF (30 mL). After stirring for 30 min at –78 °C, a solution of the aldehyde (4.6 g, 15.4 mmol) in THF (15 mL) was added dropwise with a cannula. After stirring the resulting solution for 1 h at –78 °C and 3 h at 25 °C, a solution of 10% HCl was added until pH was neutral, and the resulting mixture was extracted with Et₂O (3×25 mL). The combined organic layers were washed with H₂O and brine, dried (MgSO₄), and concentrated. Purification by chromatography (silica, 90:10 hexane/ethyl acetate) afforded the *cis*-configured ester (6.20 g, 95%) as a yellowish oil (Francesch, 1997).

cis-Alkenes can also be introduced by the choice of appropriate *cis*-configured synthons. Commercial *cis*-1-propenyl bromide (Hayashi, 1999) or *cis*-alkenyl alcohols are obtained by Michael addition of dimethyl copper lithium to ethyl propiolate (Smith, 2001).

cis-enoate-. Under argon, a suspension of freshly purified CuI (6.77 g, 35.6 mmol) in THF (140 mL) at 0 °C was treated with methyllithium (1.5 M solution in ether, 47 mL, 71.1 mmol), dropwise over 10 min. The resultant opaque mixture was stirred at 0 °C for 10 min, then cooled to –78 °C, and a solution of ynoate (–) (4.05 g, 15.5 mmol) in THF (10 mL) was introduced via cannula, dropwise over 3 min. The resultant orange slurry was stirred at –78 °C for 70 min, after which pH 7 phosphate buffer (3 mL) was added slowly, and the entire mixture was poured into saturated aqueous NH₄Cl (150 mL). The biphasic mixture was then stirred at ambient temperature for 30 min. The layers were separated, and the aqueous phase was extracted with ethyl acetate (3× mL). The combined organic extracts were washed once with aqueous NH₄Cl, dried over MgSO₄, filtered, and concentrated *in vacuo*. Gradient flash chromatography (20% hexanes/ethyl acetate → 25% hexanes/ethyl acetate → 33% hexanes/ethyl acetate) provided the corresponding enoate (3.37 g, 78% yield) as an oil.

In addition to the Grignard and Wittig reactions discussed so far, there are many more synthetic procedures, which are based on the electrophilicity of the carbonyl carbon atom. We shall discuss a few of them in Sections 1.7 (aldol) and 1.9 (Michael) as well as in Chapter 2 (Carbocycles). Now, however, we shall turn to reactions between alkenyl or aryl compounds carrying electronegative substituents (boride, bromide, iodide) and alkenes with or without labile bonds to slightly electropositive substituents (Pd, Sn, B). Electron-rich alkenes and arenes or carbon monoxide, (–)CO(+), act as donor synthons here. The alkene and aryl synthons are less polar than aldehydes, ketones, carbanions or enolate anions and the reaction conditions are therefore undemanding. Moisture, oxygen, unprotected functional groups, etc. are often allowed.

The first alkenes synthons were obtained by the reaction with aluminum hydride and did not have these properties. The resulting binary aluminum organyls (AlR₃)₂ are even more reactive than the lithium and magnesium analogs. Their main advantage is their low cost, because they are important intermediates in industrial Ziegler–Natta-type polymerization processes. Aluminum alkyls with short carbon chains appear as nonviscous fluids which are spontaneously combustible in air, react explosively with water, and are difficult to handle. Long-chain aluminum alkyls are less aggressive, but have so far not found much application. Furthermore aluminum is the only main group element, which readily abstracts β-hydrogen atoms from carbon. It did not become popular in laboratory syntheses.

Nonmetallic Lewis acids, in particular electron-deficient boron, are less dangerous and therefore more useful as activating agents for the C=C double bond. Organoboranes are obtained by hydroboration of alkenes using dialkyl borohydride dimers or diborane, B_2H_6. The reaction of diborane with unhindered alkenes is essentially instantaneous in ether solvents, which act as a catalyst, and produce some solvated monomers. Alkenes are substituted in an anti-Markovnikov manner (H to carbon with less H, boron to the less substituted carbon atom) and hydride and boron are added on the same side (*syn*-addition). Hydroborations are thus regio- as well as stereoselective. Trialkylboranes are monomeric and are stabilized by hyperconjugation effects in the B–C bonds. They constitute quite stable compounds and are easy to handle on a large scale. Caution is necessary with volatile BR_3-compounds as well as with B_2H_6, because they are spontaneously combustible. Trialkylboranes are electron-deficient Lewis acids with no carbanion character (A^0). They become donor synthons if the boron atom in a trialkyl borane is combined with an electron donating nucleophile nu to give a negatively charged boride. The partial negative charge on the carbon atom of carbon monoxide makes it sufficiently nucleophilic to load its electrons onto the boron atom. Hydroxide, enolate, or acetylide anions do the same. This negative charge now gives anionic character to the alkyl substituents and mobilizes them. Rearrangements become favorable, which lead to a neutralization of the boron atom. New C–C bonds are formed in high yield (Brown, 1975). The C–B bond may be cleaved either by hydrolysis with alkoxides and water to give a C–H bond and water-soluble boric acid or by oxidation with hydrogen peroxide to yield an alcohol and boric acid. As usual the organic compound is extracted with a solvent and the boric acid derivatives dissolve in water. The scheme below gives an impression of the variability of organoboron chemistry. Its character as Lewis acid and carbanion promotor changes, depending on reaction conditions. This is exemplified in the Scheme, as are the many functional group inversions, which can be effected on the C–B bond.

cis-Olefins or cis,trans-dienes can be obtained from alkynes in similar reaction sequences. The alkyne is first hydroborated and then treated with alkaline iodine. If the other substituents on boron are alkyl groups, a *cis*-olefin is formed (Zweifel, 1967). If they are *cis*-alkenyls, a *cis,trans*-diene results. The reactions are thought to be iodine-assisted migrations of the *cis*-alkenyl group followed by *trans*-deiodoboronation (Zweifel, 1968). *trans,trans*-Dienes are made from haloalkynes and alkynes. These compounds are added one after the other to thexylborane. The alkenyl(l-haloalkenyl)thexylboranes are converted with sodium methoxide into *trans,trans*-dienes (Negishi, 1973). The thexyl group does not migrate.

(cis)-alkenes:

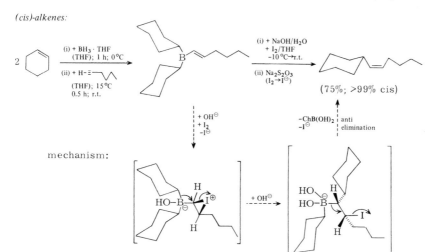

(cis)-alkenes: scheme showing

2 cyclohexene

(i) + BH₃·THF
(THF); 1 h; 0°C

(ii) + H–≡–
(THF); 15°C
0.5 h; r.t.

B

(i) + NaOH/H₂O
+ I₂/THF
–10°C→r.t.

(ii) Na₂S₂O₃
(I₂→I⁻)

(75%; >99% cis)

mechanism:

+ OH⁻
+ I₂
–I⁻

–ChB(OH)₂ | anti
–I⁻ | elimination

HO–B⁻ ... I⁺ H ... H

+ OH⁻

HO H
HO–B⁻ ... I ... H

conjugated (trans, cis)-dienes:

2

(i) + ThexBH₂
(THF); 1 h; 0°C

(ii) Me₃NO; 1 h; 0°C

B

(i) + NaOH/H₂O
+ I₂/THF
0°C

(ii) Na₂S₂O₃

(69%; 100% trans,cis)

(trans)-alkenes:

Br–≡–

+ Ch₂BH(THF)
0°C; 0.5 h; r.t.

B Br

NaOMe
(THF); 1 h; r.t.

MeO B

AcOH/H₂O
1 h; Δ

(90%; > 99% trans)

conjugated (trans, trans)-dienes:

ThexBH₂

(i) + Cl–≡–
(THF); 1 h; –25°C

(ii) + H–≡–
(THF); 1 h; –25°C

B Cl

(i) NaOMe
(THF); 1 h; r.t.

(ii) PrⁱCOOH
1 h; Δ

(63%; > 99% trans, trans)

<5% thexyl
migration

$\left(\bigcirc\!\!\!-\right)_3 B$ + LiC≡C(CH₂)₃CH₃ → $\left(\bigcirc\!\!\!-\right)_3 \overset{\ominus}{B}$–C≡C(CH₂)₃CH₃ $\xrightarrow[H_2O_2]{HCl}$ $\left(\bigcirc\!\!\!-\right)_2 \overset{}{C}(CH_2)_4CH_3$ OH

Silicon can either be considered as an enlarged carbon atom or as the mildest of metals. The Si–C single bond is not very reactive. Silyl double bonds or "silyl-ylenes" are not formed with base, because back bonding and double bond formation are weak. Silylated alkenes, silyl allenes and silyl arenes are available from commercial silylated alkyl chlorides and Grignard reagents or by hydrosilylation with silanes. Silylated carbon–carbon double bonds are good electron donors. The silicon substituent stabilizes the positive charges that occur at the β-carbon atom after attack by an electrophile. Typical examples are Friedel–Craft type allylations of carbonyl compounds in the presence of Lewis acids as well as substitutions of acetal esters. Neighboring methylene chloride groups can also be metallated (reduced) and produce exceptionally stable carbanions. To summarize: the silicon atom has no low-lying unoccupied d-orbitals, which could take up electrons in back bonding, like phosphorus. Nevertheless, silicon compounds behave as if it could. Steric hindrance by trimethylsilane groups is small, because the C–Si bond is long (1.9 Å). Furthermore Si–C bonds can be cleaved by HF to give a silicon fluoride and a hydrocarbon (Holmes, 1996; Fleming, 1997; Suginome, 2000).

The most significant progress in current syntheses comes from the application of palladium-catalyzed reactions between alkenyl chlorides, boranes, and tinorganyls and alkenes or arenes. A catalyst does not only lower the activation energy of a given reaction. If there are several possible reaction pathways it may also favor one of the pathways, leading to more product specificity. Transition metal–olefin complexes are ideal in the latter respect, because reactants are fixed in a defined ligand field. This is particularly true for homogeneous catalysis in inert solvents at low temperature. Many catalytic synthetic cycles run in the gigantic reaction vessels of industry where they are used in the polymerization, carbonylation, oxidation, and reduction of small alkenes and alkynes coming from catalytic cracking of mineral oil. There are, however, several closely related synthetic reactions, which have become very popular in the laboratory since about 1980. They are not so productive in quantity, but often allow regio- and stereoselective reactions under mild conditions and without extensive protection of functional groups. In this context the term "catalysis" does not really mean that one needs only very small amounts of palladium, what is often not the case. Rather, it indicates very mild and nondemanding reaction conditions, and a lowering of the activation energy. Several experimental details given in this section will prove this point.

Reactions of organometals are driven by two principal forces: ligand coordination and oxidation–reduction of the metal center. The oxidation state of a metal is defined as the charge left on the metal atom after all ligands, together with their electron pairs, have been removed. In metal hydrides the hydride ligand is always formally H-, although some transition metal "hydrides" are strong acids. After having calculated the formal oxidation state the number of d-electrons can be determined by referring to the periodic table. One should assume here that all outer electrons are d-electrons, which is probably correct in most of the metal complexes discussed here (Hegedus, 2000). Now the 18-electron rule can be applied, which says that transition metals try to obtain that many electrons in the outer sphere. The sum of the d-electrons and those of the ligands determines the maximum allowable number or a coordinatively saturated complex. Complexes with fewer electrons are coordinatively unsaturated. A vacant coordination site on the metal is the most important prerequisite for metal catalysis, for it allows the substrate to attach to the metal. Thermal and photochemical activation can usually be traced back to the expelling of a ligand, dissociation of bridged dimers, or $\pi \rightarrow \sigma$ rearrangement of ligands. Oxidation–reduction, on the other hand, can also lead to coordinative unsaturation by electron transfer from the ligand to the metal.

Oxidation state

$$Pd-\| \quad \Longrightarrow \quad Pd(0) \quad + \quad \|$$

$$Pd(PPh_3)_2 \quad \Longrightarrow \quad Pd(0) \quad + \quad 2\ \text{:}PPh_3$$

$$Pd(PPh_3)_4 \quad \Longrightarrow \quad Pd(0) \quad + \quad 4\ \text{:}PPh_3$$

$$R-Pd(PPh_3)_2-X \quad \Longrightarrow \quad Pd(II) \quad + \quad 2\ \text{:}PPh_3 \quad + \quad R^{\oplus} \quad + \quad X^{\ominus}$$

Number of d-electrons and vacant sites

	$Pd(PPh_3)_2$	$Pd(PPh_3)_4$	$R-Pd(PPh_3)_2-X$	$\overset{X}{\underset{\|}{H-Pd(PPh_3)}}$
d-electrons:	$10+2\times2=14$	$10+4\times2=18$	$2+8+4+2=16$	$4+8+2+2=16$
vacant sites:	2	0	1	1

Another important aspect in the use of organometals in synthesis is the possibility of choosing between homo- and heterolysis of the carbon–metal bond. Homolysis leads to radicals, which induce racemization of chiral centers and isomerization of *cis*- or *trans*-alkenes. It is, however, controllable in intramolecular reactions. Heterolysis, on the other hand, usually leads to the retention of the stereochemistry of the educts. Transition metal organyls are used to create new carbon bonds between alkyl, aryl, vinyl, or acyl groups. Homo coupling involves two like partners, while cross coupling refers to the joining of dissimilar alkyl and hydrogen, aryl, vinyl, or acyl groups.

Metal-catalyzed carbon–carbon bond formation is limited to alkenes and alkynes or other carbon ligands without β-hydrogen atoms. Soft and redox-active metals must be used in catalytic systems and transition metals show a strong tendency to first eliminate a β-hydrogen or a halide atom from neighboring sp³-hybridized carbon atoms and then to leave the carbon atom. A carbon–carbon atom double bond remains. Carbon–carbon bonds to β-carbon groups are usually not cleaved. In transition metal organyls, the activation of β-hydrogen atoms on sp³-carbons is coupled with the presence of a free coordination site on the metal, which is also responsible for carbon–carbon bond formation. Strong ligands, such as triphenylphosphine may prevent β-elimination, but also prevent catalysis.

Palladium complexes have a rich organic chemistry in two stable oxidation states, the +2 state Pd(II) and the zerovalent state Pd(0). Pd(II) complexes are electrophilic and react with electron-rich olefins and arenes in a similar way to Lewis acids in Friedel–Crafts reactions. $[PdCl_2]_n$, a rust-brown chloro-bridged oligomer is soluble in acetonitrile where it produces a monomeric $PdCl_2(CH_3CN)_2$ complex. It adds rapidly and reversibly to olefins in the order $CH_2=CH_2 > RCH= > cis$-$RCH=CHR > trans$-$CH=CHR \gg R_2C=CH_2, R_2C=CHR$. The resulting Pd(II) complexes prefer electron-rich olefins, enol ethers and enamides as ligands. Electron-poor olefins are more loosely bound, but activation towards further reaction might be strong, particularly in intramolecular carbon–carbon bond formation. Once the Pd(II)–C bond is formed, it becomes vulnerable towards nucleophilic attack. This occurs usually at the more substituted positions of the alkene and from the face opposite to the metal. Hard nucleophiles do not interfere, because they do not ligate to Pd(II). However, carbanions, thiols, thioethers, amines, etc. either reduce the Pd(II)-ion or poison it by irreversible binding.

Pd(OAc)$_2$ is directly soluble in organic solvents and used as common precursor of Pd(0) in catalyzed processes. Palladium(0) complexes are strong nucleophiles and strong bases and react with organic halides, acetates and triflates in a similar way to Mg(0) in Grignard reactions. Pd(0) complexes are usually prepared from Pd(OAc)$_2$ in presence of almost any weak reductant, e.g. methanol, and ligands such as triphenyl phosphine (PPh$_3$) or bisbenzylidene acetone (= dba). Pd(0) catalysts should be prepared freshly in small batches, as their catalytic activity may be lost on standing because they are air-sensitive.

Alkenes are also suitable reactive ligands for palladium. Ligands of palladium exchange quite slowly. Palladium is kinetically much more inert than lithium, magnesium or copper, but more active than kinetically inert platinum and ruthenium. Li and Mg bind preferentially to hard oxygen ligands, e.g of carbonyl groups, Pt and Ru can only activate loosely bound hydrogen. The highly polarizable π-electron pair of a C=C double bond, floating between the carbon atoms of medium electronegativity prefers palladium. It hardly matters whether it is Pd(0), which readily reduces C–I bonds at an alkene carbon, taking up iodide and a carbanion, or Pd(II), which slips into a C–H bond. Both reactions work, because both partners, alkenes and palladium, are soft and resemble each other. The reaction brings almost no energy gain as neither the C–C nor the H–I single bonds are more stable than the original C–H and C–I bonds. Nevertheless it runs with high yields along very low energy profiles. The importance of palladium in comparison with strong reductants such as Mg(0) and strong Lewis acids, such as Al(III) lies in the fact, that one can switch between the two types of reactivity by treatment with mild acid–base or redox systems. This is comparable to the acid–base triggered transformations between organoboranes and -borides. The following series of events can now lead to efficient carbon–carbon bond formation in the case of palladium:

1. A coordinatively unsaturated PdL$_2$(0) complex with 14 d-electrons on palladium enters a an alkene–halide bond R^{1}X to give an R^{1}PdL$_2$(II)X-σ-complex with 16 electrons on Pd(II). This complex is relatively stable toward elimination, but contains an electropositive C–Pd end and is still coordinatively unsaturated and therefore catalytically active.

2. An electronegative carbon atom of an alkene now enters this bond. The more highly substituted carbon atom of the alkene goes to the palladium atom, its larger substituent preferring a position *syn* to palladium. The less hindered carbon atom, at best a terminal CH$_2$-group, attaches to the carbon atom bearing R^1. The palladium is bound to an sp^3-hybridized carbon atom, has 16 d-electrons, and is labile because there is a hydrogen atom now at a neighboring sp^3-carbon.

3. Palladium waits until one of the β-hydrogen atoms has rotated in its plane into a *syn*-position. *syn*-β-Palladium hydride elimination now takes place, removing the metal from the carbon and releasing the *E*-configured alkene.
4. The formal palladium hydride is in reality a proton and can easily be removed by a weak base together with the halide. The original Pd(0) catalyst is thus regenerated and a catalytic cycle is started (Hegedus, 1994; Crabtree, 2001).

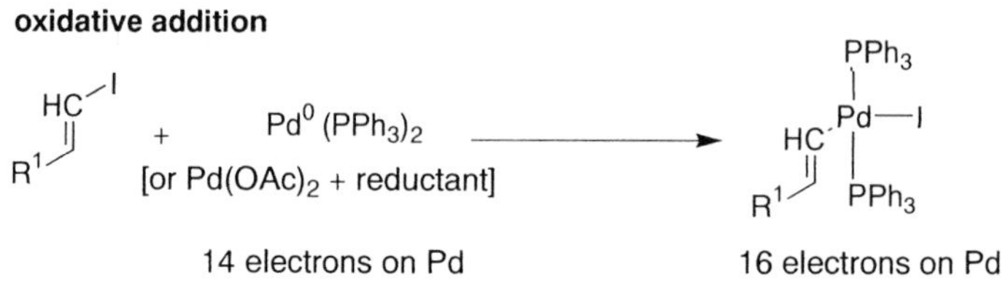

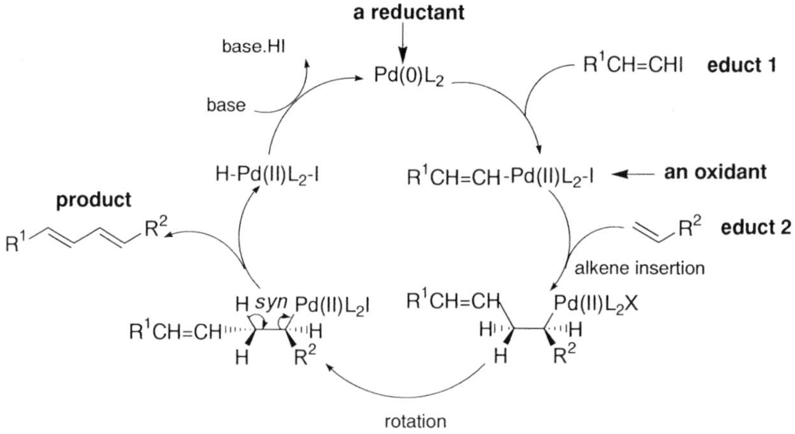

oxidative addition

14 electrons on Pd   16 electrons on Pd

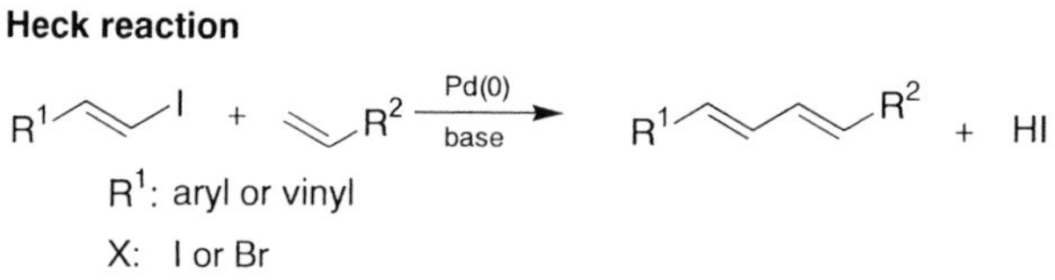

Heck reaction

R^1: aryl or vinyl

X: I or Br

Nucleophilic attack of olefins (Heck reaction) of palladium(II) thus produces first C–Pd–C bonds which rearrange to form C–C bonds. Main group organometals M (Sn, Zn, Cu, B, Si, Mg, in particular Suzuki and Stille reactions using boron or tin) often lead to a faster transfer of their alkene substituent as compared to the free alkene. The same σ-alkyl palladium(II) complexes are formed at –20 °C and the metal is removed as a salt MX (transmetallation). Upon warming, the two olefins are coupled and reductive β-hydrogen elimination removes the palladium as Pd(0). The alkylation occurs predominantly at the less substituted alkene carbon; electronic effects are usually not important. In order to start a new olefin activation, Pd(0) must be reoxidized to Pd(II). Oxygen in the presence of CuCl$_2$, quinones, or persulfate may be used, but finding experimental conditions for appropriate Pd(0)/Pd(II) equilibria is often a tedious task. The overall reaction course can be summarized as a Grignard-type reaction of an alkene iodide with Pd(0) first and a Pd(II) migration to an electronegative carbon atom as a second step. A simplified scheme is given below, and practical examples of Heck, Suzuki, and Stille reactions follow. The most important experimental advantages are (i) the nondemanding conditions of the Grignard-type reaction and (ii) the easy exchange and final cleavage of the σ-Pd–C bonds.

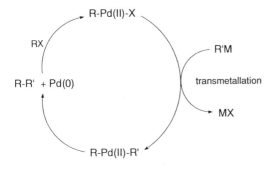

M = Mg, Zr, Zn, Sn, B, Si, Cu

Palladium-catalyzed coupling reactions of alkene halides with olefins or dienes (Heck, 1979, 1990) are broad in scope and simple to carry out. Neither anhydrous conditions nor any special techniques are required, and most functional groups are tolerated. Some alkene alkylations even occur in water. Tetraalkyl ammonium salts, e.g. Bu$_4$NBr, enhance the reaction rate (Jeffery, 1996). Aryl and vinylic bromides and iodides react with the least substituted and most electrophilic carbon atoms of activated olefins, e.g., styrenes, allylic alcohols, α,β-unsaturated esters, and nitriles. In the Heck reaction a vinylic halide is replaced by a vinyl, benzyl, or aryl group. The catalyst is a Pd π-complex formed *in situ* from palladium acetate, triethylamine, and triphenylphosphine. At first an olefin without CH-acidic β-hydrogen is added (vinyl-, aryl-, or benzyl halide) with replacement of the halide, then the reactive olefin is inserted into the Pd-C bond and immediate β-elimination of the hydrogen atom liberates the new olefin. Triethylamine then removes the proton thus producing the original palladium diphosphine and the cycle starts again. The palladium catalyzed arylation or alkenylation of alkenes is ideally suited for the construction of complex molecular frameworks. The insertion of the olefin into the C–Pd σ-bond occurs as a *syn*-addition, the organic ligand of the palladium complex migrating to the less hindered carbon atom of the olefin. The *syn*-hydride elimination described above gives the olefin.

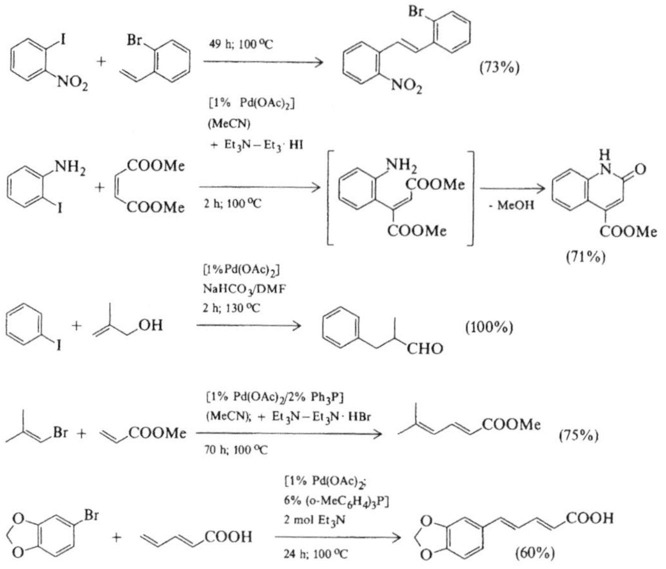

A more demanding intermolecular Heck reaction is depicted below. The alkene contains two unprotected alcohol groups and reacts directly with the vinyl iodide containing an acid-labile acetal group. Cesium carbonate is used as a base which deprotonates the reductant triethylamine. The latter produces Pd(0) needed for the palladation of the iodide, the terminal CH_2-group inserts into the Pd–C bond, PdHI(NEt_3)_2 is eliminated and loses HI, and the catalytic cycle starts again with the insertion of $Pd(NEt_3)_2$ or a similar Pd(0) complex into the vinyl iodide. The yield of 80% is remarkable for the connection of two large and polyfunctional molecules, and reaction conditions are remarkably simple. Mix at room temperature and wait – that is all. The reaction is very selective with respect to the terminal double bond (Bhatt, 2001).

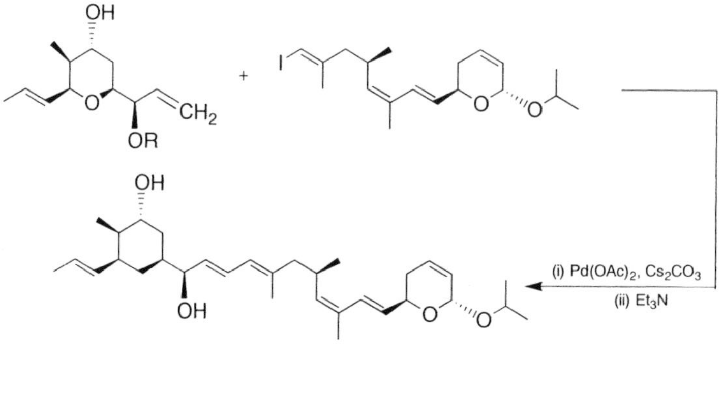

The alkene (30 mg, 0.14 mmol), the vinyl iodide (32 mg, 0.08 mmol) Bu₄NBr (25 mg, 0.08 mmol), Cs₂CO₃ (31 mg, 0.09 mmol), and Pd(OAc)₂ (18 mg, 0.08 mmol) were taken up in 0.3 mL of DMF, and Et₃N (0.012 mL, 0.08 mmol) was added. The solution was stirred for 3 h at room temperature and then chromatographed (EtOAc/hexane, 3:2) to yield the diene (31 mg, 80%).

The Sonogashira coupling is a variant of the Heck reaction using terminal alkynes instead of olefins. They react smoothly with bromoalkenes, iodoarenes, and bromopyridines and even alkene chlorides in the presence of catalytic amounts of bis(triphenylphosphine) palladium dichloride and cuprous iodide in diethylamine. The example given below shows the analogy of palladium activation with a Friedel–Crafts reaction. Direct conversion of 1,6-bis(trimethylsilyl)-hexa-3-ene-1,5-diyne into the diketone proceeded in 63% yield by reaction with acetyl chloride in the presence of AlCl₃. The *ortho*-ester of propiolic acid was coupled with *cis*-1,2-dichlorethylene in presence of copper(I) iodide by palladium(0). Iodide presumably replaces the less reactive chloride atoms. Neat solutions of both ene-diynes were not stable and polymerized within minutes (König, 2001).

1,2-Bis(3,3,3-triethoxyprop-1-ynyl)ethene. To a solution of Pd(PPh₃)₄ (831 mg, 0.72 mmol), CuI (344 mg, 1.81 mmol), 3,3,3-triethoxypropyne (4.66 g, 27.08 mmol), and 8 mL (81.2 mmol) of *n*-BuNH₂ in 80 mL of toluene at 0 °C under dinitrogen was added 1,2-dichloroethene (0.68 mL, 9 mmol) dropwise. The ice bath was removed, and the dark reaction mixture was allowed to warm to room temperature and stirred for a further 15 h. The reaction mixture was filtered through a plug of Celite, the solvent was removed *in vacuo*, and the residue was purified over silica (Et₂O with 1% NEt₃). The first fraction (R_f 0.8) gave 1.49 g (45%) of the ene-diyne as a light yellow oil (König, 2001).

The Suzuki reaction consists of a cross coupling of a 1-alkenylboron compound with a palladiumorganic electrophile. The boron compounds are usually prepared from the corresponding alkenyl- or aryl-bromides (or iodides) by reduction with butyllithium or Grignard reagents and reaction with boron triesters. The arene-

lithium salt may also be obtained by deprotonation of an activated benzene derivative. Hydroboration of alkynes is another possibility. The classical boron hydride
used is the catechol diester, which is less reactive than the borane hydrides and
leaves the boronated double bond untouched, or 9-BBN-H. The C–B bond is then
cleaved by Pt(0) in the same way as the halides in the Heck reaction and the catalytic cycle may start. The C–B bond is strong (100 kcal mol^{-1}) and apolar. In catalytic cycles it only becomes reactive in the form of borides. The boron atom must
be loaded with a negative charge by addition of hydroxide or other basic ions. If
the educt is water soluble, one may use a 10% aqueous solution of thallium(I) hydroxide.

Suzuki reaction

R^1–X + R^2‒‒BY$_2$ $\xrightarrow[\text{hydroxide or carbonate}]{\text{Pd(0)}}$ R^2‒‒R^1

R^1: alkenyl, aryl
or allylic
X: halogen

R^2‒‒BY$_2$ $\xrightarrow{\text{NaOH}}$ R^2‒‒$\overset{\ominus}{B}Y_2$ $\underset{\text{OH}}{\mid}$ $\xrightarrow{\text{R}^1\text{Pd(II)L}_2\text{X}}$ R^2‒‒$\underset{\text{L}_2}{\overset{\text{Pd}\cdot\text{R}^1}{\mid}}$ + BY$_2$OH

R^2‒‒R^1 + Pd(0)L$_2$ ⟵

a conjugated diene or a styrene

Microwave irradiation is a particularly useful energy source for Suzuki reactions
(Larhed, 2002; Blettner, 1999). α-Amino alcohols, *cis*-configured alkenes, and polyenes have been synthesized by Suzuki coupling (Suzuki, 1986; Miyaura, 1983).

HO‒‒‒‒‒B(OH)$_2$ + [Br‒‒‒‒Me]

HO‒‒‒‒‒‒‒‒‒‒

bombykol (82%)   Me

Pd(PPh$_3$)$_4$,
NaOEt,
PhH, reflux

Suzuki coupling

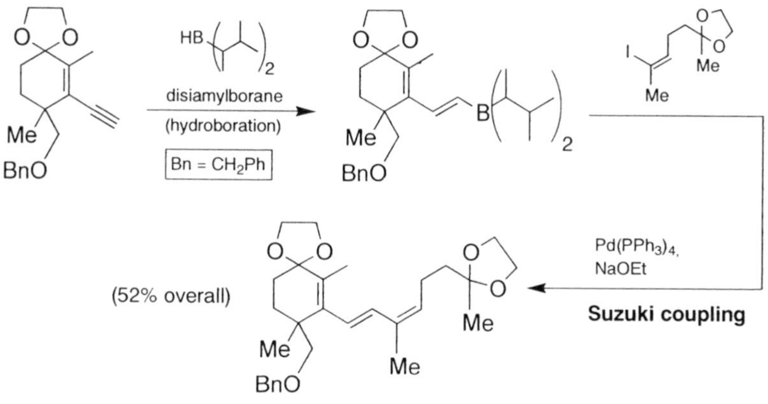

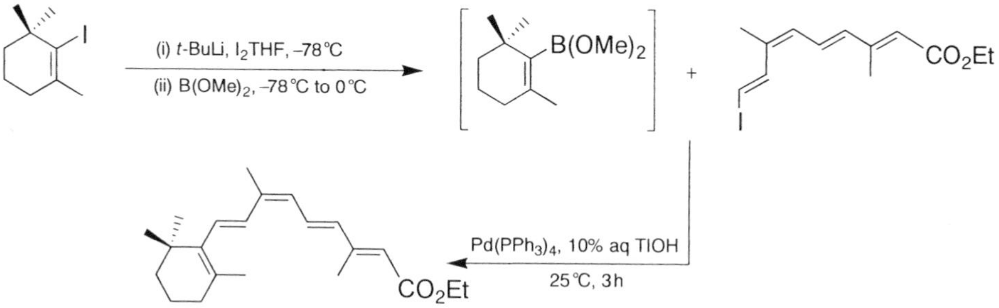

General Procedure for Hydroboration-Suzuki Coupling. To the alkene (0.22 mmol) in THF (1 mL) at 0 °C under nitrogen was added 9-BBN-H (0.5 M in THF, 0.88 mL, 0.44 mmol). The mixture was warmed to room temperature and stirred for 2 h. The flask was covered with foil, and K$_3$PO$_4$ (3 M in H$_2$O, 0.15 mL. 0.44 mmol) was added carefully (H$_2$ evolution) followed quickly by addition of the aromatic halide (0.24 mmol) in dry degassed DMF (1 mL) and finally PdCl$_2$ (dppf). CHCl$_3$ (9.4 mg, 5 mol%) under nitrogen. The reaction was stirred overnight, and then the solvent was removed *in vacuo* using an oil pump rotary evaporator. The residue was taken up in Et$_2$O (25 mL) and saturated aqueous NaHCO$_3$ (10 mL). The aqueous layer was re-extracted with Et$_2$O (25 mL), and the combined organic layers were dried, filtered, and concentrated *in vacuo* to give the crude product, which was purified by flash column chromatography, eluting with light petroleum–EtOAc mixtures to afford the Suzuki coupling product (Collier, 2002)

Two or three metal-catalyzed coupling reactions have frequently been combined in one-pot procedures The phosphonate given at the end of the following scheme carries an enol ether group at one end, which can be converted to an aldehyde by acid treatment, and a phosphonate on the other end, which can be converted to a Wittig ylene by base treatment. Both complementary functional groups survived metallation, a Pd-catalyzed Heck coupling, and an olefin σ-bond construction via a Suzuki one-pot three-component triene synthesis (Smith III, 1998).

One-pot Heck–Suzuki coupling

Triene Phosphonate. A solution of *tert*-butyllithium (1.4 M) in pentane was added dropwise to ethyl vinyl ether in THF (5 mL) at –78 °C. After 30 min at 23 °C a freshly prepared solution of $ZnCl_2$ (1.0 M in THF) was added, and after 30 min at 23 °C, the resulting solution of vinylzinc was transferred via cannula to a mixture of vinyl bromide (0.86 mmol) and bis(triphenylphosphine)palladium(II) chloride (0.034 mmol in THF). After 3 h at 23 °C, H_2O (5 mL), silver(I) oxide (3.5 mmol), and vinyl iodide (0.42 mmol) were added. The mixture was stirred vigorously at reflux for 3 h, diluted with ethyl acetate (50 mL), filtered through Celite, washed with saturated aqueous $NaHCO_3$, dried over $MgSO_4$, and concentrated. Filtration through silica gel afforded the unstable triene (70 mg. 64% yield). THF (2.5 mL) was added, and at –78 °C the solution was treated sequentially with *n*-BuLi (1.6 M in hexanes, 0.186 mL; 0.30 mmol) and methyl iodide (0.2 mL, excess). The dark red anion was replaced with an orange color during 15 min at 4 °C, after which the mixture was diluted with ethyl acetate (10 mL), washed with saturated aqueous $NaHCO_3$, and concentrated (Smith III, 1998).

Palladium organyls also react efficiently with organostannanes (Stille reaction). Tributyltin-alkenes are accessible by reactions between organolithium or -magnesium compounds of alkenyl halides and tributyltin chloride, or by the addition of tributyltin hydride to acetylenes in the presence of AiBN. The Grignard reaction gives a mixture of diastereomers, while the hydrogenation of alkyne tin compounds leads to *cis* tin alkenes (Labadie, 1983). Tributyltin chloride is more volatile and toxic than trimethyltin, but the latter becomes very reactive and nonvolatile in the presence of KF and PMHS and is even sometimes efficient in catalytic amounts (Maleczka, 2001). The catalytic cycle in which transmetallation of acyl- or alkenylpalladium chloride by the organotin reagent takes place has been characterized by changes of ^{31}P-shifts in the NMR spectra of palladium-phosphine complexes after addition of acyl or alkenyl chloride followed by tributyltin alkenyls (Labadie, 1983; Ricci, 2002). Solvated palladium and palladium(II)-tinalkenyl complexes b–d are apparent intermediates. The tin transfer order of different organic groups in the palladium-catalyzed coupling with acid chlorides, alkenyl chlorides and carbon monoxide decreases in the order CC (very fast) $> CH_2=CH > Ph > Me$, Bu. This order points to an electrophilic Pd(II)-halide complex that attaches the carbon to tin. Experimentally, one finds that derivatives of phenyl-, stryryl- or alkenyl-tributyltin transfer only the unsaturated ligand. Replacement of a second butyl group by an unsaturated ligand lowers the reactivity dramatically. This experimental finding indicates that alkenes and arenes withdraw electrons from the tin atom, in other words they are electron rich.

Stille reaction mechanism

The stereochemistry of the transmetallation reaction is not uniform. Retention of the geometry on carbon predominates with respect to *cis/trans* alkenyl-tributyltin as an educt and the resulting alkenes. If radical formation takes place, 1:1 mixtures of *cis/trans* diastereomers prevail. It is also noteworthy that Stille reactions are usually carried out on a small scale (≤ 100 mg). The first paper (Milstein, 1978) also used only millimoles of simple acid chlorides as educts. Kalesse (Bhatt, 2001) reported low yields upon scaling up. Dimerization of the stannane became dominant. The palladium(II) complex again plays the role of a Lewis acid, the tin(IV) organyl that of a nucleophile. The electroneutral character of the C–Sn bond stabilizes it against attack by protons, whereas its weak bond energy (an estimated 50 kcal mol^{-1}) favors its replacement by a more stable C–C bond (120 kcal mol^{-1}) in catalytic reaction cycles. The Stille reaction has been applied to the coupling of aryl bromides or triflates with organostannanes. Two types of reactions are encountered: direct R^1–C–C–R$_2$ coupling and coupling over a CO bridge in presence of carbon monoxide to give R^1–CO–R$_2$ ketones. Enol triflates and acid chlorides are also effective. Virtually all functional groups including aldehydes are tolerated, yields are high, and reaction conditions are mild (Stille, 1985, 1986; Echavarren, 1987). Tin organyls even survive oxidation of an alcohol to an aldehyde with manganese dioxide. The Stille reaction is, however, more sensitive to steric hindrance than the Suzuki reaction, because tin substituents are much larger than boron substituents.

(95%)

Ethyl (2*E*,4*E*,6*Z*,8*E*)-3,7-Dimethyl-9-(2-methylcyclohex-1-en-1-yl)nona-2,4,6,8-t etraenoate. A solution of Pd$_2$-(dba)$_3$ (0.010 g, 0.011 mmol) in NMP (3.2 mL) was treated with AsPh$_3$ (0.03 g, 0.11 mmol). After stirring for 5 min, a solution of triflate (0.1 g, 0.41 mmol) in NMP (0.6 mL) was added, followed, after 10 min, by a solution of the stannane (0.22 g, 0.45 mmol) in NMP (0.6 mL). The resulting mixture was stirred at 25 °C for 3 h. Saturated aqueous KF solution was added, and the mixture was stirred for 30 min and extracted with Et$_2$O. The combined organic extracts were washed with H$_2$O and saturated aqueous KF solution, dried (Na$_2$SO$_4$), and evaporated. The residue was purified by column chromatography (SiO$_2$, 90:10 hexane/EtOAc) to afford 0.12 g (95%) of the pentaene as a yellow oil.

1.6
Aldehydes, Ketones, and Carboxylic Acids

Monofunctional aldehydes and ketones are usually made by oxidation of Grignard-made alcohols. Monofunctional carboxylic acids come from synthons which contain the carboxyl group already. Nevertheless, a few carbonylations and carboxylation reactions are known, which should be familiar to the organic chemist, although most of them are rarely performed in the laboratory.

A general synthetic route to ketones uses the reaction of carboxylic acids (or their derivatives) or nitriles with Li-, Zn- or Cu-metallic compounds (Jorgenson, 1970). Lithium carboxylates react with organolithium compounds to give stable *gem*-diolates, which are decomposed by water to the ketone (Riley, 1974). Formation of tertiary alcohols, which results from the reaction of the ketone with the organometallic compound, can generally be avoided if the acid derivative is always in excess relative to the lithium organyls. Most common, however, were the *N*-methoxy-*N*-methylamide derivatives of the carboxylic acids ("Weinreb amide"). The lithium and magnesium chelate, which is formed as an intermediate, does not open to a reactive ketone under Grignard reaction conditions of low temperature (Nahm, 1981). Primary, secondary, and even tertiary lithium alkanyls can be used with generally good success, but with lithium alkenyls yields are less satisfactory. Chiral ketones may partly epimerize if the chiral center is in the *α*-position (Bartlett, 1935). This problem can be avoided if lithium alkanides or lithium dialkyl-cuprates (from Grignard reagents and copper(I) salts) are applied and the workup is done quickly at low temperature (Posner, 1975).

OH
NH₂
S ⋯H

(S)-valinol

(i) COCl₂ or Et₂CO₃
(ii) BuLi/THF
   +EtCOCl; –78 °C
(iii) CF₃SO₃BBu₂
   + EtN(Pr')₂
   (CH₂Cl₂); 0 °C

Bu   Bu
B
O   O
O   N   Z

(~70%)

(iv)   CHO
   0.5 h; –78 °C
   + 1.5 h; r.t.
(v) aq. H₂O₂/MeOH
   pH=7; 1 h; 0 °C
(vi) NaOMe/MeOH
   5 min; 0 °C

OH
COOMe

98.6% d.e.; >99% e.e

H   OH
Ph
R ⋯NH₂
H

(1S,2R)-
norephedrine

Bu   Bu
B
O   O
O   N   Z

Ph

OH
COOMe

99.6% d.e.; >99% e.e.

O   X
N   O
Bn

X = O
X = S

Bu₂BOTf/Et₃N
or
TiCl₄,
(–)-sparteine
CH₂Cl₂
3-butenal

OH O   X
N   O
Bn

X = O   63%
X = S   75%

TBSOTf
2,6-lutidine

TBSO   O   X
N   O
Bn

X = O   92%
X = S   70%

TBSO   O   X
N   O
Bn

X = O   92%
X = S   91%

L₄Ru=CHPh
CH₂Cl₂

(4R)-Benzyl-3-[(3S)-hydroxy-(2R)-vinylhex-5-enoyl]oxazolidin-2-one. A solution of the chiral acylox-azolidinone (13.17 g, 54 mmol) in 280 mL of CH₂Cl₂ was cooled to –78 °C. The borane Bu₂BOTf (14.7 mL, 59 mmol) was diluted in 15 mL of CH₂Cl₂ and then added dropwise slowly to the oxazole. The mixture was stirred for 5 min at –78 °C. Et₃N (10.5 mL, 75 mmol), was diluted with 10 mL of CH₂Cl₂ and then added dropwise slowly. After complete addition, the mixture was stirred at –78 °C for 1 h. The mixture was gradually warmed to 0 °C, stirred for 15 min, and then recooled to –78 °C. A solution of 3-butenal (277 mmol) in 100 mL of CH₂Cl₂ was cooled to –78 °C and added to the boron enolate solution via cannula. The reaction was stirred for 1 h at –78 °C, gradually warmed to 0 °C, and stirred for 1 h. Workup including an oxidation of the borane with 30% H₂O₂ and chromatography yielded 3.58 g of acyloxazolidinone and 10.7 g of the desired S-alcohol (63%; 90% based on recovered starting material).

Another useful catalyst for asymmetric aldol additions is prepared *in situ* from mono-O-(2,6-diisopropoxybenzoyl)tartaric acid-boron complex under hydrogen in propionitrile solution. Aldol reactions of ketone enol silyl ethers with aldehydes were promoted by 20 mol% of this catalyst solution. The relative stereochemistry of the major adducts was assigned as Fischer-*threo*, and predominant *Re*-face attack of enol ethers at the aldehyde carbonyl carbon atom was found with the (R,R)-enantiomer of the tartaric acid catalyst (Furuta, 1991). A chiral open-chain boron enolate was used with acetaldehyde to produce a specific stereotriad in a one-pot assembly. An *R*-chiral ethylketone was readily synthesized in three steps (71% overall yield) from commercially available methyl (*S*)-2-methyl-3-hydroxy-propionate. Enolization with (c-Hex)$_2$BCl/Et$_3$N led to the selective generation of the (*E*)-boron enolate, which underwent addition to acetaldehyde through a highly ordered chairlike transition state. This was followed by *in situ* reduction of the intermediate boron aldolate with LiBH$_4$, via axial hydride delivery, to provide, after oxidative workup, the 1,3-*syn* diol (86%, >97% ds). Such a one-pot reaction relies on the properties of the non-polar boron enolate. It allows the establishment of equilibria and subsequent addition of different reactants, which are not compatible with each other, e.g. acetaldehyde and LiBH$_4$ (Paterson, 2001).

R^1	R^2	R^3	yield	d.e.	e.e.
Et	Me	Pr^n	61%	60%	88%
Et	Me	Ph	96%	88%	96%
Ph	H	Ph	98%	—	85%

Chiral diols from chiral ketones. To a stirred solution of dicyclohexylboron chloride (8.96 mL, 41.3 mmol) in Et$_2$O (60 mL) at 0 °C Et$_3$N (6.54 mL, 46.4 mmol) was added. Ketone (*S*) (5.33 g, 25.8 mmol) in Et$_2$O (20 mL) was added via cannula and the reaction mixture stirred at 0 °C for 2 h before cooling to –78° C. Acetaldehyde (4.30 mL, 77.4 mmol) was added and stirring continued for 3 h before transferring to a freezer (–20 °C, 20 h). The reaction was recooled to –78 °C before the addition of lithium borohydride (31.8 mL of a 3 M THF solution, 95.5 mmol). After 3 h, the reaction mixture was cannulated into NH$_4$Cl (400 mL) at room temperature and the layers were separated. Workup and flash chromatography (10–50% EtOAc/hexanes) gave the diol as a colourless oil (5.52 g, 86%).

The Masamune reaction of aldehydes with boron enolates of chiral α-silyloxy ketones establishes a "double asymmetric induction", and also generates two new chiral centers with enantioselectivities >99%. It is again explained by a chair-like six-center transition state stabilized by two boron–oxygen bonds between the carbonyl groups of both educts and the borane or titanium propoxide. The repulsive interactions of the bulky cyclohexyl group with the vinylic hydrogen and the boron ligands dictate the approach of the enolate to the aldehyde (Masamune, 1981 A). The β-hydroxy-α-methyl ketones obtained are pure threo products (threo = threose- or threonine-like Fischer formula; also termed "*syn*" = planar zigzag chain with substituents on one side), and the reaction has successfully been applied to macrolide syntheses (Masamune, 1981 B). Optically pure *threo* (= "*syn*")-β-hydroxy-α-methyl carboxylic acids are obtained by desilylation and periodate oxidation (Masamune, 1981 A). Chiral O-[(*S*)-*trans*-2,5-dimethyl-1-borolanyl]ketene thioketals giving pure erythro (= "anti") diastereomers have also been developed by Masamune (1986). Stereoselectivities of 99% are also obtained by Mukaiyama-type aldol reactions of the titanium enolate of Masamune's chiral α-silyloxy ketone with aldehydes. An excess of titanium reagent (≥2 mol) must be used to prevent interference by the lithium salt formed, when the titanium enolate is generated via the lithium enolate (Siegel, 1989). The mechanism and the stereochemistry are the same as with the boron enolate. Stereoselectivities of 99% are also obtained by Mukaiyama-type aldol reactions using the titanium enolate of Masamune's chiral α-silyloxy ketone and aldehydes as educts (Yoshimitsu, 1997).

Syntheses of *anti*-configured 2-alkyl-3-hydroxycarboxylic esters using a norephedrine ester as stereocontroller and LDA-Cp$_2$ZrCl$_2$ as E-enolizer were easily scaled up to multigram quantities. The *anti/syn* ratios were approximately 9:1, the *cis* for *anti* often more than 98:2 (Kurosu, 2001).

Experimental Procedures for the Large-Scale Aldol Reaction. To a stirred solution of diisopropylamine (2.80 mL, 2 equiv.) in THF (50 mL) at 0 °C was added *n*-BuLi (1.5 M, 13.3 mL, 2 equiv.). The LDA solution was cooled to −78 °C and a THF solution of Cp$_2$ZrCl$_2$ (879 mg, 3.00 mmol, 0.3 equiv.) was added. After 15 min, the ketone (5.00 g, 10.0 mmol) in THF (10 mL) was added. The reaction mixture was stirred for 90 min, and Cp$_2$ZrCl$_2$ (7.30 g, 25.0 mmol, 2.5 equiv.) in THF (9 mL) was added. After a further 10 min at −78 °C, to the reaction was added propionaldehyde (794 mg, 11.0 mmol, 1.1 equiv.) in THF (5 mL). The reaction mixture was stirred at

1-Hydroxyalkyl substituted cyclopropanes operate as a^4-synthons. d^0-Synthons, e.g. hydroxide or halides, yield 1,4-disubstituted products (Wenkert, 1970 A). (1-Hydroxyalkyl)- and (1-haloalkyl)-cyclopropanes are rearranged to homoallylic halides, e.g. in Julia's general terpene synthesis (Julia, 1961, 1974; Brady, 1968; McCormick, 1975).

trans:cis = 75:25

Julia's terpenoid synthesis

Substituted cyclopropanes readily available from addition of carbenoids to C=C double bonds constitute another interesting class of d^3-reagents, since electron donating and withdrawing groups strongly facilitate ring opening by acids or bases. In the first example in Tab. 1.4 the d^3-synthon is generated directly by Lewis acid-promoted ring opening (Nakamura, 1977) and in the second one ring opening occurs in a homoallylic rearrangement after the addition of a cyclopropyl anion to a carbonyl compound has taken place (Corey, 1975 B).

Since cyclopropane resembles olefins in its reactivity and is thus an electron-rich carbocycle, it forms complexes with Lewis acids, e.g. TiCl$_4$, and is thereby destabilized. This effect is even more pronounced in cyclopropanone ketals. If one of the alcohols forming the ketal is a silanol, the ketal is stable and distillable. The O–Si-bond is cleaved by TiCl$_4$ and a d^3-reagent is formed. This reacts with a^1-reagents, e.g. aldehydes or ketals. Various 4-substituted carboxylic esters are available from 1-alkoxy-2-siloxycyclopropanes in this way (Nakamura, 1977). If one starts with 1-bromo-2-methoxycyclopropanes, the bromine can be selectively substituted by lithium. Subsequent treatment of this reagent with carbonyl compounds yields (2-methoxycyclopropyl)methanols, which can be transformed to β,γ-unsaturated aldehydes (Corey, 1975 B).

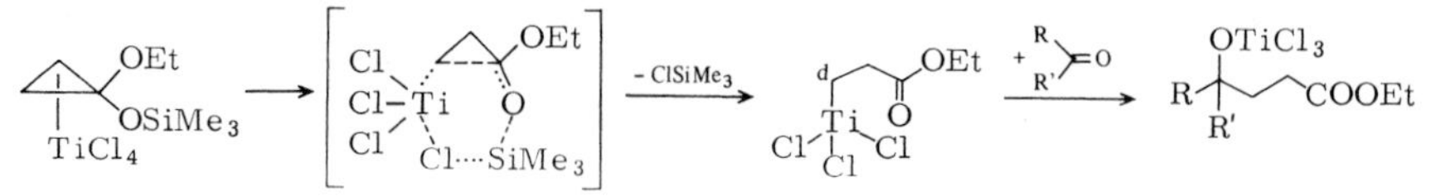

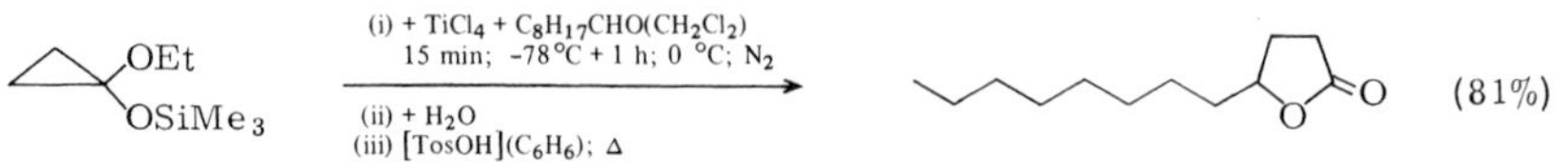

(i) + TiCl$_4$ + C$_8$H$_{17}$CHO(CH$_2$Cl$_2$)
15 min; −78 °C + 1 h; 0 °C; N$_2$

(ii) + H$_2$O
(iii) [TosOH](C$_6$H$_6$); Δ

(81%)

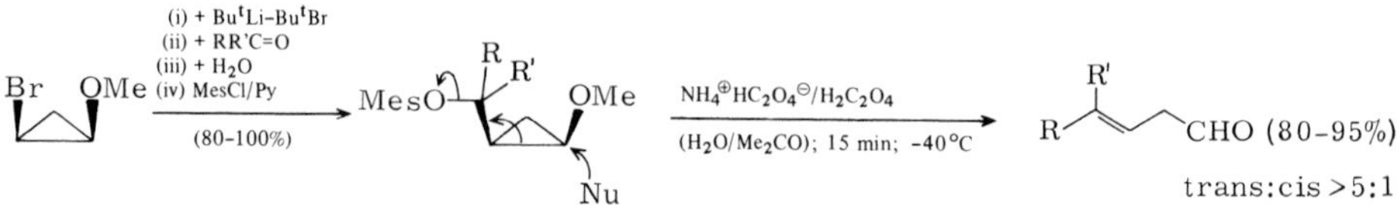

(i) + Bu^{t}Li−Bu^{t}Br
(ii) + RR'C=O
(iii) + H$_2$O
(iv) MesCl/Py

(80−100%)

NH$_4^{\oplus}$HC$_2$O$_4^{\ominus}$/H$_2$C$_2$O$_4$

(H$_2$O/Me$_2$CO); 15 min; −40 °C

R—CH=... CHO (80−95%)

trans:cis > 5:1

Tab. 1.4 Some ring opening reactions.

d^3-synthon	equivalent reagent		
	Li—△—OMe → (i) RR'C=O (ii) MesCl	→ (intermediate) $^{\ominus}$OMe	NaOMe → R'—CH=...—OMe, OMe
	Cl$_3$Ti...O...SiMe$_3$ + TiCl$_4$ − ClSiMe$_3$	[Cl$_3$Ti...O ...OR] RR'C=O →	OTiCl$_3$ R'—...—OR, O

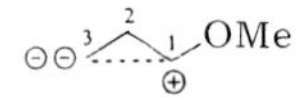

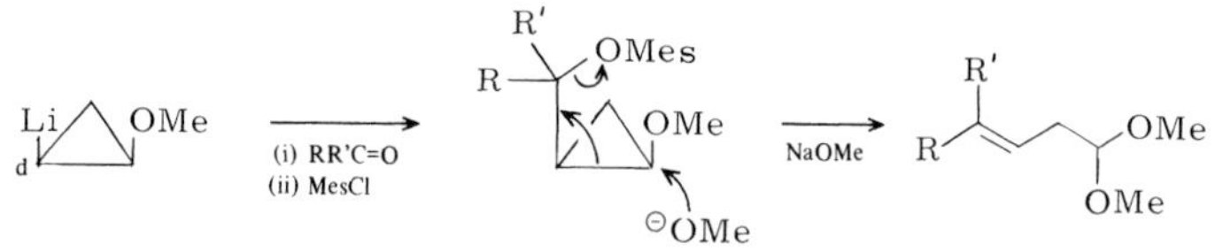

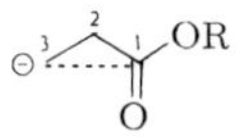

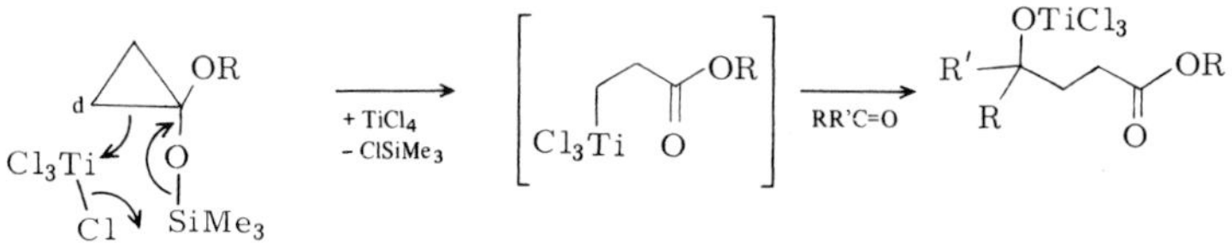

1.10
1,5-Difunctional Compounds

The Michael reaction is again of central importance here. This reaction is a vinylogous aldol addition, and most facts that have been discussed in Section 1.9 also apply here: the reaction is catalyzed by acids and by bases, and it may be made regioselective by the choice of appropriate enol derivatives. Stereoselectivity is observed in reactions with cyclic educts. An important difference to the aldol addition is that the Michael addition is usually less prone to steric hindrance. This is evidenced by the two examples given in the scheme below, in which cyclic 1,3-diketones add to α,β-unsaturated carbonyl compounds (Hiroi, 1975; Smith, 1964).

(70%)

(97%)

synthetic steroid precursor

Michael reactions, together with aldol reactions, belong to the very few synthetic steps which may lead to single stereoisomers in the synthesis of open-chain compounds. Two attractive preparations look exotic at first sight, but turn out to be quite simple and efficient experimentally. Glycolic acid and commercial (R)- or (S)-3-chloro-propane-1,2-diol yield a cylohexane-type butane diacetal lactone (BDA) chair with an acidic CH_2-group between a carbonyl group and an acetal oxygen. Upon deprotonation, this anion adds at –78 °C activated double bonds only in an equatorial position. After workup, open-chain α-hydroxy-γ-amino acids are obtained with ee's of up to 95% (Dixon, 2001). Another stereoselective Michael addition uses chiral 2-ynyl-alcohols and alkylidenemalonates as educts. At first, the ynyl-alcohol is phosphatized and the acetylene group trimethylsilylated. Treatment with a 1:2 mixture of Ti(O-i-Pr)$_4$ and i-PrMgCl converts the alkyne to an allene with a titanium(II) substituent. This electron-rich metal organyl then adds to the alkylene group via an antiperiplanar transition starter. The anti/syn ratio is consistently above 50, the enantiospecificity above 90% (Song, 2001). There are several other examples of dia- and enantio-selective Michael additions in modern literature. The activated C=C double bond is obviously suitable for the introduction of chirality at β-carbonyl carbons.

(diastereomeric ratio ~ 10-20 : 1)

Me₃Si⁓⁓⁓‴R¹ (OEt)₂(O)PO⁓H
→ Ti(O-i-Pr)₄, 2 i-PrMgCl, ether, –50 °C ~ –40 °C, 2 h →
[L₃Ti=⟨ Me₃Si / ‴R¹ / H ⟩] (L= (O-i-Pr) and/or OP(O)(OEt)₂)

R²⁓(CO₂Et)(CO₂Et), ether, –40 °C ~ r.t. °C, 1 h →

Me₃Si⁓⁓⁓ R² CO₂Et / R¹ CO₂Et

→ (i) LiCl, DMSO/H₂O, 150 °C; (ii) NaIO₄, cat. RuCl₃, CH₄/CH₃CN/H₂O →

HOOC⁓(R²)⁓COOEt / R¹ (82%)

The Lewis acid-promoted addition of an enolsilane can be made enantioselective with chiral copper(II) bisoxazoline (box) as a catalyst. (Z)-Enolsilanes give, for example, syn diastereoselection, while (E)-enolsilanes are anti selective in addition reactions to ethyl-fumaroyl-oxazolidinones (Evans, 1999).

OTMS / R¹⁓R² + EtO₂C⁓⁓C(O)—N(oxazolidinone)
R¹ = SMe or Ph
R² = Me, Et or i-Pr

→ 10 mol% cat., CH₂Cl₂, HFIP →

O / N(pyrrole) ... CO₂Et O ... Me ... O N-oxazolidinone
98:2 diastereoselectivity
97% e.e. 94% yield

Me Me / oxazoline-Cu-oxazoline / Me₃C ... CMe₃ / 2 SbF₆⊖ 2⊕
catalyst

- - - →

Me Me / oxazoline-Cu-oxazoline / Me₃C ... CMe₃ / 2X⊖ 2⊕
Rᵦ⁓⁓C(O)—N-oxazolidinone

1.11
1,6-Difunctional Compounds

Cyclohexene derivatives can be oxidatively cleaved under mild conditions to give 1,6-dicarbonyl compounds. The synthetic importance of the Diels–Alder reaction described in Chapter 2 originates to some extent from this fact.

The most common procedure is ozonolysis at –78 °C (Bailey, 1978) in methanol or methylene chloride in the presence of dimethyl sulfide or pyridine, which reduce the intermediate ozonides to aldehydes. Unsubstituted cyclohexene derivatives give 1,6-dialdehydes; enol ethers or esters yield carboxylic acid derivatives. Oxygen-substituted C=C bonds in cyclohexene derivatives, which may also be obtained by Birch reduction of aldoxyarenes, are often more rapidly oxidized than nonsubstituted bonds (Corey, 1968; Stork, 1968 A, B). Catechol derivatives may also be directly cleaved to afford conjugated hexadienedioic acid derivatives (Woodward, 1963). Highly regioselective cleavage of the more electron-rich double bond is achieved in the ozonization of dienes (Knöll, 1975). Workup with alcohols and acetic anhydride gives asymmetric aldehydes–ester mixtures, which can be cyclized to uniform heterocycles, e.g. pyrroles (Taber, 2001).

Aldehyde Methyl Esters. An ozone stream was bubbled through a suspension of the ketal (3.40 g, 18.7 mmol) and NaHCO$_3$ (6.31 g, 75.1 mmol) in a 5:1 mixture of CH$_2$Cl$_2$:MeOH (36 mL) at −78 °C until complete consumption of the ketal was observed by TLC analysis. The mixture was then flushed with nitrogen, and NaHCO$_3$ was removed by filtration. The filtrate was concentrated *in vacuo* to give the crude mixture of the esters as a colorless oil, which was taken up in CH$_2$Cl$_2$ (30 mL). The mixture was cooled to 0 °C, and acetic anhydride (8.8 mL, 93 mmol) and triethylamine (3.9 mL, 28 mmol) were added. The mixture was stirred at room temperature for 1 h and then partitioned between MTBE and, sequentially, 3 M aqueous HCl, 3 M aqueous KOH, and brine. The combined organic extract was dried (MgSO$_4$) and concentrated *in vacuo*. Chromatography of the crude product afforded 3.11 g (68%) of an inseparable mixture of the aldehydes–methylesters as a colorless oil.

Another method to achieve selectivity in oxidative splitting of C=C bonds to carbonyl groups is controlled epoxidation followed by periodate cleavage (Nagarkatti, 1973).

Most of the synthetic reactions leading to substituted carbon compounds can be reversed. *Retro*-aldol or *retro*-Diels-Alder reactions, for example, are frequently used in the degradative fragmentation of complex molecules to give simpler fragments. In synthesis, such procedures are of limited value. There are, however, some noteworthy fragmentations of easily accessible polycycles with n rings, and those lead to useful products possessing n–1 rings, but the same total number of carbon atoms. 6-Keto acids are obtained by acylation of cyclopentanone enamines with acid chlorides and subsequent base-catalyzed *retro*-aldol cleavage (Hünig, 1960). A 2-cyclohexenone derivative can be transformed into the corresponding epoxy tosylhydrazone by sequential treatment with peracid and tosylhydrazine. The elimination of nitrogen and *p*-toluenesulfinate and fragmentation after rearrangement to the 3-tosylazo allylic alcohol may occur under mild conditions. Carbonyl compounds with 5,6-triple bonds are formed in high yields (Schreiber, 1967; Tanabe, 1967). If one applies this reaction to a 9-10-epoxy-1-decalone, a ten-membered 5-cyclodecyn-1-one ring is formed (Felix, 1971). This product is an important intermediate in the perfume industry and is produced on a large scale. For this purpose Eschenmoser developed a synthesis in which the readily removed styrene was split off instead of a sulfinic acid. 1-amino-2-phenylaziridine hydrazone was used instead of a tosylhydrazone (Felix, 1968).

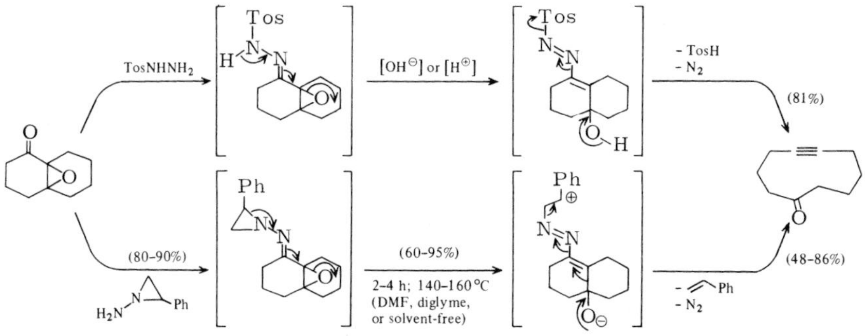

Similar fragmentations to produce 5-cyclodecen-1-ones and 1,6-cyclodecadienes have employed 1-tosyloxy-4*a*-decalols and 5-mesyloxy-1-decalyl boranes as educts. The ring-fusing carbon–carbon bond was smoothly cleaved and new π-bonds were thereby formed in the macrocycle (Wharton, 1965; Marshall, 1966). The mechanism of these reactions is probably E2, and the positions of the leaving groups determine the stereochemistry of the olefinic product.

four-bond fragmentation

five-bond fragmentation

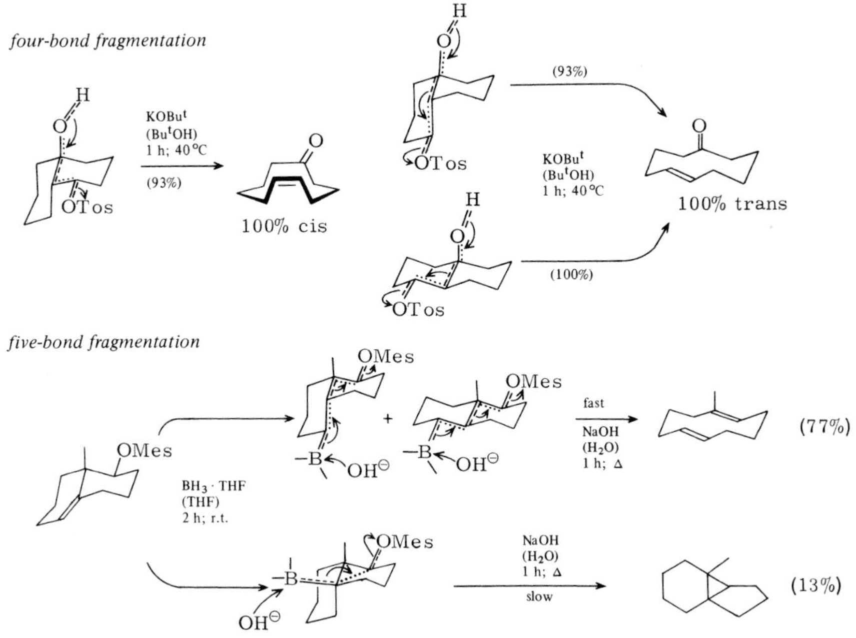

All these fragmentation reactions occur readily, when the C–X bond and the breaking C–C bond have the trans antiparallel arrangement. Olefin formation can proceed along a concerted pathway. The stereochemistry of the olefinic product is determined by the orientation of groups in the cyclic precursor. If the *trans anti-*

parallel arrangement of the breaking bonds has been attained, the relative orientation of the hydrogen atoms in the precursor is retained in the olefin.

Conventional synthetic schemes to produce 1,6-disubstituted products, e.g. reaction of a³- with d³-synthons, are largely unsuccessful. An exception is the following reaction, which provides a useful alternative when Michael type additions fail, e.g. at angular or other tertiary carbon atoms. In such cases the addition of allylsilanes catalysed by titanium tetrachloride, the Sakurai reaction, is most appropriate (Hosomi, 1977). Isomerization of the δ-double bond with bis(benzonitrile-N)dichloropalladium gives the γ-double bond in excellent yield. Subsequent ozonolysis provides a pathway to 1,4-dicarbonyl compounds. Thus 1,6-, 1,5- and 1,4-difunctional compounds are accessible by this reaction.

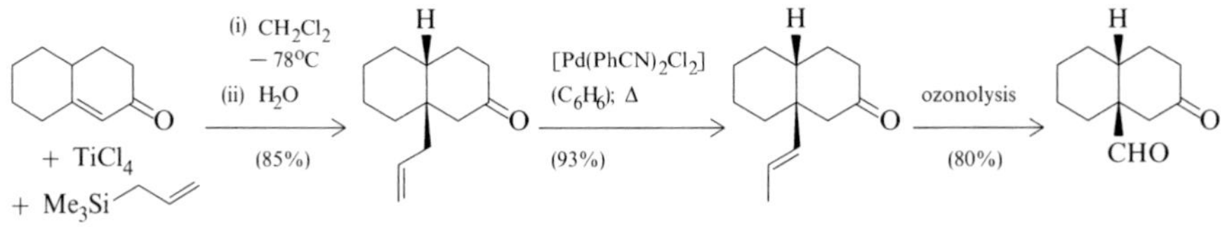

2
Carbocycles

2.1
Introduction

The chemistry of carbocycles is dominated by small-angle strain in cyclopropane and cyclobutane derivatives, by pseudorotation of the carbon atoms of cyclopentane to avoid the eclipsed positions of substituents in the planar conformer, and by the chair-like conformation of cyclohexane, which leads to the differentiation of *equatorial* and *axial* substituents conformers (Smith III, 2001). With respect to synthetic problems: (i) small-angle strain leads to thermodynamic instability and sometimes makes straightforward intramolecular reactions leading to strained carbocycles difficult; (ii) pseudorotation within the ring of cyclopentane derivatives often leads to extraordinarily highly soluble products and difficult crystallizations; (iii) in cyclohexyl products and cyclohexane-chair-like transition states, chiral stereocenters are best defined with respect to the relative positions of large and small substituents and charges. Equatorial and axial conformers exist in equilibrium, but the *equatorial* isomer persistently dominates. The difference in energy between *axial* and *equatorial* conformations of methyl cyclohexane is usually erroneously attributed to 1,3-*syn-axial* interactions between the axial methyl group and the axial hydrogens. The two *gauche*-butane interactions are present in the *axial* conformer, but absent in the *equatorial* isomer or, in more general terms, steric repulsion between the *axial* methyl group and the ring carbons is much more important (Wiberg, 1999). This general rule is as important for stereoselective synthesis as the consideration of neighboring group effects.

gauche butane

$K = 220; \quad \Delta H = -1.9 \text{ kcal mol}^{-1}, \quad \Delta G = -2 \text{ kcal mol}^{-1}$

Another general approach to predicting the outcome of cyclization reactions, namely consideration of stereoelectronic effects, is more suitable for heterocycle synthesis. It is discussed in Chapter 4.

2.2
Radicals

Radical reactions are most useful in synthesis if they occur intramolecularly. Neighbor effects make radical combinations fast and selective. Typical examples are the pinacol and acyloin coupling reactions, which are used to dimerize ketones and esters (=r^{1}synthons) to 1,2-diols or -dienols. Neighbor effects in intramolecular cyclization reactions not only allow the combination of different carbonyl groups ("heterodimerization") in high yield but also lead to stereoselectivity.

In the pinacol reduction, carbonyl groups are reduced by titanium(II) or (III) (Hashimoto, 2001) in the presence of magnesium or zinc metals. Ketyl radicals are formed and combine to give cyclic 1,2-diols (Tyrlik, 1973; Mukaiyama, 1973; Corey, 1976). The same holds true for the reductive coupling of two ester groups to form acyloins (=a-hydroxy ketones; McElvain, 1948; Bloomfield, 1976). Yields are generally better if the alkoxide anions and enediolate dianions are trapped as silyl ethers. Undesirable side-reactions caused by the strong bases, e.g. cleavage, aldol-type reactions, and dehydrations, are thus prevented (Rühlmann, 1971). 1,2-Dicarboxylic esters yield 1,2-bis(trimethylsiloxy)-cyclobutenes, which can be opened thermally. This has been used in a procedure for ring expansion of carbocycles (Mori, 1969).

pinacol coupling:

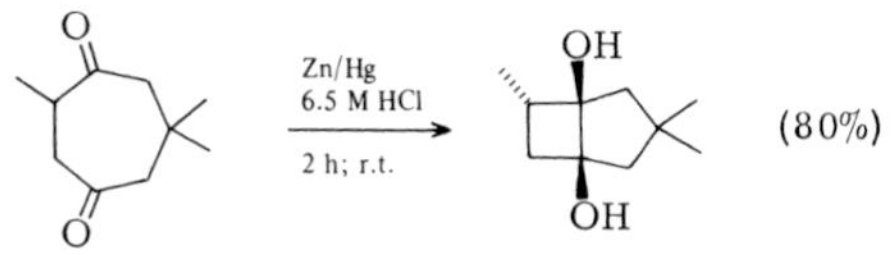

acyloin coupling

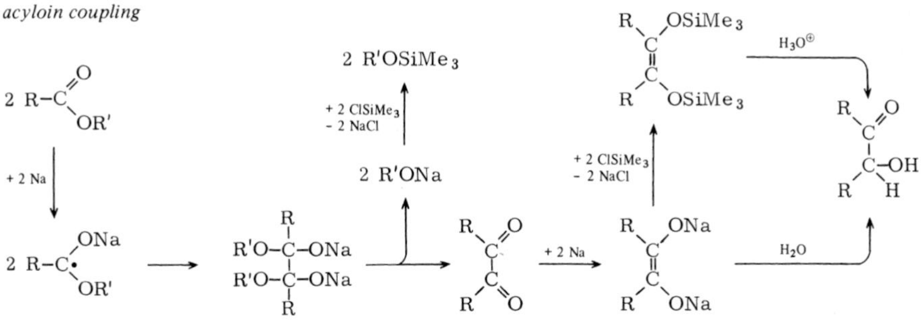

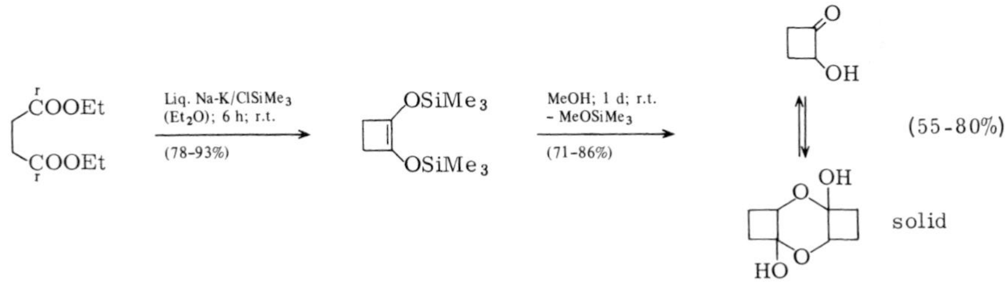

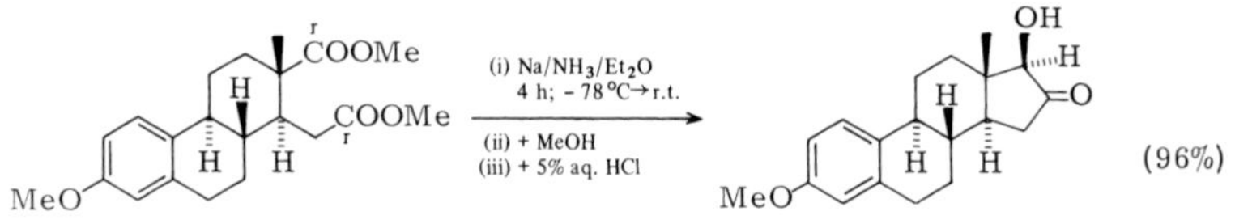

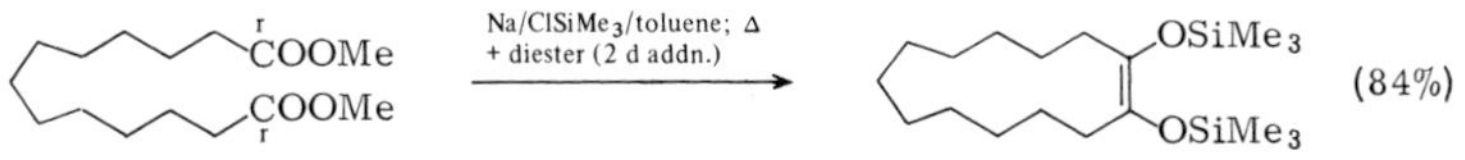

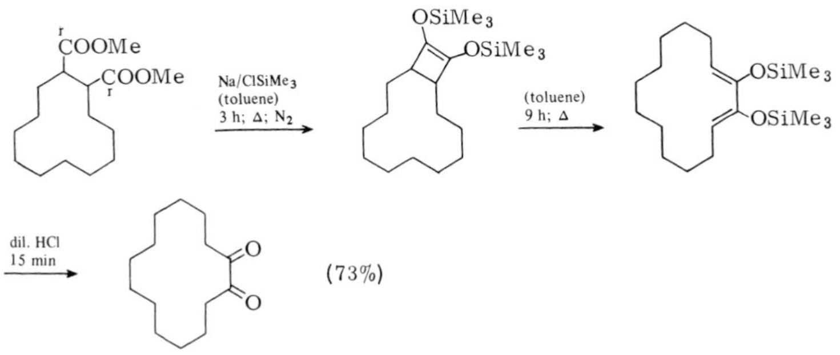

(80%)

2 R'OSiMe₃ → rendered as image

The standard reactions for radical cyclizations are reductions of alkyl or alkenyl iodides with tin organyls or silanes (Ryu, 1996; Molander, 1998). $(Me_3Si)_3SiH$ and Bu_3SnH in the presence of AIBN are particularly suitable to start reductive tandem reactions leading to medium-sized rings. Carbon monoxide can be added under high pressure and behaves then as electron acceptor. The radical, which is formed by thermal splitting of a C–Si or a C–Sn bond, reacts directly with an olefin or adds carbon monoxide first and then yields cyclopentanones or macrocycles. Both electron-rich and electron-poor double bonds react with the radical.

Organolanthanides contain unpaired electrons in f- rather than in d-orbitals. Their organometallic compounds, e.g. methyllutetium, are more stable than those of other transition metals and they are efficient catalysts for polymerizing ethylene (Watson, 1985). Samarium(II) iodide is a popular reagent to produce radicals from alkyl iodides, which may then further react with alkenes (Molander, 1996). This reaction is analogous to the pinacol and acyloin cyclizations, but uses α-iodo-ω-alkenes as substrates instead of dicarbonyl compounds. A C–C bond without heteroatomic substituents is formed. Polyhalide compounds with chloride and fluoride substituents are preferred for stepwise cyclization of medium-sized rings with SmI$_2$. Initially it reacts exclusively with alkyl iodides and adds them to carbo-nyl groups, the primary radical formed can then again react with SmI$_2$ and adds iodide to aldehydes or ketones, which are present in solution (Molander, 1998). The diastereoselectivity of most cyclization reactions is high, because the standard cyclohexane-type chair conformation is anchored on samarium with the allylic stereogenic center.

2.3
Ring Closing Metathesis

Some homogeneous metal catalysts induce the exchange of alkylidene units of al-kenes. This so-called "olefin metathesis" includes splitting and formation of dou-ble bonds in a series of [2+2] cycloadditions and cycloreversions and leads to an equilibrium mixture of alkenes and cycloalkanes. The ring closing metathesis (RCM) of a diene must escape this equilibrium in order to be productive. This is often the case, because the competing retro-reaction (ring opening metathesis, ROM) and polymerization (ROMP) are kinetically hindered. The metathesis cata-lysts often do not catalyze these reactions and the cyclization products accumulate (Fürstner, 2000).

Types of alkene metathesis:

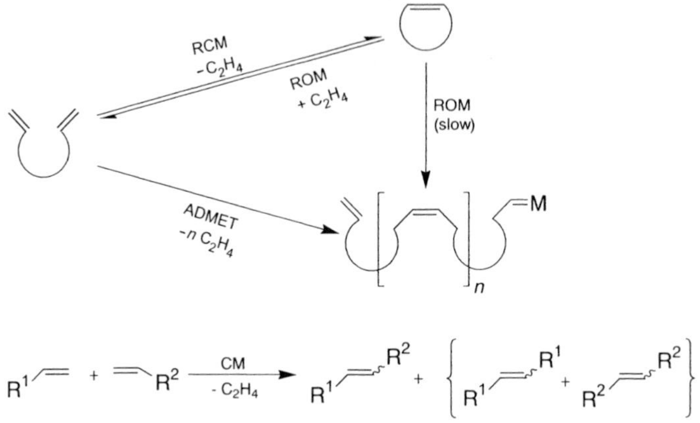

RCM: ring closing metathesis; ROMP: ring opening metathesis polymerization; ADMET: acyclic diene metathesis polymerization; CM: cross metathesis

Ring closing metathesis:

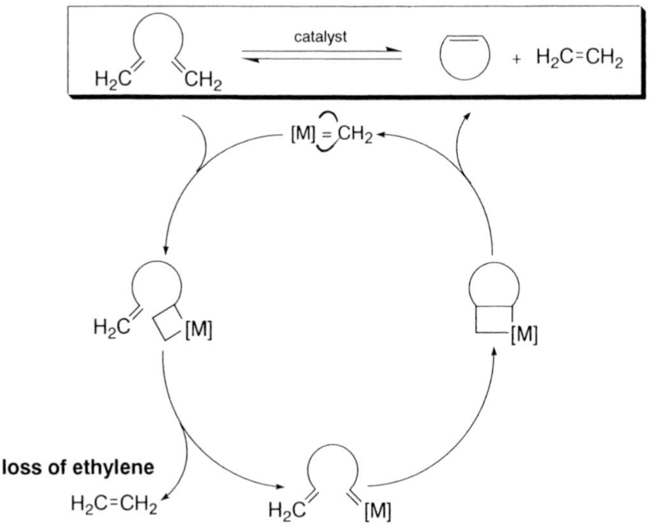

Ruthenium catalysts with a Ru=C double bond (a carbene) are particularly useful. If they contain charged phosphine ligands they may be used in water, methanol, or detergent solutions.

$$RuCl_3 \cdot 3H_2O \xrightarrow[\text{ClCH}_2\text{CH}_2\text{Cl}]{\text{H}_2,\ \text{PCy}_3,\ \text{Mg}} [RuHCl(H_2)(PCy_3)_2] \xrightarrow[\text{HCl}]{\text{HC}\equiv\text{CH}} \underset{\substack{\text{Cl} \\ \text{Cl}}}{\overset{\text{PCy}_3}{\text{Ru}}}=\text{CH}_3$$

$L = Cy_2P\text{—}\underset{}{N}^{\oplus}\ X^{\ominus}$

$L = Cy_2P\text{—}\overset{\oplus}{N}Me_3\ X^{\ominus}$

$L = Cy_2P\text{—}SO_3^{\ominus}\ \overset{\oplus}{Na}$

All these five-coordinated, 16d-electron complexes are quite stable against oxygen and water and can take up an extra olefin molecule. A ligand then dissociates off, the olefin is located cis to carbene, and a four-membered ring containing part of both olefins and ruthenium is formed (Grubbs, 1995). The rearrangements of metathesis can then take place (Fürstner, 2000).

Metathesis-type carbon–carbon bond rearrangements are neither restricted to a special ring size nor do they depend on substituents on the *a-ω*-dialkene used. Glucose, for example, was first converted to an *a,ω*-diene by a Wittig reaction and then to a tetrahydroxy-cyclohexene in 90% yield by RCM (Ackermann, 2000). *a,ω*-Dienylesters were cyclized on a solid support to a lactone using liquid carbon dioxide as a solvent. Immobilized *a,ω*-dienes were not only cyclized by RCM but also liberated, when one of the terminal vinyl groups was bound to the support. Unwanted side-products remained tagged to the carrier material (Fürstner, 2000).

$$\xrightarrow[\text{[M]=CH}_2]{\text{RCM}}$$

$$+ \quad CH_2 = CH_2$$

$$\xrightarrow[\text{[M]=CH}_2]{\text{RCM}}$$

$$+ \quad =\text{[M]}$$

RCM: ring closing metathesis

2.4
Macrocyclization

Stiff benzene units form macrocycles, if they are connected at the meta-positions. The simplest cases are the cyclophanes, in which two arene rings are connected by short alkyl chains.

Terminal acetylenes react with copper(II) salts or with copper(I) salts in the presence of oxygen to form bisacetylides, which undergo alkyne coupling (Glaser coupling) in aqueous ammonium hydroxide solutions or hot pyridine (Eglington, 1963). This reaction has been used for the synthesis of macrocycles (Sondheimer, 1963). Cu(II) probably interacts with the alkyne groups and favors a bent conformation. Nickel(0) forms a similar π-complex with three butadiene molecules at low temperature. This complex rearranges spontaneously at 0 °C to afford a bi-sallylic system, from which a large number of interesting olefins can be obtained. The scheme below and the example of the synthesis of the odorous compound muscone (Baker, 1972, 1974; Kozikowski, 1976) indicate the variability of such re-arrangements (P. Heimbach, 1970). Many rather complicated cycloolefins are synthesized on a commercial scale by such reactions and should be kept in mind as possible starting materials for a,ω-bisfunctional chains, e.g. after ozonolysis. Nickel-allyl complexes prepared from $Ni(CO)_4$ and allyl bromides are useful for the olefination of alkyl bromides and iodides (Corey, 1967 B; Kozikowski, 1976). The reaction has also been extended to the synthesis of macrocycles (Corey, 1967 C, 1972 A).

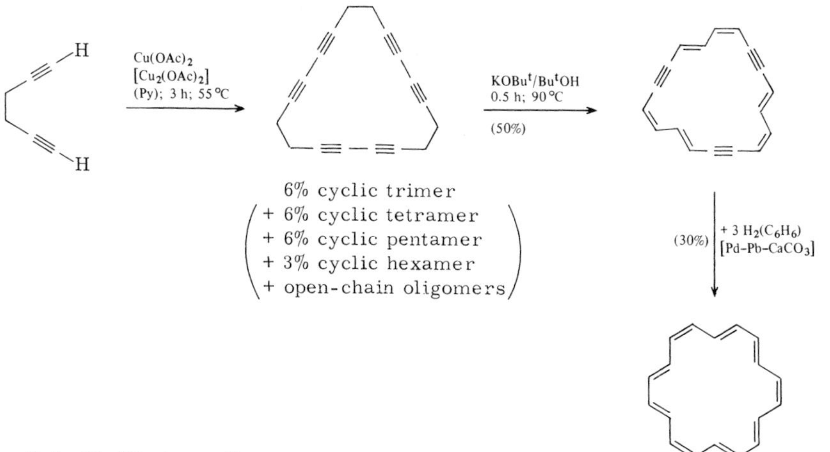

Cu(OAc)$_2$
[Cu$_2$(OAc)$_2$]
(Py); 3 h; 55 °C

KOBu^t/Bu^{t}OH
0.5 h; 90 °C

(50%)

6% cyclic trimer
+ 6% cyclic tetramer
+ 6% cyclic pentamer
+ 3% cyclic hexamer
+ open-chain oligomers

(30%) | + 3 H$_2$(C$_6$H$_6$)
[Pd-Pb-CaCO$_3$]

Cadiot-Chodkiewicz coupling:

[18] annulene

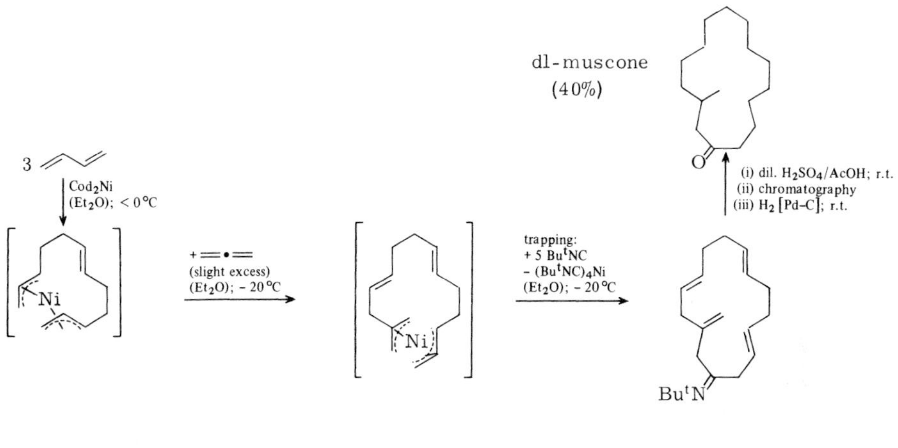

dl-muscone
(40%)

3

Cod$_2$Ni
(Et$_2$O); < 0 °C

+ ═•═
(slight excess)
(Et$_2$O); – 20 °C

trapping:
+ 5 Bu^{t}NC
– (Bu^{t}NC)$_4$Ni
(Et$_2$O); – 20 °C

(i) dil. H$_2$SO$_4$/AcOH; r.t.
(ii) chromatography
(iii) H$_2$ [Pd-C]; r.t.

Ni

Ni

Bu^{t}N

The kekulene macrocycle consists of twelve annellated benzene rings and may be considered as an [18]annulene (inside) or a [30]annulene (outside). Staab (Diederich, 1978) called it a "superbenzene", since it has the same D$_{6h}$ symmetry as benzene. 2,19-Dimethyl-6,7,9,10,23, 24,26,27-octahydro-2,19-dithionia-[3,3] (3,11)dibenz[a,j]-anthracenophane-bis(fluorosulfonate) (A) was synthesized by procedures similar to those described earlier for other cyclophanes. (A) was treated with potassium t-butoxide (Stevens rearrangement), methyl fluorosulfonate (methylation), and again potassium t-butoxide (elimination of dimethyl sulfide), to give the vinylene-bridged octahydrodibenzanthracenophane (B). Short irradiation with iodine caused cyclodehydrogenation of both cis-stilbene units. The resulting octahydrokekulene was dehydrogenated with DDQ to kekulene.

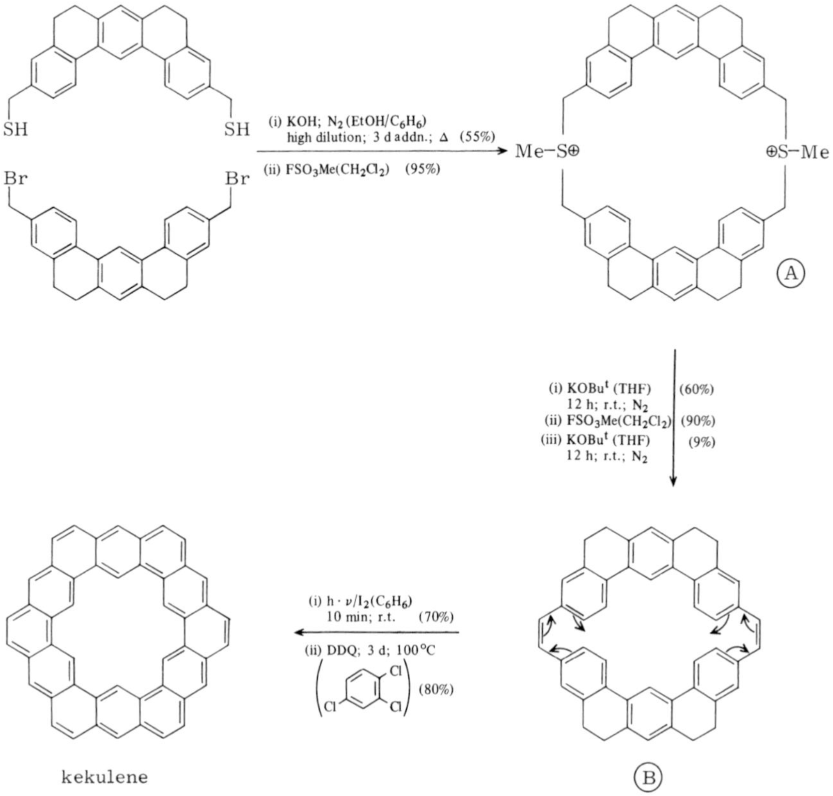
SH
SH
Br
Br
(i) KOH; N2 (EtOH/C6H6) high dilution; 3 d addn.; Δ (55%)
(ii) FSO3Me(CH2Cl2) (95%)
Me–S⊕
⊕S–Me
A
(i) KOBut (THF) 12 h; r.t.; N2 (60%)
(ii) FSO3Me(CH2Cl2) (90%)
(iii) KOBut (THF) 12 h; r.t.; N2 (9%)
(i) h · ν/I2(C6H6) 10 min; r.t. (70%)
(ii) DDQ; 3 d; 100°C
Cl
Cl
Cl
(80%)
kekulene
B

Macrocycles containing up to 24 benzene rings in a hexagon have been synthesized in multi-step Suzuki cross couplings of oligophenylene units (Hensel, 1999). Zirconocene coupling of oligophenylene-alkynes yields up to 90% of a hairpin-shaped macrocycle containing 18 benzene rings and two connecting alkyne groups. No dilution techniques are required here, because the ligands of zirconocene dichloride do not allow for polymerization, acting instead as a capping reagent (Nitschke, 2001).

R = C₆H₁₃

Two arenes connected by alkyl chains in a macrocycle are called cyclophanes. They may be effective in charge separation, if an acceptor and a donor are fixed at appropriate distances. Two efficient syntheses of strained cyclophanes indicate the synthetic potential of allyl or benzyl sulfide intermediates, in which the combined nucleophilicity and redox activity of the sulfur atom can be used. The dibenzylic sulfides from xylylene dihalides and dithiols can be methylated with dimethoxy-carbenium tetrafluoroborate (Meerwein, 1960; Borch, 1968, 1969; from trimethyl orthoformate and BF_3, 3:4). The sulfonium salts are deprotonated and rearrange to methyl sulfides (Stevens rearrangement). Repeated methylation and Hofmann elimination yields double bonds (Mitchell, 1974). Treatment of dibenzylic sulfides with triethylphosphite and UV light also led to cyclophanes in high yield (Staab, 1979). A diene-cyclophane was also obtained by a Stevens rearrangement and then aromatized to a pyrene-benzene cyclophane using a dehydrogenation with DDQ. The benzene ring was thus nestled into the concave face of a bent pyrene, indicating ring strain (Bodwell, 2001).

2.5
Cyclopropane and Cyclopropene Derivatives

The majority of preparative methods which have been used for obtaining cyclopropane derivatives involve carbene addition to an olefinic bond. Carbenes are usually applied in the form of zinc organyls. They are formally transition metal carbon compounds, but their closed d^{10}-shell renders their Grignard-type carbon compounds quite stable. They are made from alkyl iodides or Grignard compounds and zinc-copper alloys and always occur as stable dimers in THF solution. They are less reactive than the lithium and magnesium analogs and are occasionally used to transform esters to alkylated ketones (Reformatsky reaction). More important is their reaction with diiodomethane or other *gem*-diiodides (Nishimura, 1969; Wendisch, 1971; Denis, 1972; Simmons, 1973; Girard, 1974). Both iodine atoms are left in a monozinc adduct which then converts alkenes to cyclopropane derivatives. If acetylenes are used in the reaction, cyclopropenes are obtained. Heteroatom-substituted or vinyl cyclopropanes come from alkenyl bromides or enol acetates (de Meijere, 1979; Corey, 1975 B; Wenkert, 1970 A). The carbenes needed for cyclopropane syntheses can also be obtained *in situ* by a-elimination of hydrogen halides with strong bases (Köster, 1971; Corey, 1975 B), or by copper-catalysed decomposition of diazo compounds (Wenkert, 1970 A; Burke, 1979; Turro, 1966), or by reductive elimination of iodine from *gem*-diiodides (Nishimura, 1969; Wendisch, 1971; Denis, 1972; Simmons, 1973; Girard, 1974).

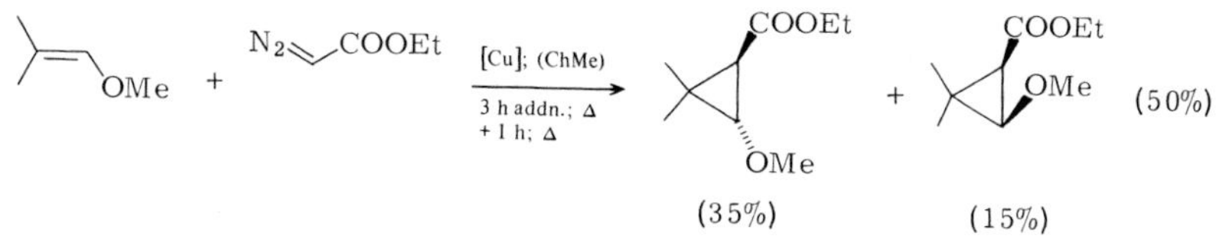

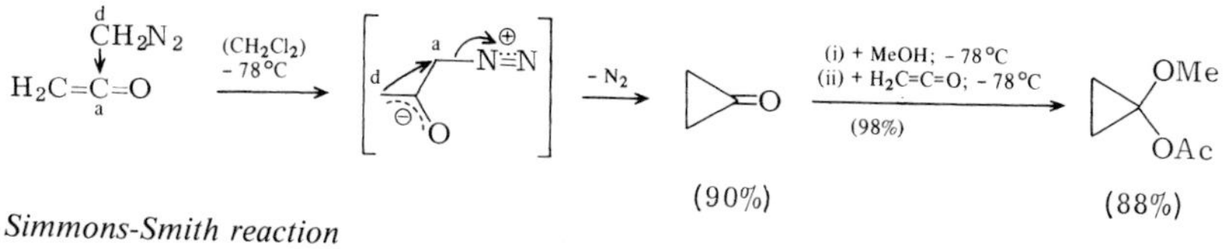

$$N_2\diagup COOEt$$

$$\underset{\substack{3\text{ h addn.; }\Delta \\ +1\text{ h; }\Delta}}{\xrightarrow{\text{[Cu]; (ChMe)}}}$$

(35%) (15%) (50%)

(90%) (98%) (88%)

Simmons-Smith reaction

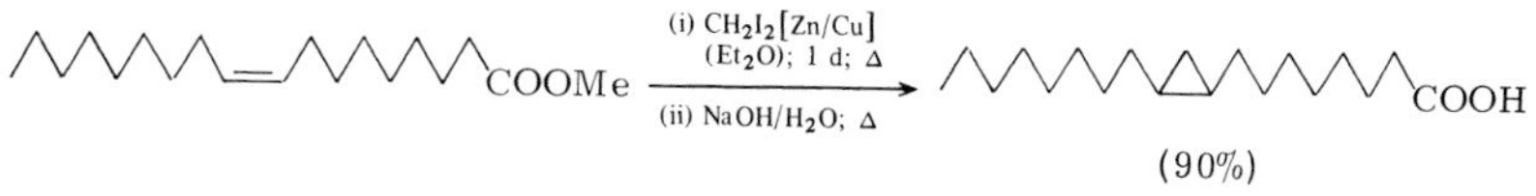

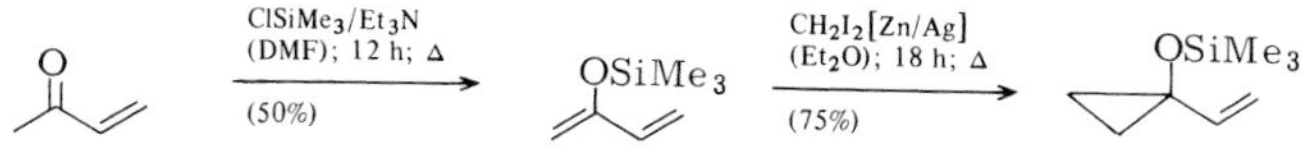

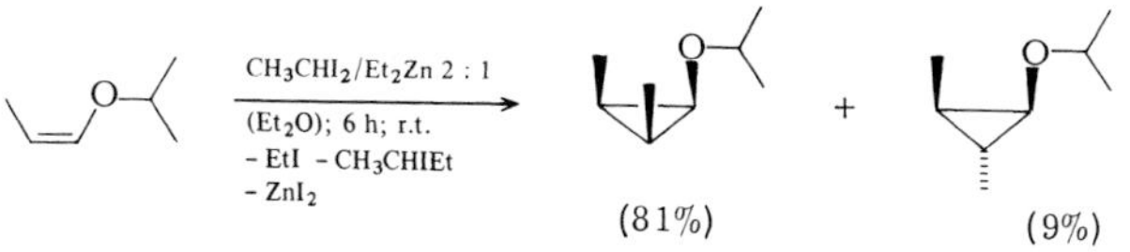

(i) CH₂I₂[Zn/Cu]
(Et₂O); 1 d; Δ
(ii) NaOH/H₂O; Δ

(90%)

ClSiMe₃/Et₃N
(DMF); 12 h; Δ
(50%)

CH₂I₂[Zn/Ag]
(Et₂O); 18 h; Δ
(75%)

CH₃CHI₂/Et₂Zn 2 : 1
(Et₂O); 6 h; r.t.
– EtI – CH₃CHIEt
– ZnI₂

(81%) (9%)

Another widely used route to cyclopropanes involves the addition of sulfonium ylides to *a,β*-unsaturated carbonyl compounds (Landor, 1967; Sowada, 1971; Johnson, 1973B, 1979; Trost, 1975A). Nonactivated double bonds are not attacked. Steric hindrance is of little importance in these reactions because the C–S bond is extraordinarily long and the sulfur atom is relatively far away from the ylide carbon bound to it. The last example shown below illustrates control of regio- and stereoselectivity in an intramolecular reaction (Matthew, 1971).

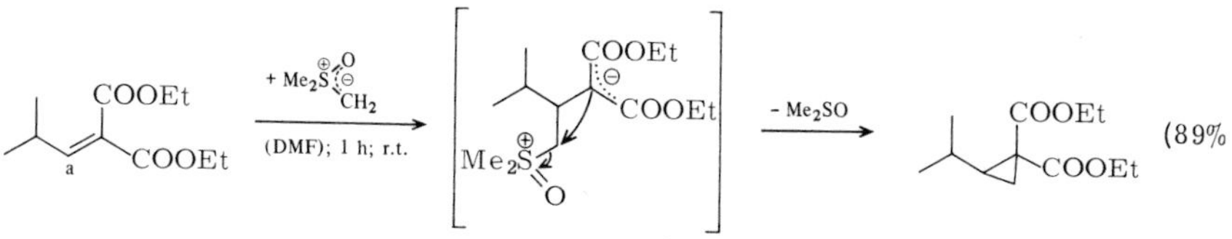

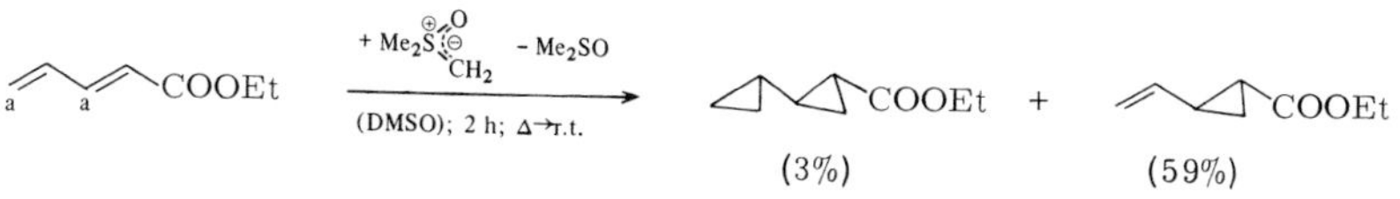

(89%)

(3%) (59%)

Cyclopropane derivatives are synthetic intermediates, which behave like olefins (de Meijere, 1979). Cyclopropane derivatives with one or two activating groups are easily opened, often in a highly regio- and stereoselective manner (Wenkert, 1970 A, B; Corey, 1956 A, B, 1975). Many appropriately substituted cyclopropane derivatives yield 1,4-difunctional compounds under mild nucleophilic or reductive reaction conditions. Such compounds are especially useful in syntheses of cyclopentenone derivatives and of heterocycles.

(Bromomethyl)- or (hydroxymethyl)cyclopropane derivatives undergo acid-catalyzed homoallylic rearrangements to yield *trans*-olefins (McCormick, 1975; Brady, 1968; Julia, 1974). This rearrangement is the basis of Julia's terpene synthesis.

R = alkyl, aryl   R' = H   : 95–100% trans
R' = alkyl : ⩽75% trans

A dioxaborolane ligand derived from (*R*,*R*)-(+)-*N*,*N*,*N*,*N*-tetramethyltartaric acid diamide works as an efficient controller for the enantioselective conversion of allylic alcohols to substituted cyclopropylmethanols. Its activity relies on the presence of an acidic boron and a basic amide site that allows the simultaneous complexation of acidic halomethylzinc reagents and the basic allylic metalalkoxides (Nicolaou, 2001).

Me₂NOC　　　CONMe₂

An azetidine-ligated dirhodium(II) catalyst that possesses an *l*-menthyl ester attachment provided significant diastereocontrol (3:1) and high enantiocontrol for the formation of *cis*-cyclopropane products from reactions of substituted styrenes with diazo esters (Hu, 2002).

The reaction between an (*R*)-epichlorohydrin and phenylsulfonylacetonitrile in the presence of NaOEt in EtOH followed by treatment with acid gives chiral cyclopropane lactones with 98% e.e. in 82% yield. They were converted into both the *cis*- and *trans*-chiral cyclopropane units via reductive desulfonylation with Mg/ MeOH. The corresponding enantiomers were prepared starting from (*S*)-epichlorohydrin (Kazuta, 2002).

Three-dimensional cyclopropane derivatives have also been synthesized. 2-Butyne trimerizes in the presence of aluminum chloride to give hexamethyl "Dewarbenzene" (Schäfer, 1967). Its irradiation leads not only to aromatization but also to hexamethylprismane (Lemal, 1966). Highly substituted prismanes may also be obtained from the corresponding benzene derivatives by irradiation with 254 nm light. The rather stable prismane itself was synthesized via another C_6H_6 hydrocarbon, namely benzvalene, a labile molecule (Katz, 1971, 1972).

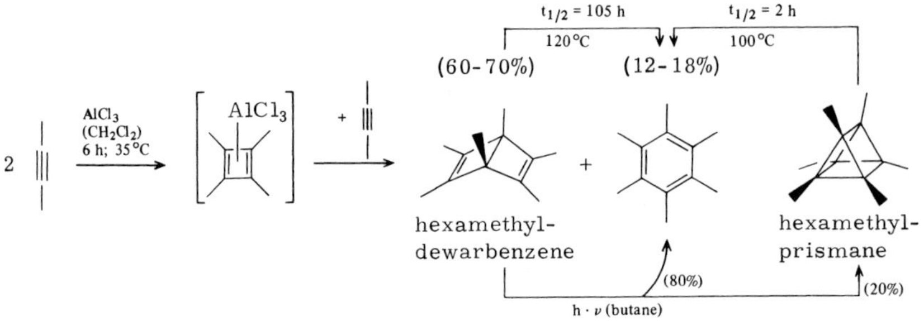

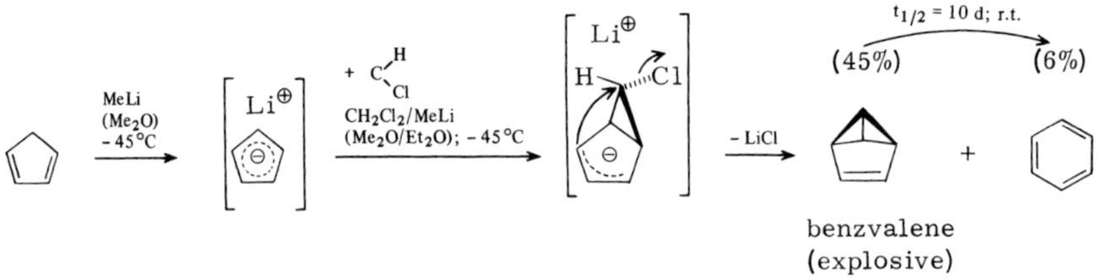

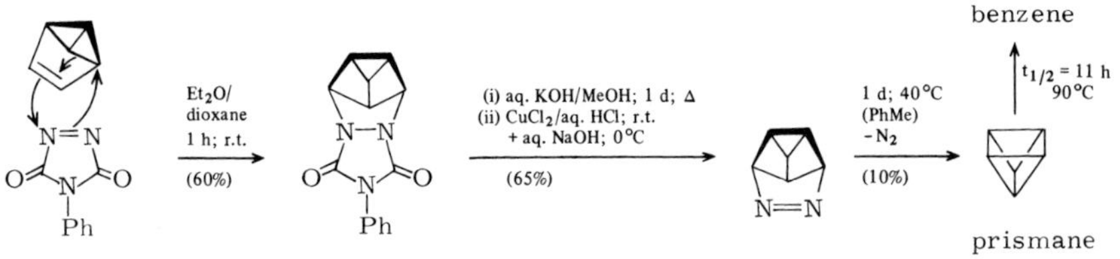

hexamethyl-dewarbenzene

hexamethyl-prismane

benzvalene (explosive)

benzene

prismane

Benzvalene (tricyclo[3.1.0.0^{2,6}]hex-3-ene) was first prepared by photoirradiating benzene, but the amount that could be made in this way was tiny. The combination of cyclopentadiene, methylene chloride, and methyllithium has made benzvalene available easily and in quantity and has allowed it to be used as the starting material for the synthesis of a variety of its derivatives and of the parents of a number of other skeletons (Katz, 1999). The preparative procedure is included here as an example for a modestly explosive compound.

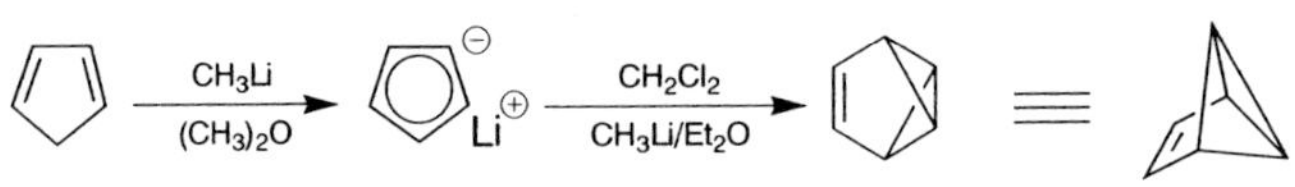

A flame-dried 2 L round-bottomed flask was fitted with an airtight mechanical stirrer, dropping funnel, and low-temperature thermometer and was connected via a branched inlet to a source both of gases and (through a tube filled with Drierite) aspirator vacuum (the "gas inlet") and to a stopcock sealed with a serum-bottle cap. Methyllithium in diethyl ether (400 mL, 1.6 M, containing 0.4% lithium chloride) was introduced by syringe. The aspirator was then used to remove the ether. While the flask was cooled in a bath of dry ice and acetone, dimethyl ether (ca. 1 L) was distilled from $LiAlH_4$ into the flask through the gas inlet. The temperature was raised to $-35\,°C$, and cyclopentadiene (52 mL, 640 mmol) was added slowly by syringe (at ca. $1\,mL\,min^{-1}$). After the vigorous evolution of methane had subsided, methylene chloride (60 mL, 940 mmol) was added to the slurry again at ca. $1\,mL\,min^{-1}$). Methyllithium in ether (440 mL, 1.6 M, containing 0.4% lithium chloride) was then added in drops from the dropping funnel. (The slurry was now yellow.) While the apparatus was flushed with nitrogen, the stopcock with the serum-bottle cap was replaced by a distillation head, condenser, and receiver, all of which had been washed with aqueous ammonia and dried. The dimethyl ether was distilled at atmospheric pressure and ambient temperature into a receiver that was cooled in dry ice–acetone. When the pot temperature reached $20\,°C$, the receiver was replaced by a 500 mL flask, which was cooled in the dry ice–acetone mixture. The product, a solution of benzvalene in ether, was distilled at aspirator vacuum into this flask. The yield was ca. 400 mL of 0.7 M benzvalene in ether. The benzvalene was contaminated with only 5–10% of benzene. In seven experiments, the yields of benzvalene ranged from 30 to 59% and averaged 45 (±5%). **Caution**: The distillation residue should be destroyed cautiously, for explosions have occurred when the residue was simply exposed to air or water. While the flask was being flushed with nitrogen, the distillation head was replaced by a stopcock sealed with a serum-bottle cap connected to a Teflon tube through which diethyl ether (100 mL) was syringed. The mixture was stirred, and wet ether (20 mL) was added in drops, followed by methanol (100 mL) and, very slowly, water (100 mL). Pure benzvalene can be isolated by GLPC, but it explodes when scratched.

2.6
Cyclobutane Derivatives

Only relatively few examples of interesting target molecules containing C_4 rings are known. These include caryophyllene (Corey, 1963 A, 1964) and cubane (Barborak, 1966). The photochemical [2+2]-cycloaddition applied by Corey yielded mainly the *trans*-fused isomer, but isomerization with base led via an enolate to the formation of the more stable *cis*-fused caryophyllene.

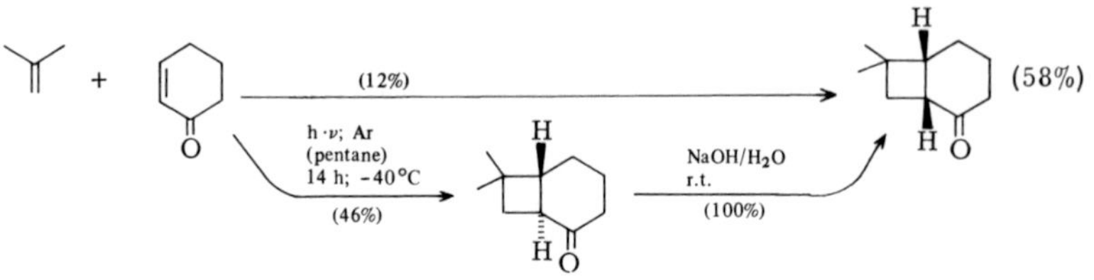

The syntheses of C-12-C-13-cyclobutyl aldehydes were carried out as shown below. Starting from a monoacetate, readily available through enzymatic group-selective saponification of the corresponding diacetate, *cis*-aldehyde was prepared by Dess–Martin periodinane oxidation (95% yield), while the corresponding *trans*-aldehyde was conveniently available by base-catalyzed epimerization (88%; Nicolaou, 2001).

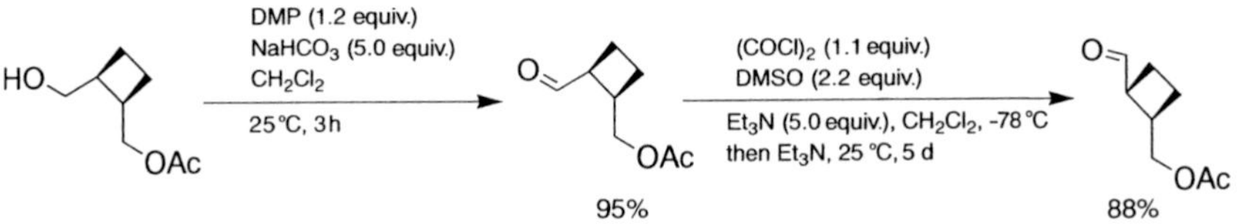

Cycloaddition of two Michael-systems by electrochemical reduction yielded the carbonyl-substituted cyclobutanes (Roh, 2001). Intramolecular Michael addition did the same (Takasu 2001).

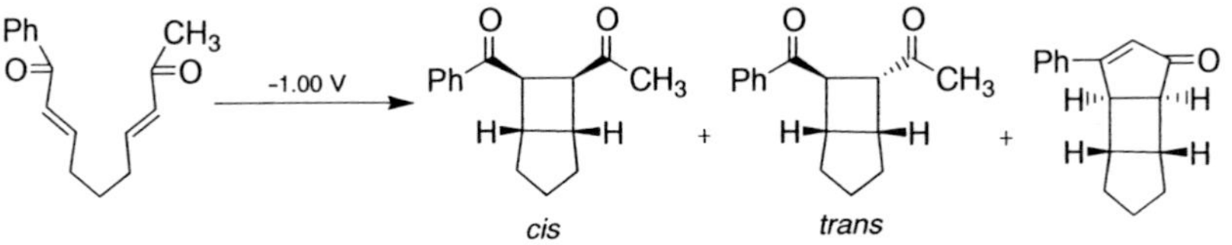

Within the cubane synthesis the initially produced cyclobutadiene moiety was only stable as an iron(0) complex (Avram, 1964; Emerson, 1965; Cava, 1967). When this complex was destroyed by oxidation with cerium(IV) in the presence of a "dienophilic" quinone derivative, the cycloaddition took place immediately. Irradiation lead to a further cyclobutane ring closure. The cubane synthesis also exemplifies another general approach to cyclobutane derivatives. This starts with cyclopentanone or cyclohexanedione derivatives, which are brominated and treated with strong base. A Favorskii rearrangement then leads to ring contraction (Barborak, 1966).

Ketones and aldehydes have been cyclized with diphenylsulfonium cyclo-propylide followed by Li⁺-catalyzed rearrangement to produce cyclobutanone derivatives (Trost, 1973). These compounds react on heating with *tert*-butoxy-bis(dimethylamino)methane to give vinylogous amides (Bredereck, 1963, 1965, 1968). Reaction with 1,3-propanedithiol ditosylate in buffered ethanol leads to replacement of the dimethylaminomethylene unit by the 1,3-dithiane unit. The yield of this reaction sequence is often close to quantitative and correlates with the molecular size of the carbonyl educt because small vinylogous amides are much more sensitive to decomposition than are larger ones. The resulting spirobicyclic systems may then be opened by nucleophiles, and the 1,3-dithiane can be hydrolyzed. 1,4-Dicarbonyl butanones are formed. The versatility of the dithiane group as an acyl anion equivalent allows various modifications of this reaction scheme. Condensation reactions of the 1,4-dicarbonyl functions lead to cyclopentenone rings. Thus a method for spiroannelation of such a ring on a carbonyl-containing ring is given (Trost, 1975 B). Ring-contraction of vinylfuranoses with samarium diiodide has also been applied for the preparation of substituted cyclobutanes (Aurrecoechea, 2000). The samarium ion adds to the vinyl as well as to the aldehyde group of the carbohydrate and reduction by SmI₂ leads to a biradical, which closes to form 1-vinyl-2-ols.

Highly strained polycycloalkanes consisting of several interconnected cyclopropane or cyclobutane rings undergo thermal and hydrogenolytic ring opening with great ease (Kaufmann, 1979). Those bonds which release the largest amount of strain energy are cleaved selectively.

Simple cyclobutanes do not readily undergo such reactions, but cyclobutenes do. Benzocyclobutene derivatives tend to open to give extremely reactive dienes, namely *ortho*-quinodimethanes (examples of syntheses see in sections 2.8.2 and 4.5). Benzocyclobutenes and related compounds are obtained by high-temperature elimination reactions of bicyclic benzene derivatives such as 3-isochromanone (Spangler, 1973, 1976, 1977), or more conveniently in the laboratory, by Diels–Alder reactions (Thummel, 1974) or by cyclizations of silylated acetylenes with 1,5-hexadiynes in the presence of (cyclopentadienyl)dicarbonylcobalt (Aalbersberg, 1975; Thummel, 1980).

2.7
Cyclopentane Derivatives

Direct a-dialkylation of a ketone enolate anion with an a,ω-dihalide leads preferentially to five- or six-membered rings. Neither dialkylation nor the formation of larger or smaller rings are generally competitive (Mousseron, 1957; Johnson, 1963). Intramolecular reactions are usually much faster than intermolecular ones, and the reversibilty of enol alkylations under strongly basic conditions leads to the thermodynamically most stable and entropically most likely products.

Cyclopentanone derivatives are classically made by base-catalyzed intramolecular aldol or ester condensations (Ellison, 1973). An important example is 2-methylcyclopentane-1,3-dione. It was prepared by intramolecular acylation of diethyl propionylsuccinate dianion followed by saponification and decarboxylation. This cyclization only worked with potassium *t*-butoxide in boiling xylene (Bucourt, 1965). Faster routes to this diketone start with succinic acid or its anhydride. A Friedel–Crafts acylation with 2-acetoxy-2-butene in nitrobenzene or with propionyl chloride in nitromethane leads to acylated adducts, which are deacylated in aqueous acids (Grenda, 1967; Schick, 1969). Another versatile route to substituted cyclopent-2-enones makes use of intermediate 5-nitro-1,3-diones. A lithium enolate first reacts with an acid chloride bearing an ω-nitro group. Under mildly basic conditions the ketone is recovered and the deprotonated methylene group next to the nitro group becomes nucleophilic and adds to the carbonyl double bond. The nitro group leaves the molecule to release α,ω-unsaturated ketones (Seebach, 1977).

Cyclopentene-1-carboxaldehydes are obtained from cyclohexene precursors by the sequence cyclohexene → cyclohexane-1,2-diol open-chain dialdehyde → cyclopentane aldol. The main advantage of this ring contraction procedure is, that the regio- and stereoselectivity of the Diels–Alder synthesis of cyclohexene derivatives can be transferred to cyclopentane synthesis (Stork, 1953; Büchi, 1968).

A closely related synthesis of the cyclopentanone ring in steroids also introduced the angular methyl group (Johnson, 1957). The α-methylene group of a cyclohexanone precursor is protected by aldol condensation with an aromatic aldehyde, the tertiary α'-carbon atom methylated with methyl iodide, and oxidative cleavage with alkaline H_2O_2 converted the 2-(arylmethylene)cyclohexanone into a 1,6-dioic acid. Dieckmann condensation of the corresponding diester, saponification, and decarboxylation finally yielded a cyclopentanone ring.

Efficient cyclopentannelation procedures are based on the addition of a malonic ester anion to a cyclopropane-1,1-dicarboxylic ester followed by a Dieckmann condensation (Danishefsky, 1974) or addition of β-ketoester anions to a (1-phenylthiocyclopropyl)phosphonium cation followed by intramolecular Wittig reaction (Marino, 1975). Another procedure starts with a [2+2]-cycloaddition of dichloroketene to alkenes followed by regioselective ring expansion with diazomethane. The resulting 2,2-dichlorocyclopentanones can be converted to a large variety of cyclopentane derivatives (Greene, 1979; Déprés, 1980).

R O + a SPh PPh₃⊕ → [R a O SPh PPh₃ d COOEt] → R SPh R' COOEt HCl/H₂O (dioxane) 60 °C → R O R' COOEt

R' d COOEt (THF); 3 d; Δ R' COOEt – Ph₃PO (75–80%) (85–90%) (64–72%)

R R' + [O Cl Cl] (85–90%) → R O R' Cl Cl (80–90%) CH₂N₂/Et₂O 0.5 h; r.t. → R O R' Cl Cl (80–90%) Zn/AcOH 1.5 h; 75 °C → R O R' (55–75%)

Et₃N; Δ (pentane) – HCl

Zn-Cu/POCl₃ (Et₂O); Δ – ZnCl₂

O Cl Cl Cl

O Cl Cl Cl Cl

Cyclopentanones may also be synthesized from α,β-unsaturated ketones and diiodomethane. The ketone is converted to the O-silyl enol, and carbene is added to the enol double bond using the Simmons–Smith reaction. Thermal rearrangement of the resulting 1-siloxy-1-vinylcyclopropane and acid-catalyzed hydrolysis of the silyl enol ether leads to cyclopentanones in excellent yields (C. Girard, 1974). Very high temperatures, however, are needed, and this obviously limits the generality of this rearrangement reaction. A condensation of an acid chloride with an enolate anion under anhydrous conditions yields ketones. Allylsilanes can be decomposed with ether-soluble fluorides to give an alkylanion, which undergoes an ester condensation with a neighboring ester group (Hirose, 2002).

O → Et₃N/ClSiMe₃ (DMF); 12 h; Δ (70%) → CH₂I₂/Zn-Ag (Et₂O); 18 h; Δ (70%) → OSiMe₃ (i) 0.5 h; 330 °C vapor phase (ii) dil. HCl; r.t. (> 95%) → H O H (47%)

TBSO N H CO₂Me + COCl Pr TMS → LDA → TBSO N H MeO₂C O Pr TMS (88%) → TBAT DMF → HO N H O O O (52%) (+)-Madindoline A

Another popular route to cyclopentanes leads over the ring contraction of 2-bromo-cyclohexanones by a Favorskii rearrangement to give cyclopentanecarboxylic acids. If α,β-dibromoketones are used, ring opening of the intermediate cyclopropanone leads selectively to β,γ-unsaturated carboxylic acids (S. A. Achmad, 1963, 1965; J. Wolinsky, 1965). The resulting methylenecyclopentanes are easily modified, e.g. by ozonolysis, hydroboration, etc., and thus a large variety of interesting cyclopentane derivatives is accessible.

Stereoselective allylic alkylations have been carried out with the aid of palladium catalysts. The 17-(Z)-ethylidene groups of steroids (obtained from the ketones by Wittig olefination) form π-allyl palladium complexes, and addition of equimolar amounts of copper(II) salts recovers palladium(II) from palladium(0) (Trost, 1974, 1976). Their alkylation with dimethyl malonate anions in the presence of 1,2-ethanediylbis(diphenylphosphine) (=diphos) gives a reaction exclusively at the side chain and only the (20S) products. If one starts with the endocyclic 16,17 double bond and replaces an (S)-20-acetoxy-group by using terakis(triphenylphosphine)palladium, the substitution occurs with complete retention of configuration, resulting from two complete inversions (Trost, 1976).

Versatile [3+2]-cycloaddition pathways to five-membered carbocycles involve the "trimethylenemethane" (= 1-methylene-propanediyl) synthons (Trost, 1986). Palladium(0)-induced 1,3-elimination at suitable reagents generates a reactive η^3-2-methylene-1,3-propanediyl complex which reacts highly diastereoselectively with electron-deficient olefins.

Palladium-catalysed cycloisomerizations of 6-en-1-ynes lead most readily to five-membered rings. Palladium binds exclusively to terminal $C \equiv C$ triple bonds in the presence of internal ones and induces cyclizations with high chemoselectivity. Synthetically useful bis-exocyclic 1,3-dienes have been obtained in high yields, which can, for example, be applied in Diels–Alder reactions (Trost, 1989).

A highly diastereoselective alkenylation of cis-4-cyclopentene-1,3-diols has been achieved with O-protected (Z)-1-iodo-1-octen-3-ols and palladium catalyst (Torii, 1989). The (E)-isomers yielded 1:1 mixtures of diastereomeric products. The (Z)-alkenylpalladium intermediate is thought to undergo syn-addition to the less crowded face of the prochiral cyclopentene followed by syn-elimination of a hydropalladium intermediate.

R = SiBu^{t}Me$_2$, CMe$_2$(OMe)

cis-Cyclopentane-1,3-dialdehyde has been made by dihydroxylation and periodate treatment of norbornene. This chiral dialdehyde is an ideal precursor for U-shaped receptor molecules (Hubbard, 2001). Ring-opening metathesis (ROM) of norbornyl ethers with styrene gives substituted cyclopentyl dienes in >98% ee and trans-olefin selectivities (La, 2001).

Ring opening of carbonyl or 1-hydroxyalkyl substituted cyclopropanes, which operate as a^4-synthons, yield 1,4-disubstituted products with d^0-Synthons, e.g. hydroxide or halides (Wenkert, 1970A). (1-Hydroxyalkyl)- and (1-haloalkyl)-cyclopropanes are rearranged to homoallylic halides, e.g. in Julia's method of terpene synthesis (Julia, 1961, 1974; Brady, 1968; McCormick, 1975).

The reductive coupling of aldehydes or ketones with α,β-unsaturated carboxylic esters by >2 mol samarium(II) iodide (Soderquist, 1991) provides a convenient route to γ-lactones (Otsubo, 1986). Intramolecular coupling of this type may produce *trans*-2-hydroxycycloalkaneacetic esters with high stereoselectivity, if the educt is an (*E*)-isomer (Enholm, 1989 A, B).

α-Alkylidene lactones with halogenated side chains were prepared from unsaturated diols (not shown). Treatment with tin hydrides produced radicals and preferentially yielded trisubstituted cyclopentane derivatives and smaller amounts of cyclohexane isomers (Kim, 2001).

R = Me, *n*-Bu, Ph or SiMe$_3$

Regioselective Michael additions occur with phosphine catalysts if one of two competing Michael acceptors is less electrophilic than the other. Phosphine adds, for example, to the C–C double bond of an enone rather than to an enoate. Many highly substituted cyclopentene derivatives are thus accessible from hexadienes with terminal electron withdrawing groups (EWGs) (Frank, 2002).

(80%) (regiosel. 95:5)

Zirconium(II) Cp_2 is a source of a 14d-electron species with two valence-shell empty orbitals. They may form π-complexes with two alkyne or alkene units. At least one filled nonbonding orbital is available for σ-carbometallation (Negishi, 1994). $ZrCp_2$ generated from Cp_2ZrCl_2 and Mg, $HgCl_2$ or n-BuLi is effective for converting enynes or diynes to zirconia bicycles, which loose the zirconium upon simple treatment with water (Kasatkin, 2000).

In allylic ethers such cycles rearrange in the presence of BF_3, and allylic zirconium intermediates are formed. Protected 6-deoxy-6-vinyl-glucose, for example, gives a zirconium organyl, which is also ligated by the aldehyde group at C-1. The pyrane then rearranges stereoselectively to a cyclopentane ring with three OH-groups in the original glucose configuration and the ethyl groups cis to the 1-OH and trans to the 5-OH groups (Ito, 1993; Nitschke, 2001). Samarium diiodide has a similar effect (Aurrecoechea, 2000).

Several interesting polycyclic $C_{10}H_{10}$ hydrocarbons have been obtained from cyclooctatetraene and maleic anhydride. Thermal cycloaddition, photochemical [2+2]-cycloaddition, and oxidative decarboxylation yield basketene, which may be rearranged in almost quantitative yield to snoutene. Irradiation with UV light and silver(I)-catalyzed rearrangement give triquinacene, a tricyclic triene which was made with the anticipation that it could be dimerized to dodecahedrane (Eaton, 1979; Paquette, 1977, 1979).

basketene

snoutene   diademane   triquinacene

Spherical, pentagonal dodecahedrane is the thermodynamically most stable $C_{20}H_{20}$ polycycloalkane. It is the so-called "$C_{20}H_{20}$ stabilomer". It should therefore be available by thermodynamically controlled, e.g. acid-catalyzed, isomerization of less stable $C_{20}H_{20}$ isomers. Experiments along this line, e.g. treatment of the basketene photodimer with Lewis acids, were, however, unsuccessful (Eaton, 1979) until the "pagodane route" was discovered (Fessner, 1987 A). Attempts to dimerize triquinacene, $C_{10}H_{10}$, or to combine a C_{15}-unit ("peristylane") with a C_5-unit (Paquette, 1979; Grahn, 1981) have so far failed, because the carbon–carbon bonds that have to be formed are all in the unfavourable *endo* position. Strong preference for *exo,exo* carbon–carbon bond formation was evident in reductive coupling reactions (Paquette, 1979).

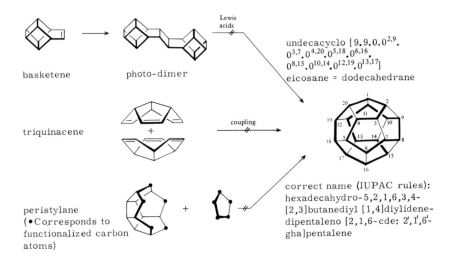

basketene   photo-dimer

Lewis acids

undecacyclo [9.9.0.0^{2,9}.0^{3,7}.0^{4,20}.0^{5,18}.0^{6,16}.0^{8,15}.0^{10,14}.0^{12,19}.0^{13,17}] eicosane = dodecahedrane

triquinacene

+

coupling

correct name (IUPAC rules): hexadecahydro-5,2,1,6,3,4-[2,3]butanediyl [1,4]diylidene-dipentaleno [2,1,6-cde: 2',1',6'-gha]pentalene

peristylane (●Corresponds to functionalized carbon atoms)

+

The synthesis of 1,16-dimethyldodecanedrane starts with a "pincer Diels–Alder reaction" between 9,10-dihydrofulvalene and dimethyl acetylenedicarboxylate. The reaction involves initial intermolecular cycloaddition of the dienophile to the first 1,3-diene moiety and subsequent reaction of the residual double bond of the acetylenic dienophile in intramolecular bond formation with the second diene (Paquette, 1974). The resulting diester (**C**) hydrolyzed to the diacid. Its iodolactonization proceeded with high efficiency to give (**D**). Cleavage with methanolic sodium methoxide at room temperature, Jones oxidation, and reductive deiodination with a zinc-copper couple furnished isomerically pure diketoester (**E**) (Paquette, 1976, 1979). To the dimethyl ester of this "cross-corner oxidized" C_{14}-frame, six more carbon atoms were added symmetrically by spiroannelation with cyclopropyldiphenylsulfonium ylide. The dispiro-diketodiester (**F**) therefore contains all twenty carbon atoms of the dodecahedrane molecule. Sixteen more steps were performed to build up all the five-membered rings necessary for the final closure of the polycyclic sphere. In this sequence of delicate reactions it was necessary to replace two acidic hydrogen atoms neighboring carboxylic ester groups by methyl substituents. The final acid-catalyzed rearrangement of olefin (**G**) gave 1,16-dimethyldodecahedrane which was characterized by an X-ray structure determination (Paquette, 1982; Christoph, 1982).

Pagodane (Fessner, 1983, 1987 B), another interesting $C_{20}H_{20}$ hydrocarbon, rearranges to dodecahedrane on heating with hydrogenation catalysts (Fessner, 1987 A).

The strained undecacyclic pagodane framework was obtained in a series of 14 one-pot operations with an overall yield up to 24% from commercial isodrin. The key steps are (i) a benzene–benzene [6+6]photocycloaddition, and (ii) a domino Diels–Alder reaction. As a consequence of the rigid face-to-face orientation, there are strong electronic interactions between the benzene rings in the dibenzo-anellated isodrin derivative. Irradiation with 254 nm UV light gave rise to a 7:3 equilibrium mixture of the educt with the [6+6]cycloaddition isomer. At an irradiation wavelength of 300 nm the cycloaddition was completely reversed.

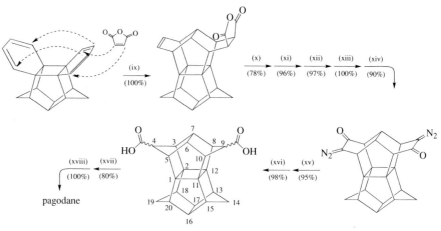

With the acquisition of the first birdcage, the stage was set for a domino Diels–Alder addition of an acetylene equivalent, with the two *syn*-oriented, 1,3-cyclohexadiene units to produce the other half of the framework. With maleic anhydride, a classic dienophile, this was easily verified. The yellow color of a charge-transfer complex instantly developed in benzene solution and slowly faded on boiling. A single product with the pagodane topology formed in quantitative yield. Pagodane was obtained in a few steps after removal of functional groups and ring contraction.

(i) 2 d; 110 °C
(ii) conc. H$_2$SO$_4$; –CO; 7h; 80→140 °C
(iii) Li/Bu^{t}OH/THF; 15 h; Δ
(iv) = (i) xylene; 12 h; Δ
(v) Na/Bu^{t}OH/THF; 1 d; Δ
(vi) aq. TsOH; –CO (MeOH/butanone); 16 h; Δ
(vii) [10% Pd-C] 0.5 h; 250 °C
(viii) h · v (254 nm) (2.2.4-trimethylpentane); 16 h; N$_2$

(ix) (C$_6$H$_6$) 12 h; Δ
(x) Cu$_2$O/2,2´-bipyridyl (quinoline) + H$_2$O; 40 h; Δ; –2 CO$_2$ –2H
(xi) BH$_3$/THF; 0 °C + 4 h; r.t.; + aq. NaOH/H$_2$O$_2$; 1 h; 0 °C + 10 h; r.t.
(xii) CrO$_3$/H$_2$SO$_4$ (aq. acetone); 2 h; r.t.
(xiii) HCOOMe/NaH (THF) 4 h; r.t. + 16 h; 40 °C
(xiv) TsN$_3$/Et$_3$N (THF); 8 h; r.t.
(xv) h · v; MeOH (CH$_2$Cl$_2$); 6 h
(xvi) KOH/EtOH; 16 h; Δ
(xvii) h · v; I$_2$/Pb(OAc)$_4$ (CCl$_4$); Δ→ 4,9-diiodopagodane
(xviii) Na-K/Bu^{t}OH/THF; 12 h; r.t.

2.7.1

Prostaglandins

Prostaglandins (= PG) constitute a class of hormones that are present in almost all human tissues and fluids in minute concentrations and play a dominant role in the control of pregnancy, hypertension, osteoporosis, ulcers, asthma, and pain. Some PGs play a dangerous role in inflammation processes of heart attacks, arthritic rheumatism and chemotherapy. The principal structures of prostaglandins are shown in the scheme below. The capital letters A, B, C, E, F denote the state of oxidation and the position of double bonds in the cyclopentane or cyclopentene ring; the numeral subscript refers to the number of double bonds in the side-chains: 1 = (*trans*)-13; 2 = (*cis,trans*)-6,13; 3 = (*cis,trans,cis*)-6,13,17; α(= below) and β(= above) indicate the position of the hydroxyl group on C-9.

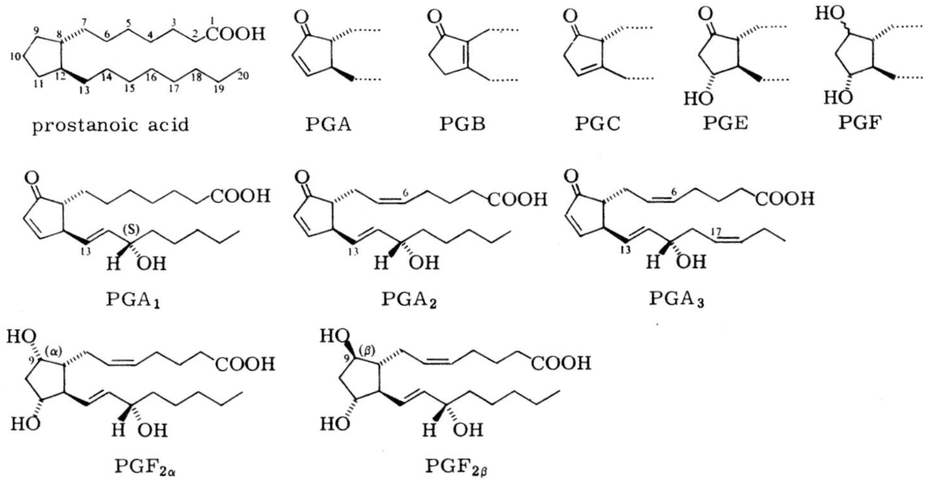

prostanoic acid   PGA   PGB   PGC   PGE   PGF

PGA$_1$ PGA$_2$ PGA$_3$

PGF$_{2\alpha}$ PGF$_{2\beta}$

The partial or full structures given below indicate the following difficulties which one has to face in synthesis of pharmacologically interesting PGE and PGF derivatives: (i) The 11-alcohol group of PGE$_{1-3}$ (β-ketol grouping) is sensitive to acids and bases which will cause dehydration (= PGA$_{1-3}$ formation) and isomerization (= PGB$_{1-3}$) formation. The β-ketol should therefore be introduced at the end of the synthesis.

(ii) PGFs will be relatively stable towards acids and bases, but their natural 9*a*-configuration produces an extra problem not present in the other prostaglandins.

We shall concentrate in our short account of PG syntheses on solutions to both problems as provided in the literature.

PGA$_2$ derivatives can be obtained in relatively large amounts from gorgonian corals. Since PGAs are of little pharmaceutical interest, they are converted commercially to the highly active PGF$_{2\alpha}$ and PGE$_2$ (Bundy, 1971, 1972). The regioselective hydration of the endocyclic double bond between C-10 and C-11 in the presence of two exocyclic double bonds can be achieved by utilizing the activating effect of the keto group. Hydrogen peroxide in the presence of small amounts of potassium hydroxide attacks only the electrophilic β-carbon atom of the conjugated enone system. The ratio of the desired a-epoxide to the β-isomer is approximately 7:3. Reductive opening of the epoxide mixture by chromium(II) acetate in buffered solution gives PGE$_2$ derivatives regioselectively, presumably because the Cr(II) ion is first bound by the epoxyketone, and reduction leads to the enol. The alcohol can be protected as the trimethylsilyl ether and the carbonyl group on C-9 reduced with NaBH$_4$. The desired 9*a*-compound, PGF$_{2\alpha}$, is formed as the major product since the large a-trimethylsilyl ether group directs the hydride reagent to the β-position (Bundy, 1972).

15-O-acetyl-PGA$_2$
methyl ester

$\alpha:\beta \approx 7:3$

Cyclopentene derivatives with carboxylic acid side-chains can be stereoselectively hydroxylated by the "iodolactonization" procedure (Corey, 1969, 1970). To the trisubstituted cyclopentene a large iodine cation is added stereoselectively to the less hindered β-side of the 9,10 double bond. Lactone formation occurs on the intermediate iodonium ion specifically at C-9a. Later the iodine is reductively removed with tri-*n*-butyltin hydride. The cyclopentane ring now bears all oxygen and carbon substituents in the right stereochemistry, and the carbon chains can be built starting from the C-8 and C-12 substituents.

(opt. pure)

$PGF_{2\alpha}$ precursor

Another synthesis (Dimsdale, 1977) leads in three highly stereoselective steps from a cyclopentadiene-ketene adduct to a bridged cyclohexanone, which can be transformed to $(\pm)$-(15 R/S)-$PGF_{2\alpha}$ by Baeyer–Villiger oxidation and chain elongation. The first step involves a highly regio- and stereoselective formation of a bromohydrin with N-bromoacetamide. The bromonium cation is only formed at the less hindered β-side of the C=C double bond and opened selectively at the less hindered 9a-position. After protection of the alcohol group with *t*-butyldimethylsilyl chloride, only one enolate is formed with the non-nucleophilic base potassium *tert*-butoxide. The enolate to the bridgehead carbon does not form (Bredt's rule). The carbanion then substitutes the bromine atom, and a tricycloheptanone is formed in the second step. This compound is susceptible to homo-Michael-type attack at the most labile bond of the highly strained cyclopropane. The cuprate reagent with the strongly nucleophilic alkenyl substituent and the less nucleophilic acetylide react at −78 °C to give the desired norbornanone in the third step. The following conventional steps of the synthesis involve Baeyer–Villiger oxidation of the ketone, DIBAL reduction of the lactone, and Wittig reaction with the resulting aldehyde.

Highly enantioselective copper-catalyzed conjugate addition reactions of diorgano-zinc reagents to enones have been reported using chiral ligands, e.g. phosphorami-dite. There is nearly complete stereocontrol in the reaction of (functionalized) dia-lkylzinc (R_2Zn) reagents with six-, seven- and eight-membered cycloalkenones. The catalytic enantioselective 1,4-addition to 2-cyclopentenone in a prostaglandin synthesis starts with an enone, an aldehyde and a functionalized zinc reagent. In the presence of 3 mol% of catalyst two carbon chains were added in 60% yield. A mixture of diastereomers (ratio 83:17), which differed in the configuration at the exocyclic stereocenter bearing the hydroxy functionality, was formed. This one-pot procedure is carried out with an enone and an enal. To differentiate between these, the unsaturated aldehyde was equipped with a removable silyl substituent, exploiting the fact that β-disubstituted enones are not reactive in 1,4-addition under these conditions. Reduction of the ketone moiety proceeded with 95% stereoselectiv-ity using $Zn(BH_4)_2$ in ether at $-30\,°C$. The prostaglandin was isolated after chroma-tography as a single isomer in 63% yield with an e.e. of 94% (Arnold, 2001).

A conceptually surprising route to prostaglandins (Johnson, 1988) involves the simple idea of adding alkenylcopper reagents stereoselectively to a protected chiral 4,5-dihydroxy-2-cyclopenten-1-one and completing the synthesis of the trisubstituted cyclopentanone by stereoselective allylation of the resulting enolate.

Two difficulties with this idea were overcome in unconventional ways: The en-one enantiomer mixture given below was obtained by photooxygenation of cyclopentadiene and *in situ* reduction to the diol (Kaneko, 1974), diacetylation dihydroxylation of the alkene followed by ketalization with acetone, and enantioselective esterase-catalyzed mono-deacetylation. The diol monoacetate underwent oxidation and β-elimination to the enone in one step with Jones' reagent. The enolate equilibration to the undesired position 10 was suppressed by dipole repulsion and angle strain in the acetonide. The surplus oxygen next to the 9-oxo group was finally removed with aluminium amalgam (not shown).

Unique chemistry is associated with the cyclopentenone: all five carbon atoms can be functionalized, and the *endo*-methyl groups of the acetonide assure clean stereoselective addition of the alkenylcopper reagent from the convex side. The use of the acetonide group to control enolate regioselectivity and to mask alcohols should be generally applicable.

The upper reaction scheme shows:

$(Bu_3P)_2Cu$ [13] (S) 15, 20, $OSiBu^tMe_2$

+

H, H, 10, 9, O, O, 8, 11, 12, H_3C, O, CH_3

(THF); 10 min; -78°C + 1 h; -30°C (i) (46%)

[HMPTA] 3 h; -30°C (7%)

+

I [7] $\overline{}$ 1 COOMe

[NaOAc] (MeOH) 3 h; Δ

(ii) 50% aq. HF (Py/MeCN) 10 min; 0°C + 4.5 h; r.t. (78–90%)

(iii) Al–Hg (aq. THF) 1.5 d; r.t. (89–98%)

9, 8, 1, COOMe, 20, 11, 12, OH, OH

(−)-PGE$_2$ methyl ester

COOMe 8β-epimer (10%)

Synthesis of a prostaglandin intermediate by a Suzuki coupling is outlined in the scheme below. The upper chain bearing a benzyl-protected hydroxylfunctionality was installed via a Suzuki cross coupling reaction between a 9-BBN derivative generated *in situ* from benzyl 3-butenyl ether and an iodoenone catalyzed by PdCl$_2$-(dppf). The coupling product was isolated in 72% yield. Conjugate addition of the lower chain component was successfully achieved by treatment of the enone with a high-order cuprate generated by lithiation of an iodide with *t*-BuLi in combination with lithium 2-thienylcyanocuprate. Reduction of the ketone with L-Selectride at −78 °C proceeded smoothly to provide the *R*-alcohol with good enantioselectivity (92:8 *R/S*). Protection of the three hydroxyl groups as TBS ethers under standard conditions and subsequent purification by silica gel chromatography provided the enantiopure PGE-intermediate (Gu, 2001).

Reaction scheme:

TBSO, O, I

Benzyl 3-butenyl ether (1.5 equiv.)
9-BBN (0.5 M in THF, equiv.)
PdCl$_2$(dppf) (5 mol%)
K$_3$PO$_4$ (2 equiv.), DMF-THF-H$_2$O, r.t.

→ TBSO, O, OBn (72%)

t-BuLi (2.7 equiv.)
ThCu(CN)Li (1.1 equiv.)
Pentane-Et$_2$O
−78 to −20 °C

→ TBSO, O, OBn, OTBS (83%)

L-selectride (1.2 equiv.)
THF, −78 °C
TBSCl, imidazole, DMF, r.t.

→ OTBS, TBSO, OBn, OTBS (86%)

The two open-chain educts 1 and 2 below were obtained as follows. Ozonolysis of cyclohexane and treatment with acetanhydride/triethylamine gave 6-carbethoxy-hexanals. A Wittig Reaction with a *trans*-butenol phosphonium salt converted it to 1. The azo-compound Z was made from a diketone and p-nitrobenzenesulfonyl-azide (p-NBSA) and DBU. The potassium enolate of the diazoketone also reacted with the dienal in the presence of triethylchlorosilane (TESCl) in toluene to give the TES-protected aldol together with the free aldol. The diazoketone was deazotized and cyclized with rhodium octanoate catalyst in CH$_2$Cl$_2$ to provide the diastereomeric bicyclic ketones in a ratio of 65:35. **7** was isolated by chromatography and treated with 3 equiv. of thiophenol and 4 equiv. of BF$_3$ · OEt$_2$ in CH$_2$Cl$_2$ (0.2 M in **7**) at –45 °C for 8 h . The cyclopropane ring opened and the desired thioether in was obtained in 55% yield. The labile *tert*-butyldimethylsilyl group survived this treatment (Taber, 2002).

2.8
Cyclohexane and Cyclohexene Derivatives

We begin with a discussion of intramolecular reactions, the usual approach to cyclic compounds. An example of a regioselective Dieckmann condensation (Schaefer, 1967) used an educt with two ester groups, of which only one could form an enolate. Regioselectivity was dictated by the structure of the educt.

The Michael reaction is a vinylogous aldol addition, catalyzed by acids and by bases, and it is made regioselective by the choice of appropriate enol derivatives. Stereoselectivity is also observed in reactions with cyclic educts. An important difference from the aldol addition is that the Michael addition is usually less prone to steric hindrance. This is evidenced by the two examples given below, in which cyclic 1,3-diketones add to *a,β*-unsaturated carbonyl compounds (Hiroi, 1975; Smith, 1964).

synthetic steroid precursor

Torgov introduced an important variation of the Michael addition: allylic alcohols are used as vinylogous a^3-synthons and 1,3-dioxo compounds as d^2-reagents (Ananchenko, 1962, 1963; Smith, 1964; Rufer, 1967). Mild reaction conditions have been successful in the addition of 1,3-dioxo compounds to vinyl ketones. Potassium fluoride can act as a weakly basic, non-nucleophilic catalyst in such Michael additions under essentially nonacidic and nonbasic conditions (Kitahara, 1964).

The addition of large enolate synthons to cyclohexenone derivatives via Michael addition leads to equatorial substitution. If the cyclohexenone conformation is fixed, e.g. as in decalones or steroids, the addition is highly stereoselective. This is also the case with the δ-addition to conjugated dienones (Abe, 1956). Large substituents at C-4 of cyclic a^3-synthons direct incoming carbanions to the *trans*-position at C-3 (Battersby, 1960). The thermodynamically most stable products are formed in these cases, because the addition of 1,3-dioxo compounds to activated double bonds is essentially reversible.

Michael addition of enolates to vinyl ketones yields saturated 1,5-diketones, which may be again enolized and may undergo a further aldol addition. This sequence is important in the synthesis of polycyclic compounds and is called Robinson annelation (*anellus*, Lat. small ring). Some representative examples are summarized in the scheme below (Johnson, 1956; Marshall, 1968; Zurflüh, 1968; Coates, 1968). The regio- and stereoselectivities which are indicated can all be understood if one considers the arguments given in the discussions of aldol and Michael addition reactions as well as the relative reactivity of enolates.

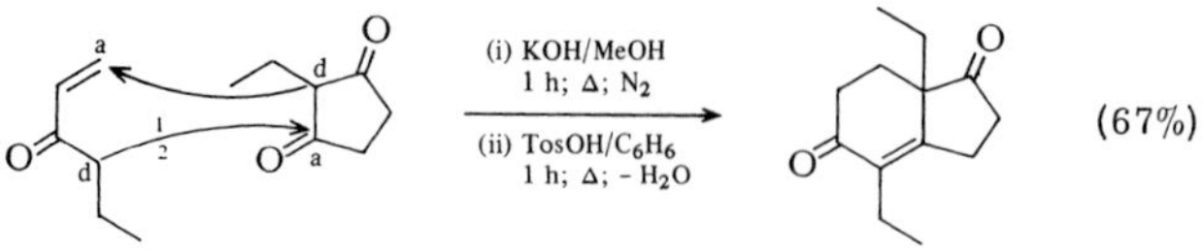

(i) KOH/MeOH
1 h; Δ; N₂

(ii) TosOH/C₆H₆
1 h; Δ; – H₂O

(67%)

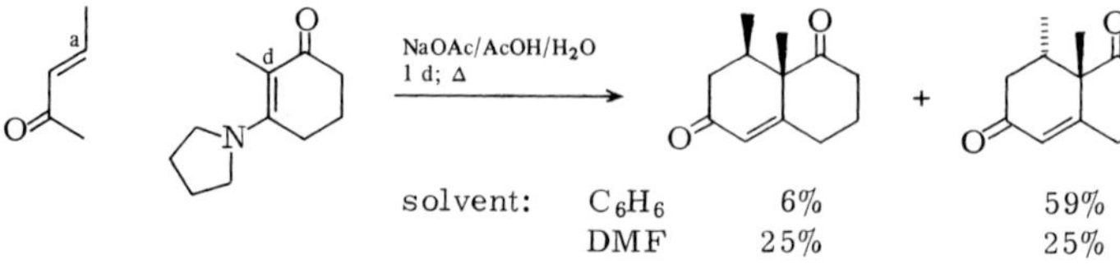

NaOAc/AcOH/H₂O
1 d; Δ

| solvent: | C₆H₆ | 6% | 59% |
| | DMF | 25% | 25% |

If a Michael reaction uses an unsymmetrical ketone with two CH-groups of similar acidity, the enol or enolate is first prepared in pure form. To avoid equilibration one has to work at low temperatures. The reaction may then become slow, and it is advisable to further activate the carbon–carbon double bond. This may be achieved by the introduction of an extra electron withdrawing silyl substituent at C-2 of an a³-synthon. Treatment of the Michael adduct with base removes the silicon, and may also lead to an aldol addition (Stork, 1973, 1974 B; Boeckman, Jr., 1974).

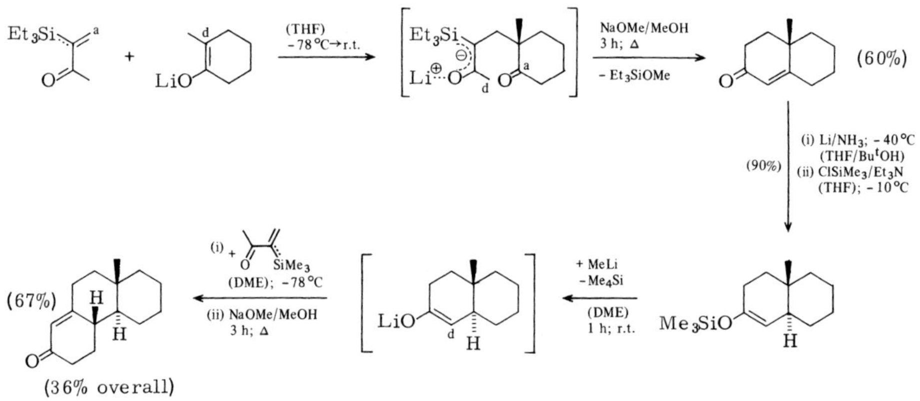

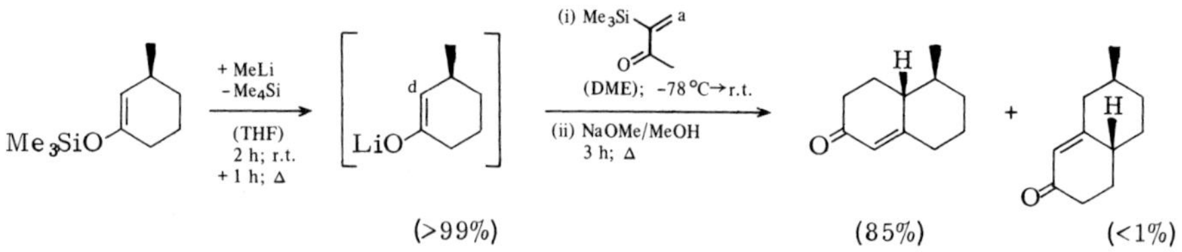

Et₃Si

+

LiO

(THF)
–78 °C→r.t.

Et₃Si

Li⊕O

NaOMe/MeOH
3 h; Δ

– Et₃SiOMe

(60%)

(90%)

(i) Li/NH₃; – 40 °C
(THF/BuᵗOH)
(ii) ClSiMe₃/Et₃N
(THF); – 10 °C

(67%)

(i) +

O   SiMe₃

(DME); –78 °C

(ii) NaOMe/MeOH
3 h; Δ

LiO

+ MeLi
– Me₄Si

(DME)
1 h; r.t.

Me₃SiO

(36% overall)

Me₃SiO

+ MeLi
– Me₄Si

(THF)
2 h; r.t.
+ 1 h; Δ

LiO

(i) Me₃Si

O

(DME); –78 °C→r.t.

(ii) NaOMe/MeOH
3 h; Δ

(>99%)

(85%)

(<1%)

Enantioselective and highly regioselective Michael-type alkylations of 2-cyclohex-en-1-one occurred with alkylcuprates bearing chiral auxiliary ligands, e.g. *N*-[2-(di-methylamino)ethyl]ephedrines (Corey, 1986), 2-(methoxymethyl)pyrrolidine (from prolines; Dieter, 1987) or "chiramt" (= (*R,R*)-*N*-(1-phenylethyl)-7-[(1-phenyl-ethyl)imino]-1, 3,5-cycloheptatrien-1-amine); Villacorta, 1988). Enantioselectivities of up to 95% have been reported.

R	yield	e.e.
Et	66-90%	92-95%
Bu^n	90%	89%
CH$_2$OBu^t	85%	85%

The synthesis of spiro compounds from ketones and methoxyethynyl propenyl ketone exemplifies some regioselectivities of the Michael addition. The electrophi-lic triple bond is attacked first, and then the 1-propenyl group. The conjugated keto group is usually least reactive. The ethynyl starting material has been ob-tained from the addition of the methoxyethynyl anion to the carbonyl group of crotonaldehyde (Stork, 1962 B, 1964 A).

dl-griseofulvin

Thermal [2+4]-cycloaddition of an olefin, preferably with electron withdrawing substituents, to a diene yields cyclohexene derivatives in high yields (Carruthers, 1978). High temperatures must be used with electron-deficient dienes, the activa-tion energy is ≥ 22 kcal mol^{-1} (Huang, 2001). This so-called "Diels–Alder" reaction is stereoselective: the olefin adds suprafacially to the diene, and so the configura-tion of both components is retained. With cyclic dienes oligocyclic products are formed, which have preferentially the *endo*-configuration (Arrieta, 2001). The Diels–Alder reaction is one of the most powerful synthetic reactions of all since its application can result simultaneously in an increase of (i) the number of rings,

(ii) the number of asymmetric centers, and (iii) the number of functional groups (Corey, 1967 A). A few examples illustrate the wide range of applications of this classical electrocyclization reaction. The first examples demonstrate the effect of substituents on regio- and stereoselectivity (Büchi, 1966 B; Woodward, 1952). All reaction schemes show the synthesis of complex polyfunctional molecules from relatively simple, symmetric precursors (Büchi, 1968; Evans, 1972; Ruden, 1974), which would be difficult to synthesize by other means.

A major difficulty with the Diels–Alder reaction is its sensitivity to steric hindrance. Tri- and tetrasubstituted olefins, or dienes with bulky substituents at the terminal carbons, react only very slowly. Therefore bicyclic compounds with angular substituents are often obtained in low yields, and polar reactions are more suitable for such target molecules, e.g. steroids. There exist, however, several exceptions, e.g. a reaction of a tetrasubstituted alkene with a 1,1-disubstituted diene to produce a cyclohexene intermediate containing three contiguous quaternary carbon atoms (Danishefsky, 1979). This reaction was assisted by large polarity differences between the electron-rich diene and the electron-deficient ene component.

"Tandem" Diels–Alder reactions (Winkler, 1996) indicate two reactions, which follow one another. Acetylenic *bis*-dienophiles react, for example, twice with *trans*-dienes to yield polycyclic hexadienes (Goldberg, 1993). Elimination of carbon monoxide may generate a second diene equivalent in a Diels–Alder cascade (Fray, 1967) and a symmetric divinyl ketone reacts twice with an asymmetric bis-*trans* diene to yield a nonsymmetric tricycle (Winkler, 1996; Nicolaou, 1994). Lewis acid catalyzed reactions between tethered enes and dienes produce bicycles (Yakelis, 2001; Tantillo, 2001).

Dramatic rate accelerations of [4+2] cycloadditions were observed in an inert, extremely polar solvent, namely in 5 M solutions of lithium perchlorate in diethyl ether (corresponds to 532 g L^{-1} LiClO$_4$!). Diels–Alder additions requiring several days, 10–20 kbar of pressure, and/or elevated temperatures in apolar solvents are achieved in high yields in some hours at ambient pressure and temperature in this solvent (Grieco, 1990; Kumar, 2001). Also several other reactions, e.g., allylic rearrangements and Michael additions, can be drastically accelerated by this magic solvent. The diastereoselectivities of the reactions in apolar solvents and in LiClO$_4$/Et$_2$O are often different or even complementary and become thus steerable.

Bridged Diels–Alder products occur preferentially in the *endo*-form, because there is a strong two-electron interaction between atomic orbitals at 3 and 1', which are not involved in the formation of σ-bonds ("secondary orbital interactions"; Arrieta, 2001). Diels–Alder enes with a chiral EWG react with achiral cyclopentadiene to give racemic mixtures of norbornene derivatives. Upon addition of planar and chiral bisoxazoline copper(II) complexes, however, stereoselectivity was observed. It is thought, that the chiral ene is also bound to the copper(II) ion and blocks one face of it in the Diels–Alder cyclization (Sibi, 2001).

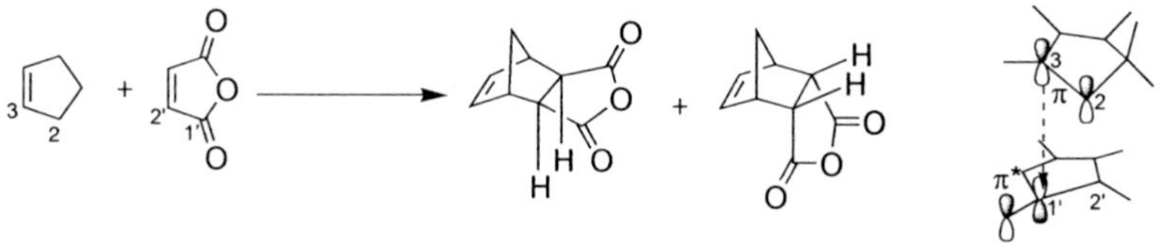

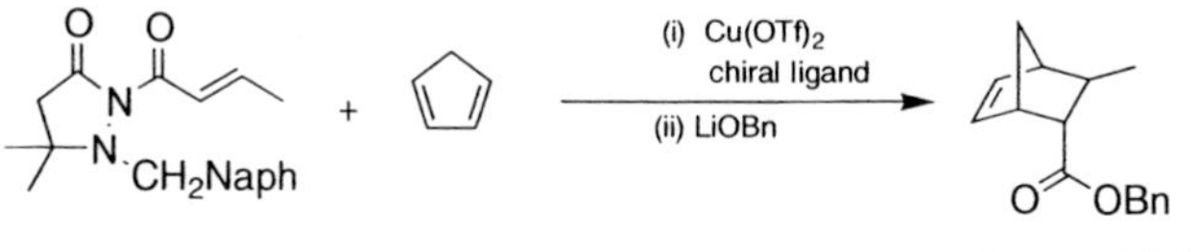

endo:exo = 91:9

A highly successful route to stereoisomers of substituted 3-cyclohexene-1-carbox-ylates runs via Ireland–Claisen rearrangements of silyl enolates of ω-vinyl lactones. The rearrangement proceeds stereospecifically through the only possible boat-like transition state, in which the connecting carbon atoms come close enough (Danishefsky, 1980; Nakatsuka, 1990).

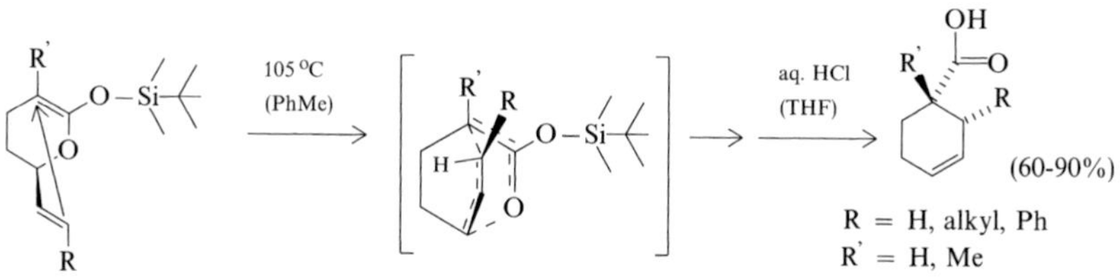

(60-90%)

R = H, alkyl, Ph
R' = H, Me

Several substituted cyclohexane derivatives may be obtained by the reduction of a benzenoid precursor. Partial reduction of resorcinol, for example, and subsequent methylation yields 2-methylcyclohexane-1,3-dione, which is frequently used in steroid synthesis (Newman, 1960; see also section 2.8.1). From lithium-ammonia reduction of alkoxybenzenes 1-aldoxy-1,4-cyclohexadienes are obtained (Corey, 1968 D).

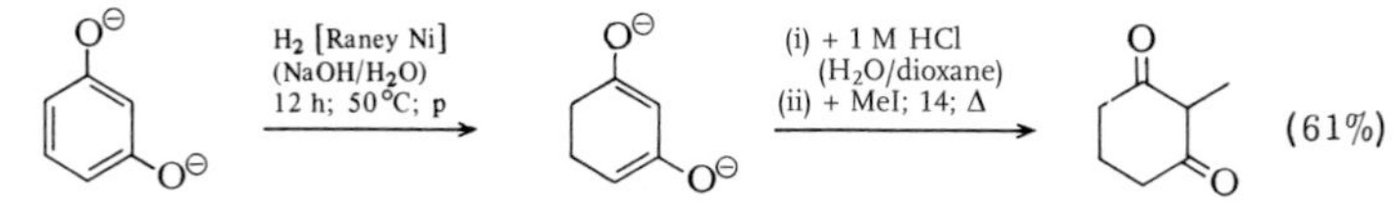

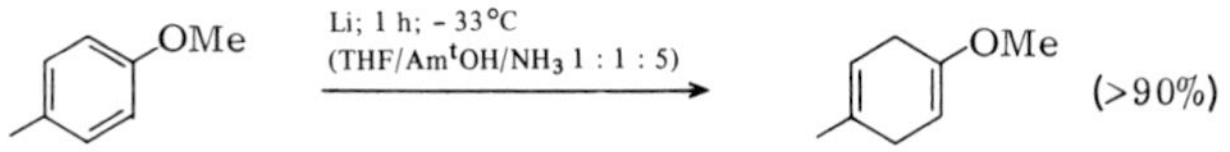

Substitution and dimerization reactions of cyclohexanone derivatives are selective and have been investigated in great detail.

The phosphorus ylides of the Wittig reaction are preferably replaced by trimethylsilylmethyl-carbanions (Peterson reaction). These silylated carbanions add to carbonyl groups and can easily be eliminated with base to give olefins. The only byproducts are volatile silanols. They are more easily removed than the phosphine oxides or phosphates of the more conventional Wittig or Horner reactions (Peterson, 1968). The Peterson reaction has two more advantages over the Wittig reaction: (i) it is sometimes less vulnerable to steric hindrance, and (ii) groups that are susceptible to nucleophilic substitution are not attacked by silylated carbanions. The introduction of a methylene group into a sterically hindered ketone (Boeckman Jr., 1973) and the syntheses of olefins with sulfur, selenium, silicon, or tin substituents (Seebach, 1973; Gröbel, 1974, 1977) illustrate useful applications. The reaction is, however, more limited and time-consuming than the Wittig reaction, since metallated silicon derivatives are difficult to synthesize and their reactions are rarely stereoselective (Chan, 1974).

epoxide mixture of
gipsy moth pheromone

trans:cis = 1:1

(50%)

The Julia–Lythgoe olefination operates by addition of alkyl sulfone anions to carbonyl compounds and subsequent reductive deoxysulfonation (Kocienski, 1985). In comparison with the Wittig reaction, it has several advantages: sulfones are often more readily available than phosphorus ylides, and it is often successful when the Wittig olefination fails. The elimination step yields exclusively or predominantly the more stable trans olefin stereoisomer.

Syntheses of alkenes with three or four bulky substituents, e.g. two cyclohexane, benzyl or *tert*-butyl groups connected by a double bond, cannot be achieved with an ylide or by a direct coupling reaction. Steric hindrance of substituents presumably does not allow the direct contact of polar or radical carbon synthons in the transition state. A generally applicable principle formulated by Eschenmoser indicates a possible solution to this problem: If an intermolecular reaction is complex or slow, it is advisable to change the educt in such a way, that the critical bond formation can occur intramolecularly (Eschenmoser, 1970). The problem of the synthesis of highly substituted olefins from ketones according to this principle was solved by Derek Barton. The ketones are first connected to azines by hydrazine and secondly treated with hydrogen sulfide to yield 1,3,4-thiadiazolidines. In this heterocycle the substituents of the prospective olefin are too far from each other to produce problems. Mild oxidation of the hydrazine nitrogens produces Δ^3-1,3,4-thiadiazolines. The decisive step of carbon–carbon bond formation is achieved in a thermal reaction: a nitrogen molecule is cleaved off and the biradical formed recombines immediately since its two reactive centers are hold together by the sulfur atom. The thiirane (episulfide) can be finally desulfurized by phosphines or phosphates, and the desired olefin is formed. With very large substituents the 1,3,4-thiadiazolidines do not form with hydrazine. In such cases, however, direct thiadiazoline formation from thiones and diazo compounds is often possible, or a thermal reaction between alkylideneazinophosphoranes and thiones may be successful (Barton, 1972, 1974, 1975).

Alkylation of cyclohexane is most easily achieved by a hydroboration procedure of cyclohexene.

(cis)-alkenes:

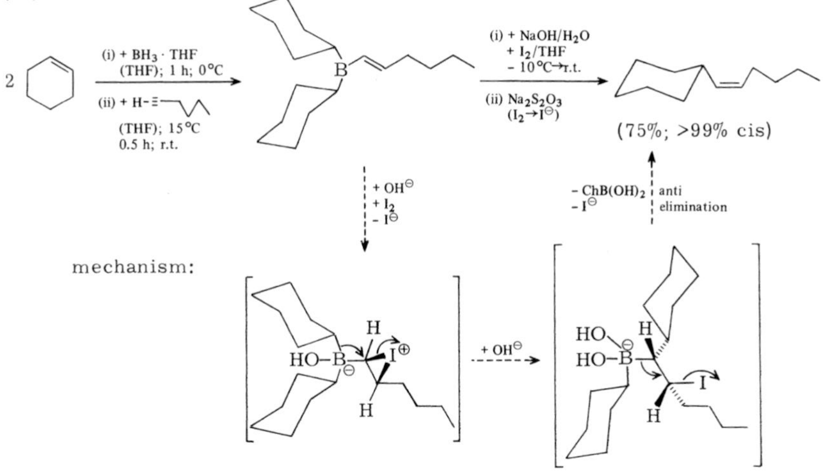

2

(i) + BH$_3$ · THF
(THF); 1 h; 0°C

(ii) + H–≡
(THF); 15°C
0.5 h; r.t.

B

(i) + NaOH/H$_2$O
+ I$_2$/THF
– 10°C→r.t.

(ii) Na$_2$S$_2$O$_3$
(I$_2$→I$^\ominus$)

(75%; >99% cis)

+ OH$^\ominus$
+ I$_2$
– I$^\ominus$

– ChB(OH)$_2$ | anti
– I$^\ominus$ | elimination

mechanism:

HO–B   H   I$^\oplus$
H

+ OH$^\ominus$

HO   H
HO–B
I
H

(trans)-alkenes:

Br—≡—

+ Ch$_2$BH(THF)
0°C; 0.5 h; r.t.

B   Br

NaOMe
(THF); 1 h; r.t.

MeO   B

AcOH/H$_2$O
1 h; Δ

(90%; > 99% trans)

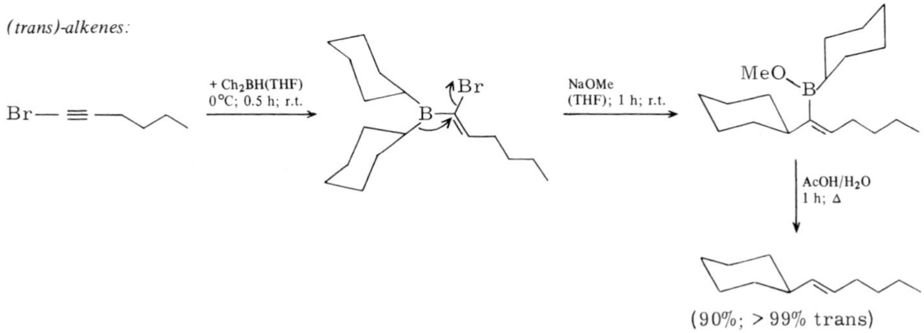

2.8.1
Steroids

About one third of prescribed pharmaca in the western world are steroid hormones. They steer protein synthesis as anabolica or catabolica. Pharmaceutically useful steroids are obtained in a few cases by total synthesis (most important: norgestrel). Much more important are functional group conversions of inexpensive natural steroids, in particular androstadiene-dione, which is made by microbial degradation of cholesterol and stigmasterols. Both approaches will be discussed in this section (Langecker, 1977; Blickenstaff, 1974).

Classical syntheses of steroids consist of the stepwise formation of the four rings with or without angular alkyl groups and the final construction of the C-17 side-chain. The most common reactions have been described in the two previous sections, e.g. Diels–Alder and Michael additions, the Robinson annelation and the Torgov reaction, Dieckmann condensation, and regioselective alkylation. Although many of the classical syntheses are stereoselective, they yield only racemic products, and separation of enantiomers at an early stage is mandatory.

The commercially most successful steroid total synthesis is that of "norgestrel", the gestagen of most contraceptives. The 13-ethyl group causes an extremely high efficiency of the drug. The synthesis starts with the stereospecific microbiological reduction of the Torgov adduct of tetralone, vinylmagnesium chloride, and 2-ethyl-cyclopentane-1,3-dione. Both "prochiral" carbon atoms C-13 and C-17 become chiral in this enzyme-catalyzed reaction, and since all following reactions are highly stereoselective only (+)-norgestrel is formed. The acid-catalyzed cyclization of the enone does not produce new asymmetric centers but hydrogenation of the 14,15-double bond occurs on the less hindered a-side. Birch reduction of the 8,9-double bond and of the aromatic ring produces the thermodynamically most stable *trans*-configuration of rings B and C. The 17-alcohol is oxidized to the ketone, and the chiral center is introduced by ethynylation of the prochiral keto group. Acid-catalyzed cleavage and rearrangement of the enol ether produces norgestrel in high yield (H. Smith, 1964; Rufer, 1967; Langecker, 1977).

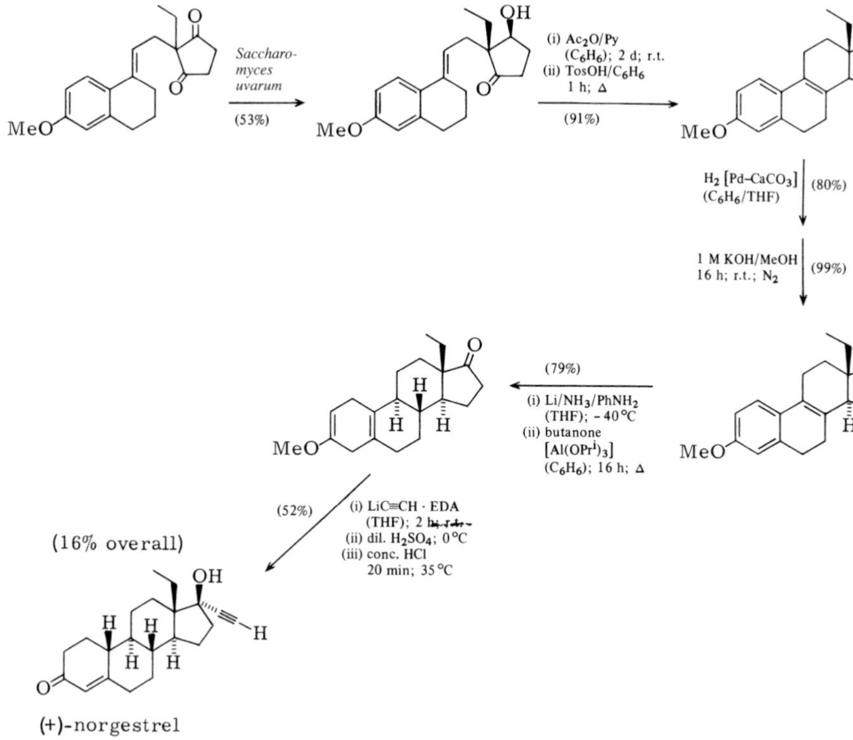

Biomimetic syntheses of steroids apply strategies in which open-chain or mono-cyclic educts with appropriate side-chains are stereoselectively cyclized in one step to a tri- or tetracyclic steroid precursor. These procedures mimic the biochemical scheme where acylic, achiral squalene is first oxidized to a 2,3-epoxide containing one chiral carbon atom and then enzymatically cyclized to lanosterol with no less than seven asymmetric centers (Johnson, 1968, 1976; van Tamelen, 1968). Non-en-zymatic cyclizations of educts containing chiral centers can lead to products with additional "asymmetric" centers. The underlying effect is called "asymmetric in-duction". Its systematic exploration in steroid syntheses started when G. Saucy dis-covered in 1971 that a chiral carbon atom in a cyclic educt induces a stereoselec-tive Torgov condensation several carbon atoms away (Rosenberger, 1971, 1972).

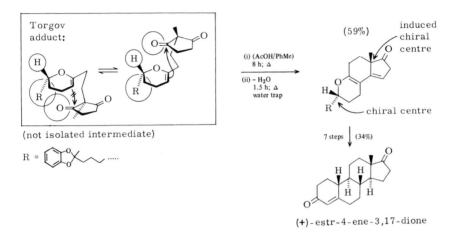

(+)-estr-4-ene-3,17-dione

Proton-catalyzed olefin cyclization of open-chain educts may give tri- or tetracy-clic products but low yields are typical (van Tamelen, 1968, 1977). More useful are cyclizations of monocyclic educts with appropriate side-chains. The chiral center to which the chain is attached may direct the steric course of the cyclization, and several asymmetric centers may be formed stereoselectively since the cyclizations usually lead to *trans*-fused rings. The following acid-catalyzed cyclizations leading to steroid hormone precursors exemplify some important facts: an acetylenic bond is less nucleophilic than an olefinic bond; acetylenic bonds tend to form cyclopen-tane rather than cyclohexane derivatives, if there is a choice; in proton-catalyzed olefin cyclizations the thermodynamically most stable trans connection of cyclo-hexane rings is obtained selectively; electroneutral nucleophilic agents such as eth-ylene carbonate can be used to terminate the cationic cyclization process, forming stable enol derivatives which can be hydrolysed to carbonyl compounds; without this nucelophile and with trifluoroacetic acid the corresponding anol ester may be obtained (Gravestock, 1978 A, B; Peterson, 1969).

Two approaches to *convergent* steroid syntheses are based on the thermal opening of benzocyclobutenes to the *o*-quinodimethane derivatives (Oppolzer, 1978 A) and their stereoselective intramolecular Diels–Alder cyclizations. Kametani (1977 B, 1978) obtained (+)-estradiol in a six-step synthesis. The final Diels–Alder reaction occurred regio- and stereoselectively in almost quantitative yield, presumably because the *exo* transition state given in the scheme below is highly favored over the *endo* state, in which rings A and D would sterically interact.

ergosterol
(provitamin D_2)

tachysterol
(previtamin D_2)

ergocalciferol
(vitamin D_2)

H
R

H
R

H
R

H

H
H

H

H

9

H
H

9

light induced
conrotatory
ring opening

thermal
antarafacial
1,7 H-shift

H

H
H

8

H
H

H

HO

$h\nu$

1 10
2
3 5 6
4

7

HO

Δ

6
5
4
10 19
HO 3 1
2

8
7 H

D_2: R =

H

Hydroxy vitamin D was synthesized by palladium-catalyzed coupling of the A-ring synthons with the CD-ring portion, followed by deprotection of the silylated hydroxyl groups with camphorsulfonic acid (CSA) in MeOH to give the target 2R-alkyl and 2R-hydroxyalkyl steroids (Suhara, 2001).

OH

OH

Br

+

(i) (dba)$_3$Pd$_2$, CHCl$_3$, TPP, TEA/toluene

(ii) CSA, MeOH

TBSO

OTBS

HO

OH

$(\ )_n$
R

$(\ )_n$
R

Open-chain 1,5-polyenes (e.g. squalene) and some oxygenated derivatives are the biochemical precursors of cyclic terpenoids (e.g. steroids, carotenoids). The enzymic cyclization of squalene 2,3-oxide, which has one chiral carbon atom, to produce lanosterol introduces seven chiral centers in one totally stereoselective reaction. As a result, organic chemists have tried to ascertain whether squalene or related olefinic systems could be induced to undergo similar stereoselective cyclizations in the absence of enzymes (Johnson, 1968, 1976).

A simple acid-catalyzed cyclization transforms ψ-ionone into α-ionone (Kimel, 1957, 1958). Further treatment with protic acids transforms the α-ionone to the thermodynamically more stable β-ionone.

pseudoionone
(ψ-ionone)

α-ionone

β-ionone

Lewis acid: (i) BF$_3$/C$_6$H$_6$; 10 min; 0→10 °C
(ii) neutralization: dil. NaOH; < 10 °C

protic acid: H$_2$SO$_4$/AcOH 7 : 3; 20 min; 10 °C
+ 20 min; r.t.

	α-ionone	β-ionone
Lewis acid	92 : 8 (70%)	(6%)
protic acid	10 : 90 (7%)	(65%)

$-A = -\overset{\ominus}{B}F_3, -H$

The achiral triene chain of (*all-trans-*)-3-demethyl-farnesic ester as well as its (6-*cis*-)-isomer cyclize in the presence of acids to give the decalol derivative with four chiral centers whose relative configuration is well defined (Stadler, 1957; Eschenmoser, 1959; Johnson, 1968, 1976). A monocyclic diene is formed as an intermediate (Stork, 1955). With more complicated 1,5-polyenes, such as squalene, oily mixtures of various cyclization products are obtained. The 18,19-glycol of squalene 2,3-oxide, however, cyclized in modest yield with picric acid catalysis to give a complex tetracyclic natural product with nine chiral centers. Picric acid acts as a protic acid of medium strength whose conjugated base is non-nucleophilic. Such acids activate oxygen functions selectively (Sharpless, 1970).

The early Eschenmoser–Stork results indicated that stereoselective cyclizations may be achieved if monocyclic olefins with 1,5-polyene side-chains are used as substrates in acid treatment. This assumption has now been justified by many syntheses of polycyclic systems. A typical example synthesis is given in the last reaction of the scheme below. The cyclization of a trideca-3,7-dien-11-ynyl-cyclopentenol leads in 70% yield to a 17-acetyl A-norsteroid with correct stereochemistry at all ring junctions. Ozonolysis of ring A and aldol condensation gave *dl*-progesterone (Gravestock, 1978).

(60–70%)

2,4,6-trinitrophenol
(MeNO$_2$); 1 d; r.t.

(erythro-18,19-glycol)

7% dl-malabaricanediol
7% dl-18,19-epimalabaricanediol

TFA/ O=⟨O/O⟩; N$_2$
(DCE); 3 h; 0°C

K$_2$CO$_3$
(H$_2$O/MeOH)
1 h; r.t.

(70%)

2.9
Bridged Carbocycles

In polycyclic hydrocarbon skeletons the adjacent rings may have more than one common bond ("fused ring systems"). Bridges, which consist of one or more carbon atoms, may connect two tertiary ring carbon atoms, which are termed bridgeheads: Bridged structures are found in many natural products from plants (terpenes, alkaloids) and transpose the three-dimensionality of the carbon tetrahedron into larger structures. Whereas fused cyclic systems are more or less planar, often with kinks from *cis*-linked rings, bridged systems may approach the space filling of a sphere or a hemisphere.

The syntheses of bridged carbon compounds start from monocyclic or fused polycyclic compounds with appropriate donor and acceptor centers. We start again with the most simple and straightforward synthetic reaction. This is a Diels–Alder-type combination of dienes and alkenes in which the diene or both components are cyclic. As with fused ring systems this reaction is again stereoselective. The double bond of the cyclohexene boat, which is formed in the reaction, and the substituents on the carbon atoms of the educt olefin lie on the same side (= endo) in the major product. Sterically strained cyclopropene rings are particularly reactive as "ene"-components (Deem, 1972). With electron-deficient dienes, reactions are still selective but they give low yields and require high temperatures (Zimmerman, 1969A; Cookson, 1956). Benzyne (Wittig, 1956; Chapman, 1973, 1975; Levin, 1978; Hoffmann, 1967) has also been used as the "ene"-component.

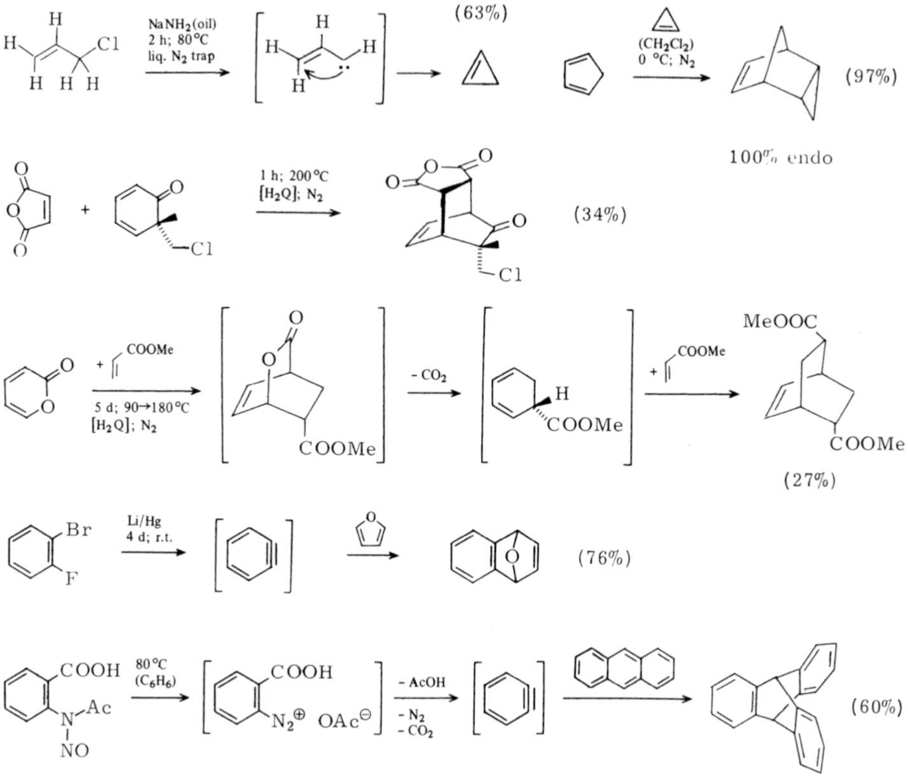

(63%)

(97%)

100% endo

(34%)

(27%)

(76%)

(60%)

Intramolecular reactions between donor and acceptor centers in fused ring systems provide a general route to bridged polycyclic systems. The *cis*-decalone mesylate given in the scheme below contains two d^2-centers adjacent to the carbonyl function and one a^1-center. Treatment of this compound with base leads to reversible enolate formation, and the C-3 carbanion substitutes the mesylate on C-7 (Gauthier, 1967; Bélanger, 1968).

In an intramolecular aldol condensation of a diketone, many products are conceivable, since four different enols can be made. Five- and six-membered rings, however, will be formed preferentially. Kinetic or thermodynamic control or different acid–base catalysts may also induce selectivity. In the Lewis acid-catalyzed aldol condensation given in the scheme, the more substituted enol is formed preferentially (Corey, 1963 B, 1965 B).

Another synthesis of a bridged hydrocarbon takes advantage of high electron release from the para-position of phenolate anions, which may be used to transform the phenol moiety into a substituted cross-conjugated cyclohexadienone system (Masamune, 1961, 1964).

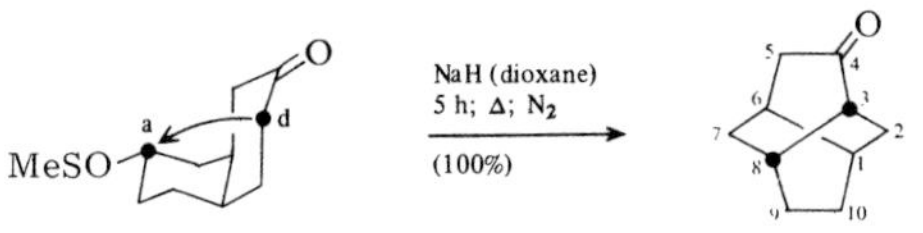

NaH (dioxane)
5 h; Δ; N₂

(100%)

tricyclo[4.4.0.0³,⁸]-
decan-4-one
= 4-twistanone

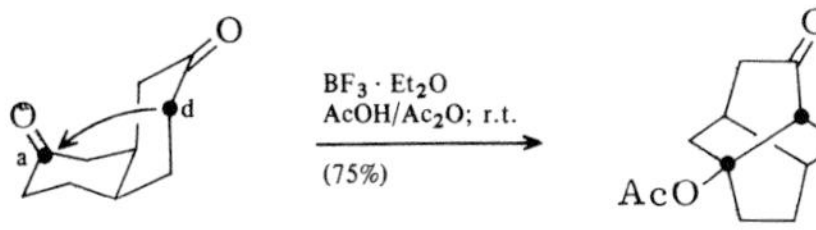

BF₃·Et₂O
AcOH/Ac₂O; r.t.

(75%)

8-acetoxy-
-4-twistanone

epimerizable

Lewis
acids

+

+

(Robinson annelation
products)

BF₃(CH₂Cl₂); 16 h; r.t.: (32%)   (8%)
SnCl₄(C₆H₆); 45 min; Δ: (28%)   (28%)

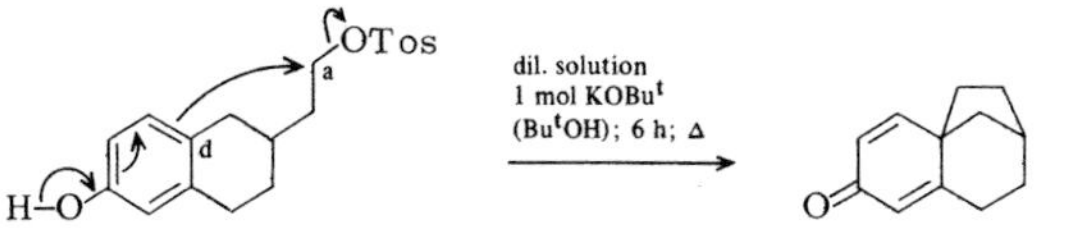

dil. solution
1 mol KOBuᵗ
(BuᵗOH); 6 h; Δ

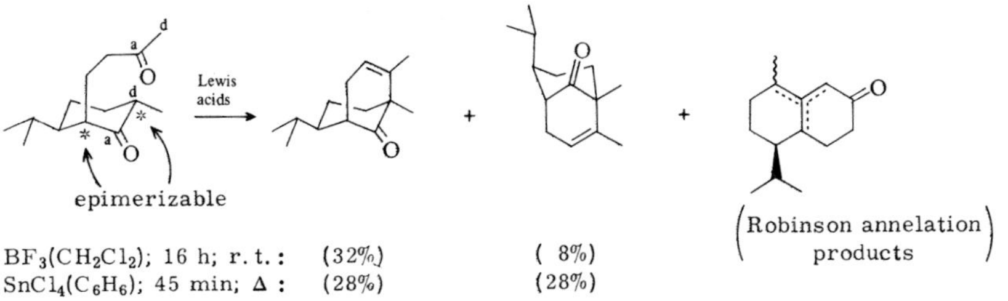

(90%)

As final examples, the intramolecular cyclopropane formation from cycloolefins with diazo groups (Burke, 1979), intramolecular cyclobutane formation by photochemical cycloaddition, and intramolecular Diels–Alder reactions are remarkable. The application of these three cycloaddition reactions has led to an enormous variety of exotic polycycles (Corey, 1967 A).

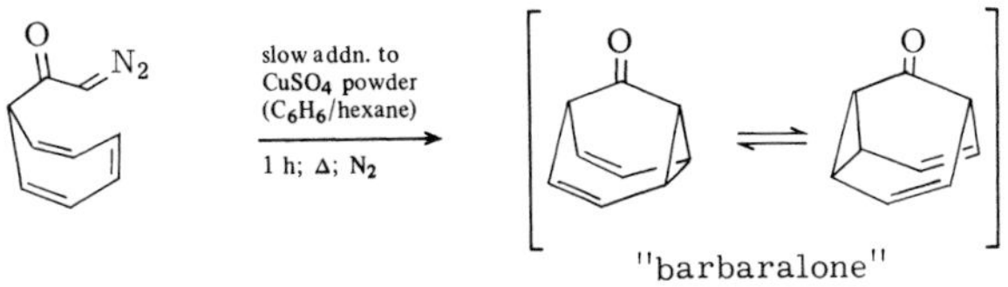

slow addn. to
CuSO₄ powder
(C₆H₆/hexane)

1 h; Δ; N₂

(30%)

"barbaralone"

Cu (hexane)
12 h; Δ; N₂

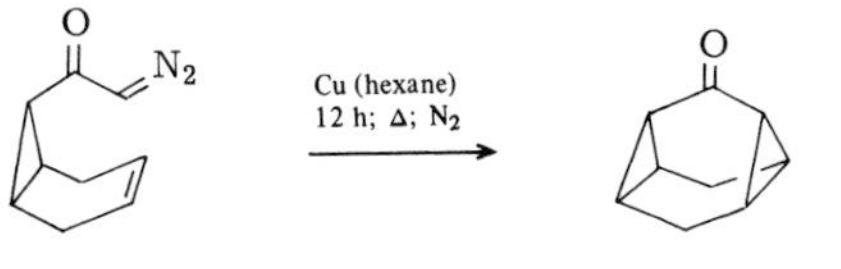

(73%)

"triasteranone"

Bicyclo [1.1.1]-pentanes are stiff units, can be polymerized, and have some interesting physical properties. They are best synthesized by reduction of halogenated dimethyl cyclopropane or cyclopropene derivatives (Levin, 2000).

A simple Diels–Alder reaction between TMS-protected 2,6-dihydroxyanthracene and maleic anhydride gave a dicarboxylate-bridged bis-phenol, which forms molecular complexes with acetylated arginine and lysine derivatives in water (Thompson, 2002). Such bridged systems with a hydrophobic and a hydrophilic face are probably the most promising three-dimensional carbocycles for future work.

3
Functional Group Interconversion (FGI)

3.1
Introduction

Construction of a carbon skeleton seldom yields the target molecule directly. Almost inevitably some of the functional groups are in the wrong oxidation state, contain the wrong hetero atoms, or are undesired derivatives. The functional groups have then to be manipulated. Since most target (or intermediate) molecules contain several functional groups, these interconversions have to be selective. Selectivity may arise from the specific electronic structure of the functional group (e.g. C=C double bonds are more nucleophilic than C=O double bonds) or from its steric environment (e.g. a sterically hindered ketone does not react with bulky boranes). In this part of the book selected FGIs, which are synthetically "realistic", will be discussed. "Realistic" means that good control of single reactions in complicated molecules has been achieved.

Site- and stereoselective interconversions of functional groups are employed to achieve one of the following objectives:

- change of the oxidation state of specific carbon atoms,
- introduction, removal, or substitution of hetero atoms,
- connection of monomers or cyclization by formation of new C–X–C or C–X=C bonds,
- reversible protection of reactive sites.

The major problem in FGIs is the choice of the right reagents, namely oxidants, reductants, acids, bases, or nucleophiles. They must be only as strong as necessary to be efficient for the particular FGI, they must be as bulky as possible to allow regio- and stereoselective attack, they must be soluble and allow for easy workup procedures. The development of all the reagents that are successfully applied today was tedious and led by experience only.

3.2
Reduction

3.2.1
General Considerations

Of all synthetic operations, hydrogenation of organic molecules is probably the most highly developed. Knowledge of selectivities is comparatively far advanced. Common reducing agents are hydrogen in the presence of solid or dissolved catalysts (e.g. Ni, Pd, Pt, Ru, Rh), hydrides (e.g. alanes, boranes, $LiAlH_4$, $NaBH_4$), reducing metals (e.g. Li, Na, Mg, Ca, Zn), and low-valent compounds of nitrogen (e.g. N_2H_4, N_2H_2), phosphorus (e.g. triethyl phosphite, triphenylphosphine), and sulfur (e.g. SFSi, sodium formaldehyde sulfoxylate = $HO–CH_2–SO_2Na$ (SFS), sodium dithionite = $Na_2S_2O_4$).

Catalytic hydrogenation is mostly used to convert carbon–carbon triple bonds into double bonds or alkenes into alkanes, or to replace allylic or benzylic hetero atoms by hydrogen (Kropft, 1980). Simple theory postulates *cis-* or *syn-*addition of hydrogen to the C–C triple or double bond with heterogeneous (Augustine, 1965, 1968, 1976; P. N. Rylander, 1979) and homogeneous (Birch, 1976) catalysts. Sulfur functions can be removed with reducing metals, e.g. with Raney nickel (Pettit, 1962 A). Heteroaromatic systems may be reduced with the aid of ruthenium on carbon.

Hydrides are available in many molecular sizes (see table 8.2) and possessing different reactivities. $LiAlH_4$ reduces most unsaturated groups except alkenes and alkynes. $NaBH_4$ is less reactive and reduces only aldehydes and ketones, but usually not carboxylic acids or esters (Gaylord, 1956; Hajós, 1979).

The conversion of carboxylic acid derivatives (halides, esters and lactones, tertiary amides and lactams, nitriles) into aldehydes can be achieved with bulky aluminum hydrides (e.g. DIBAL = diisobutylaluminum hydride, lithium trialkoxyalanates). Simple addition of three equivalents of an alcohol to $LiAlH_4$ in THF solution produces those deactivated and selective reagents, e.g. lithium triisopropoxyalanate, $LiAlH(OPr^i)_3$ (Malek, 1972).

Diborane or alkylboranes are used for reduction of alkenes and alkynes via hydroboration followed by hydrolysis of the borane with acetic acid (Brown, 1975).

Halides and tosylates are reduced by $LiAlH_4$, if S_N2 displacement is easy (Gaylord, 1956; Hajós, 1966, 1979; Brown, 1951). Tin hydrides and samarium iodide are especially reactive to these substrates (Brady, 1970).

They work in one-electron (= radical) steps. Reducing metals (Na, Na/LiHg, Zn, Fe) are also applied to produce anion radicals, e.g. in acyloin and pinacol coupling reactions or in Birch reduction of arenes and carbonyl-conjugated carbon–carbon multiple bonds (Akhrem, 1972).

Low-valent nitrogen and phosphorus compounds are used to remove hetero atoms from organic compounds. Important examples are the Wolff–Kishner-type reduction of ketones to hydrocarbons (Augustine, 1968; Todd, 1948; Hutchins, 1973B) and Barton's olefin synthesis, both using hydrazine derivatives.

Tab. 3.1 Reactivity of reducing agents towards functional groups (adapted from: J. B. Hendrickson, D. J. Cram, and G. S. Hammond, *Organic Chemistry*, 3rd ed., McGraw-Hill 1970, with modifications).

Educt	Product	H_2 with catalyst	$NaBH_4$	$LiAlH_4$	$AlH(OR)_3^-$ or R_2AlH	B_2H_6 or $(R_2BH)_2$	Li, Na	Other reagents
alkene (A,C / B,D)	saturated (A,C,H / H,B,D)	+++ S	–	–	+++ S	+++ S	(+)	N_2H_2 = diimine (extremely mild) chiral Wilkinson-type catalysts S
A–≡–B	cis alkene (A,B / H,H)	+++ S	–	–	++– S	++– S	–	
	trans alkene (A,H / H,B)	–	–	–	–	–	+++ S	
phenol (Ar–OR)	cyclohexadienol (–OR)	–	–	–	–	–	+++ S	
$R–CH_2–X$ (i), $R_2CH–X$ (i)	$R–CH_3$, R_2CH_2	+++	–	+++	–	–	+++	Bu_3SnH
$R_3C–X$ (i), $Ar–X$ (i)	R_3CH, $Ar–H$	+++	–	–	–†	–	++–	Bu_3SnH
ROH, ROR	R–H	–	–	–	–	–	–	convert to olein or tosylate

Tab. 3.1 (cont.)

Educt	Product	H_2 with catalyst	$NaBH_4$	$LiAlH_4$	$AlH(OR)_3^-$ or R_2AlH	B_2H_6 or $(R_2BH)_2$	Li, Na	Other reagents
$Ar\text{-}\overset{\mid}{\underset{\mid}{C}}\text{-}Y$ (ii)	$Ar\text{-}\overset{\mid}{\underset{\mid}{C}}\text{-}H$	+++	–	–	–	–	+++	
epoxide	$H\overset{\diagup}{\underset{\diagdown}{C}}\text{-}\overset{\diagup}{\underset{\diagdown}{C}}\text{-}OH$	+++	–	+++	(+)	+++	+++	
R–SH, R–S–R	R–H	+++ Raney Ni	–	–	–	–	+++	Raney nickel without H_2
R–NO$_2$	R–NH$_2$	+++	–	(+)[iii]	–	–	+++	Sn^{2+}, Ti^{2+} etc.
R–CHO, R$_2$C=O	R–CH$_3$, R$_2$CH$_2$	(+–)	(+–)	(+–)	(+–)	(+–)	(+–)	(i) N$_2$H$_4$ (ii) KOBut/DMSO or (i) TosNHNH$_2$ (ii) NaBH$_4$ or (i) HSC$_2$H$_4$SH (ii) Raney Ni or Clemmensen red.
	R–CH$_2$OH, R$_2$CH–OH	(+)	+++ S	+++	+++ S	+++	+++	
R$_2$C=NOH	R$_2$CH–NH$_2$	(+)	–	+++	–	–	+++	Sn, Zn, Ti^{2+}

Tab. 3.1 (cont.)

Educt	Product	H_2 with catalyst	$NaBH_4$	$LiAlH_4$	$AlH(OR)_3^-$ or R_2AlH	B_2H_6 or $(R_2BH)_2$	Li, Na	Other reagents
R–COOH	R–CH$_2$OH	–	–	+++	–	+++	–	
R–COOR′	R–CH$_2$OH	(+)	–	+++	–	(+)	(+)	Bu$_3$SnH, Na$_2$Fe(CO)$_4$
R–COCl	R–CHO	(+–)	(+–)	(+–)	+++	–	–	DIBAL
R–CONR′$_2$	R–CHO	–	+++	(+–)	+++	–	–	DIBAL
R–CONR′$_2$	R–CH$_2$–NR′$_2$	–	–	+++	–	+++	–	NaBH$_4$ + CoCl$_2$ or RCOOH (i) Et$_3$O$^+$BF$_4^-$ (ii) NaBH$_4$
R–C≡N	R–CH$_2$NH$_2$	+++	–	+++	+++	+++	+++	DIBAL

+++ Syntehtically useful reaction, ++– rate of secondary reactions can be kept comparatively small, (+) slow reaction or complex product mixture, (+–) reaction does not stop at or does not reach the desired oxidation state, – no reaction, S Regio- and/or stereoselectivity has been achieved in syntheses of complex molecules, (i) X=Cl, Br, I, OMes, OTos, (ii) Y=OH, OR, NR$_2$, (ii) complex mixtures, if R=aryl

Tab. 3.2 Reactivites of functional groups towards different reducing agents

Reactivity	Catalytic hydrogenation	Complex hydrides	Boranes
High	$-C \equiv C-$ $\text{C}=\text{C}$ $-COCl$ $-C \equiv N$	$-COCl$ $-CHO$ $\text{C}=\text{O}$ $\text{C}=\text{N}$	$-C \equiv C-$ $\text{C}=\text{C}$ $-COOH$ $-CONR_2$ $-C \equiv N$
Medium	$-CHO$ $\text{C}=\text{O}$ $-NO_2$ $Ar-\overset{|}{\underset{|}{C}}-OR$ $\overset{|}{\underset{|}{C}}\overset{}{-}\overset{|}{\underset{|}{C}}-OR$ epoxide $\overset{O}{C-C}$ $\text{C}-X$ cyclopropane $\overset{C}{C-C}$ Heteroarenes	epoxide $\overset{O}{C-C}$ $\overset{|}{\underset{|}{C}}\overset{}{}\overset{|}{\underset{|}{C}}-OR$ $Ar-\overset{|}{\underset{|}{C}}-OR$ $-COOR$ $-CONR_2$ $-C \equiv N$ $\text{C}-X$ $-NO_2$ $-COO^{\ominus}$	$-CHO$ $\text{C}=\text{O}$ $-CONR_2$ $-C \equiv N$ epoxide $\overset{O}{C-C}$
Low	$-COOR$ $-CONR_2$ Arenes	Pyridines	$-COOR$ Heteroarenes
Very low	$-COOH$ $-COO^{\ominus}$	$-C \equiv C-$ $\text{C}=\text{C}$ Arenes	Arenes

Tab. 3.1 gives a broad summary of the reactions of the common classes of reducing agents. The following sections contain some typical examples of synthetically useful reductions (in the educt order given in the table) together with some more sophisticated methods of stereoselective hydrogenations.

If it is necessary to reduce one group in a given molecule without affecting any other unprotected reducible group, the reactivity orders given in Tab. 3.2 for "ease of reduction" toward catalytic hydrogenation, LiAlH$_4$, and diborane may serve as a guideline.

3.2.2
Hydrogenation of Carbon–Carbon Multiple Bonds and Cyclopropane Rings

The carbon–carbon triple bond is the most readily hydrogenatable functional group with several reagents. Its full or partial hydrogenation can take place in the presence of nearly all other functionalities. Most important is its selective conversion into cis-double bonds. Cis-hydrogenation is readily accomplished using diisobutylaluminum hydride (DIBAL; Winterfeldt, 1975; Gensler, 1963) or hydrogen (Baker, 1955; Sondheimer, 1962) with palladium catalysts (e.g. 1–2% Pd on BaSO$_4$ or CaCO$_3$, poisoned by quinoline and/or lead(II) acetate; Lindlar, 1973). The less hindered trans-olefins may be obtained by reduction with lithium or sodium metal in liquid ammonia or amine solvents (Birch reduction). This reagent, however, attacks most polar functional groups (except for carboxylic acids; Dear, 1963; Fried, 1968), and their protection is necessary. Terminal alkynes are only reduced in the presence of proton donors, e.g. ammonium sulfate, because the acetylide anion does not take up further electrons. If, however, an internal C≡C triple bond is to be hydrogenated without any reduction of a terminal one, it is advisable to add sodium amide to the alkyne solution first. On catalytic hydrogenation the less hindered triple bonds are reduced first (Dobson, 1955, 1961). The reduction of medium-size cycloalkynes, however, always yields considerable amounts of the less strained cis-cycloalkenes (Cope, 1960A; Svoboda, 1965). Cyclodecyne, for example, is reduced almost exclusively to cis-cyclodecene. In polycyclic systems the Birch reduction of C=C double bonds is also highly stereoselective, e.g. in the synthesis of the thermodynamically favored trans-fused steroidal skeletons.

(i) DIBAL (heptane)
1 h; 0 °C + 12 h; r.t.

(ii) MeOH/PE; 1 h; < 40 °C
(iii) 20% aq. H$_2$SO$_4$

(82%)

H$_2$ [Lindlar cat.]
(C$_6$H$_6$); r.t.

(≈ 25%)

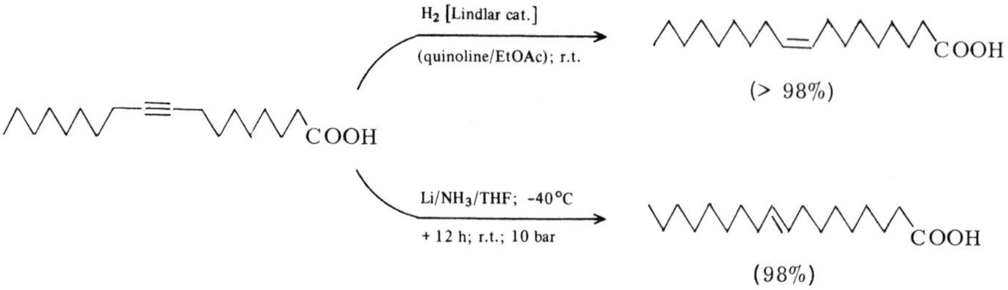

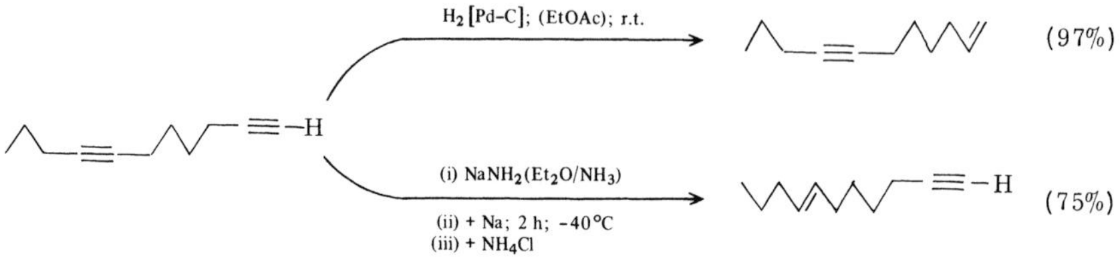

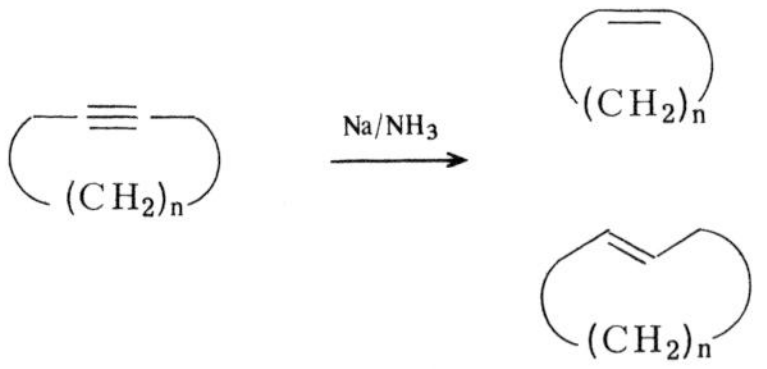

H₂ [Lindlar cat.]

(quinoline/EtOAc); r.t.

(> 98%)

Li/NH₃/THF; –40 °C

+ 12 h; r.t.; 10 bar

(98%)

H₂ [Pd–C]; (EtOAc); r.t.

(97%)

(i) NaNH₂(Et₂O/NH₃)

(ii) + Na; 2 h; –40 °C
(iii) + NH₄Cl

(75%)

	n = 7	n = 8	n = 9	n = 10
	19%	94%	47%	9%
	71%	2%	53%	38%

Na/NH₃

Carbon–carbon double bonds are usually reduced using hydrogen and a hetero-geneous catalyst. The activity of hydrogenation catalysts decreases in the order Pd > Rh > Pt > Ni > Ru. Catalysts other than Pd are especially chosen to minimize migration of hydrogen, e.g. if one wants to deuterate a C=C double bond (Ni, Ru) or if hydrogenolysis of sensitive groups is to be prevented (Rh). The ease of hydro-genation is inversely proportional to the number and size of substituents at the C–C multiple bond (Newhall, 1958). Hydrogenation of tetrasubstituted double bonds is strongly retarded and occurs only if the double bond shifts to a less hin-dered position in the presence of catalysts. It may also happen, however, that a di- or trisubstituted double bond isomerizes to a tetrasubstituted double bond, which resists reduction. Such isomerizations are particularly fast in the presence of pro-tic acids (Barton, 1956).

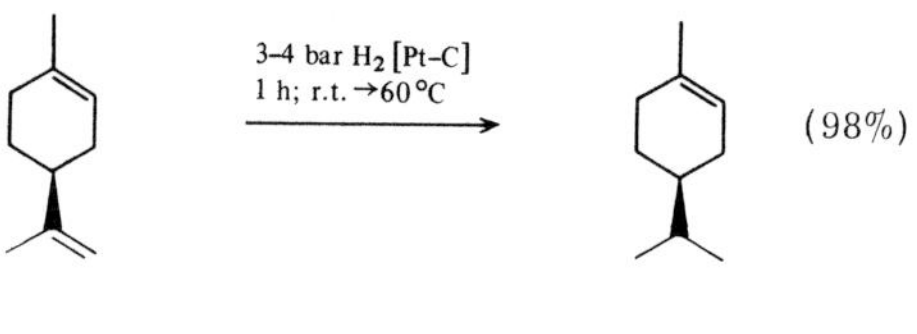

3–4 bar H₂ [Pt–C]
1 h; r.t. →60 °C

(98%)

The commonly accepted mechanism of heterogeneously catalysed hydrogenation involves activation of both the hydrogen and the C–C multiple bond adsorbed on the metal surface. First one hydrogen atom is transferred to the least hindered position of the multiple bond to give a half-hydrogenated adsorbed species. This reaction is fully reversible and accounts for double-bond shifts and cis-trans isomerizations as well as for hydrogen scrambling, when deuterium is used for reduction. Under conditions of high hydrogen availability (high pressure, rapid agitation) another hydrogen atom adds to the same side of the half-hydrogenated bond as the first (Augustine, 1976). If the hydrogenation proceeds more slowly, this syn or cis selectivity is largely lost.

Considering the properties of the substrates, the highest stereoselectivities in cis-hydrogenations are observed with chiral cycloalkenes of rigid conformation. The hydrogenation of the tricyclic system given below, for example, led selectively to the a-hydrogenated trans-fused product because the aromatic ring keeps the methyl groups rigidly in the axial position (Stork, 1962). Very selective cis-hydrogenations are also achieved by reduction with diimine (N_2H_2, Hünig, 1965; Miller, 1965; Pasto, 1991). The reagent can be used at low temperatures and has been employed in the selective reduction of C=C double bonds, e.g. in the presence of a sensitive peroxidic function (Adam, 1978).

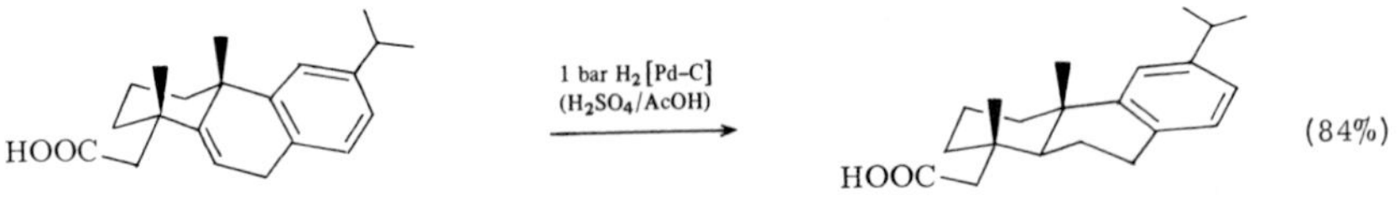
HOOC
1 bar H2 [Pd–C]
(H2SO4/AcOH)
HOOC
(84%)

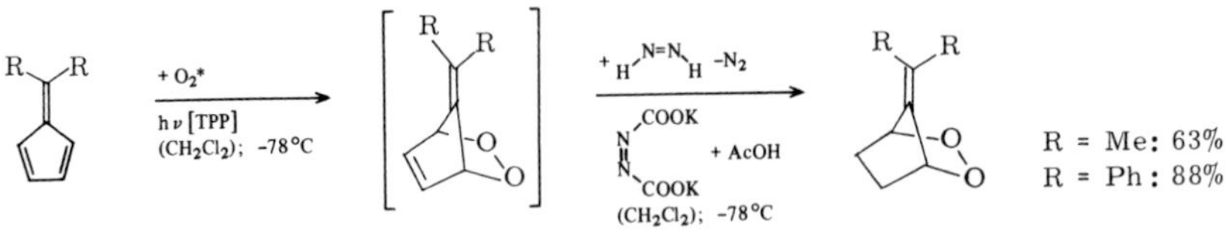
R R
+ O2*
hν [TPP]
(CH2Cl2); -78°C
R R
O
O
+ HN=NH -N2
COOK
N=N
COOK
+ AcOH
(CH2Cl2); -78°C
R R
O
O
R = Me: 63%
R = Ph: 88%

Asymmetric hydrogenation has been achieved with dissolved Wilkinson-type catalysts (Birch, 1976; Valentine, Jr., 1978; Kagan, 1978). The (*R*)- and (*S*)-[1,1′- bi-naphthalene]-2,2′-diylbis[diphenylphosphine] (="binap") complexes of ruthenium (Miyashita, 1980) and rhodium (Miyashita, 1984; Noyori, 1987) have been prepared as pure atropisomers and used for the stereoselective "Noyori hydrogenation" of *α*-(acylamino) acrylic acids and, more significantly, *β*-keto carboxylic esters. In the latter reaction enantiomeric excesses of more than 99% are often achieved (see also Nakatsuka, 1990). The commercial syntheses of aspartame and DOPA both occur with a Noyori reduction of an enamine acetate of a *trans*-cinnamic acid derivative.

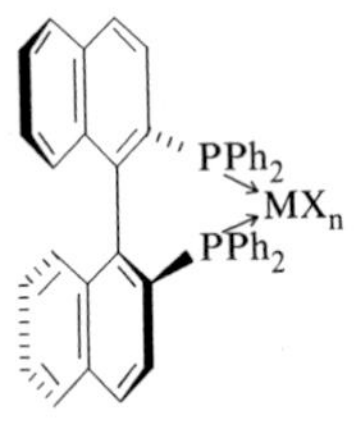
PPh2
MXn
PPh2

(S)-binap complexes:

A: M = Rh, X_n = ClO_4 — [Rh((S)-binap)ClO_4]

B: M = Ru, X_n = 2 Cl — [Ru_2Cl_4((S)-binap)$_2$]

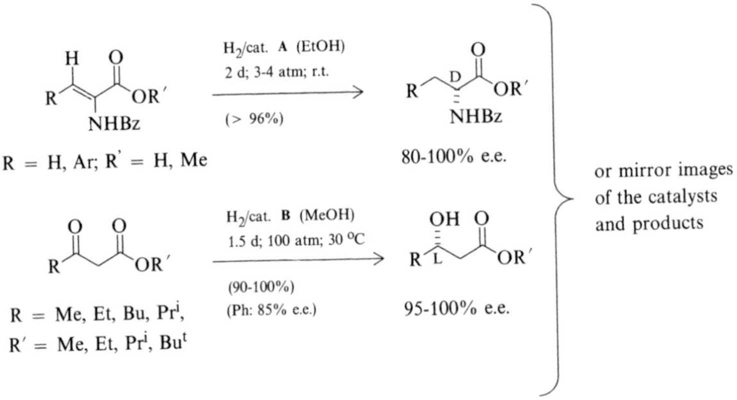
H O
R OR′
NHBz
H2/cat. A (EtOH)
2 d; 3-4 atm; r.t.
(> 96%)
O
R D OR′
NHBz
R = H, Ar; R′ = H, Me
80-100% e.e.
O O
R OR′
H2/cat. B (MeOH)
1.5 d; 100 atm; 30 °C
(90-100%)
(Ph: 85% e.e.)
OH O
R L OR′
R = Me, Et, Bu, Pri,
R′ = Me, Et, Pri, But
95-100% e.e.
or mirror images
of the catalysts
and products

(i) (R,R)-PNNP-Rh(I) (cat.),
H₂, EtOH (83% e.e.)
(catalytic asymmetric hydrogenation)

(ii) H⊕, MeOH

(S)-phenylalanine methyl ester
(97% e.e. after recrystallization)

(R,R)-PNNP=

Birch reductions of C=C double bonds with alkali metals in liquid ammonia or amines obey other rules than do the catalytic hydrogenations (Caine, 1976). In these reactions regio- and stereoselectivities are mainly determined by the stabilities of the intermediate carbanions. If one reduces, for example, the α,β-unsaturated decalone shown in the scheme below with lithium, a dianion is formed, for which three different conformations (A), (B), and (C) are conceivable. Conformation (A) is the most stable, because repulsion disfavors the cis-decalin system (B) and in (C) the conjugation of the dianion is interrupted. Thus, protonation yields the trans-decalone system (Stork, 1964 B). Similar rules hold true for the Birch reduction of substituted benzene rings to give 1,4-dihydro derivatives (Akhrem, 1972; Birch, 1972). In the first step an anion radical is formed, which is selectively protonated at the site of highest electron density. The resulting pentadienyl radical is further reduced to the corresponding anion and selectively protonated para to the first proton. Alkoxy- and dialkylamino-substituted benzene rings are reduced in this way to produce the corresponding 2,5- or 3,6-dihydro derivatives, whereas benzoic acid derivatives are hydrogenated at 1,4-positions. Enol ethers from alkoxyarenes may be converted to ketones by acidic hydrolysis.

(i) Li; –30°C
(Et₂O/EtOH/NH₃)
(ii) + NH₄Cl

(iii) CrO₃/H₂SO₄
(Me₂CO); 5 min; 0°C

(94%)
less stable isomer

(< 1%)
more stable isomer

transition states:

A
most stable
transition state

>

B
enhanced axial-
axial repulsion

≫

C
zero overlap
of dianion

Selective reduction of a benzene ring (Grimme, 1970) or a C=C double bond (Cole, 1962) in the presence of protected carbonyl groups (acetals or enol ethers) has been achieved by Birch reduction. Selective reduction of the C=C double bond of an α,β-unsaturated ketone in the presence of a benzene ring is also possible in aprotic solution, because the benzene ring is reduced only very slowly in the absence of a proton donor (Caine, 1976).

The phenol rings of calyx[4] arene have been reduced to the enol state with hydrogen in the presence of rhodium chloride and Aliquat 336 (= tricapryl methyl-ammonium chloride). Workup with water yielded the calyx[4] cyclohexanone with eight stereocenters in statistical arrangement. Treatment with ethoxide gave a multianion in which all carbons neighboring the same carbonyl groups became mirror images. A bowl form was enforced leading to the formation of a single stereoisomer of the molecule, which was retained after reduction of the carbonyl groups to the alcohols (Columbus, 1998).

Calix[4]arene   Calix[4]cyclohexanone

Calix[4]cyclohexanol

Triple bonds are best reduced by the conjugate addition reactions using chlorobis-(η^5-cyclopentadienyl) hydridozirconium, Cp$_2$ZrClH. At first the reagent is added in THF, then the C–Zr bond is either cleaved oxidatively with iodine to produce vinyliodides (see Section 3.8) or followed directly by a metal catalyzed coupling (Sun, 1992). Another procedure uses tributyltinhydride (Bhatt, 2001).

Cyclopropane rings are opened hydrogenolytically, e.g. over platinum on platinum dioxide (Adam's catalyst) in acetic acid at 1–4 bar hydrogen pressure. The bond that is best accessible to the catalyst and most activated by conjugated substituents is cleaved selectively (Irwin, 1968; Augustine, 1976). Synthetically this reaction is useful as a means to hydromethylate C=C double bonds via carbenoid addition (Majerski, 1968; Woodworth, 1968).

3.2.3
Reduction of Aldehydes, Ketones, and Carboxylic Acid Derivatives

These polar functional groups are mostly reduced to the corresponding alcohols with hydride reagents (Hajós, 1966, 1979). The general selectivities are indicated in Tabs. 3.1 and 3.2 and a few specific examples will be given here.

Hydroborates with an electron withdrawing cyano group reduce protonated C=O and C=N double bonds selectively. Aldehydes and ketones are therefore only reduced at pH < 3, whereas imines are hydrogenated at pH 6. The bulky salt tetrabutylammonium cyanoborate reduces aldehydes much faster than ketones in slightly acidic HMPTA solutions (Hutchins, 1973 A). Hydrotris(alkylthio)borates show the same selectivity (Maki, 1977). Bulky aluminium hydrides, e.g. diisobutylaluminium hydride (DIBAL), convert esters or amides into the corresponding aldehydes or hemiacetals (Corey, 1975 A). Lithium tri-sec-butylhydroborate (= L-Selectride®) can be used to reduce lactones to cyclic hemiacetals in the presence of acyclic ester groups (Nakatsuka, 1990). α,β-Unsaturated carbonyl compounds are reduced with alane (AlH$_3$, from 3 LiAlH$_4$ + AlCl$_3$; Wigfield, 1973, Dauben, 1973) or with NaBH$_4$ in the presence of cerium(III) ions (Luche, 1978) to give allylic alcohols. LiAlH$_4$ itself tends to reduce the C=C double bond.

Bu$_4$N$^{\oplus}$ BH$_3$CN$^{\ominus}$; r.t.
(H$_2$SO$_4$/HMPTA; pH = 4)

Na$^{\oplus}$ BH(SBu^t)$_3$$^{\ominus}$
(THF); 0.5 h; r.t.

R–CHO, R'–C=O → R–CH$_2$OH, R'–C=O ($\approx$ 90%) and R–CH$_2$OH, R'–CH–OH ($\approx$ 5%)

R = divalent hydrocarbon radical

DIBAL(CH$_2$Cl$_2$)
0.5 h; –78 °C
(> 94%)

MeO OMe ... COOMe, OSi(CH$_2$Ph)$_3$ → [MeO OMe ... CHO, OSi(CH$_2$Ph)$_3$]

Wittig reaction → . . . .

$$3 \; LiAlH_4 + AlCl_3 \rightleftharpoons 2 \; Al_2H_6 + 3 \; LiCl$$

LiAlH$_4$/AlCl$_3$ 3 : 1
(Et$_2$O); 1 h; r.t.; N$_2$
$\alpha : \beta \approx 1 : 1$

HO— (93%)   HO— (2%)

NaBH$_4$/CeCl$_3$ · 6 H$_2$O
(MeOH); 5 min; r.t.

—OH (100%)

An important aspect of the reduction of carbonyl compounds is the stereoselectivity of these reactions. With open-chain compounds such selectivity is difficult to achieve, although there exists some preferences. Cram's rule (1952), for example, predicts the steric course of hydride and carbanion addition to open-chain aldehydes and ketones containing a chiral carbon atom with three substituents of different size (large, medium, and small, l,m,s). The rule has been substantiated and revised several times (Morrison, 1971). A version by Chérest et al. (1968) explains the observed diasteromeric product ratios with a sterically favored approach of the hydride donor or carbanion anti to l. In the preferred transition state the small carbonyl oxygen is near to m, the bulky alkyl group R near to s. The extent of asymmetric induction is, however, usually unsatisfactory, and therefore such reactions are of limited value in complex organic syntheses.

hydride donor →

preferred staggered transition state

	threo product	
R	Ch	Ph
Me	62%	74%
Et	67%	76%
Pr^i	80%	83%
Bu^t	62%	98%

Synthetically useful stereoselective reductions have been possible with cyclic carbonyl compounds of rigid conformation. Reduction of substituted cyclohexanone and cyclopentanone rings by hydrides of moderate activity, e.g. NaBH$_4$ (Luche, 1978), leads to alcohols via hydride addition to the less hindered side of the carbonyl group. Hydrides with bulky substituents are especially useful for such regio- and stereoselective reductions, e.g. lithium hydroxi-*t*-butoxyaluminate (Kuo, 1968) and lithium or potassium tri-*sec*-butylhydroborates or hydrotri-*sec*-isoamylborates (= L-, K-, LS- and KS-Selectrides®) (Brown, 1972 B; Brown, 1973; Krishnamurthy, 1976).

Cyclohexanone derivatives are reduced to the equatorial alcohol in almost quantitative yield by treatment with isopropanol as reductant (a variant of the Meerwein–Ponndorf reaction) and samarium iodide as a catalyst (Collin, 1986; Molander, 1992). As with most radical reactions, steric hindrance is negligible.

Another possibility for asymmetric reduction is the use of chiral complex hydrides derived from LiAlH$_4$ and chiral alcohols, e.g. N-methylephedrine (Jacquet, 1974), or 1,4-bis(dimethylamino)butanediol (Seebach, 1974). But stereoselectivities are mostly below 50%. (Percent stereoselectivity = % stereoisomer A − % stereoisomer B in the product mixture A + B.) Attempts to form chiral alcohols from ketones are usually less successful than the asymmetric reduction of C=C double bonds via hydroboration or hydrogenation with Wilkinson-type catalysts (Zweifel, 1963; Kagan, 1978).

In cases where Noyori's reagent and other enantioselective reducing agents are not successful, (+)- or (−)-chlorodiisopinocamphenylborane (Ipc$_2$BCl) may help. This reagent reduces prochiral aryl and *tert*-alkyl ketones with exceptionally high enantiomeric excesses (Chandrasekharan, 1985; Brown, 1986). The initially formed boron moiety is usually removed by precipitation with diethanolamine. Ipc$_2$BCl has, for example, been applied to synthesize polymer-supported chiral epoxides with 90% e.e. from Merrifield resins (Antonsson, 1989).

Tab. 3.3 The formation of chiral alcohols.

Reducing agent	Example		R:S
	O (phenyl propyl ketone)	(Et₂O); 2 h; 0°C (> 90%; 89% e.e.) → HO H (phenyl propyl carbinol)	94:6
	O (phenyl butyl ketone)	(Et₂O); 4 h; r.t. (92%; 47% e.e.) → HO H (phenyl butyl carbinol)	73:27
	COOMe (cyclopentadiene acetate)	(i) (THF); 5 h; –78°C +16 h; 0°C (ii) H₂O₂/OH⁻; r.t. (Et₂O/H₂O) → HO COOMe 98% (R,R)	–

Enantioselective addition of dibutylzinc in the presence of a chiral bisulfonamide titanium complex based on 1S,2S-diamino cyclohexane yielded an open-chain tertiary alcohol with 94% e.e. (Arredondo, 1999). Azide displacement of the free hydroxyl groups under Mitsunobu conditions (see p. 217 f.) followed by LAH reduction to the amine leads to amines with inversion of configuration.

$n\text{-}C_3H_7$ CHO → [SO_2CF_3 / N-Ti(O^{i}Pr)$_2$ / N / SO_2CF_3; n-Bu₂Zn; –60 to –20 °C; toluene] → $n\text{-}C_3H_7$ ... OH (70% yield, 94% e.e.) → [(i) Ph₃P, DEAD, DPPA, 25 °C; (ii) LiAlH₄, Et₂O] → $n\text{-}C_3H_7$ ... NH₂ (69% yield)

The direct four-electron reduction of the carbonyl group to the methylene group is possible directly with zinc (Clemmensen reductions; Vedejs, 1975) or, more conveniently, via the corresponding hydrazones or thioacetals (Asinger, 1970). The drastic conditions required in the original Wolff–Kishner reduction of hydrazones (alkali hydroxide, above 200 °C; Cole, 1962; potassium t-butoxide in boiling toluene; Grundon, 1963) can be avoided, if potassium t-butoxide in DMSO is used (Cram, 1962). Another possibility is the use of hydrazone derivatives bearing good leaving groups, e.g. tosyl. They may eliminate molecular nitrogen on reduction with hydrides (Hutchines, 1973 B, 1975; Taylor, 1976). Hydrazones of α,β-unsaturated carbonyl compounds yield rearranged olefins. Desulfurization of thioacetals and thioketals occurs with Raney nickel (Pettit, 1962 A; Asinger, 1970). Several examples are known for the direct reduction of the keto group to a methylene group in a,β-unsaturated ketones (Barton, 1968).

Esters can usually not be reduced directly to the corresponding ethers. Efficient conversion with the $NaBH_4$-BF_3 reagent is only possible if the alcohol component is tertiary (Pettit, 1962 B).

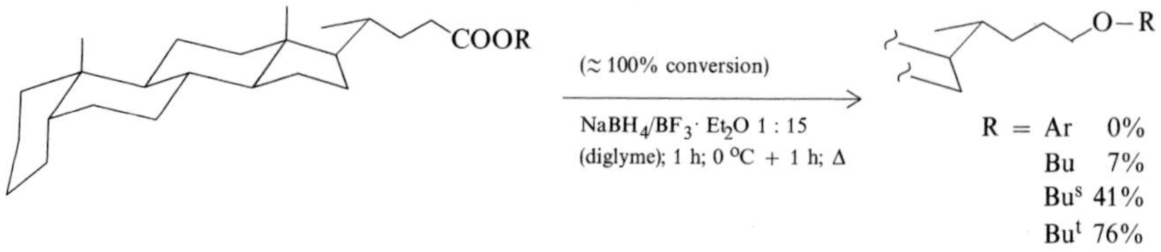

($\approx$ 100% conversion)

NaBH$_4$/BF$_3$· Et$_2$O 1 : 15
(diglyme); 1 h; 0 °C + 1 h; Δ

R = Ar   0%
Bu   7%
Bu^s 41%
Bu^t 76%

Noyori's catalyst is effective in the asymmetric isomerization of allylic amines to give enamines. This reaction has been used to produce (*R*)-citronellal from achiral geranyl diethylamine and, after a Grignard reaction and asymmetric hydrogenation, the side-chain of tocopherol (Takaya, 1987). The chiral BINAP catalyst differentiates between the *a*-hydrogen atoms of the amine. The same isomerization–hydrogenation sequence also occurs in a commercial (–)-menthol synthesis. The starting material comes from cheap turpentine oil (*β*-pinene). The BuLi-catalyzed amination of 1,3-dienes is known as telomerization (Nicolaou, 1996).

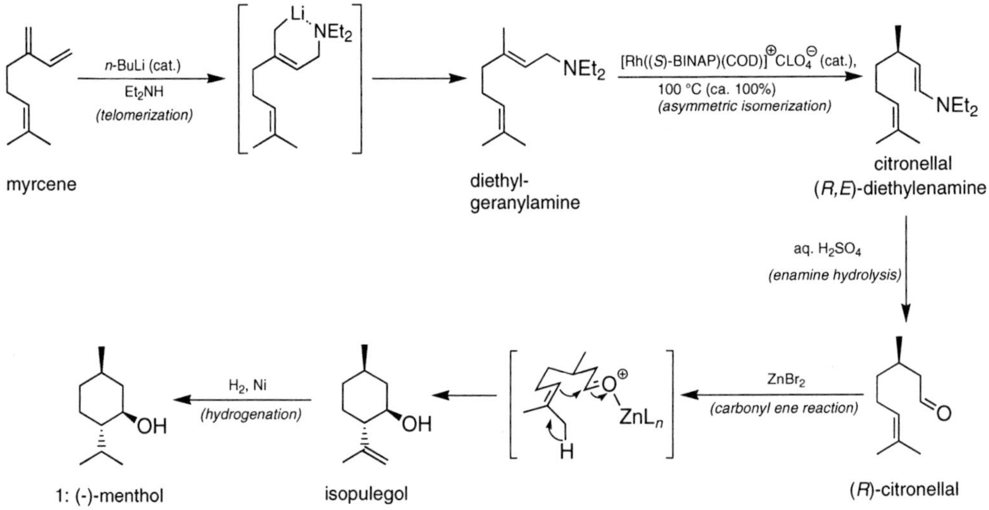

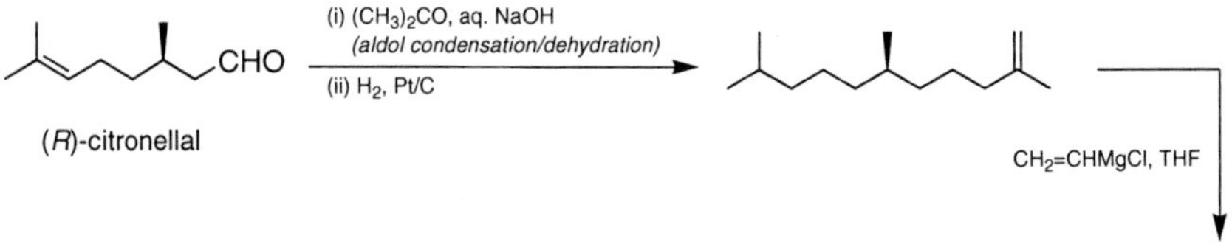

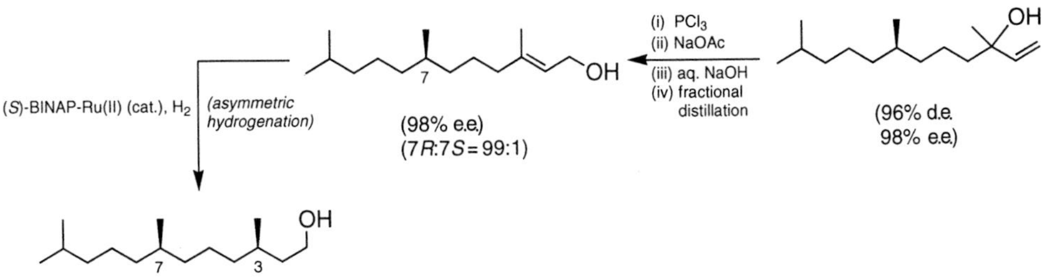

(98% e.e.)
(7R:7S = 99:1)

(96% d.e.
98% e.e.)

Sterically hindered α,α-disubstituted amino methylesters are reduced to the amino alcohols by the BuLi/DIBAL-ate complex. The latter is formed by mixing a DIBAL solution in hexane/THF with BuLi in hexane. If the methyl ester cannot be produced because of steric hindrance, persilylation with trimethylsilyl N,N-diethyltrimethylsilylamine (TMSDEA) and subsequent reduction with LAH can be used (Glunz, 1999). Diethylamine is removed by simple evaporation.

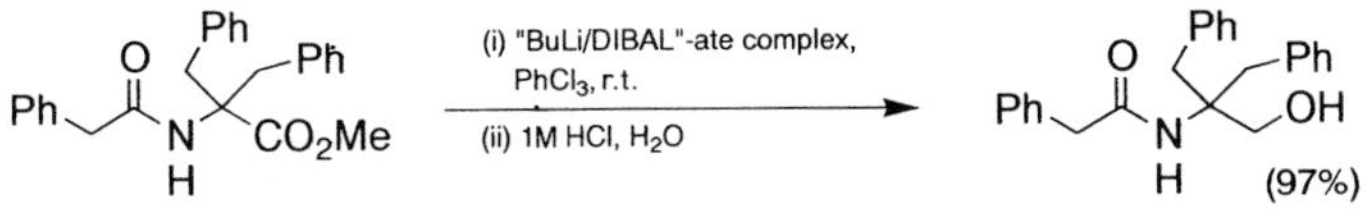

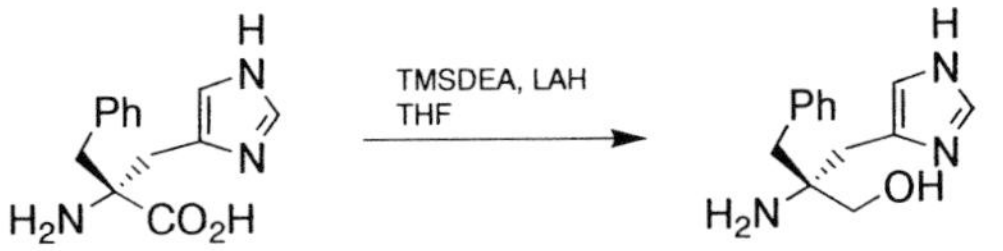

Reducing methylenation of the ester carbonyl group with Tebbe's reagent and the conversion of thionolactones to cyclic thioketals with subsequent reduction have been performed using electrophilic titanium or tantalum "ylides" (Tebbe, 1978 [footnote 20]; Schrock, 1976). Vinyl ethers were obtained in high yields with μ-chlorobis(η^5-2,4-cyclopentadien-1-yl)   (dimethylaluminum)-μ-methylenetitanium (Pine, 1980; Barrett, 1989).

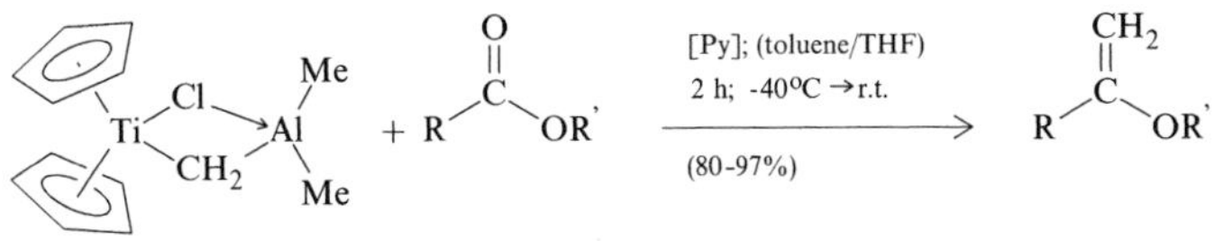

Tebbe's reagent

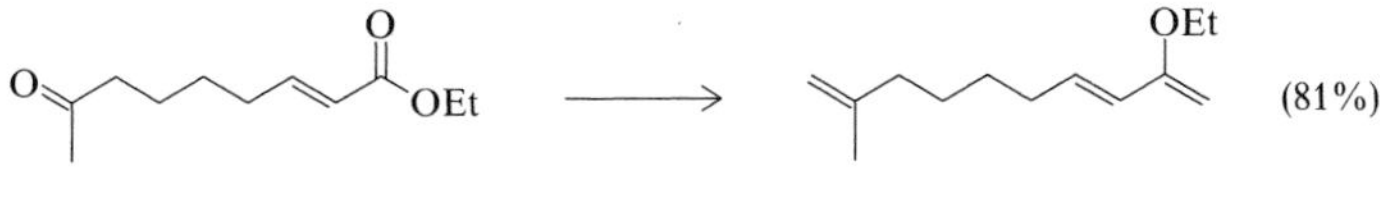

(69%)

Lactones were efficiently converted to a-alkylated cyclic ethers by successive ox-othioxo exchange, C-alkylation, and reductive desulfurization. The lactone thioxo-nation is most commonly executed by a 1,3,2,4-dithiadiphosphetane 2,4-disulfide, a Lawesson-type reagent. Nucleophilic addition of a wide range of organolithium reagents proceeds smoothly at −78 °C to provide cyclic thioketals upon trapping by S-alkylation with iodomethane. Reductive desulfurization of the thioketals is ac-complished with triphenylstannane. At bicyclic or further substituted ring systems the hydrogen usually stereoselectively attacks the least hindered side of the most stable ring conformation of the intermediate radical (Nicolaou, 1990 A).

Deoxygenation of allylic alcohols may lead to dehydration instead. Activation of the alcohol by xanthate formation with CS_2 in the presence of sodium hydride and reduction with triphenyltin and AIBN was found to be the mildest procedure to lead to reduction only. Nitro groups are easily removed from a-nitroketones with Ph_3SnH in the presence of ACN. Reduction of the ketone with borohydride, prepa-ration of xanthates and again treatment with Ph_3SnH/ACN removed the oxygen atom (Jeong, 1999).

Trialkyltin radicals show a strong affinity to thiocarbonyl compounds. This is applied in the reduction of alcohols to hydrocarbons, in particular for sterically hindered alcohols. In the Barton–McCombie reaction thioacetylation of an alcohol is achieved by a thioester chloride and reduction occurs at relatively high temperatures. Thiohydroxamates can be used under milder conditions to convert acid chlorides into alkanes or halides (Barton, 1975, 1986).

3.2.4
Reduction of Nitrogen Compounds

Amides can be reduced to amines with LiAlH$_4$, although the reaction proceeds more slowly than the reduction of most other functional groups (Harrison, 1961; Ayer, 1968), which have to be reoxidized afterwards if desired. Diborane is also useful and does not attack ester groups, but C=C double bonds (Brown, 1964). Sodium tetrahydroborate reduces amides only in the presence of acidic catalysts, e.g. CoCl$_2$ (prim. and sec. amides only; Satoh, 1969) or carboxylic acids (Umino, 1976). Secondary and tertiary amides are O-alkylated with Meerwein's reagent (Et$_3$O$^+$BF$_4^-$), and the resulting carbenium ions are reduced in high yield with NaBH$_4$ in ethanol (Borch, 1968). In all these cases the C–N linkages remain intact after reduction. Cleavage into amines and alcohols (from the reduction of the acyl moiety) occurs only occasionally. Esters, in contrast, are almost always cleaved on reduction because alkoxide ions are easily cleaved from the intermediate hemiacetal anions. In amides the carbonyl oxygen bound to boron or aluminium is removed much more easily than an amide anion. Nitriles are converted into aldehydes by several reducing agents, e.g. DIBAL, complex hydrides, or catalytic hydrogenation (Winterfeldt, 1975).

R—CONH$_2$ $\xrightarrow{\begin{array}{c}\text{NaBH}_4/\text{CoCl}_2 \ 1:2\\ \text{(H}_2\text{O); 1 h; r.t.}\end{array}}$ R—CH$_2$NH$_2$ ($\approx 70\%$)

$\xrightarrow{\begin{array}{c}\text{NaBH}_4/\text{AcOH} \ 1:11\\ \text{(dioxane); 2 h; } \Delta\end{array}}$ (93%)

$\xrightarrow{\begin{array}{c}\text{Et}_3\text{O}^\oplus\text{BF}_4^\ominus\\ \text{(CH}_2\text{Cl}_2\text{); 1 d; r.t.}\end{array}}$ $\left[\right]$ BF$_4^\ominus$ $\xrightarrow{\begin{array}{c}\text{NaBH}_4\\ \text{(EtOH); 1 d; r.t.}\end{array}}$ (92%)

$\xrightarrow{\begin{array}{c}\text{(i) DIBAL(C}_6\text{H}_6\text{)}\\ \text{1 h; r.t.; N}_2\\ \hline \text{(ii) + dil. aq. H}_2\text{SO}_4\end{array}}$ (75-85%)

Ketones may also be converted into amines, if they are first reacted with ammonium salts or methoxyamine and then reduced with sodium trihydrocyanoborate at pH 7, where carbonyl groups are not attacked (Boutigue, 1973).

$\xrightarrow{\begin{array}{c}\text{excess NH}_4^\oplus \text{AcO}^\ominus\\ \text{NaBH}_3\text{CN(MeOH)}\\ \text{1 h; r.t.}\\ \hline \text{(100\%)}\end{array}}$

$\alpha:\beta \ = 1:9$

Nitro groups are efficiently reduced with hydrogen over Raney nickel catalyst (I. Felner, 1967), with hydrides, or with metals. As mentioned above, nonaromatic C=N double bonds are easily reduced by hydrides (Stork, 1963). AlH$_3$ reduces activated pyridine rings (Ferles, 1970). Mild and selective reduction of pyridine rings, e.g. with NaBH$_4$ or dithionite, necessitates quaternization or protonation of the nitrogen atom and/or electron withdrawing substituents on a carbon atom (Eisner, 1972; Dyke, 1972). Catalytic hydrogenation of pyridine rings also works best with 3,5-disubstituted electron-deficient pyridinium cations, but further reduction to tetra- and hexahydropyridines is a side-reaction in all reductions (Eisner, 1972; Ferles, 1970; Acheson, 1976A). The complete hydrogenation of neutral pyridine rings occurs in catalytic hydrogenation, e.g. with rhodium on carbon or Adam's catalyst (Pt/PtO$_2$) (Freifelder, 1962, 1963; Coppola, 1978).

Reaction: starting material (pyrrolidinone with NO2 and COOMe) → 100 bar H2, [Raney Ni], (MeOH); r.t. → product (88%)

Reaction: indole alkaloid (MeO-substituted) → NaBH4, (MeOH); r.t. → product (100%)

Reaction: N-methyl dimethylpyridinium iodide → LiAlH4/AlCl3 3:1, (Et2O); 6 h; Δ → two products (73%) and (9%)

Reactions: pyridinium CONH2 with OPr, Cl⊖ →
- Na2S2O4/NaHCO3, (H2O/Et2O); 4 h; 0 °C → product (80%)
- NaBH4/NaHCO3, (H2O/Et2O); 1 d; r.t. → product (80%)

Reaction: pyridyl propanol → 2.7 bar H2 [Rh–C], (EtOH); 1 h; 55 °C → piperidine propanol (77%)

Reaction: quinoline-COOH → 3.5 bar H2 [Pt–PtO2], (EtOH); 2 h; r.t. → tetrahydroquinoline-COOH (87%)

3.2.5
Reductive Cleavage of Carbon–Heteroatom and Heteroatom–Oxygen Bonds

Single-bond cleavage with molecular hydrogen is termed hydrogenolysis. Palladium is the best catalyst for this purpose, platinum is not useful. Desulfurizations are most efficiently performed with Raney nickel (with or without hydrogen; Pettit, 1962 A) or with alkali metals in liquid ammonia or amines. The scheme below summarizes some classes of compounds most susceptible to hydrogenolysis.

Tab. 3.5 (cont.)

Oxidative rearrangement or cleavage of ketones		
		Tl^{3+}/H_2O, MeOH
		peracids
		(i) NH_2OH (ii) H^+ (Beckmann rearr.)
		CrO_3, $KMnO_4$; (i) SeO_2 (ii) HIO_4

tials. The E_0 values for the CH_4/CH_3OH and C_2H_6/C_2H_5OH couples are at +0.59 V and 0.52 V, respectively. The oxidation of alcohols and aldehydes corresponds to E_0 values around 0.0 V (Latimer, 1952). Therefore all applied oxidants are, in thermodynamic terms, able to oxidize or to dehydrogenate all hydrocarbons and all oxidizable functional groups of organic molecules.

Specificity in synthetic oxidation procedures is only possible if one C–H, C–C, or C–X bond reacts much more rapidly than all other bonds. Therefore it depends much more on the structure of the organic substrate than on the oxidant used, and large excesses of oxidants should always be avoided. The rationale for the use of a large variety of oxidizing reagents in organic syntheses is largely based on empirical results with specific oxidations of specific substrates (Augustine, 1969, 1971; Wiberg, 1965; Trahanovsky, 1973, 1978; Arndt, 1975). It has been shown frequently, that supposedly "general" rules previously developed for the use of different oxidants in fact depend on the nature of the substrates. Nevertheless, nowadays such rules constitute the only basis for the choice of an oxidant, and some of them will therefore be outlined within this section. Initial guidelines to the preferential uses of oxidants may be deduced from Tab. 3.5. Very few of the oxidants, however, are selective, and other easily oxidizable groups generally have to be protected (see Section 3.9).

3.3.2
Oxidation of Non-Functional Carbon Atoms

Chromic acid, potassium permanganate, and chlorine attack alkanes under vigorous conditions. The relative rates of chromic acid oxidation of primary, secondary, and tertiary C–H bonds, for example, are approximately 1:100:10000 (Mareš, 1961; Roček, 1957, 1959). Oxidations of complex saturated hydrocarbons with external oxidants are, however, of limited synthetic use because complicated mixtures of products are generally obtained in low yield. A classic attempt in this field was the oxidative degradation of the side-chain on C-17 of steroids (e.g. cholesterol, stigmasterol). Such degradation converted inexpensive steroids into precious steroidal hormones. Simple chromic acid oxidation led to a useful androstenolone derivative in 7–8% overall yield from cholesterol (Schering process, 1957, 1977; Fieser, 1959). This reaction sequence became unimportant with the discovery of bacteria that were able to produce androstene diones from cholesterol. Fluorination of steroids has been shown to be quite selective for tertiary C–H bonds at the ends of the side-chain (Barton, 1976). This reaction, however, has not been exploited either, because both fluorination sites are not of medical interest and it is difficult to cleave-off HF. If the oxidizing agent, e.g. (dichloroiodo)benzene, is covalently bound to the steroid, yields of regio- and stereoselective chlorination may be as high as 60% (Breslow, 1977). This "remote oxidation" approach has not been tested either on a large scale. A less esoteric method is biological oxidation with the aid of specialized microorganisms. Such biotechnological procedures allow highly regio- and stereoselective hydroxylations of complex organic molecules even on an industrial scale. The enormous data collection of Kieslich (1976) and introductory texts should be consulted (Bonse, 1978; Johnson, 1978; Roberts, 1994; Wackett, 2001) for the scope of selective biological transformations of organic substrates.

cholesterol

$Li^{\oplus}AlH(OBu^t)_3^{\ominus}$

Schering process

(i) Ac$_2$O (DCE); 5 h; Δ
(ii) PyBr$^{\oplus}$ Br$_3^{\ominus}$ (DCE); 12 h; –15 °C } protection
(iii) CrO$_3$/H$_2$SO$_4$/AcOH (DCE/SiO$_2$);
    12 h; 12 °C + 12 h; r.t. } oxidation
(iv) Zn/AcOH/H$_2$O; 9 h; r.t. } deblocking
(v) removal of carboxylic acids
(vi) H$_2$NNHCONH$_2$/AcOH(EtOH); 2 h; Δ } purification
(vii) CH$_3$COCOOH(NaOAc/H$_2$O/AcOH);
    10 min; Δ

androstenolone acetate
(7–8%)

(i) F$_2$/NaOTFac/TFA
    (CFCl$_3$/CH$_2$Cl$_2$)
    [inh.: PhNO$_2$]
    0.5 h; – 25 °C

(ii) Zn/AcOH/H$_2$O; Δ

17α-fluoro :   40%
17α,25-difluoro : 20%

Conventional regioselective oxidation of C–H bonds require neighboring activating groups. The oxidation of allylic methylene or methyl groups to yield allylic alcohols or α,β-unsaturated carbonyl compounds is very common in synthesis. The most successful reagents are derivatives of selenium(IV) and chromium(VI). Microorganisms are also used. The primary reaction is the electrophilic attack of the highly electropositive metal atoms upon the less hindered side of the C=C double bond. Selenium dioxide is added to olefins in an "ene" reaction to yield allyl-selenic acids, which rearrange to allyl selenoxylates (Arigoni, 1973; Sharpless, 1972, 1977). These are then cleaved thermally or hydrolytically. The original allylic methyl, methylene, or methane group is oxidized by this sequence to give a carbonyl or alcohol group. The oxidation proceeds regioselectively trans to the largest substituents of the C=C double bond (Sharpless, 1972; Arigoni, 1973; reviews: Trachtenberg, 1969; Reich, 1978; Rabjohn, 1949, 1976).

Allylic oxidation with derivatives of chromic acid (Wiberg, 1965; Bosche, 1975), e.g. di-*t*-butyl chromate (Fujita, 1961), CrO$_3$/pyridine (Dauben, 1969), or K$_2$CrO$_4$ (Cuilleron, 1970), gives similar products in occasionally very high yields. Steric effects are important with the large oxidant, and rigid rings are less easily attacked than flexible ones.

Alkyl groups attached to aromatic rings are oxidized more readily than the ring itself in alkaline media. Complete oxidation to benzoic acids usually occurs with nonspecific oxidants such as $KMnO_4$, but activated tertiary carbon atoms can then also be oxidized to the corresponding alcohols (Stewart, 1965; Arndt, 1975). With mercury(II) acetate, allylic and benzylic oxidations are also possible. It is most widely used in the mild dehydrogenation of tertiary amines to give enamines or heteroarenes (Shamma, 1970; Arzoumanian, 1971; Friedrich, 1975).

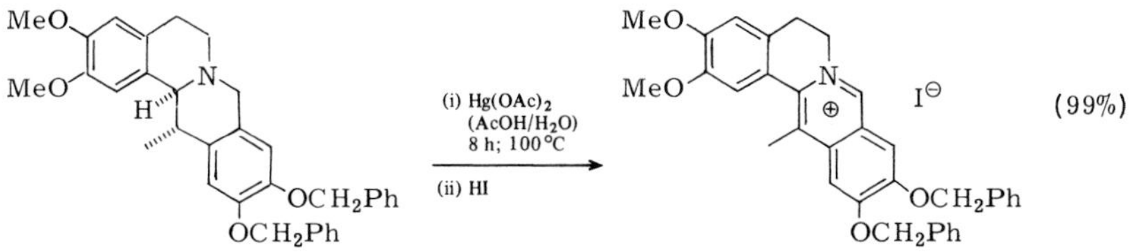
MeO
MeO
H
N
OCH2Ph
OCH2Ph
(i) Hg(OAc)2
(AcOH/H2O)
8 h; 100 °C
(ii) HI
MeO
MeO
N
⊕
I⊖
OCH2Ph
OCH2Ph
(99%)

The α-oxidation of carbonyl compounds may be performed by addition of molecular oxygen to enolate anions and subsequent reduction of the hydroperoxy group, e.g. with triethyl phosphite (Bailey, 1962; Gardner, 1968 A, B). If the initially formed α-hydroperoxide possesses another enolizable α-proton, dehydration to the 1,2-dione occurs spontaneously, and further oxidation to complex product mixtures is usually observed.

free-radical chain oxygenation:

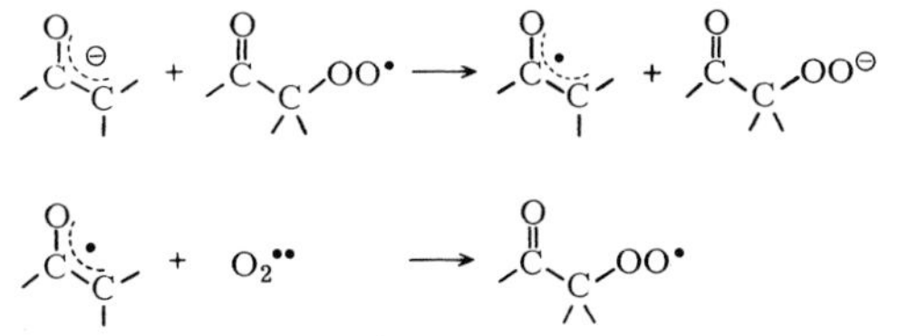

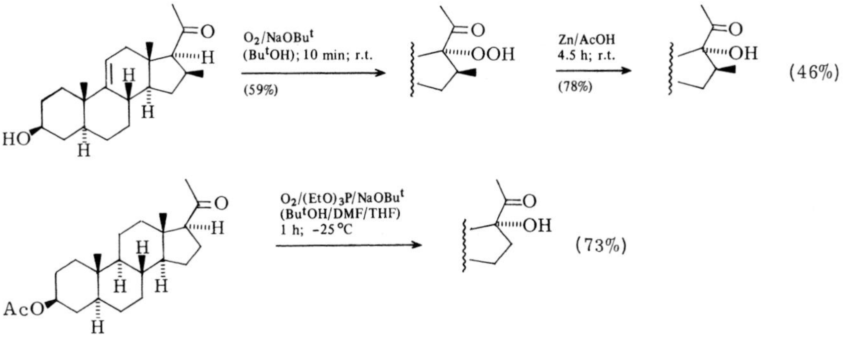
HO
H
H
H
O
H
O2/NaOBut
(ButOH); 10 min; r.t.
(59%)
O
OOH
Zn/AcOH
4.5 h; r.t.
(78%)
O
OH
(46%)
AcO
H
H
H
H
O
H
O2/(EtO)3P/NaOBut
(ButOH/DMF/THF)
1 h; -25 °C
O
OH
(73%)

SeO_2 oxidizes ketones and aldehydes to 1,2-dioxo compounds (Rabjohn, 1949, 1976; Trachtenberg, 1969). Some systems give α,β-unsaturated carbonyl compounds as by-products or even as the major product, especially if the β-carbon site is activated (Wiesner, 1958). The reaction sequence is similar to that described for the allylic oxidation of olefins with SeO_2, except that the C=O double bond is attacked instead of a C=C double bond. For the dehydrogenation of C–C single bonds activated by carbonyl, vinyl, or aryl groups, various other reagents, e.g. phenylselenium bromide or quinines, have also been used frequently.

3.3.3
Oxidation of Carbon Atoms in Carbon–Carbon Multiple Bonds

Oxidation of olefins and dienes provides the classic means for syntheses of 1,2- and 1,4-difunctional carbon compounds. The related cleavage of cyclohexene rings to produce 1,6-dioxo compounds has already been discussed in Section 1.11. Many regio- and stereoselective oxidations have been developed within the enormously productive field of steroid syntheses. Our examples for regio- and stereoselective C=C double bond oxidations as well as the examples for C=C double bond cleavages are largely selected from this area.

Peracids or mixtures of hydrogen peroxide and formic acid have been frequently used to transform olefins into epoxides, alcohols, or *trans*-glycols. The reactivities of the peracids parallel the acidities of the corresponding acids (Swern, 1953, 1970; Lewis, 1969; Plesničar, 1978). In cycloalkenes the less hindered side of the C=C double bond is epoxidized selectively (Brown, 1970). If two different C=C double bonds are present within a molecule, the bond with the higher electron density reacts faster (electron releasing alkyl groups, no electron withdrawing groups; Hückel, 1955; Woodward, 1958; Knöll, 1975).

Epoxide opening with nucleophiles occurs at the less substituted carbon atom of the oxirane ring. Catalytic hydrogenolysis yields the more substituted alcohol. The scheme below contains also an example for *trans*-dibromination of a C=C double bond followed by dehydrobromination with strong base for overall conversion into a conjugated diene. The bicyclic tetraene then isomerizes spontaneously to the aromatic 1,6-oxido[10]annulene (Vogel, 1964). Conjugated dienes form 2,3-unsaturated 1,4-diols by treatment with peracids first and hydroxide thereafter (Woodward, 1957).

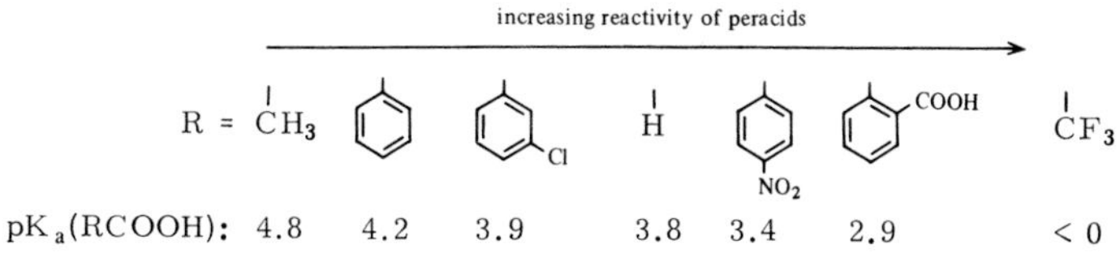

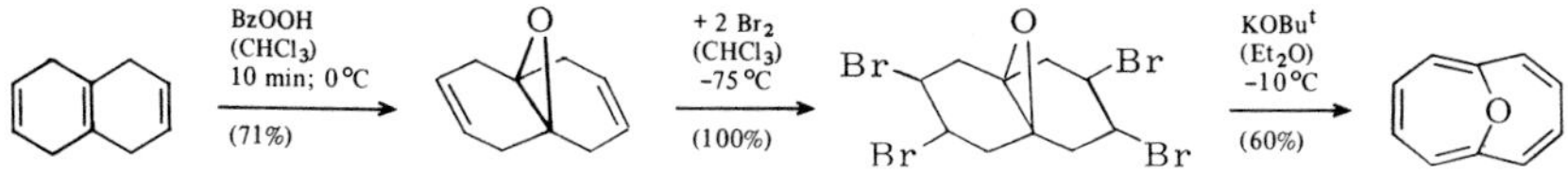

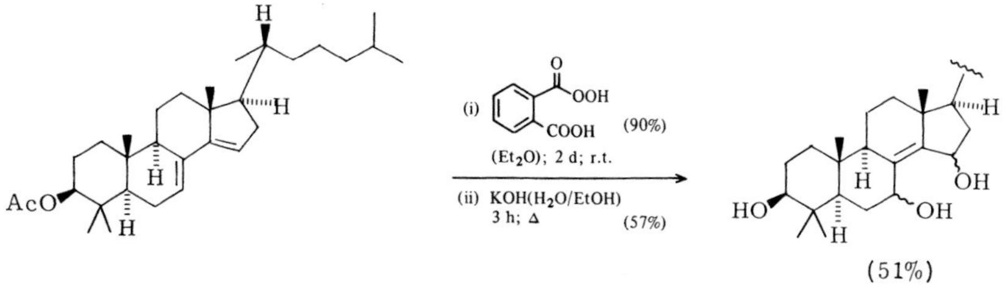

$$R-\overset{O}{\underset{O-O-H}{C}} \quad + \quad \overset{}{\underset{}{C}}=\overset{}{\underset{}{C} } \quad \longrightarrow \quad R-\overset{O}{\underset{OH}{C}} \quad + \quad \overset{}{\underset{}{C}}\overset{O}{-}\overset{}{\underset{}{C}}$$

increasing reactivity of peracids →

$R = \overset{|}{CH_3}$ (phenyl)   (Cl-phenyl) $\overset{|}{H}$ (NO_2-phenyl)   (COOH-phenyl) $\overset{|}{CF_3}$

$pK_a(RCOOH)$: 4.8 4.2 3.9 3.8 3.4 2.9 < 0

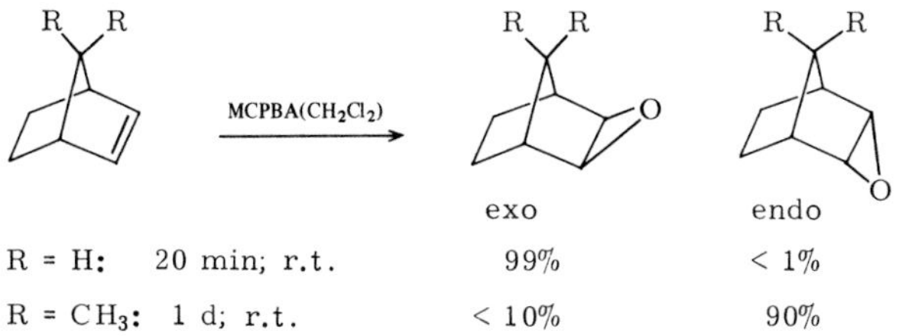

MCPBA(CH₂Cl₂)

exo   endo

R = H: 20 min; r.t.   99%   < 1%

R = CH₃: 1 d; r.t.   < 10%   90%

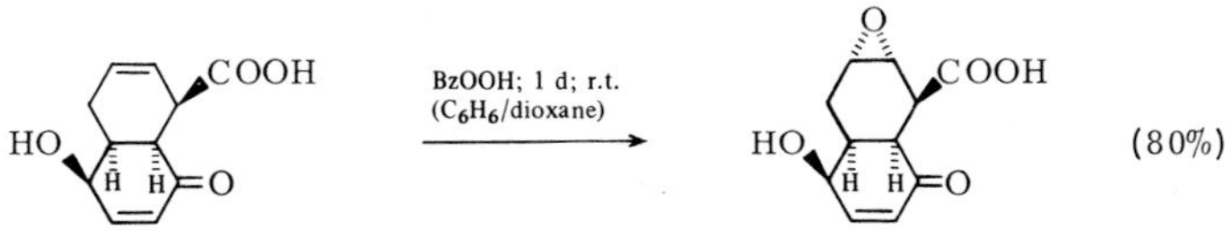

BzOOH; 1 d; r.t.
(C₆H₆/dioxane)

(80%)

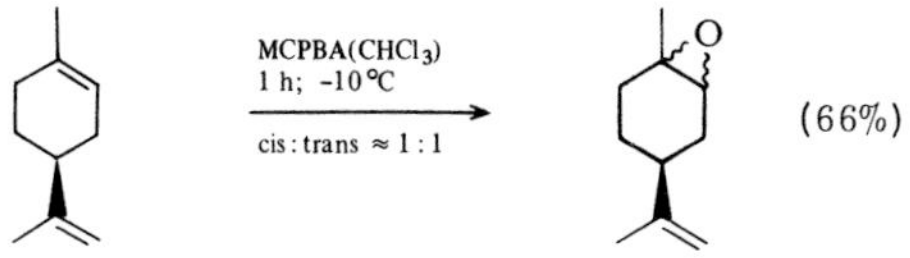

MCPBA(CHCl₃)
1 h; −10 °C

cis : trans ≈ 1 : 1

(66%)

BzOOH
(CHCl₃)
10 min; 0 °C

(71%)

+ 2 Br₂
(CHCl₃)
−75 °C

(100%)

KOBuᵗ
(Et₂O)
−10 °C

(60%)

(i) (2-carboxyperbenzoic acid structure) (90%)
(Et₂O); 2 d; r.t.

(ii) KOH(H₂O/EtOH)
3 h; Δ   (57%)

(51%)

The first practical method for asymmetric epoxidation of primary and secondary allylic alcohols was developed by Sharpless in 1980 (Katsuki, 1980; Sharpless, 1983 A, B, 1986; see also Hoppe, 1982). Tartaric esters, e.g., "DET" and "DIPT"

(=diethyl and diisopropyl (+)- or (–)-tartrates), are applied as chiral auxiliaries, titanium tetrakis(2-propanolate) as a catalyst and *tert*-butyl hydroperoxide (=TBHP, *t*-BuOOH) as the oxidant. If the reaction mixture is kept absolutely dry, catalytic amounts of the dialkyl tartrate titanium(IV) complex are sufficient, which largely facilitates workup procedures (Gao, 1987). Depending on the tartrate enantiomers used, either one of the 2,3-epoxy alcohols may be obtained with high enantioselectivity. The titanium probably binds to the diol grouping of one tartrate molecule and to the hydroxy groups of the bulky hydroperoxide and of the allylic alcohol.

Several structures of the transition state have been proposed (Williams, 1984; Jørgensen, 1987; Corey, 1990; Takano, 1991). They are compatible with most data, such as the observed stereoselectivity, NMR measurement (Finn, 1985), and X-ray structures of titanium complexes with tartaric acid derivatives (Williams, 1984). The models, e.g. Jørgensen's and Corey's, are, however, not compatible with each other. One may predict that there is no single dominant Sharpless transition state.

The catalyst is sensitive to pre-existing chirality in the substrate: the epoxidation of racemic secondary allylic alcohols often proceeds rapidly with only one of the enantiomers:

The method has been applied in asymmetric and regioselective syntheses of several natural compounds. Two simple examples are the commercial syntheses of the gipsy moth hydrophobic sex attractant, disparlure (Rossiter, 1981, 1985) and a mono-epoxidation of a diene in a leukotriene B_4 synthesis (Mills, 1983).

Sharpless epoxidations can also be used to separate enantiomers of chiral allylic alcohols by "kinetic resolution" (Martin, 1981; Sharpless, 1983 B). In this procedure the epoxidation of the allylic alcohol is stopped at 50% conversion, and the desired alcohol is either enriched in the epoxides fraction or in the nonreacted allylic alcohol fraction.

Very high degrees of enantiomeric excess are frequently obtained by reactions of enantioselective reagents with achiral, symmetrical oligofunctional educts containing a prochiral atom (or prochiral atoms) between a pair of enantiotopic reactive groups. Examples are enzyme-catalyzed hydrolyses of symmetrical diol diacetates to chiral monoacetates (Wand, 1984), hydroborations of 5-substituted 1,3-cyclopentadienes (Kagan, 1978), and the Sharpless epoxidation of divinylmethanols (Schreiber, 1987). In each of these reactions one of the two enantiotopic groups reacts rapidly, the other slowly, in an asymmetric synthesis step (*enantiotopic group selectivity*). The slowly forming stereoisomer, however, reacts preferentially in a subsequent "kinetic resolution" step at the remaining, rapidly reacting group to give the double-reaction product which is easily separated from the desired product. Thus, at high conversion, the e.e. theoretically and experimentally approaches 100%, if the ratio $k_{fast}:k_{slow}$ is sufficiently large (Wang, 1984; Schreiber, 1987).

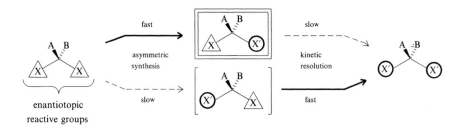

In the Sharpless epoxidation of divinylmethanols only one of four possible stereoisomers is selectively formed. In this special case the *diastereotopic face selectivity* of the Sharpless reagent may result in diastereomeric by-products rather than the enantiomeric one, e.g., for the L_g-(+)-DIPT-catalyzed epoxidation of (*E*)-*a*-(1-propenyl)cyclohexanemethanol to [*S*(*S*)]-, [*S*(*R*)]- and [*R*(*R*)-*trans*]-*a*-cyclohexyl-3-methyloxiranemethanol the ratio of rate constants is 971:19:6:4 (see above; Schreiber, 1987). This effect may strongly enhance the e.e. in addition to the "kinetic resolution" effect mentioned above, which finally reduces further the amount of the enantiomers formed.

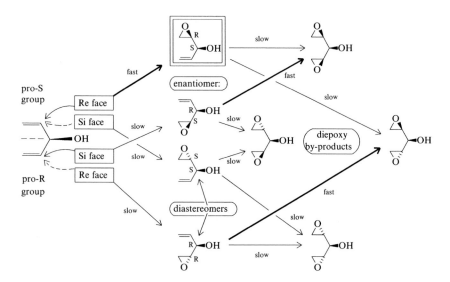

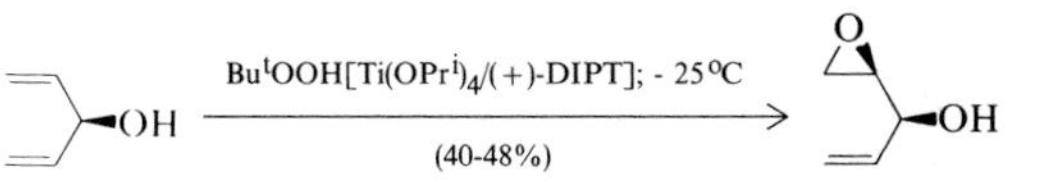

Bu^tOOH[Ti(OPr^i)$_4$/(+)-DIPT]; - 25 °C
(40-48%)

time	% e.e.	% d.e.
3 h	84	92
24 h	93	99.7
140 h	$\geq$ 97	> 99.7

As has been shown above, in cyclic systems the epoxy group is introduced on the less hindered side, and its hydrolysis leads to *trans*-diols. Another dihydroxylation procedure works with acyl hypoiodites, which are prepared *in situ* from iodine and silver(I) carboxylates in dry aprotic solvents. First *O*-acyl iodohydrins are formed. From these an iodide anion is removed by Ag$^+$, and the slightly nucleophilic carboxyl oxygen adds to the carbocation to form a *cis*-fused cyclic dioxolenium cation, which is opened by carboxylate to give *trans*-diol derivatives (Prévost reaction; Wilson, 1957; Criegee, 1979). The selectivity and the stereochemistry depend strongly on solvent, metal ions, substrate, and temperature, but *diaxial* glycols are obtainable in good yield (Cambie, 1977).

Prévost reaction:

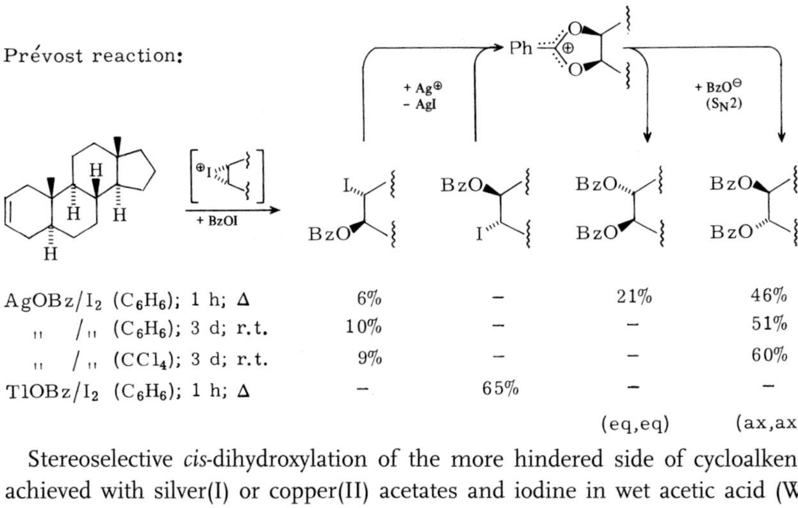

+ Ag$^\oplus$
- AgI

+ BzO$^\ominus$
(S$_N$2)

AgOBz/I$_2$ (C$_6$H$_6$); 1 h; Δ	6%	–	21%	46%
" / " (C$_6$H$_6$); 3 d; r.t.	10%	–	–	51%
" / " (CCl$_4$); 3 d; r.t.	9%	–	–	60%
TlOBz/I$_2$ (C$_6$H$_6$); 1 h; Δ	–	65%	–	–

(eq,eq)   (ax,ax)

Stereoselective *cis*-dihydroxylation of the more hindered side of cycloalkenes is achieved with silver(I) or copper(II) acetates and iodine in wet acetic acid (Woodward glycolization; Siddall, 1966; Mangoni, 1973; Criegee, 1979) or with thallium(III) acetate via organothallium intermediates (Glotter, 1976). In these reactions the intermediate dioxolenium cation is supposed to be opened hydrolytically, not by an S$_N$2 reaction.

The less hindered side of C=C double bonds is best *cis*-dihydroxylated with osmium tetroxide. Cyclic osmate diesters are isolable, often crystalline, intermediates. The very expensive and very toxic OsO_4 may be used in catalytic amounts with $KMnO_4$, H_2O_2, or N-oxides (Criegee, 1979) to regenerate the OsO_4 *in situ*. OsO_4 and especially its bis(pyridine) complex are bulky reagents and attack more accessible and more electron-rich C=C double bonds stereo- and regioselectively (Bernstein, 1956; Corey, 1964 B). The glycolization with $KMnO_4$ is largely limited to molecules with no other oxidizable groups (Criegee, 1979).

(i) OsO₄Py₂(C₆H₆/Py)
   5 d; r.t.; N₂
(ii) Na₂SO₃/KHCO₃/H₂O
   (MeOH/C₆H₆); 8 h; r.t.

(67%)

(i) OsO₄Py₂(Et₂O/Py)
   1 h; −20→0 °C + 1 d; r.t.
(ii) NaHSO₃(H₂O/Py); 0.5 h; r.t.

(100%)

A catalytic enantio- and diastereoselective dihydroxylation procedure without the assistance of a directing functional group (like the allylic alcohol group in the Sharpless epoxidation) has also been developed by Sharpless (Jacobsen, 1988; Kwong, 1990; Kim, 1990; Waldmann, 1992). It uses osmium tetroxide as a catalytic oxidant (as little as 20 ppm to date) and two readily available cinchona alkaloid diastereomers, namely the 4-chlorobenzoate esters or bulky aryl ethers of dihydroquinine and dihydroquinidine, as stereosteering reagents (for structures of the Os complexes: see Pearlstein, 1990). The transformation lacks the high asymmetric inductions of the Sharpless epoxidation, but it is broadly applicable and insensitive to air and water.

[dihydroquinidine 4-chlorobenzoate ester/OsO₄]
K₃Fe(CN)₆/BuᵗOH/H₂O; 1 d; r.t.

(80-95%)

[dihydroquinine 4-chlorobenzoate ester/OsO₄]
aq. acetone; 1 d; 0 °C

70-99% e.e.
styrene: 73% e.e.
5-decene: 74% e.e.
stilbene: 99% e.e.

20-80% e.e.
(not optimized)

The stoichiometric dihydroxylation of olefins with OsO₄ is a most reliable reaction; even tetrasubstituted alkenes react. In order to save on the expensive and poisonous oxidant, it is best to employ only 0.2–1% of the stoichiometric amount and use the Upjohn catalyst N-methylmorpholine N-oxide (NMO). Most important, however, is the binding of OsO₄ to chiral ligands, which allows catalytic and stereoselective reactions. Commercial phthalazine (PHAL) and pyrimidine (PYR) connected dihydroquinine (DHQ) and dihydroquinidine (DHQD) dimers fulfil this task. The ligands bind tightly enough to accelerate the reaction with substrates, but not too tightly to slow down subsequent steps. They are mixed in water/*tert*-butanol 1:1 with the olefin and give a stereoselective dihydroxylation overnight (Jacobsen, 1988; Berrisford, 1995).

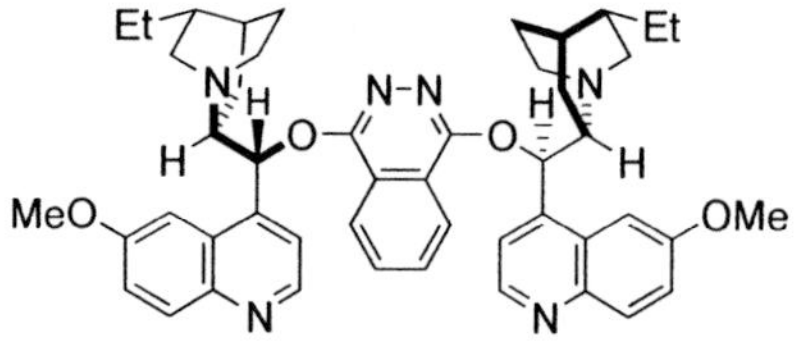

(DHQ)₂PHAL

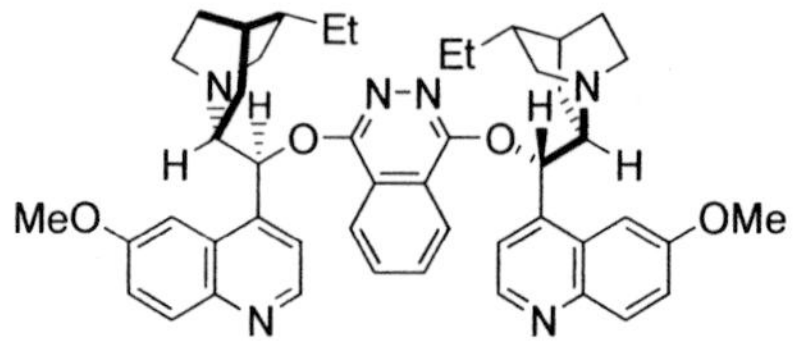

(DHQD)₂PHAL

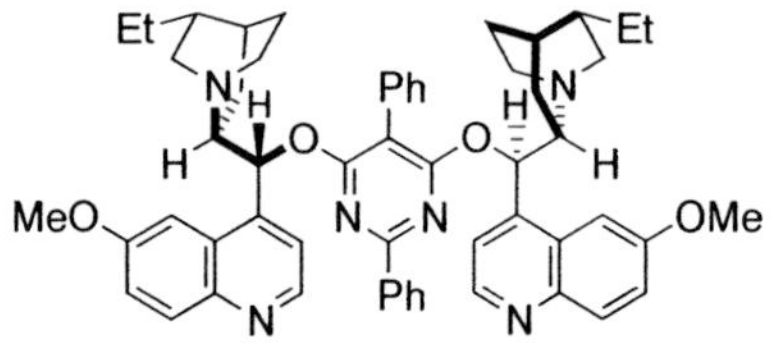

(DHQ)₂PYR

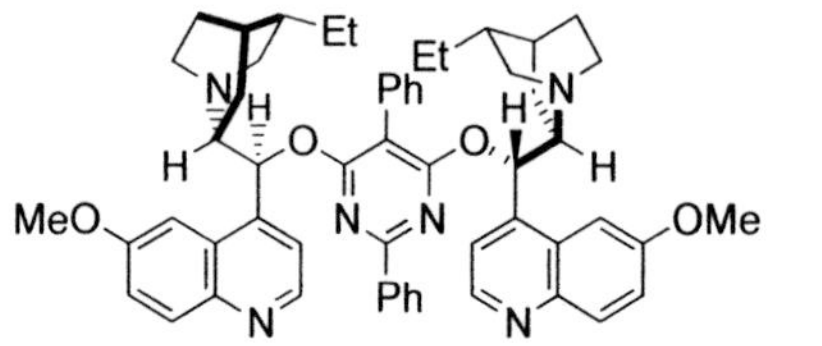

(DHQD)₂PYR

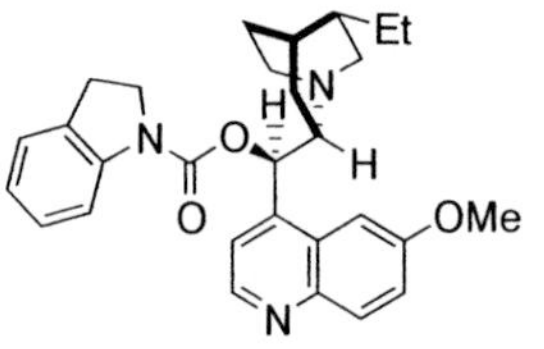

(DHQ)IND

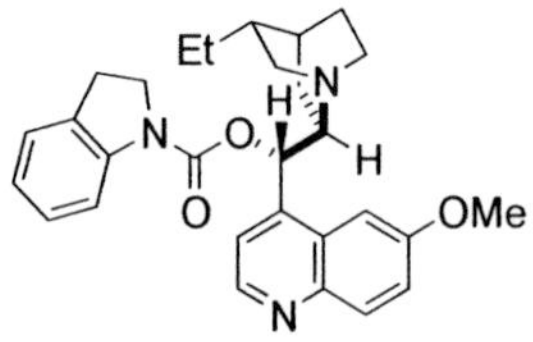

(DHQD)IND

PHAL catalysts are the first choice for substituted olefins, PYR catalysts are superior for terminal alkenes, while IND ligands are chosen for the most difficult to differentiate *cis*-olefins. The stereoselective rules obeyed are summarized for the example given in the scheme below (Kim, 1993).

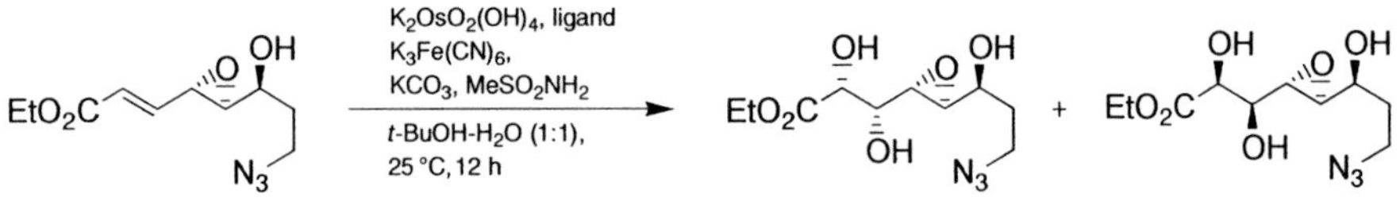

no ligand - OsO₄ used	1	2
ligand = (DHQD)₂PHAL	< 1	20
ligand = (DHQ)₂PHAL	10	1

Asymmetric dihydroxylation of squalene is not regioselective, but treating it in three rounds with the AD-mix followed by ketalization to improve solubility of the products, leads to a dodeca-hydroxylation in 79% yield with an average of 98% e.e. (Crispino, 1993).

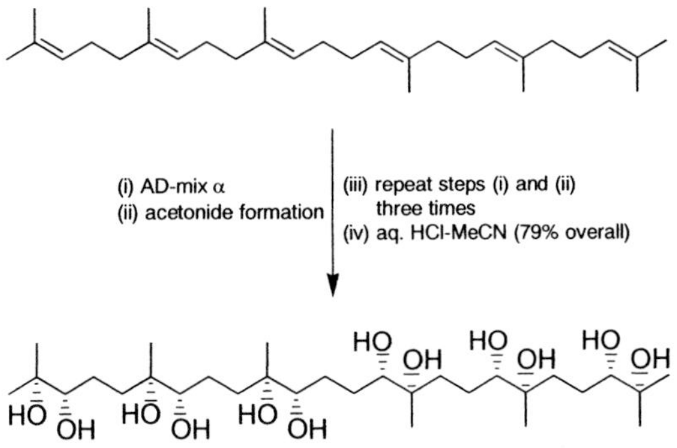

(i) AD-mix α
(ii) acetonide formation

(iii) repeat steps (i) and (ii)
three times
(iv) aq. HCl-MeCN (79% overall)

Sharpless AD-mixes were also repeatedly used in the functionalization of substituted vinyl groups of multifunctional steroids (Jeong, 1999).

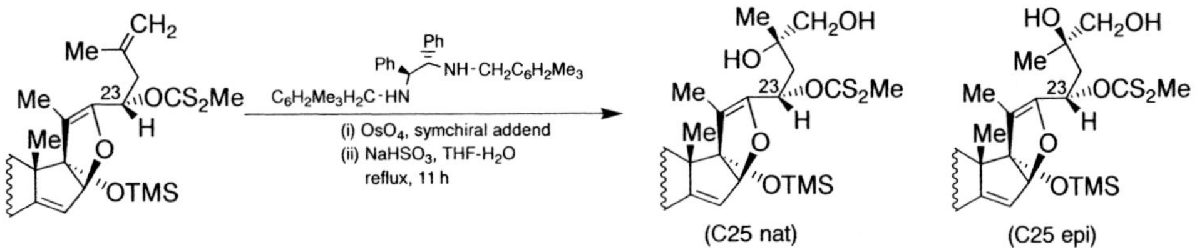

(i) OsO₄, symchiral addend
(ii) NaHSO₃, THF-H₂O
reflux, 11 h

(C25 nat)   (C25 epi)

Sharpless catalysts are also useful in asymmetric amino hydroxylation reactions (Li, 1996; Kolb, 1994; O'Brien, 1999). So far this reaction has been most successful on styrene derivatives, where the benzylic carbon is stereoselectively aminated and the terminal carbon is preferentially hydroxylated. Cinnamic acid derivatives are also good substrates. As an oxidant one uses freshly prepared sodium or lithium salts of the N-halogenated amides given at the end of the following scheme, potassium osmate and the chiral pyridine or quinone catalysts, which are also depicted in the scheme.

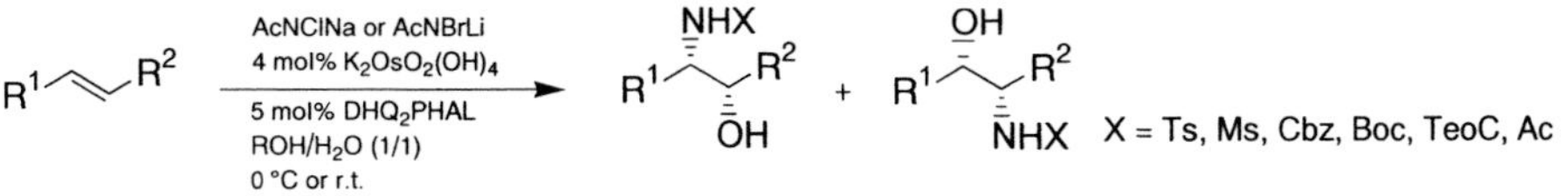

AcNClNa or AcNBrLi
4 mol% K₂OsO₂(OH)₄
5 mol% DHQ₂PHAL
ROH/H₂O (1/1)
0 °C or r.t.

X = Ts, Ms, Cbz, Boc, TeoC, Ac

OAlk*
(Alk*)₂PHAL

(Alk*)₂AQN

Alk* = DHQ

Alk* = DHQD

$$Ph\diagdown CO_2R \xrightarrow[\text{aminohydroxylization}]{\text{asymmetric}} Ph\diagdown^{\text{NHX}}_{\overline{\text{OH}}}CO_2R \;+\; Ph\diagdown^{\text{OH}}_{\overline{\text{NHX}}}CO_2R$$

$Ph\diagdown^{\text{NHTs}}_{\overline{\text{OH}}}CO_2Me$	$Ph\diagdown^{\text{NHMs}}_{\overline{\text{OH}}}CO_2iPr$	$Ph\diagdown^{\text{NHCbz}}_{\overline{\text{OH}}}CO_2Me$	$Ph\diagdown^{\text{NHTeoC}}_{\overline{\text{OH}}}CO_2iPr$	$Ph\diagdown^{\text{NHAc}}_{\overline{\text{OH}}}CO_2$
(60%, 82% e.e)	(65%, 94% e.e)	(65%, 94% e.e.)	(70%, 99% e.e)	(81%, 99% e

Thallium(III) acetate reacts with alkenes to give 1,2-diol derivatives while thallium(III) nitrate leads mostly to rearranged carbonyl compounds via organothallium compounds (Taylor, 1970, 1976; Ouelette, 1973; Rotermund, 1975; Criegee, 1979). Very useful reactions in complex syntheses have been those with olefins and ketones containing conjugated aromatic substituents, e.g. porphyrins (Kenner, 1973; Smith, 1975).

$$R^1\diagdown R^2 \;\; R^3\diagdown R^4 \xrightarrow[{-H^\oplus}]{\substack{+\,Tl^{3\oplus}\\ +\,ROH}} \cdots \xrightarrow[{-Tl^\oplus}]{} \cdots \xrightarrow[{-H^\oplus}]{+\,ROH} \cdots$$

Wagner-Meerwein rearrangement

$$\xrightarrow[{-H^\oplus}]{+\,ROH}$$

(i) Tl(NO₃)₃/MeOH; 10 min; r.t.

(ii) 1 M aq. H₂SO₄; 5 min; r.t.

(52%)

(28%)

(i) Tl(NO₃)₃/MeOH; 1 min; r.t.

(ii) 1 M aq. H₂SO₄; 5 min; r.t.

—CHO   (85%)

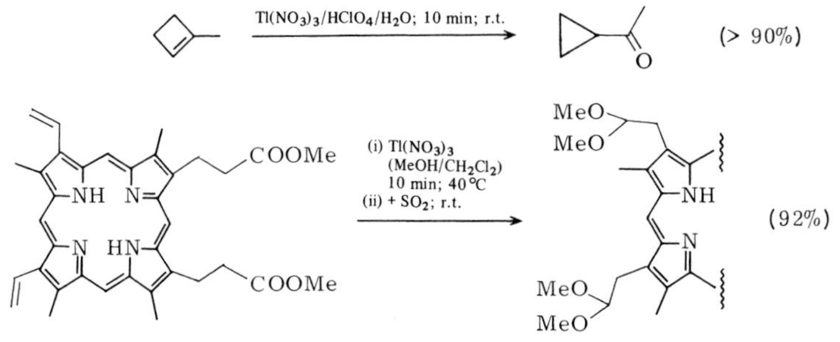

(> 90%)

(i) Tl(NO₃)₃
(MeOH/CH₂Cl₂)
10 min; 40 °C
(ii) + SO₂; r.t.

(92%)

Regio- and stereoselective monohydroxylations of C=C double bonds are achieved in high yield by successive hydroborations and oxidative cleavage of the borane with H₂O₂ (Zweifel, 1963; Brown, 1975). This procedure is much more suitable for laboratory use than a similar one with organoaluminum compounds first discovered by Ziegler (1960). Boranes will add to the least hindered side of the C=C double bond, and oxidative cleavage proceeds with retention of configuration (Fuhrer, 1970). Upon reaction of thexylborane with dienes cyclic boranes may be formed, which yield diols on oxidative cleavage. Treatment of the cyclic boranes with acids leads to hydrogenation of the less hindered carbon atom (Brown, 1972 A).

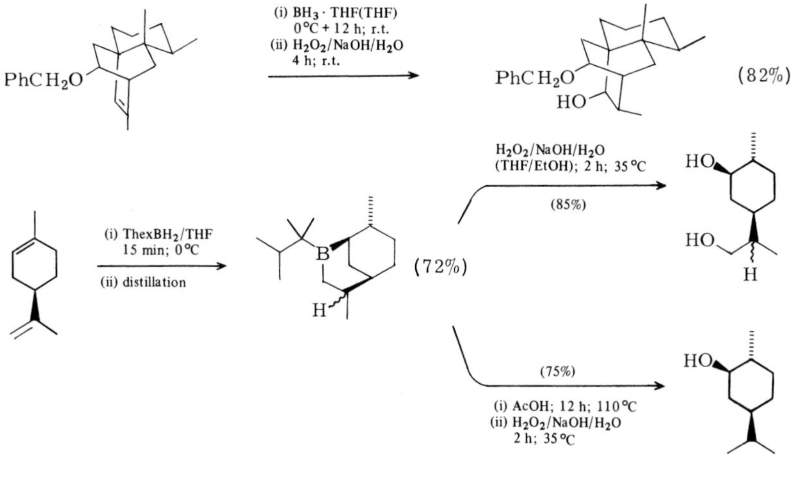

In the hydroboration/oxidation reactions of 2-cyclohexen-1-ol derivatives the regio- as well as the stereoselectivities can be steered in either of two directions by use of different reagents and reaction conditions. Dialkylboranes, e.g. 9-BBN, give mainly *trans*-1,2-diols, whereas with 1,3,2-benzodioxaborole ("catecholborane") in the presence of the rhodium hydrogenation catalyst Rh(PPh₃)₃Cl the *trans*-1,3-diols become predominant. The *cis*-1,2-diols are usually only formed in trace amounts, but the *cis*-1,3-diols are always produced as by-products in 10–20% yield (Evans, 1988).

R	product ratio							yield
H	83	:	5	:	2	:	10	86%
CH$_2$Ph	68	:	13	:	0	:	19	73%
SiBu^{t}Me$_2$	74	:	13	:	0	:	13	70%
H	18	:	72	:	1	:	9	84%
CH$_2$Ph	7	:	72	:	8	:	13	87%
SiBu^{t}Me$_2$	2	:	86	:	1	:	11	79%

C≡C triple bonds are "hydrated" to yield carbonyl groups in the presence of mercury(II) ions or by successive treatment with boranes and H$_2$O$_2$. The first procedure gives preferentially the most highly substituted ketone, the latter the complementary compound with high selectivity (Gibson, 1969).

The stereospecific and regioselective hydrobromination of alkynes with chlorobis(η^5-cyclopentadienyl)hydrozirconium and NBS produces (*E*)-vinylic bromides in good yields. The bromine atom usually adds regioselectively to the carbon atom that bears the smaller substituent and stereoselectively trans to the larger substituents (Hart, 1975; Schwartz, 1976; Corey, 1978; Nakatsuka, 1990).

R = H, CH$_3$; R' = *n*-alkyl, tert-alkyl, cycloalkyl, etc.

X = Cl, Br, I

Internal alkynes are oxidized to acyloins by thallium(III) in acidic solution (McKillop, 1973; Rotermund, 1975) or to 1,2-diketones by permanganate or by *in situ*-generated ruthenium tetroxide (Lee, 1969, 1973; Gopal, 1971). With these oxidants terminal alkynes undergo oxidative degradation to carboxylic acids with loss of the terminal carbon atom.

3.3.4
Oxidation of Alcohols to Aldehydes, Ketones, and Carboxylic Acids

The oxidation of primary alcohols to give aldehydes requires mild oxidants and careful control of reaction conditions. Oxidation of primary alcohols with pyridinium dichromate in acetic acid gives the aldehyde in about 70% yield. Further oxidation to carboxylic acids is generally a problem. Di-t-butyl chromate (Bosche, 1975), MnO$_2$ (Arndt, 1975), or Ag$_2$CO$_3$ (Manegold, 1975) in aprotic solvents, lead tetraacetate in pyridine (Rotermund, 1975), copper(II) acetate (Nigh, 1973), and DDQ (Walker, 1967; Stechl, 1975) are more reliable oxidants. MnO$_2$ and high-potential quinones oxidize allylic alcohols with high selectivity, whereas copper(II) acetate has often been used for the synthesis of sensitive α-keto aldehydes.

$$CH_3(CH_2)_{14}CH_2OH \xrightarrow[\text{(ii) } H_2C_2O_4/H_2O]{\substack{\text{(i) } CrO_2(OBu^t)_2 \\ \text{(PE); 14 d; r.t.}}} CH_3(CH_2)_{14}CHO \quad (95\%)$$

MnO$_2$(hexane); 0.5 h; 0 °C → (> 90%)

MnO$_2$(Et$_2$O); 1 h; r.t. → (64%)

Ag$_2$CO$_3$(C$_6$H$_6$); 12 h; Δ → (95%)

(steroid-17β) — CH$_2$OH → Cu(OAc)$_2$/O$_2$ (MeOH); 15 min; r.t. → (steroid-17β) (80-95%)

ortho-TCQ (CCl$_4$); 15 h; r.t. → (100%)

The widely used Moffatt–Pfitzner oxidation works with in situ-formed adducts of dimethyl sulfoxide with dehydrating agents, e.g. DCC, Ac$_2$O, SO$_3$, P$_4$O$_{10}$, COCl$_2$ (Pfitzner, 1965; Fenselau, 1966; Joseph, 1967; Moffatt, 1971; Martin, 1971) or oxalyl dichloride (Swern oxidation; Nakatsuka, 1990). A classical procedure is the Oppenauer oxidation with ketones and aluminum alkoxide catalysts (Djerassi, 1951; Lehmann, 1975). All of these reagents also oxidize secondary alcohols to ketones but do not attack C=C double bonds or activated C–H bonds.

Moffatt-Pfitzner oxidation:

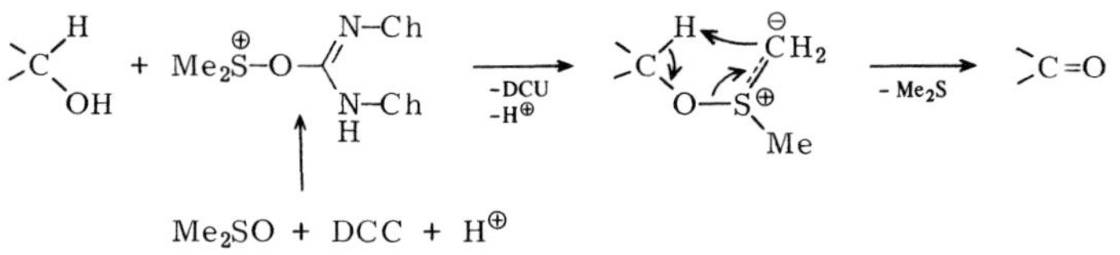

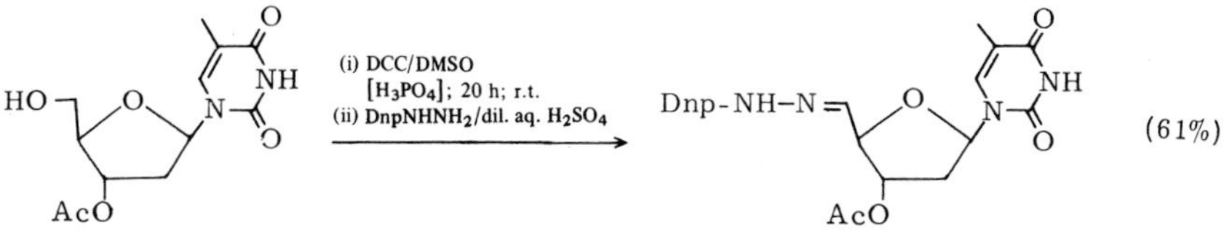

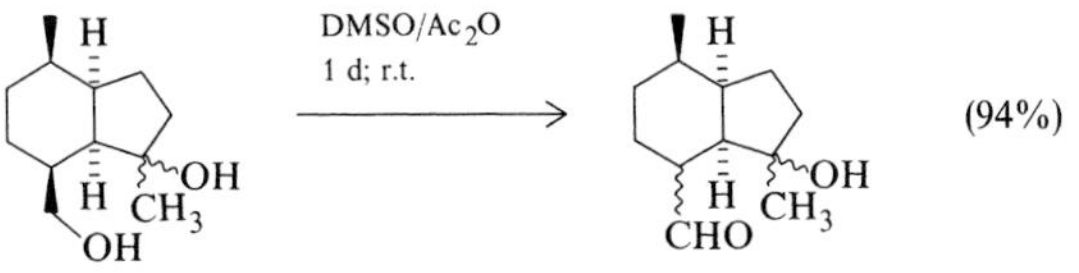

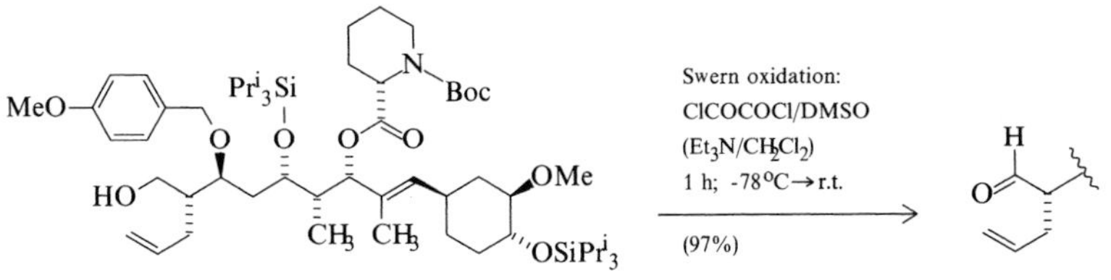

(61%)

(94%)

Swern oxidation:
ClCOCOCl/DMSO
(Et₃N/CH₂Cl₂)
1 h; -78 °C → r.t.

(97%)

Oppenauer oxidation:

Al(OPr^i)₃/Ph₂CO
dist. < 1 mbar

(73%)

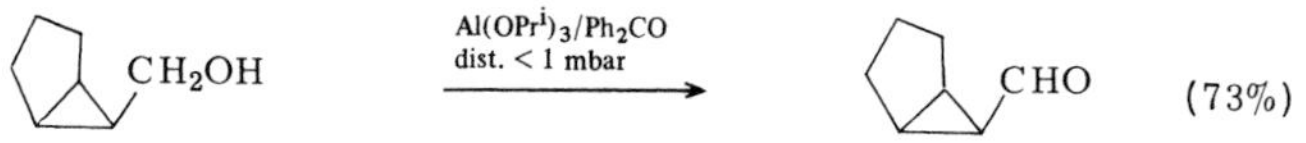

The Dess–Martin periodinane ("DMP") reagent, 1,1,1-tris(acetyloxy)-1,1-dihydro-1,2-benziodoxol-3(1H)-one, has been used in complex syntheses for the oxidation of primary or secondary alcohols to aldehydes or ketones (Nakatsuka, 1990). It is prepared from 2-iodobenzoic acid by oxidation with bromic acid and acetylation (Dess, 1983).

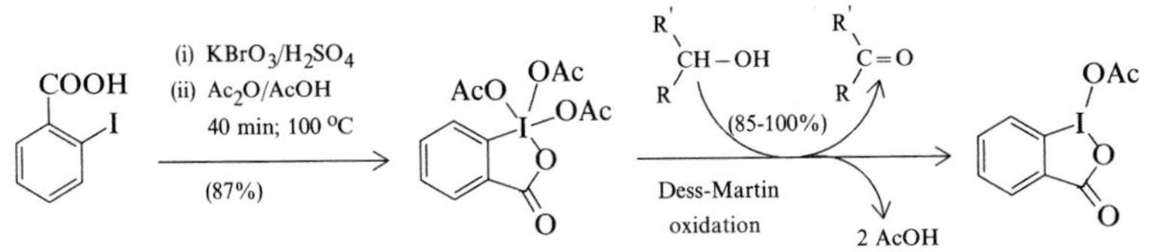

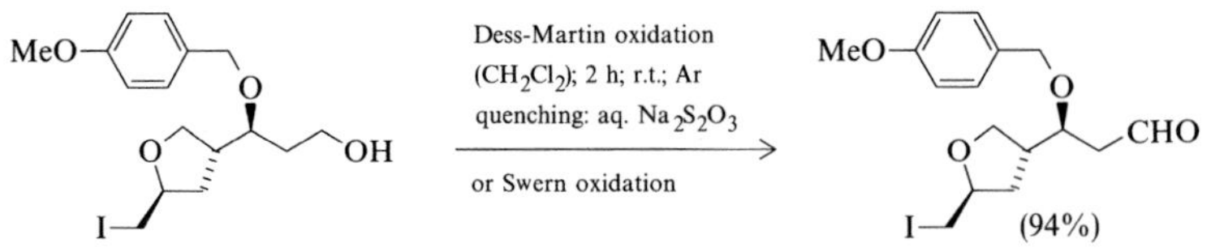

The conversion of primary alcohols and aldehydes into carboxylic acids is generally possible with all strong oxidants. Silver(II) oxide in THF/water is particularly useful as a neutral oxidant (Corey, 1968A). The direct conversion of primary alcohols into carboxylic esters is achieved with MnO_2 in the presence of hydrogen cyanide and alcohols (Corey, 1968A, D). The remarkable smooth oxidation of ethers to esters by ruthenium tetroxide has been employed quite often (Lee, 1973). Dibutyl ether affords butyl butanoate, and tetrahydrofuran yields butyrolactone almost quantitatively. More complex educts also give acceptable yields (Wolff, 1963).

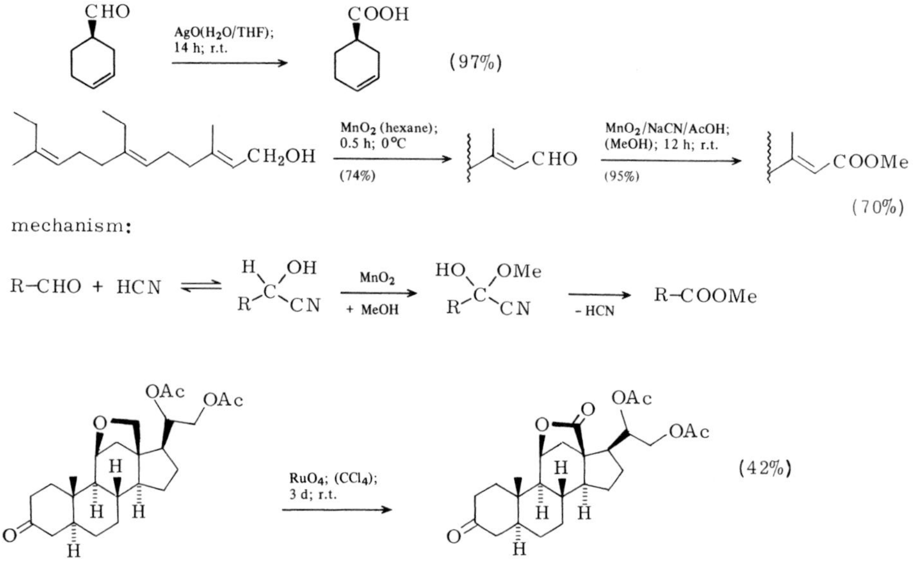

mechanism:

A high yielding and environmentally benign procedure to oxidize primary alcohols to carboxylic acids uses stoichiometric amounts of $NaClO_2$ and catalytic amounts of both 1-oxy-2,2,6,6-tetramethyl-piperidine (TEMPO) and NaOCl. No racemization is observed for substrates with labile chiral centers. The hypochlorite presumably oxidizes the primary alcohol to the aldehyde, the chlorite oxidizes the aldehyde to the carboxylate and regenerates hypochlorite (Zhao, 1999).

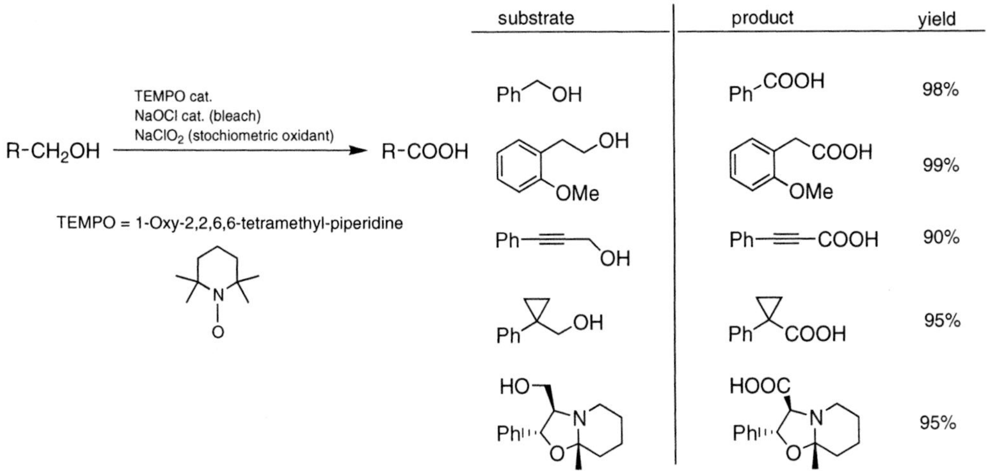

	substrate	product	yield
TEMPO cat. NaOCl cat. (bleach) NaClO$_2$ (stochiometric oxidant)	Ph⌒OH	Ph⌒COOH	98%
	(2-OMe-phenyl)-CH$_2$CH$_2$OH	(2-OMe-phenyl)-CH$_2$COOH	99%
TEMPO = 1-Oxy-2,2,6,6-tetramethyl-piperidine	Ph≡⌒OH	Ph≡—COOH	90%
	Ph-cyclopropyl-CH$_2$OH	Ph-cyclopropyl-COOH	95%
	HO-/Ph bicyclic oxazolidine	HOOC-/Ph bicyclic oxazolidine	95%

R–CH$_2$OH $\xrightarrow{\text{TEMPO cat., NaOCl cat., NaClO}_2}$ R–COOH

Selective oxidation of secondary alcohols to ketones is usually performed with CrO$_3$/H$_2$SO$_4$ 1:1 in acetone (Jones' reagent) or with CrO$_3$Py$_2$ (Collin's reagent) in the presence of acid-sensitive groups (Bosche, 1975; Djerassi, 1956; Allen, 1954). As mentioned above, a,β-unsaturated secondary alcohols are selectively oxidized by MnO$_2$ (Lee, 1969; Arndt, 1975) or by DDQ (Walker, 1967; Stechl, 1975).

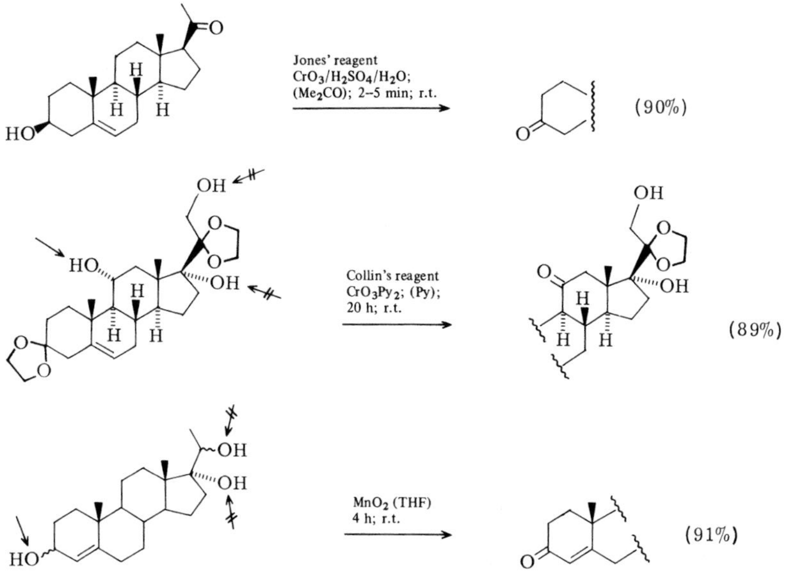

Jones' reagent
CrO$_3$/H$_2$SO$_4$/H$_2$O;
(Me$_2$CO); 2–5 min; r.t. → (90%)

Collin's reagent
CrO$_3$Py$_2$; (Py);
20 h; r.t. → (89%)

MnO$_2$ (THF)
4 h; r.t. → (91%)

(75%)

Tertiary alcohols are usually degraded unselectively by strong oxidants. Anhydrous chromium trioxide leads to oxidative ring opening of tertiary cycloalkanols (Fieser, 1948).

3.3.5
Oxidative Rearrangement and Cleavage of Ketones and Aldehydes

Ketones are rearranged oxidatively by reactions of the corresponding enols with thallium(III), e.g. to yield pyrroleacetic acids from acetyl pyrroles (Kenner, 1973 B; Rotermund, 1975).

Peracids convert ketones into esters in high yield. The peracid adds to the carbonyl group, and one carbon substituent migrates to the positively polarized peroxy oxygen (Hassall, 1957; Emmons, 1955; Plesničar, 1978). The migration tendency follows the order *t*-alkyl > *s*-alkyl > benzyl > phenyl > *n*-alkyl > methyl, if no steric effects are counteracting. This Baeyer–Villiger oxidation is particularly useful with bridged polycyclic ketones. From these compounds lactones are formed, in which the new ester bond replaces a C–C single bond with complete retention of configuration. Thus, upon hydrolysis of the lactone, a *cis*-disubstituted ring results (Meinwald, 1960; Sauers, 1961). Aldehydes are converted into sensitive formate esters by selenous peracid (Nakatsubo, 1970). Ketones may also be converted into amides or lactams via the similar Beckmann rearrangement of ketoximes (Heldt, 1960).

$(MeO)_3P$
1 d; 110 °C; N_2

(91%)

$Pb(OAc)_4$; N_2
(C_6H_6/Py); 2 h; Δ

COOH
COOH
COOEt

COOEt

(71%)

$Pb(OAc)_4/Py$
2 h; 65 °C

(30%)

$h\nu$; r.t.
(C_6H_6); r.t.

(40%)

COOH
COOH

$OOBu^t$
$OOBu^t$

basketene

3.4
Synthesis of Carboxylic Acid Derivatives

In synthetic target molecules, esters, lactones, amides, and lactims are the most common carboxylic acid derivatives. In order to synthesize them from carboxylic acids one has generally to produce an activated acid derivative, and an enormous variety of activating reagents is known, mostly developed for peptide syntheses (Bodanszky, 1976). In actual syntheses of complex esters and amides, however, only a small selection of these is used, and we shall mention only generally applicable methods. The classic means of activating carboxyl groups are the acyl azide method of Curtius and the acyl chloride method of Emil Fischer.

The most practical route to acyl azides starts with the hydrazinolysis of esters. The usually rather poorly soluble and poorly stable hydrazides are dissolved in mixtures of organic solvents (e.g. THF, DMF, AcOH) and strong acids (e.g. HCl, TFA) and then mixed with an equimolar amount of sodium nitrite or amyl nitrite at −10 °C to yield the azide almost instantaneously. The coupling step with amines at room temperature may require several days. A great advantage of the acyl azide method is the lack of α-racemization. The acyl chloride method is quicker and may also be applied for the preparation of esters and amides. Here the free acid is used as starting material, and aprotic solvents (CHCl$_3$, DMF, pyridine) must be applied in the chlorination. Thionyl chloride and oxalyl chloride are the most common agents, and low temperatures are again advantageous. Nitrosation or chlorination of activated CH groups by nitrites or SOCl$_2$, respectively, are sometimes troublesome side-reactions. Acyl azides may lose N$_2$ on heating and rearrange to isocyanates (Curtius rearrangement), which may be solvolyzed. Some of the possibilities of classical carboxyl conversions are exemplified in the schemes below, which are taken from a triquinacene synthesis (Russo, 1971; Mercier, 1973) and the ergotamine synthesis of Hofmann (1963). In these cases the acyl azides formed have been used to prepare amines via Curtius rearrangement. The acyl chloride or azide intermediates can, however, also be reacted with amines or alcohols to form amides or esters.

As a catalyst for ester and amide formation from acyl chlorides or anhydrides, 4-(dimethylamino)pyridine has been recommended (DMAP; Höfle, 1978). In the presence of this agent highly hindered hydroxyl groups, e.g. of steroids and carbohydrates, are acylated under mild conditions, which is difficult to achieve with other catalysts.

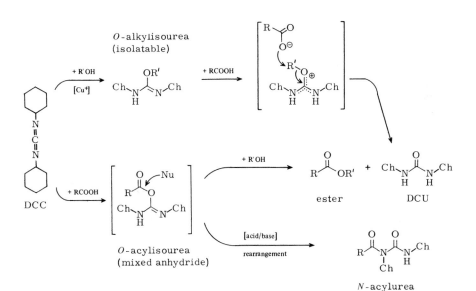

Popular routes to esters and amides employ *N,N'*-dicyclohexylcarbodiimide (DCC) as coupling reagent. This reagent functions as a mild dehydrating agent and converts acids into active *O*-acylisoureas or alcohols into similarly active *O*-alkylisoureas, if Cu(I) is added as a catalyst. The pathway via *O*-alkylisoureas is more reliable than the one via *O*-acylisoureas because the latter easily rearrange to form inactive *N*-acylureas (Mathias, 1979). Condensations using DCC are also subject to catalysis by DMAP (Höfle, 1978). The fundamental synthetic advantage of DCC and related compounds as dehydrating agents is the possibility to work in aprotic solvents. There are, however, several water-soluble carbodiimides, which work also in protic solvents, e.g. EDCI. Acid activation occurs as acid chloride made with thionyl- or oxalylchloride or with EDCI (= 1-ethyl-3-(3-diethylamino)propyl carbodiimide) in the presence of DMAP (4-dimethylamino-pyridine). The esterification of a phenol given below worked with an acid chloride and base catalysis, but when the acid-labile Mom-protective group was present, the acid chloride was not formed even in the presence of amines. The EDCI procedure was then applied successfully (Matsumoto, 1992; Hosoya, 1994).

In peptide syntheses, where partial racemization of the chiral *α*-carbon centers is a serious problem, the application of 1-hydroxy-1*H*-benzotriazole ("HBT") and DCC has been very successful in increasing yields and decreasing racemization (König, 1970; Windridge, 1971; Bosshard, 1973), 1-(Acyloxy)-1*H*-benzotriazoles or 1-acyl-1*H*-benzotriazole 3-oxides are formed as reactive intermediates. If carboxylic or phosphoric esters are to be formed from the acids and alcohols using DCC, 4-(pyrrolidine-1-yl)pyridine ("PPY"; Hassner, 1978; Patel, 1979) and HBT are efficient catalysts even with *tert*-alkyl, cholesteryl, aryl, and other unreactive alcohols as well as with highly bulky or labile acids.

3.5
Ethers

The Williamson synthesis using an alkoxide and a halide or sulfonate is commonly used to produce ethers. Optically pure chiral ethers are best synthesized by alkylation of chiral thallium alkoxides with primary alkyl halides at low temperature. No racemization was observed, for example, with lactic or tartaric ethylesters. Acidic hydroxy groups, such as the enediol group in ascorbic acid, react readily with diazomethane to give methyl ethers.

The central feature of the chemistry of ethers is their ability to bind and deactivate protons and to form coordination complexes with Lewis acids. Ethers are hard bases and interact more strongly with hard Lewis acids (H^+, Li^+, M_g^{2+}). Water-labile magnesium porphyrins can even be mixed with 0.1 M HCl in diethyl ether solution without demetallation.

Macrocyclic polyethers are synthesized by stepwise reaction of a,ω-dihalides or a,ω-disulfonate esters with a,ω-diols. In 1,2-dimethoxyethane or DMSO solvent and using potassium *tert*-butoxide as base, cyclization yields of up to 90% have been reported without recourse to high dilution techniques. The most versatile chiral macrocyclic polyether is based on the atropisomers of 2,2'-dihydroxy-1,1'-binaphthyl (BINAP). They have been applied for the separation of enantiomeric amines and as ligands for metals used in homogeneous catalysis of asymmetric hydrogenation and coupling reactions of olefins. Many other chiral crown ethers based, for example, on mannitol and tartaric acid have been prepared, but have not been found so useful in chiral recognition and catalysis as BINAP.

The condensation of bis-phenols with dichloride strongly depends on the addition of hard metal cations, M^+. If potassium hydroxide is used instead of sodium hydroxide in the synthesis of a crown-6, the yield is substantially increased, whereas *tert*-Bu$_4$NOH drastically decreased the yield. Potassium ions fit comfortably into the crown-6 ring to be synthesized, while sodium ions are too small, and *tert*-butyl ammonium ions too large.

Intermolecular etherification of two alcohol molecules is not feasible using the Mitsonobu reaction. Intramolecular reactions, however, yield three- to seven-membered rings. A tertiary alcohol acts as nucleophile with activated secondary alcohols stereoselectively (e.e. 99%; Chikashita, 1993). If both alcohols are secondary, both OH groups may act as leaving groups. Selectivity has been achieved in ribonucleoside synthesis from tetrol acetonides. Electronically the allylic OH group at C-1 is more vulnerable to substitution, as is exemplified by the furane derivative given in the scheme below. Mitsonobu treatment then gives the α-configured tetrahydrofuran ring. However, if the allylic alcohol is sterically hindered by a hydrogen-bonded indole derivative, then the less hindered alcohol reacts and the β-stereoisomer is formed (Yokohama, 1994).

Old procedures to obtain amines include nitration of double bonds and reduction of nitroalkanes or nitroarenes, substitution of bromide with phthalimide and treatment with hydrazine (Gabriel reaction), rearrangement and decarboxylation of amides (Hoffmann degradation), rearrangement of carboxylic azides (Curtius degradation) and reduction of azides. All these reactions are still applied routinely and are discussed in general textbooks of organic chemistry.

Palladium(II) acetate catalyzes the substitution of halides on double bonds by amines (Chakraborty, 1998), whereas rhodium-stabilized carbenes open a route to hindered tertiary amines (Yang, 2001). Palladium catalysis was also used to arylate pyrroles and benzophenone imide at the nitrogen atom (Mann, 1998).

Oligoamines are accessible by alkylation of an *a,ω*-diamine with an *a,ω*-chloro-

alkanol in refluxing butanol. The amino groups may then be protected with (BOC)$_2$O, the alcohol mesylated and again substituted with another amine (Lebreton, 1999 A). If side-chains are needed, one uses ketones as substrates to form substituted imines. They are then reduced with borohydride (Lebreton, 1999 B).

Guanidine derivatives are available from *N,N*-bis(*tert*-butoxycarbonyl)-*S*-methyl-isothiourea (Bergeron, 1984; 1987; Shildneck, 1943) and nucleophilic substitution of the thiomethyl unit with an amine (Bergeron, 1984; Lebreton, 1999).

In surface FGIs it is often mandatory to exchange functional groups at water–membrane interfaces. Michael additions of amines to the C=C double bonds of acrylamides (Skupin, 2001) make the surface more polar, while coupling reactions of diazonium head groups with amines (Fuhrhop 1981, 1983) make them less polar.

3.6
Phosphonates

Phosphanes have been already introduced with the Wittig reaction (see Section 1.5). Phosphonates are obtained from halides and trialkylphosphites via the intermediate phosphonium salt (Michaelis–Arbuzov reaction). A direct C–P bond is formed, which is quite stable to hydrolysis. Phosphites also rearrange to phosphonates upon irradiation with UV light ("photo-Michaelis–Arbuzov", Ganapathy, 1999).

Arbuzov reaction

$$(P(OR)_3 \ + \ R'X \ \longrightarrow \ [(RO)_3PR']X \ \xrightarrow{\Delta} \ (RO)_2\overset{\overset{\displaystyle O}{\|}}{P}-R'$$

$$(MeO)_2P-OCH_2Ph \ \xrightarrow{h\nu} \ (MeO)_2\overset{\overset{\displaystyle O}{\|}}{P}-CH_2Ph$$

The P–C bond of β-ketophosphonates is cleaved by dephosphonylation of the lithium enolates with LiAlH$_4$ followed by quenching with aqueous H$_2$SO$_4$. Regioselective alkylation as with β-keto esters or amide synthesis with isocyanates is possible (Lee, 1999).

3.7
Sulfides and Sulfonates

Hydrosulfides are accessible from halides by treatment with thiourea in DMSO and subsequent basic hydrolysis of the S-alkyl-isothiourea (Cram, 1988). The preparation of thiols is often plagued by competing disulfide or thioether formation and oxidation of the thiol. The use of lithium *tert*-butyldimethylsilanethiolate (prepared *in situ* from hexamethyl-cyclotrisilathiane and *t*-BuLi) removes these problems (Kraus, 1991). Competing Michael additions are also avoided with this bulky reagent. The resulting silyl thioethers can be quantitatively hydrolysed with hydrofluoric acid in acetonitrile (Newton, 1979).

Sulfonates can be obtained from arenes with "acetyl sulfate" which is freshly produced *in situ* from acetic anhydride and concentrated sulphuric acid. An example for polystyrene chains on silicon surfaces is given below (Tran, 2001).

3.8
Halides

Alcohols give alkyl chlorides in high yields on treatment with *N*-chlorosuccinimide and triphenylphosphine (Cram, 1988) or with diethyl diazenedicarboxylate, triphenylphosphine and zinc chloride. This Mitsunobu reaction is generally the method of choice.

Alkenyl and arenyl iodides are the best educts for palladium-catalyzed synthetic reactions (Heck, Suzuki etc.; see Chapters 1 and 2). N-Iodosuccinimide (NIS) is a potent source for the iodonium ion. It introduces iodine into activated arenes, in particular phenols and phenol ethers, as well as into alkyne derivatives. Silyl-protected ynyl-alcohols give the diiodides in the presence of silver nitrate (Gao, 2000). Treatment of phenol ethers with BuLi removes an ortho-proton and iodide yields monoiodides (Hosoya, 1994).

Another common route to vinyl iodides is the treatment of aldehydes with chromium(II) chloride and iodoform (Takai reaction; Takai, 1986; Humphrey, 1996; Andrus, 1997). The necessary aldehydes are often obtained by reduction of carboxyl esters or amides.

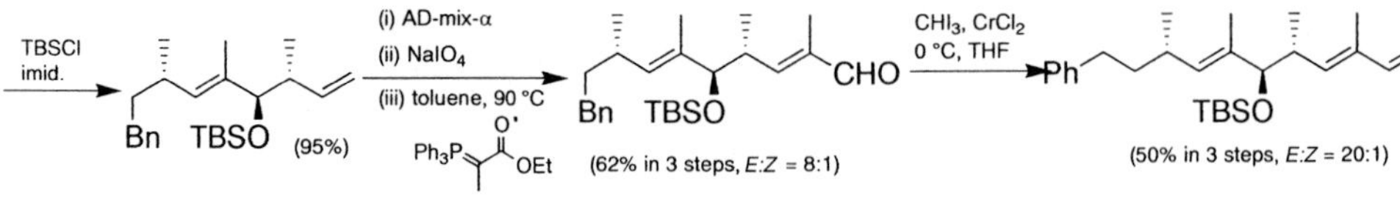

The addition of halides to alkynes using zirconium catalysis is discussed in Sections 3.2.2 and 3.3.3. Grignard reactions between terminal alkynes and alkylmagnesium bromides in the presence of copper(I) bromide and subsequent iodonolysis also produce terminal iodides (Hoffmann, 1990; Kasatkin, 1999).

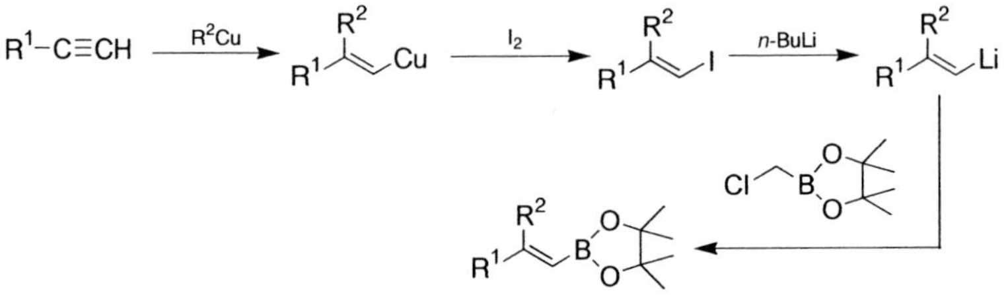

Transition metal-catalyzed cross-coupling reactions often use aryl or vinyl triflates as substrates. Aryl triflates are usually prepared by treatment of phenols with triflic anhydride in the presence of pyridine or triethylamine (Subramanian, 1976). Enol triflates are made from ketones and *N*-(2-pyridyl)triflate (Commins, 1992). Very mild conditions for triflation of phenols are sufficient, if nitrophenol triflates are applied (Neuville, 1999).

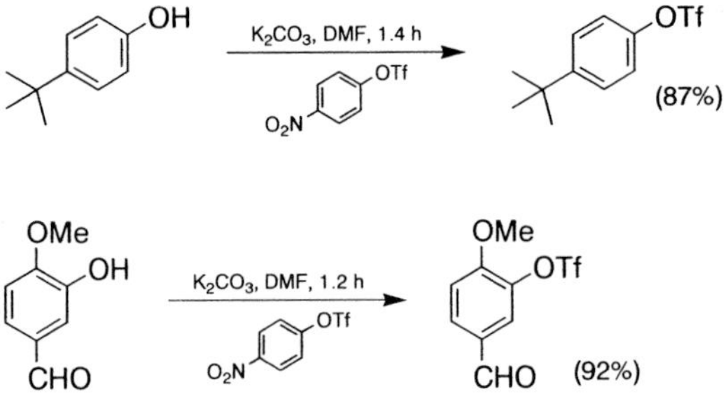

3.9 Protection of Functional Groups

The synthesis of complex organic molecules demands the availability of a variety of protective groups to ensure the survival of reactive functional groups during synthetic operations. An ideal protective group combines stability under a wide range of conditions with susceptibility to facile removal ("deblocking") by a specific, mild reagent. It is also desirable that the introduction of the "blocking" group is easy and that its reactions are complementary to other protecting groups. A large variety of new protective groups capable of removal under exceptionally mild and/or highly specific conditions has been developed mainly in syntheses of natural oligomers, e.g. peptides, nucleotides, and saccharides, which has been summarized in books (McOmie, 1973; Robertson, 2000). We shall discuss only a few protecting agents of common functional groups, which have proven to be useful in synthesis, and indicate some possible sequences for working with these blocking agents.

3.9.1
Reactive Carbon–Hydrogen and Carbon–Carbon Bonds

Only highly reactive CH groups, e.g. of terminal alkynes or of carbonyl compounds, have sometimes to be efficiently protected. Terminal alkynes are mostly used in organometallic syntheses and in oxidative couplings. Protection of the terminal C–H bond is required if another group is to be metallated, or if the oxidative coupling is to be performed with diterminal diynes. Without protection, rearranged organometallic compounds or polymers would be formed in uncontrollable side-reactions.

The most useful protecting group in these reactions is the trimethylsilyl group. It may be introduced by reaction of the acetylide anion with chlorotrimethylsilane and be removed by treatment with silver nitrate in quantitative yield. On the other hand, the C–Si bond remains intact during reactions with organometallic reagents (Corey, 1968 D) as well as during oxidative coupling of alkynes under controlled conditions (Sondheimer, 1970).

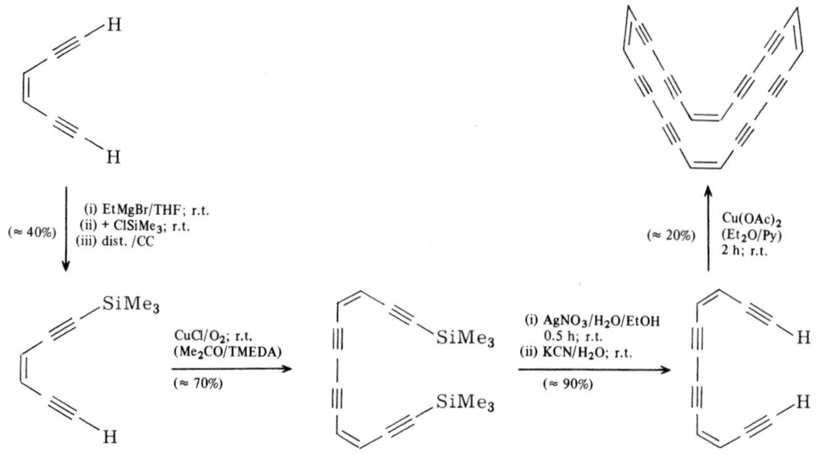

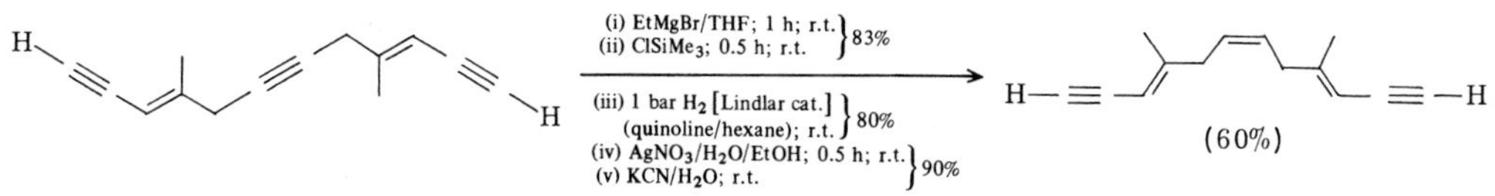

α'-Methylene groups of ketones containing a tertiary α-carbon atom, whose selective alkylation is desired, are protected by successive treatment with ethyl formate and 1-butanethiol. The butylthiomethylene derivatives formed are then deprotonated and alkylated at the tertiary α-carbon atom. The protecting group is finally removed as butanethiolate and formate by treatment with KOH in boiling diethylene glycol. A useful application is the angular alkylation of 1-decalones (Ireland, 1962). For the alkylation of α,β-unsaturated keto-steroids at C^{α} of the double bond Woodward (1957, 1971) devised another methylene-protecting group. The α'-methylene group is again first condensed with ethyl formate, and the hydroxymethylene group is then replaced by a trimethylene dithioketal unit using 1,3-bis(tosylthio)propane. The usual alkylation of the dienolate is then carried out, and finally the dithioketal is removed by reductive desulfurization with Raney nickel. This dithioketal can then be applied for a formal shift of a keto group to the neighboring carbon atom (Marshall, 1969).

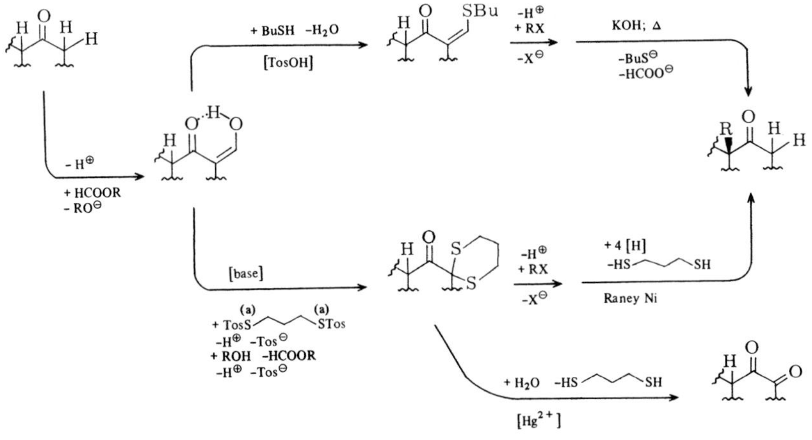
+ BuSH -H2O
[TosOH]
SBu
-H⊕
+ RX
-X⊖
KOH; Δ
-BuS⊖
-HCOO⊖
- H⊕
+ HCOOR
- RO⊖
[base]
(a) (a)
+ TosS STos
-H⊕ -Tos⊖
+ ROH -HCOOR
-H⊕ -Tos⊖
-H⊕
+ RX
-X⊖
+ 4 [H]
-HS SH
Raney Ni
+ H2O -HS SH
[Hg2+]

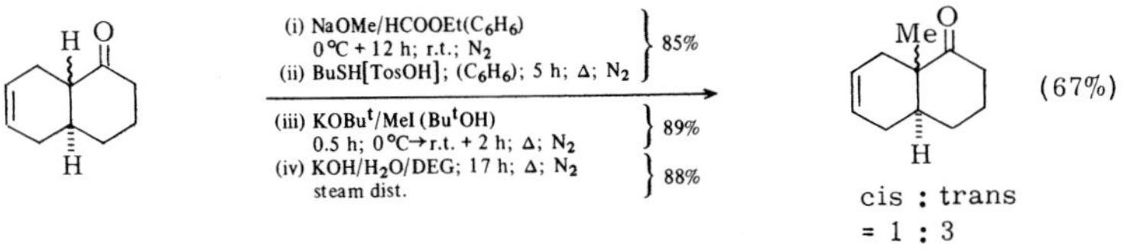
(i) NaOMe/HCOOEt(C6H6)
0 °C + 12 h; r.t.; N2
(ii) BuSH[TosOH]; (C6H6); 5 h; Δ; N2
85%
(iii) KOBut/MeI (ButOH)
0.5 h; 0 °C→r.t. + 2 h; Δ; N2
89%
(iv) KOH/H2O/DEG; 17 h; Δ; N2
steam dist.
88%
Me
(67%)
cis : trans
= 1 : 3

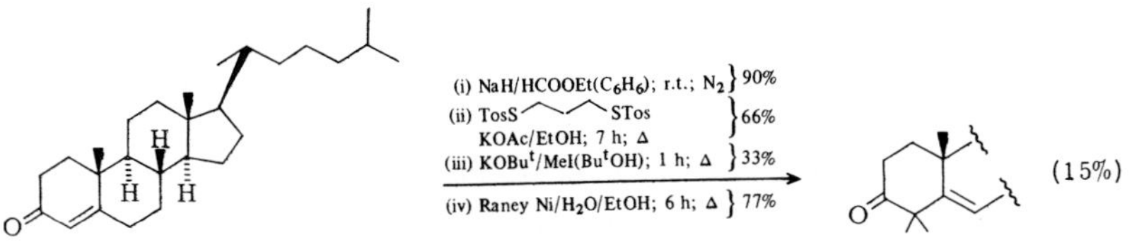
(i) NaH/HCOOEt(C6H6); r.t.; N2 90%
(ii) TosS STos
KOAc/EtOH; 7 h; Δ 66%
(iii) KOBut/MeI(ButOH); 1 h; Δ 33%
(iv) Raney Ni/H2O/EtOH; 6 h; Δ 77%
(15%)

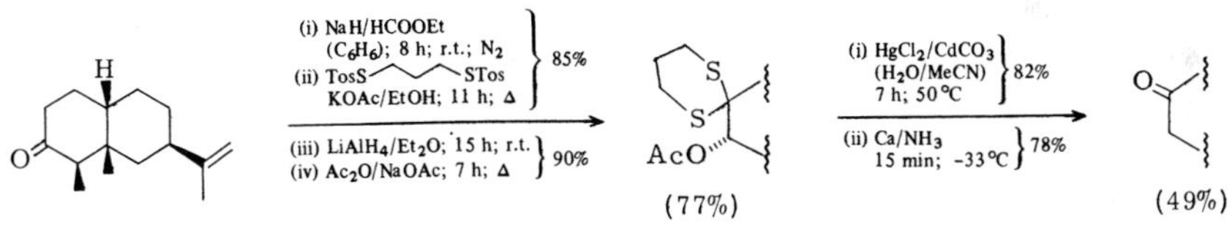
(i) NaH/HCOOEt
(C6H6); 8 h; r.t.; N2
(ii) TosS STos
KOAc/EtOH; 11 h; Δ
85%
(iii) LiAlH4/Et2O; 15 h; r.t.
(iv) Ac2O/NaOAc; 7 h; Δ
90%
AcO
(77%)
(i) HgCl2/CdCO3
(H2O/MeCN)
7 h; 50 °C
82%
(ii) Ca/NH3
15 min; -33 °C
78%
(49%)

C=C double bonds may be protected against electrophiles by epoxidation and subsequent removal of the oxygen atom by treatment with zinc and sodium iodide in acetic acid (Edwards, 1972; Knöll, 1975). Halogenation has been used for protection, too. The C=C double bond is again regenerated with zinc (Barton, 1976).

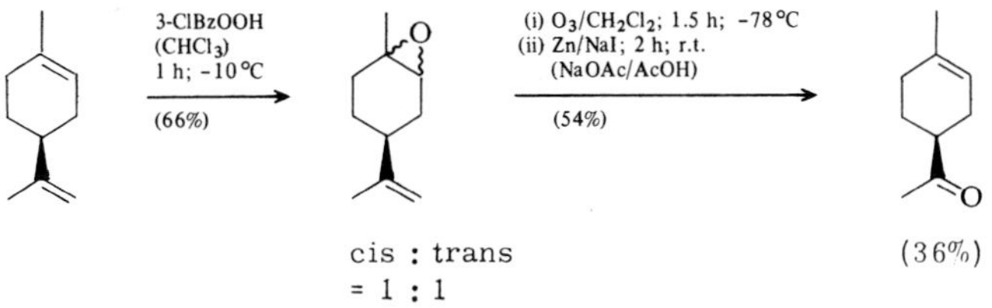
3-ClBzOOH
(CHCl3)
1 h; -10 °C
(66%)
(i) O3/CH2Cl2; 1.5 h; -78 °C
(ii) Zn/NaI; 2 h; r.t.
(NaOAc/AcOH)
(54%)
cis : trans
= 1 : 1
(36%)

(i) Cl₂(CH₂Cl₂); –78 °C
(ii) F₂/TFA [inh.: PhNO₂]; N₂
(CFCl₃/CH₂Cl₂); 0.5 h; r.t.

(iii) Zn/AcOH/H₂O; Δ

(60%)

3.9.2
Alcoholic Hydroxyl Groups

Seven protective groups for alcohols may be removed successively and selectively. (Corey, 1972 B; Nicolaou, 1996). A hypothetical heptahydroxy compound with hydroxy groups 1 to 7 protected as (1) acetate, (2) 2-methoxyethoxy methyl, (3) 2,2,2-trichloroethyl carbonate, (4) benzyl ether, (5) dimethyl-*t*-butylsilyl ether, (6) 2-tetrahydropyranyl ether, and (7) methyl ether may be unmasked in that order by the reagents (1) K₂CO₃ or NH₃ in CH₃OH, (3) Zn in CH₃OH or AcOH, (4) H₂ over Pd, (5) F⁻, (6) wet acetic acid, and (7) BBr₃. The groups may also be exposed to the same reagents in the order 4,5,2,1,3,6. The (4-methoxyphenyl)methyl group (=MPM = *p*-methoxybenzyl, PMB) can be oxidized to a benzaldehyde derivative and thereby be removed at room temperature under neutral conditions (Oikawa, 1982; Johansson, 1984; Fukuyama, 1985). Tab. 3.7 lists a number of reagents and deblocking agents for the protection of alcohols.

We shall describe a specific synthetic example for each protective group given above. Regioselective protection is generally possible if there are hydroxyl groups of different steric hindrance (prim < sec < tert; equatorial < axial). Acetylation has usually been effected with acetic anhydride. The acetylation of less reactive hydroxyl groups is catalyzed by DMAP. Acetates are stable toward oxidation with chromium trioxide in pyridine and have been used, for example, for protection of steroids (Loewenthal, 1959), carbohydrates (Wolfrom, 1963; Williams, 1967), and nucleosides (Michelson, 1963). The most common deacetylation procedures are ammonolysis with NH₃ in CH₃OH and methanolysis with K₂CO₃ or sodium methoxide.

Ac₂O/NaOAc/AcOH
2 h; 60 °C + 2 d; r.t.

(95%)

MEM can be cleaved off with zinc bromide in methylenedichloride in the presence of a ketal and an imine group or by cerium(III) chloride (Sabitha, 2001).

Tab. 3.7 Protecting groups for alcohol R–OH.

Protecting agent	Structure of protected alcohol	Reagents	Deblocking
Ac	$R-O-\overset{\displaystyle O}{\overset{\|}{C}}-CH_3$	$Ac_2O/Py[DMAP]$	$NaOMe/MeOH$ or $K_2CO_3/MeOH$ or $NH_3/MeOH$
MEM	$R-CH_2-O-CH_2-CH_2-O-CH_3$	$MEMCl/Py$	$ZnBr_2/CH_2Cl_2$
Tceoc	$R-O-\overset{\displaystyle O}{\overset{\|}{C}}-O-CH_2-CCl_3$	$TceocCl/Py$	$Zn-Cu/AcOH$
Tceoc	$R-O-\overset{\displaystyle O}{\overset{\|}{C}}-O-CH_2-CBr_3$	$Tbeoc/Py$	
Bzl, Bn	R-O-CH_2- C₆H₅	$PhCH_2Cl/Py$ $PhCH_2Br+Ag_2O$ or NaH/Bu_4NI	$H_2[Pd]$ or Na/NH_3
PMB	R-O-CH_2- C₆H₄$-O-CH_3$	$MeOC_6H_4CH_2Br$ $[NaH/Bu_4NI/THF]$	$DDQ/H_2O/CH_2Cl_2$ or aq. $(NH_4)_2Ce(NO_3)_6/MeCN$
SiBu^{t}Me$_2$ ("TBS")	$R-O-Si(CH_3)_2-C(CH_3)_3$	$Bu^t\,Me_2SiCl$ [imidazole]	$Bu_4N^{\oplus}F^{\ominus}$ or $H_2O/AcOH$
ThP	$R-O-$ (tetrahydropyranyl)	= DHP [TosOH]	aq. AcOH or 0.1 M aq. HCl
Me	$R-O-CH_3$	Me_2SO_4 [NaOH, Ba(OH)$_2$]	BBr_3 or BCl_3

2,2,2-Trichloro- and 2,2,2-tribromoethoxycarbonyl (Tceoc and Tbeoc) protecting groups are introduced with the commercially available 2,2,2-trihaloethyl chloroformates. These derivatives are stable towards CrO_3 and acids, but can smoothly be cleaved by reduction with zinc in acetic acid at 20 °C to yield 1,1-dihaloethene and CO_2. Several examples in lipid (Pfeiffer, 1968, 1970) and nucleotide syntheses (Cook, 1968) have been described.

The benzyl group has been widely used for the protection of hydroxyl functions in carbohydrate and nucleotide chemistry (McCloskey, 1957; Reese, 1965; Griffin, 1966). A common benzylation procedure involves heating with neat benzyl chloride and strong bases. A milder procedure is the reaction in DMF solution at room temperature with the aid of silver oxide (Reinefeld, 1971). Benzyl ethers are not affected by hydroxides and are stable towards oxidants (e.g. periodate, lead tetraacetate), $LiAlH_4$, and weak acids. They are, however, readily cleaved in neutral solution at room temperature by palladium-catalyzed hydrogenolysis (Tejima, 1963) or by sodium in liquid ammonia or alcohols (Reist, 1964).

HO—O OH

(i) H$_2$SO$_4$/MeOH; 5 h; r.t.
(ii) PhCH$_2$Cl/KOH
(THF); 12 h; Δ
(iii) HCl/H$_2$O/AcOH; 1 h; 65 °C
(iv) Py; r.t. (mutarot.)

(48%)

PhCH$_2$O

(89%)

(i) BzCl/Py
1 h; –15 °C
+18 h; r.t.
(ii) H$_2$ [Pd–C]
0.5 h; r.t.

(43%)

HO

OBz

(i) Na/NH$_3$; 10 min; –33 °C
(ii) + NH$_4$Cl

–3 PhCH$_3$

HO—O adenine

HO

(94%)

An active hydrogen atom on oxygen, nitrogen and sulfur can be protected as a silyl derivative (ROH, RCOOH, NH, CONH, SH). Trimethyl silyl (TMS) ethers are readily hydrolyzed by potassium carbonate or HOAc/MeOH. Primary TMS ethers react faster by a factor of 25 than secondary alcohol TMS derivatives, and TMS derivatives react in the following order: CH$_2$OTMS > NTMS > CONTMS > ArOTMS > STMS. TMS is introduced with Me$_3$SiCl (THF, Et$_3$N, 25 °C, 8 h) or hexamethyldisilazane, (Me$_3$Si)$_2$NH (Py, 25 °C, 5 min). Triethyl silyl ethers (TET) are more stable and more soluble (Et$_3$SiCl, Py, 60 °C, 30 min). The most stable and therefore most common silyl ethers are, however, isopropyldimethylsilyl (DMPES) and *tert*-butyl-dimethylsilyl ethers (TBDMS). They are quite resistant to weak acids and bases, but are rapidly cleaved by fluoride anions, e.g. from *n*-Bu$_4$NF, or by 80% HOAc at 90 °C (Nelson, 1996).

Trimethylsilyl ethers are too susceptible to solvolysis in protic media to be useful in syntheses. They are too also widely used as volatile derivatives of alcohols in gas chromatography. The bulky dimethyl-*t*-butylsilyl ethers are about 104 times more stable. Primary and secondary alcohols are readily blocked with chlorodimethyl-*t*-butylsilane in DMF solution and imidazole as catalyst. These silyl ethers are stable toward water, CrO$_3$, hydrogenolysis, and mild reduction procedures. Upon treatment with tetrabutylammonium fluoride in THF at room temperature, they are rapidly cleaved into alcohols and the fluorosilane. Corey (1972 B) demonstrated the highly selective cleavage of either of the benzyloxy and dimethyl-*t*-butylsiloxy groups at a prostaglandin precursor.

(i) Si—Cl
[imidazole]
(DMF); 10 h; 35 °C   96%

(ii) DIBAL (toluene)
10 min; –78 °C; N$_2$ 94%

HO OBzl

Me$_2$Bu^{t}SiO OBzl

+ H$_2$ [Pd–C]
1 h; r.t.

(96%)

Me$_2$Bu^{t}SiO OH

(92%)

Bu$_4$N$^{\oplus}$F$^{\ominus}$/THF
5 min; 0 °C + 1 h; r.t.

HO OBzl

3,4-dihydro-2*H*-pyran undergoes smooth acid-catalyzed addition to alcohols to give 2-tetrahydropyranyl ethers (Thp ethers; W.E. Parham, 1948). CHCl₃, ethers, and DMF may be used as solvent, and traces of HCl or *p*-tolulenesulfonic acid as catalyst. This acetal system is stable toward bases, organometallic compounds, LiAlH₄, and alkylating agents, but it undergoes acid-catalyzed hydrolysis under very mild conditions. Aqueous acetic acid or 0.1 M HCl are sufficient. The example in the scheme below is taken from a nucleotide synthesis by H.G. Khorana (Smith, 1959). It also involves selective tritylation of a primary alcohol group, which is often applied in carbohydrate and nucleoside chemistry.

Moffatt's reagent, *a*-acetoxyisobutyryl chloride, reacts with isolated hydroxy groups to form 2-alkoxy-2,5,5-trimethyl-1,3-dioxolan-4-ones, which are then replaced, again under mildly alkaline or acidic conditions. With *cis*-1,2-cycloalkanediols, *trans*-2-chlorocycloalkanol acetates are formed in high yields; *trans*-1,2-diols give complex product mixtures (Greenberg, 1973; Russell, 1973). The chloroacetoxylation reaction applies also to acyclic 1,2- and 1,3-diols. In this case, primary alcohol groups are preferentially substituted for chloride to give the 2- or 3-acetoxy-1-chloro compounds (Nakatsuka, 1990). Treatment with methanolate converts the chlorohydrin acetates into epoxides.

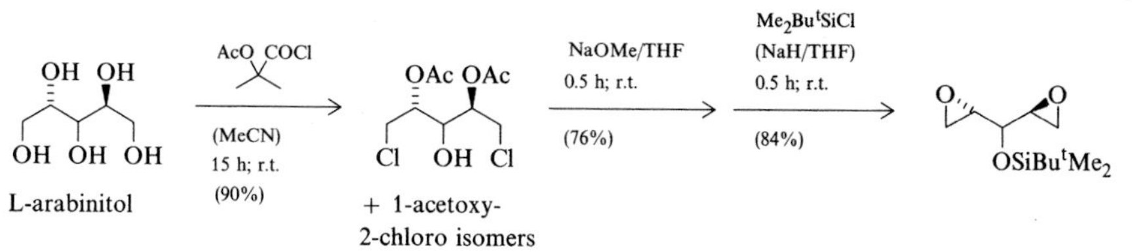

OH OH
OH OH OH
L-arabinitol
AcO COCl
(MeCN)
15 h; r.t.
(90%)
OAc OAc
Cl OH Cl
+ 1-acetoxy-
2-chloro isomers
NaOMe/THF
0.5 h; r.t.
(76%)
Me2ButSiCl
(NaH/THF)
0.5 h; r.t.
(84%)
OSiButMe2

A very convenient method for activation and S_N2 substitution of alcoholic hydroxy groups is the Mitsunobu reaction. In this reaction a zwitterion resulting from addition of a phosphine or a phosphite ester to diethyl diazenedicarboxylate converts the hydroxy groups into good leaving groups. In the presence of zinc halides or of imides, efficient conversions with nearly complete inversion of configuration to alkyl halides (Ho, 1984) or to N-alkylated imides (Mitsunobu, 1972) are observed. With carboxylic acids an "inversion–esterification" of the alcohol is achieved. Phosphoric acid derivatives and carboxamides act equally well as nucleophiles. In the case of the highly labile stereocenter at the α-carbon of the carboxylic ester in the synthesis of the β-lactam ring of 3-amino-nocardicinic acid ("ANA"), only triethyl phosphite was useful. Triphenylphosphine led to racemization (Salituro, 1990).

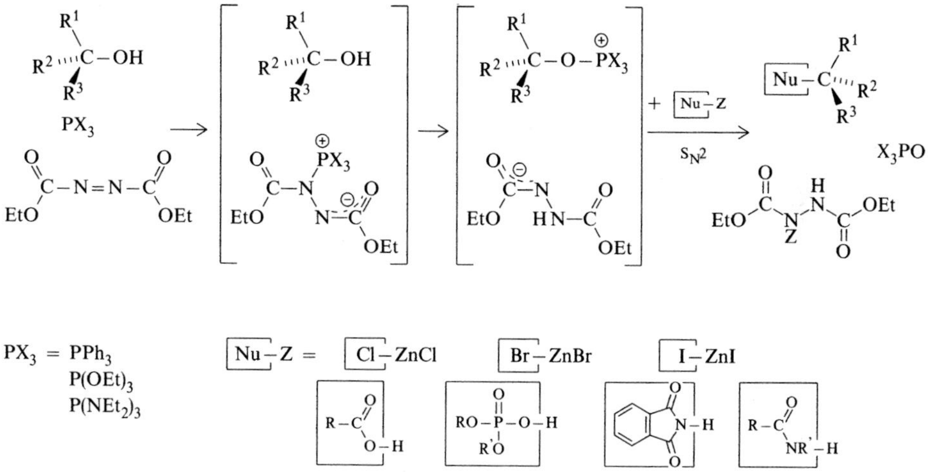

PX3
+ Nu–Z
SN2
X3PO
PX3 = PPh3
P(OEt)3
P(NEt2)3
Nu–Z =
Cl–ZnCl
Br–ZnBr
I–ZnI

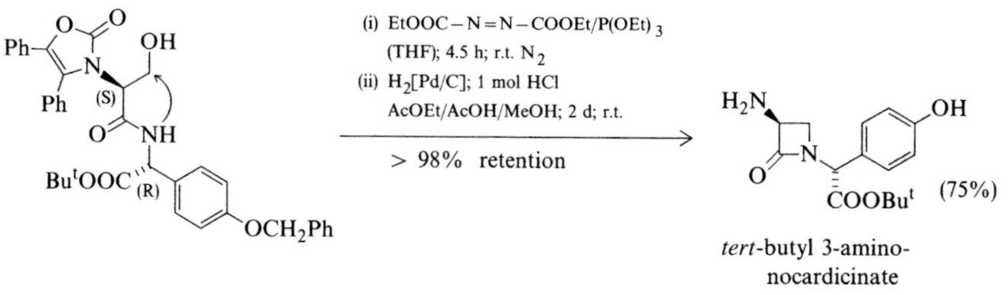

(i) EtOOC–N=N–COOEt/P(OEt)3
(THF); 4.5 h; r.t. N2
(ii) H2[Pd/C]; 1 mol HCl
AcOEt/AcOH/MeOH; 2 d; r.t.
> 98% retention
COOBut (75%)
tert-butyl 3-amino-
nocardicinate

The most versatile Mitsunobu reagent, especially for sterically congested secondary alcohols is a mixture of N,N',N'',N'''-tetramethylazodicarboxamide (=TMAD) and tributylphosphine. In dry benzene solutions the resulting dihydro-TMAD crystallizes out and is removed by filtration. Complete inversion is obtained, whenever the acylation proceeds. *p*-Methoxybenzoic acid is most suitable for sterically hindered alcohols (Tsunoda, 1995).

The most stable protected alcohol derivatives are the methyl ethers. These are often employed in carbohydrate chemistry and can be made with dimethyl sulfate in the presence of aqueous sodium or barium hydroxides in DMF or DMSO. Simple ethers may be cleaved by treatment with BCl_3 or BBr_3, but generally methyl ethers are too stable to be used for routine protection of alcohols. They are more useful as volatile derivatives in gas-chromatographic and mass-spectrometric analyses. So the most labile (trimethylsilyl ether) and the most stable (methyl ether) alcohol derivatives are useful in analysis, but in synthesis they can be used only in exceptional cases. In synthesis, easily accessible intermediates of medium stability are most helpful.

3.9.3
Amino Groups and Thiols

Primary and secondary amines are susceptible to oxidation and replacement reactions involving the N–H bonds. Within the development of peptide synthesis numerous protective groups for N–H bonds have been found (Bodanszky, 1976; Carpino, 1973), and we shall discuss five of the more general methods used involving the reversible formation of amides. A hypothetical pentaamino compound could have been blocked with the following protective reagents at all nitrogen atoms 1 to 5: (1) trifluoroacetic anhydride, (2) *t*-butyl azidoformate, (3) 2,2,2-trichloroethyl chloroformate, (4) benzyl chloroformate (="carbobenzoxy chloride"), and (5) N-ethoxycarbonylphthalimide.

In all of these reagents a leaving group ($CF3COO^-$, N_3^-, $H_2NCOOEt$) is smoothly substituted by the nucleophilic amino group under mild conditions, and an amide is formed in high yield. The amino groups may be unmasked in the given order by treatment with (1) aqueous $Ba(OH)_2$, (2) 25% CF_3COOH in $CHCl_3$, (3) Zn in acetic acid, (4) H_2 over Pd, and (5) HBr in acetic acid or aqueous N_2H_4. The five protecting agents will be discussed successively in the order given in Tab. 3.8.

Tab. 3.8 Protecting groups for amine R–NH$_2$

Tfac	R–NH–C(=O)–CF$_3$	TFac$_2$O/Py	aq. Ba(OH)$_2$ or aq. NAHCO$_3$ or aq. NH$_3$ or aq. HCl or NaBH$_4$/MeOH
Boc ("box")	R–NH–C(=O)–O–C(CH$_3$)$_3$	N$_3$–CO–O–Bu^t	TFA/CHCl$_3$ or aq. HF: –CO$_2$↑ –CH$_2$=C(CH$_3$)$_2$↑
Tceoec	R–NH–C(=O)–O–CH$_2$–CCl$_3$	Tceoc	Zn/AcOH or cathodic redcution: –CO$_2$↑ –CH$_2$=CCl$_2$↑
Cbz ("Z")	R–NH–C(=O)–O–CH$_2$–C$_6$H$_5$	Cl–CO–OCH$_2$Ph	HBr/AcOH or H$_2$[Pd]
Phth	R–N(phthalimide)	PhthN–CO–OEt	HBr/AcOH or N$_2$H$_4$·H$_2$O
Ox	R–N(oxazolidinedione, Ph Ph)	Ph Ph (oxazolidinedione)	H$_2$[Pd] or Na/liq. NH$_3$

Trifluoroacetamides are more stable toward nucleophiles than the corresponding esters and are easily formed from trifluoroacetic anhydride and the amine. The trifluoroacetyl group (Tfac) is slowly cleaved by aqueous or methanolic HCl, NH$_3$, or Ba(OH)$_2$ solutions as well as by NaBH$_4$ in methanol (Wolfrom, 1967).

The *t*-butoxycarbonyl group (Boc, "*t*-box") has been extensively used in peptide synthesis, and Boc derivatives of many amino acids are commercially available. The customary reagent for the preparation from the amine is *t*-butyl azidoformate in water, dioxane/water, DMSO, or DMF. The cleavage by acids of medium strength proceeds with concomitant loss of isobutene and carbon dioxide (Carpino, 1957, 1973).

2,2,2-Trihaloethoxycarbonyl (Tbeoc, Tceoc) derivatives of amines are prepared with the corresponding chloroformates and have been used in the synthesis of β-lactam antibiotics (see Section 4.3). They are cleaved by zinc in acetic acid or by electrolytic reduction (Semmelhack, 1972). Benzoxycarbonyl (Cbz, "Z") derivatives of amines are formed with benzyl chloroformate in alkaline aqueous solution or other polar solvents. They are more stable towards reducing agents than 2,2,2-tri-haloethoxycarbonyl derivatives. Many Cbz-protected amino acids and peptides are commercially available. The most widely used deblocking procedures are solvolysis with HBr (Losse, 1968; Olah, 1970) and hydrogenolysis over palladium catalyst. The ester bond of the Cbz group is remarkably stable towards alkaline hydrolysis.

The phthaloyl (Phth) derivatives of amines, formed from amines and N-ethoxy-carbonylphthalimide (Nefkens, 1960), are acid-resistant imides, which can be easily deblocked by nucleophilic reagents, most conveniently by hydrazine.

The cyclic carbonate of benzoin (4,5-diphenyl-1,3-dioxol-2-one, prepared from benzoin and phosgene) blocks both hydrogen atoms of primary amines; after dehydration acid-stable, easily crystallizable "Sheehan oxazolinones" are formed, which are also called "Ox" derivatives. The amine is quantitatively deblocked by catalytic hydrogenation in the presence of 1 equivalent of aqueous acid (Sheehan, 1972, 1973; Miller, 1983). An intelligent application to syntheses of acid-labile β-lactams is given in the previous section.

Allylic acetoxy groups can be substituted by amines in the presence of Pd(0) catalysts. At substituted cyclohexene derivatives the diastereoselectivity depends largely on the structure of the palladium catalyst. Polymer-bound palladium often leads to amination at the same face as the acetoxy leaving group with regioselective attack at the sterically less hindered site of the intermediate η^3-allyl complex (Trost, 1978).

Tab. 3.9 Protecting groups for thiol R–SH

Protecting agent	Structure of protected thiol	Deblocking
	$-SR^1$	aq. acid
$PhCH_2SCH_2Cl$	$R^1SCH_2SCH_2Ph$	aq. acidHg(OAc)$_2$
Ph_3CCl	R^1SCPh_3	AgNO$_3$ + pyridine in MeOh
p-MeOC$_6$H$_4$CH$_2$OCOCl	$R^1SCOOCH_2C_6H_4OMe$-p	MeOH
EtNCO	$R^1SCONHEt$	pH above 8

Tab. 3.9 summarizes methods of thiol protection based on reversible acetal or thioacetal formation, and alkylation. Acyl groups are difficult to remove, and the use of carbonates is recommended. Dimerization to S–S disulfides also reduces the S-nucleophilicity and is reversed by sodium borohydride (Barrett, 1979).

3.9.4
Carboxyl Groups

The blocking and deblocking of carboxyl groups occurs by re actions similar to those described for hydroxyl and amino groups. The most important protected derivatives are *t*-butyl, benzyl, and methyl esters, These may be cleaved in this order by trifluoroacetic acid, hydrogenolysis, and strong acid or base (McOmie, 1973). 2,2,2-Trihaloethyl esters are cleaved electrolytically (Semmelhack, 1972) or by zinc in acetic acid like the Tbeoc- and Tceoc-protected hydroxyl and amino groups.

Furthermore methylesters are relatively acid-stable but survive aqueous base for a few minutes only. Silylesters are resistant to amines and to oxidation and are readily cleaved with ethanol. *tert*-Butylesters are exceptionally resistant to base-induced hydrolysis, but are readily cleaved by acids. Phenacyl esters are cleaved by zinc in acetic acid and trichloroethyl esters are also labile toward reduction. Oxazolines are stable to Grignard reagents and hydrides, but are vulnerable to acids. *o*-Nitrophenacyl derivatives can be cleaved with UV light.

3.9.5
Aldehyde and Keto Groups

The most commonly used protected derivatives of aldehydes and ketones are 1,3-dioxolanes and 1,3-oxathiolanes. They are obtained from the carbonyl compounds and 1,2-ethanediol or 2-mercaptoethanol, respectively, in aprotic solvents and in the presence of catalysts, e.g. BF$_3$ or TosOH (Fieser, 1954; Wilson, Jr., 1968), and water scavengers, e.g. orthoesters (Doyle, 1965). Acid-catalyzed "exchange dioxolanation" with dioxolanes of low boiling ketones, e.g. acetone, which are distilled during the reaction, can also be applied (Dauben, Jr., 1954). Selective monoketalization of diketones is often used with good success (Mercier, 1973). Even from diketones with two keto groups of very similar reactivity monoketals may be obtained by repeated acid-catalyzed equilibration (Johnson, 1962; Hortmann, 1969). Most aldehydes are easily converted into acetals. The ketalization of ketones is more difficult for steric reasons and often requires long reaction times at elevated temperatures. α,β-Unsaturated ketones react more slowly than saturated ketones. 2-Mercaptoethanol is more reactive than 1,2-ethanediol (Romo, 1951; Djerassi, 1952; Wilson, Jr., 1968).

1,3-Dioxolanes and 1,3-oxathiolanes are stable under most alkaline and neutral reaction conditions. They may, however, be cleaved by organometallic compounds. Deblocking is achieved with strong acids, preferably in the presence of acetone. The ease of acidic cleavage parallels the ease of formation. Ketals of highly hindered ketones may need boiling in mineral acids. The 1,3-oxathiolanes are again more reactive and may be cleaved even in neutral or slightly basic acetone or alcohol solutions, or most efficiently using Raney nickel (Djerassi, 1952, 1958; Pettit, 1962 A). 1,3-Dithianes, which are used in the "umpolung" and removal of carbonyl groups, may also be used as protected derivatives (Seebach, 1969).

The only acid-resistant protective group for carbonyl functions is the dicyanomethylene group formed by Knoevenagel condensation with malononitrile. Friedel–Crafts acylation conditions, treatment with hot mineral acids, and chlorination with sulfuryl chloride do not affect this group. They have, however, to be cleaved by rather drastic treatment with concentrated alkaline solutions (Basts, 1963, 1963; Fischer, 1932; Woodware, 1960, 1961).

3.9.6
Phosphate Groups

If the three esterifiable OH groups of phosphoric acid have to be esterified successively with different alcohols, they have to be protected.

A widely used protecting agent is 2-cyanoethanol (= 3-hydroxypropanonitrile, hydracrylonitrile), which is condensed with phosphates with the aid of DCC. The 2-cyanoethyl (Ce) group is quantitatively removed as acrylonitrile by treatment with weak bases (Khorana, 1965)

2,2,2-Trichloroethanol may be used analogously. The 2,2,2-trichloroethyl (Tce) group is best removed by reduction with copper–zinc alloy in DMF at 50 °C (Eckstein, 1967). For specific protection in the strategy of nucleic acid synthesis see Section 5.2.

4
Heterocycles and Arenes

4.1
Introduction

Organic chemistry is largely composed of arene and heterocyclic chemistry. Because of the size of the subject, any single volume is bound to omit more systems than it includes (Sammes, 1979; Acheson, 1985; Eicher 1995). The major interest in the synthetic nitrogen heterocycles and substituted arenes lies in their extensive usefulness as drugs. Only the pyrroles are different. They are most useful as starting materials for the synthesis of porphyrins and other dyes. Only a few systems are treated here. We attempt to convey a general feeling for condensation reactions leading to nitrogen heterocycles, nitrogen-substituted arenes, lactones and lactams, porphyrins, and alkaloids. A small section treats monosaccharide synthesis and regioselective protection as an introduction to complex oxygen heterocycles. These molecules will then be treated again in Chapter 6 (Combinatorial Mixtures and Selection) and/or in Chapter 7 (Molecular Assemblies).

4.2
Nitrogen-Containing Heterocycles and Arenes in Drugs

The fundamental reasons for the applicability of arenes and heterocycles as drugs lie in their rigidity, their subtle balance between hydrophobic and hydrophilic behavior, their tendency to form strong directed hydrogen bonds and their variable dipole moments. Benzene rings are important in providing better binding to protein surfaces. They intrude into narrow hydrophobic gaps and thus help in transport through membranes. Halogen substituents make them even more membrane-soluble and retard both digestion by microorganisms and cytochrome P450-catalyzed oxidation. There is an enormous variation in these all-important binding parameters with ring systems containing a few nitrogen and oxygen atoms in different positions and combinations. Natural compounds demonstrate the importance of heterocycle variation in nucleic acids, co-enzymes (vitamins), aromas, and poisons (alkaloids). In this chapter we concentrate on systems that have proven to be useful in medicine or have been applied in light-conversion systems (see also Section 7.4).

Most syntheses of nitrogen heterocycles involve substitution and/or condensation reactions of nitrogen nucleophiles with difunctional halides or carbonyl compounds. Common nitrogen reagents are:

N_1: NH_3, $R-NH_2$, NH_2OH, $R-SO_2NH_2$

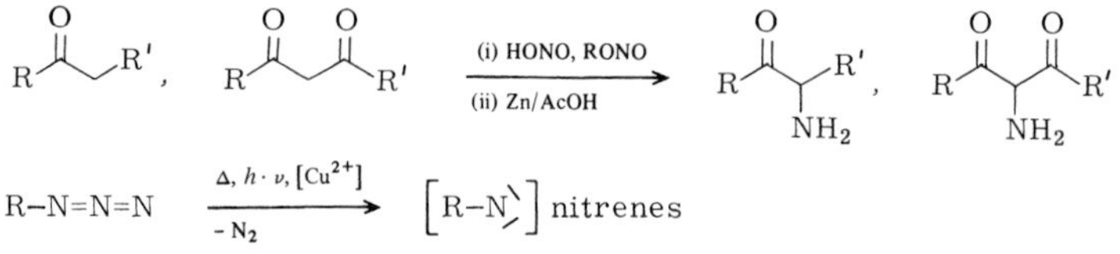

$$R-N=N=N \xrightarrow[-N_2]{\Delta,\ h\cdot\nu,\ [Cu^{2+}]} \left[R-N\right]\ \text{nitrenes}$$

N_2: H_2N-NH_2, $R-NH-NH_2$, $Tos-NH-NH_2$,

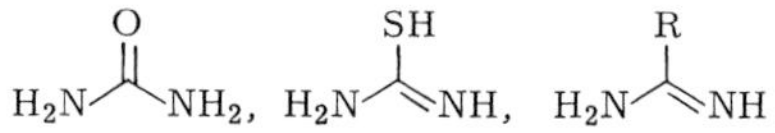

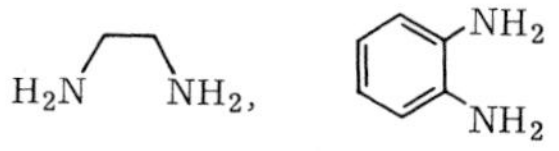

N_3: $R-N=N=N$ azides

The longest carbon chain within a heterocycle indicates possible open-chain precursors. We use this chain as a basis to classify heterocycles as 1,2- to 1,6-difunctional systems.

1,1

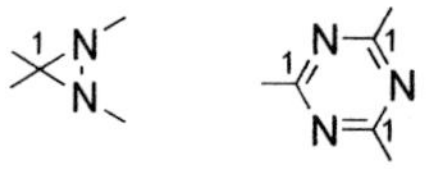

(diaziridines)   (sym-triazines)

1,2

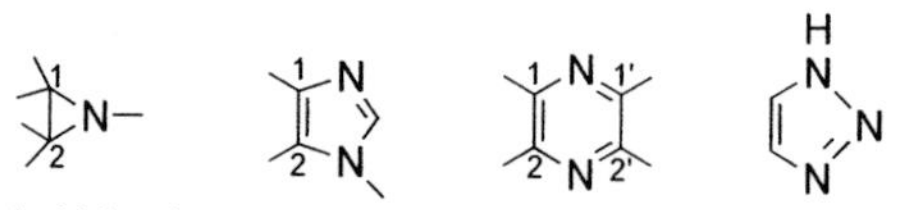

(aziridines)   (imidazoles)   (pyrazines)   (triazoles)

1,3

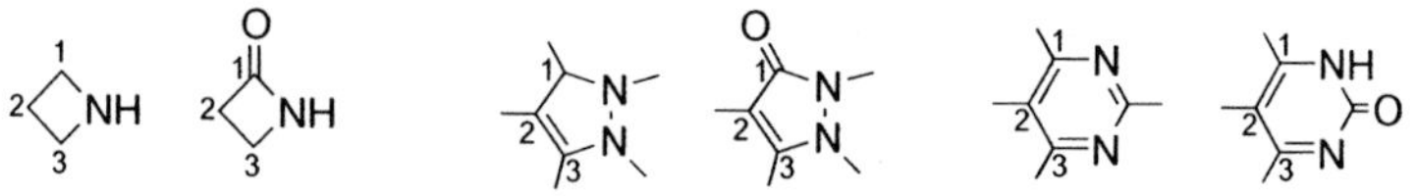

(azetidines, -ones)   (pyrazoles, -olones)   (pyrimidines, -ones)

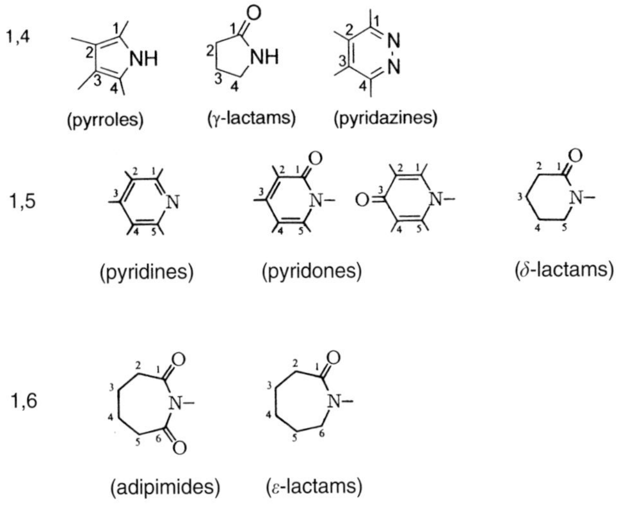

1,4

(pyrroles)   (γ-lactams)   (pyridazines)

1,5

(pyridines)   (pyridones)   (δ-lactams)

1,6

(adipimides)   (ε-lactams)

Since 1,2- to 1,6-difunctional open-chain compounds can be synthesized by general procedures (see Chapter 1), it is useful to consider them as possible starting materials for syntheses of three- to seven-membered heterocycles: 1,2-heterocycles can be made from 1,2-difunctional compounds, e.g. olefins or 1,2-dibromides. 1,3-Difunctional compounds, e.g. 1,3-dibromides or 1,3-dioxo compounds, can be converted into 1,3-heterocycles etc.

For some heterocycles straightforward syntheses can be proposed employing the addition of electrophilic azides to C=C or nucleophilic amines to C=O bonds. Regioselectivity is not a problem if both, the nitrogen reagents and the carbon skeletons are non-chiral. The oxidation states of the ring atoms in the heterocycle to be synthesized are determined by functional groups of the educts. The formation of heteroaromatic systems (e.g. imidazoles, pyrroles, pyridines) is thermodynamically favored and dehydration therefore often occurs spontaneously under acidic or oxidative reaction conditions. Heterocyclic phenols usually rearrange to lactams.

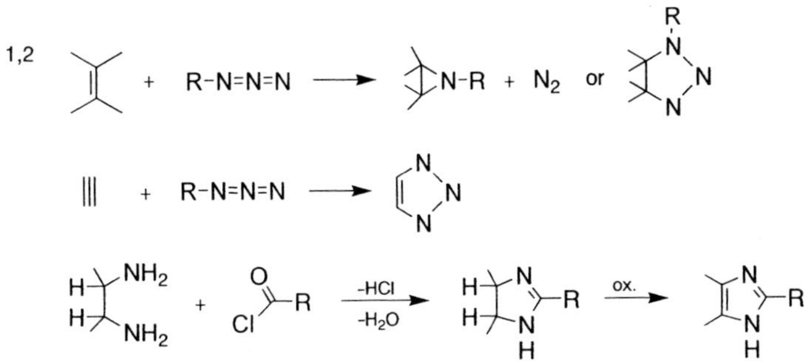

1,2

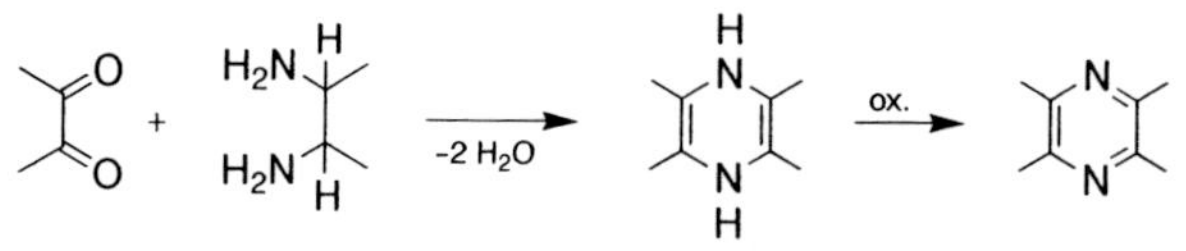

1,3

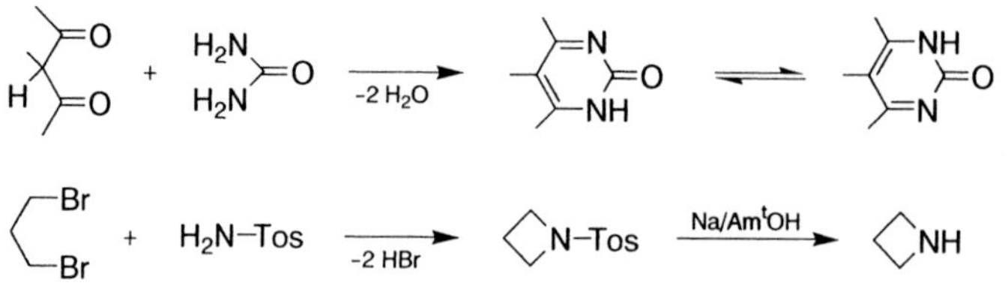

1,4

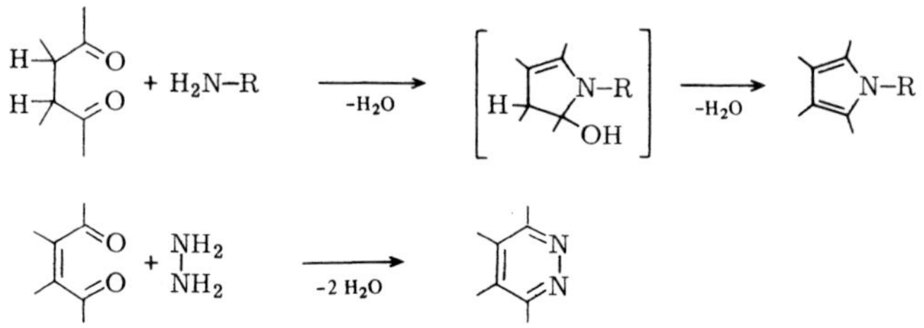

1,5

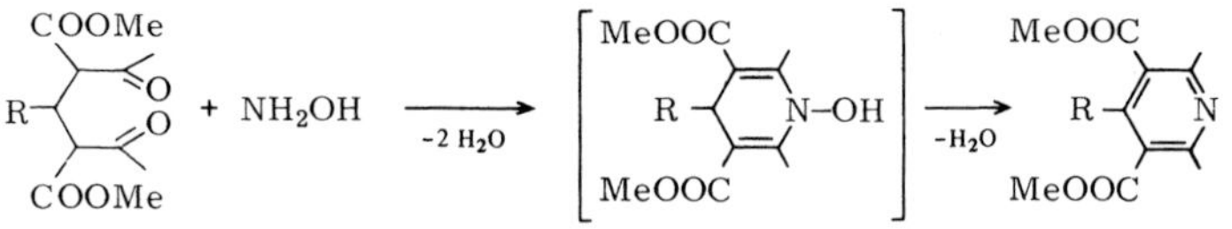

1,6

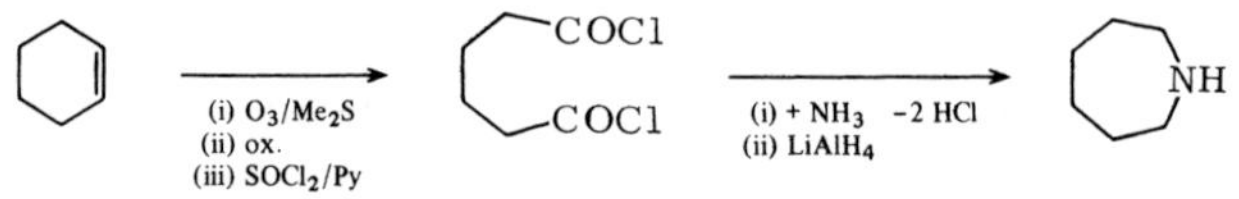

(i) O₃/Me₂S
(ii) ox.
(iii) SOCl₂/Py

(i) + NH₃ −2 HCl
(ii) LiAlH₄

Many saturated nitrogen heterocycles are commercially available from industrial processes, which involve, for example, nucleophilic substitution of hydroxyl groups by amino groups under conditions far from laboratory use, e.g. heating to 220 °C at 200 bar.

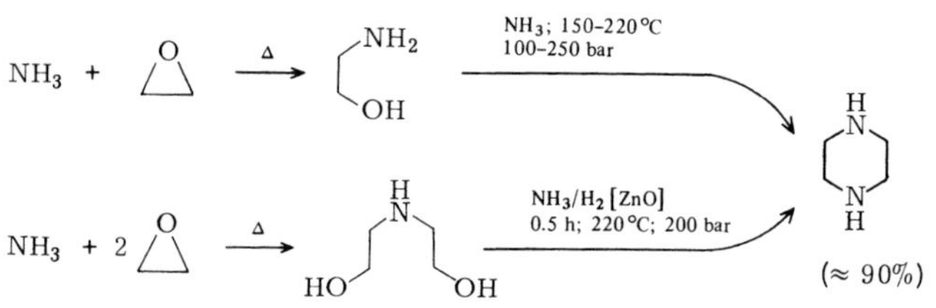

NH_3 + [epoxide] $\xrightarrow{\Delta}$ [aminoethanol]

NH_3; 150–220 °C
100–250 bar

NH_3 + 2 [epoxide] $\xrightarrow{\Delta}$ [diethanolamine]

NH_3/H_2 [ZnO]
0.5 h; 220 °C; 200 bar

(≈ 90%)

Regioselectivity becomes important if unsymmetric difunctional nitrogen components are used. In such cases two different reactions of the nitrogen nucleophile with the open-chain educt may be possible, one of which must be faster than the other. Hydrazone formation of a ketone, for example, occurs more readily than hydrazinolysis of an ester. In the second example the amide is formed very rapidly from the acyl chloride, and again only one cyclization product is observed.

$$R^1\text{-}C(=O)\text{-}COOEt \;+\; NH_2\text{-}NH\text{-}R^2 \;\xrightarrow{-H_2O}\; \left[R^1\text{-}C(=N\text{-}NH\text{-}R^2)\text{-}COOEt \right] \;\xrightarrow{-EtOH}\; \text{pyrazolone}$$

$$\text{(amino ketone, }NH_2) \;+\; Cl\text{-}C(=O)\text{-}CH_2\text{-}CN \;\xrightarrow{-HCl}\; \left[\text{intermediate} \right] \;\xrightarrow{-H_2O}\; \text{dihydropyridinone (CN, OEt)}$$

Many successful regioselective syntheses of heterocycles, however, are more complex than the examples given so far. They employ condensation of two different carbonyl or halide compounds with one nitrogen base, or the condensation of an amino ketone with a second difunctional compound. Such reactions cannot be rationalized in a simple way, and the literature must be consulted for individual target compounds.

The problems involved are exemplified here by Knorr's pyrrole synthesis (A. Gossauer, 1974). It has been known for almost a century that a-amino ketones (C_2N-heterocycles) react with 1,3-dioxo compounds (C_2 components) to form pyrroles (C_4N-heterocycles). A side-reaction is the cyclodimerization of the a-amino ketones to yield dihydropyrazines (C_4N_2), but this can be minimized by keeping the concentration of the a-amino ketone low relative to the 1,3-dioxo compound. The first step in Knorr's pyrrole synthesis is the formation of an enamine. This depends critically on the pH of the solution. The nucleophilicity of the amine is lost on protonation, whereas the carbonyl groups are activated by protons. An optimum is found around pH 5, where yields of about 60% can be reached. At pH 4 or 6 the yield of the pyrrole may approach zero. The ester groups of β-keto esters do not react with the amine under these conditions. If a more reactive 1,3-diketone is used, this has to be symmetrical, otherwise mixtures of two different imines are obtained. The imine formed rearranges to an enamine, which cyclizes and dehydrates to yield a 3-acylpyrrole as the "normal" Knorr product (Gossauer, 1974; Kenner, 1973 B).

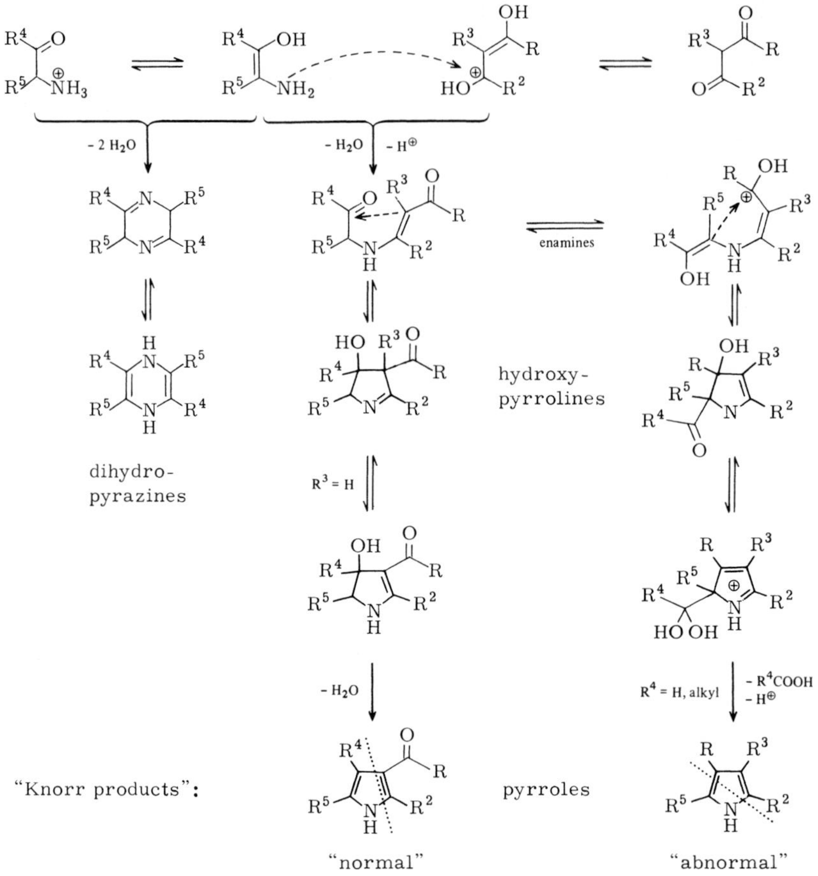

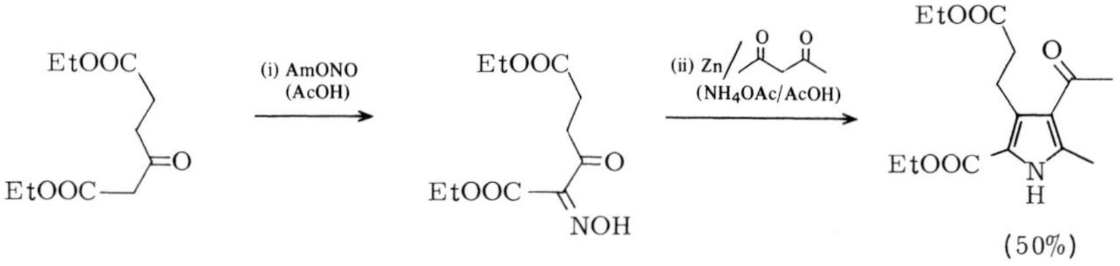

$- 2 H_2O$

$- H_2O$ $- H^\oplus$

enamines

R^4 N R^5 / R^5 N R^4

$R^3 = H$

dihydro-
pyrazines

hydroxy-
pyrrolines

"Knorr products":

$- H_2O$

pyrroles

"normal"

$R^4 = H, alkyl$ $- R^4COOH$ $- H^\oplus$

"abnormal"

EtOOC ... EtOOC—

(i) AmONO
(AcOH)

EtOOC ... EtOOC— NOH

(ii) Zn/ (NH$_4$OAc/AcOH)

EtOOC ... EtOOC N H

(50%)

The simple reaction scheme accounts for three experimental observations:

(i) The amino component may be a mono- or a dicarbonyl compound, but the other component must be a 1,3-dioxo compound because the enamine would not form at pH 5 from an imine with a non-activated neighboring methylene group. Therefore only 3-carboxy- or 3-acylpyrroles can be made efficiently. (ii) "Abnormal" products can be formed from 1,3-diketones or β-keto-aldehydes because an acyl or formyl group, respectively, may be lost by *retro*-aldol-type cleavage when the final dehydration step takes place. (iii) Yields seldom exceed 60%. Several side products are observed in these complicated reactions because several intermediates may react in different ways.

The usefulness of the Knorr synthesis arises from the fact that 1,3-dioxo compounds and α-amino ketones are much more easily accessible in large quantities than "rational" 1,4-difunctional precursors. Such "practical" syntheses are known for several important heterocycles. They are usually limited to certain substitution patterns of the target molecules.

Other interesting pyrroles are accessible by unsymmetrical ozonolysis of cyclohexene derivatives yielding 1-formyl-6-peroxycarboxylates followed by cyclization with ammonium carbonate. The necessary keto group on C-4 can easily be introduced with the ene of 1-butene-3-one in a Diels–Alder cyclization and is then protected with ethylene glycol. α-Formylation can be achieved with trimethyl *ortho*-formate in the presence of trifluoroacetic acid (Taber, 2001).

Enamine cyclization processes also occur in several other successful heterocycle syntheses, e.g. in the Fischer indole synthesis. In this case, however, a labile N–N bond of a 1-aryl-2-vinylhydrazine is cleaved in a [3,3]-sigmatropic rearrangement, followed by cyclization and elimination of ammonia to yield the indole (Robinson, 1963, 1969; Sundberg, 1970). Regioselectivity is only observed if R^2 does not contain enolizable hydrogen, otherwise two structurally isomeric indoles are obtained. Other related cyclization reactions are found in the Pechmann synthesis of triazoles (Gilchrist, 1974) and in Bredereck's (1959) imidazole synthesis (Grimmett, 1970).

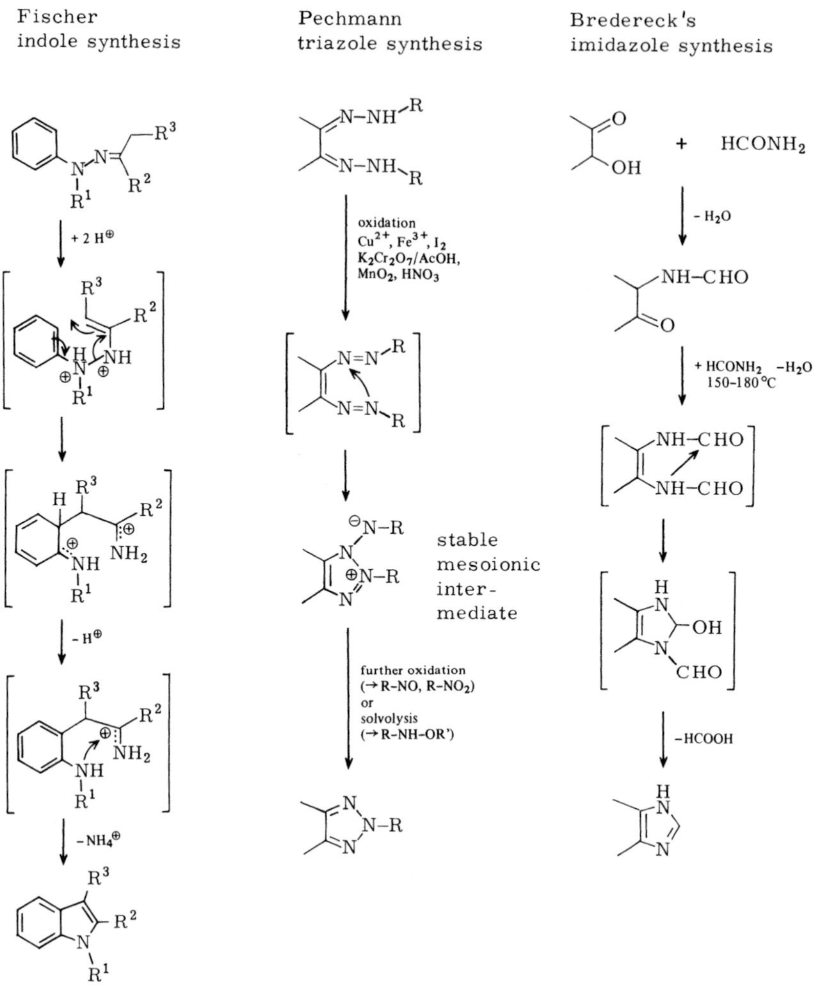
Fischer
indole synthesis
Pechmann
triazole synthesis
Bredereck's
imidazole synthesis
+ 2 H⊕
- H⊕
- NH4⊕
oxidation
Cu2+, Fe3+, I2
K2Cr2O7/AcOH,
MnO2, HNO3
stable
mesoionic
inter-
mediate
further oxidation
(→ R-NO, R-NO2)
or
solvolysis
(→ R-NH-OR')
HCONH2
- H2O
+ HCONH2 -H2O
150-180 °C
-HCOOH

One may also take advantage of the electrophilicity of nitrogen-containing 1,3-dipolar compounds rather than of the nucleophilicity of amines or enamines. 1,3-Dipoles add to multiple bonds, e.g. C=C, C≡C, C=O, in a [2+3]-cycloaddition to form five-membered heterocycles.

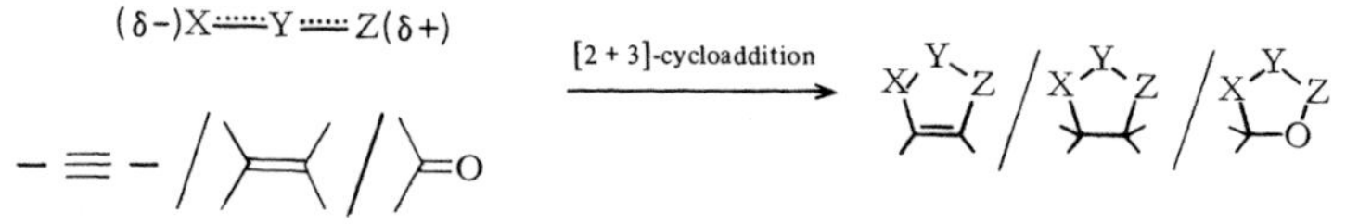
(δ-)X⋯Y⋯Z(δ+)
[2 + 3]-cycloaddition

A few typical examples indicate the large variety of five-membered heterocycles, which can be synthesized efficiently by [2+3]-cycloadditions. [2+2]-Cycloadditions, on the other hand, are useful in the synthesis of four-membered heterocycles (Ulrich, 1967), e.g. of β-lactams (Malpass, 1977). Diels–Alder-type [2+4]-cycloadditions use hetero-"ene" components (Malpass, 1977; Martin, 1980) or highly reactive o-quinodimethanes as diene components (Oppolzer, 1978A).

| | products of [2 + 3]-cycloadditions with | |
common 1,3-dipolar reagents	alkenes	carbonyl groups
R–I $\xrightarrow[-I^{\ominus}]{+N_3^{\ominus}}$ $\overset{\delta+}{N}\cdots\overset{}{N}\equiv\overset{\delta-}{N}$ azides	1,2,3-triazolines	—
imidic acid chlorides $\xrightarrow[-HCl]{base}$ $R-\overset{\delta+}{C}\equiv\overset{}{N}\cdots\overset{\delta-}{C}$ nitrile ylides	pyrrolines	oxazolines
hydroxamic acid chlorides $\xrightarrow[-HCl]{base}$ $R-\overset{\delta+}{C}\equiv\overset{}{N}\cdots\overset{\delta-}{O}$ nitrile oxides	isoxazolines	1,4,2-dioxazoles
R–CHO + R'NHOH $\xrightarrow[-H_2O]{}$ nitrones	isoxazolidines	—

Azide groups are introduced by conjugate addition of azide anions. Some peptides that adopt a β-turn structure, entrap such azide anions and then lead to enantioselective Michael addition reactions. The added azide may then add in a [2+3]-cyclization reaction to double bonds to give triazolines or to triple bonds to yield triazoles. Stereoselective addition of such heterocycles to Michael double bonds have been achieved with chiral hosts, which bind to the oxide (Guerin, 2002). N-oxides also allow 1,3-electrocyclization reactions with alkenes.

Proline, the only natural amino acid in which the chiral α-carbon atom lies in a ring, has been α-alkylated without loss of optical activity (Seebach, 1983). Proline was condensed with pivalaldehyde to give a pure bicyclic diastereomer with the bulky *tert*-butyl group on the *exo* face, i.e. cis to the angular hydrogen. LDA produces a chiral enolate which reacts with both alkylating reagents and carbonyl compounds attacking its *Re*-side, i.e. cis relating to the *tert*-butyl group. Thus this reaction sequence reproduces the original configuration with complete retention. Furthermore, aldol-type additions occur with diastereoselectivities up to 95%, i.e. probably due to lithium chelation control. The allylation of proline by this method has been applied in a complex alkaloid synthesis (Williams, 1990). It constitutes a valuable example of the stereocontrolled formation of a quarternary chiral center.

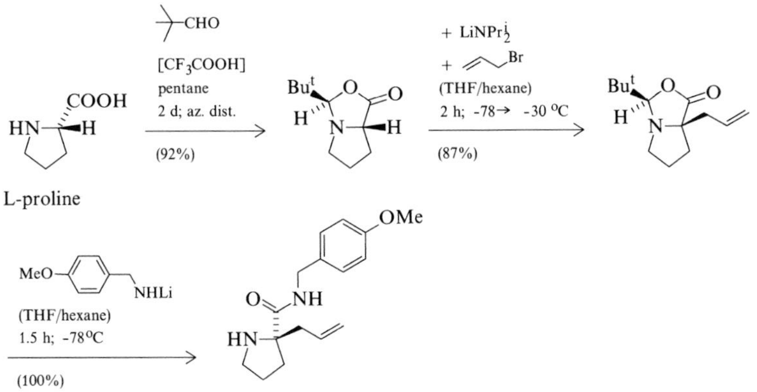

CHO

[CF$_3$COOH]
pentane
2 d; az. dist.

(92%)

COOH

HN——H

L-proline

+ LiNPr$_2^i$

+ Br

(THF/hexane)
2 h; -78 → -30 °C

(87%)

MeO—⟨⟩—NHLi

(THF/hexane)
1.5 h; -78 °C

(100%)

OMe

O NH

HN—

Alkyl azides are decomposed upon heating or irradiating to yield nitrenes, which may also undergo [1+2]-cycloaddition reactions to yield highly strained azir-idenes (Hormann, 1972). Sulfonium ylides may then be added to C=N double bonds to yield aziridines in a formal [1+2]-cycloaddition. Double HBr-elimination leads to similar bicycles.

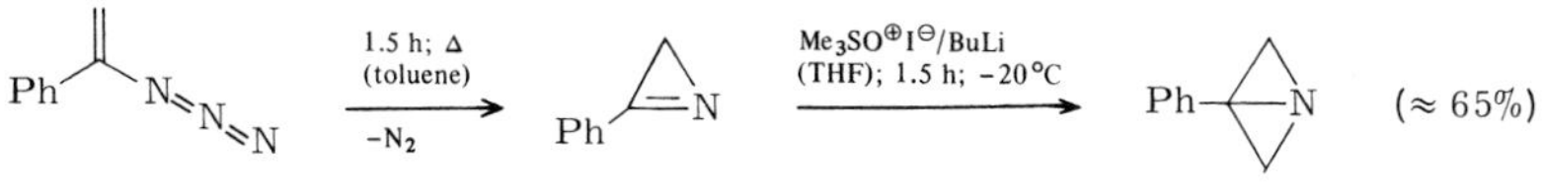

Ph
N=N=N

1.5 h; Δ
(toluene)
—N$_2$

Ph—=N

Me$_3$SO$^⊕$I$^⊖$/BuLi
(THF); 1.5 h; -20 °C

Ph—⟨N⟩   (≈ 65%)

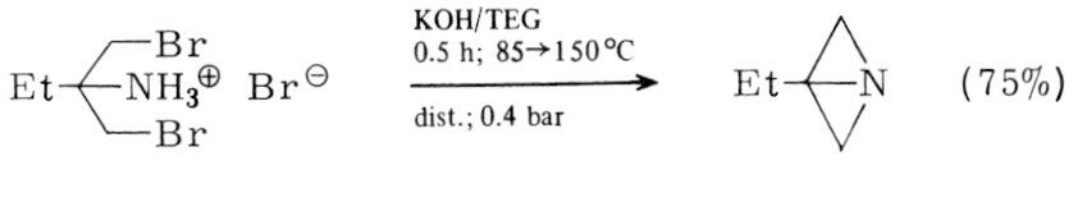

Et—⟨−Br / −NH$_3^⊕$ / −Br⟩ Br$^⊖$

KOH/TEG
0.5 h; 85 → 150 °C
dist.; 0.4 bar

Et—⟨N⟩   (75%)

The structure of drugs in which heteroycles are coupled with substituted phenyl rings often look complicated, but their synthesis is usually simple and straightfor-ward. We start with syntheses in which benzylic anions or halides are the key syn-thons. Carbanions and amines are routinely applied in multiple substitution reac-tions of saturated bromides and chlorides as well as of benzylic OH-groups. Aro-matic chlorides do not react, but halides on pyridine usually do.

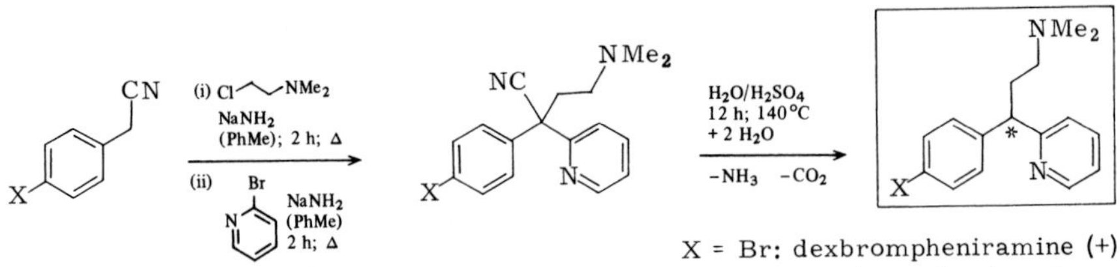

X = Br: dexbrompheniramine (+)
rac-brompheniramine
X = Cl: rac-chlorpheniramine
(antihistaminics)

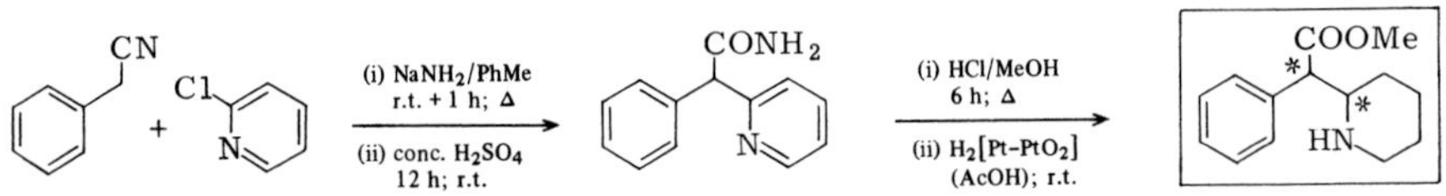

rac-methylphenidate
(central stimulant)

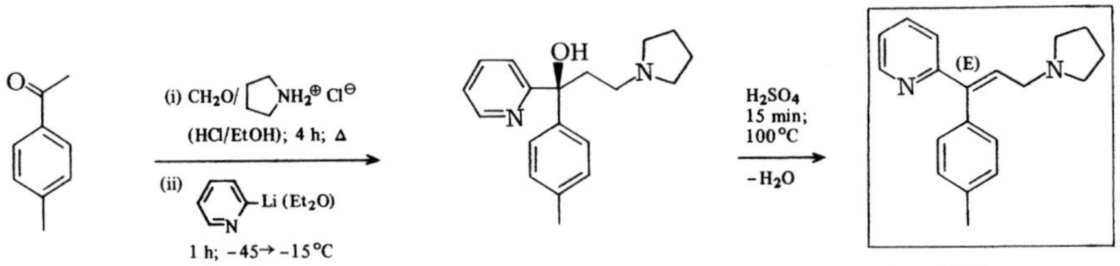

triprolidine
(antihistaminic)

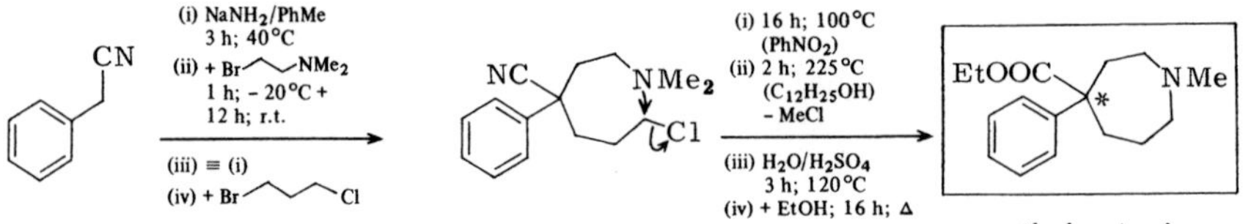

rac-ethoheptazine
(analgesic)

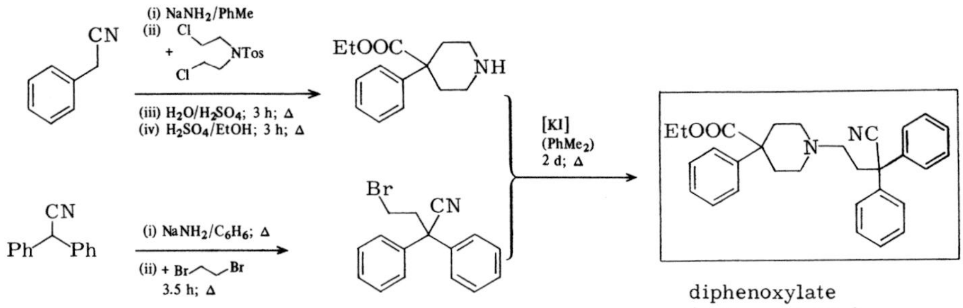

diphenoxylate
(antiperistaltic)

(i) SOCl$_2$

(ii) HN—NAc

(i) H$_2$SO$_4$/H$_2$O

(ii) Cl—

rac-meclizine = meclozine
(antinauseant)

COOMe

BzlNMe$_3$$^\oplus$ OH$^\ominus$
(dioxane)
1 h; 75–90 °C

CN COOMe

(i) KOH/MeOH
1 h; Δ; N$_2$

(ii) H$_2$SO$_4$/AcOH
10 min; 100 °C

rac-glutethimide
(sedative, hypnotic)

The following examples give more insight into typical reactions of commercial heterocycle–benzene combinations used as drugs. Nitrobenzene forms radicals on reduction with metals, e.g. iron powder, and the dimerization product hydrazobenzene can be isolated in high yield under appropriate conditions. Cyclization of the hydrazine with alkylated malonic esters yields useful drugs, e.g. phenylbutazone. Electron withdrawing substituents such as carboxyl and sulfamoyl activate the benzene ring towards nucleophilic substitution. Chlorine may then be replaced by primary amines. The regioselectivity observed in the synthesis of furosemide may be caused by steric effects of the large sulfamoyl group. The exclusive chloromethylation of carbon atom 3 in the synthesis of oxymetazoline is also caused by the strong steric hindrance exerted by the bulky *tert*-butyl group. This is especially severe in planar cyclic compounds. Finally, the benzene nucleus stabilizes the diazonium group. Such salts are useful reactants in electrophilic substitutions. Azo compounds are generally used as components of dye synthesis, but a few drug examples also exist.

PhNO$_2$

H$_2$[Pd–C]
(KOH/H$_2$O)
1 bar; r.t.

HN–Ph
HN–Ph

COOEt
COOEt

NaOEt/EtOH
(PhMe$_2$); 12 h; Δ; N$_2$

phenylbutazone
(anti-inflammatory)

rac-oxyphenbutazone
(anti-inflammatory)

furosemide
(diuretic, antihypertensive)

oxymetazoline
(adrenergic, vasoconstrictor)

phenazopyridine
(urinary analgesic)

The nucleophilicity of the nitrogen atom survives in many different functional groups, although its basicity may be lost. The reactions of nonbasic but nucleophilic urea nitrogens, for example, provide an easy entry to sleeping pills (barbiturates) as well as to stimulants (caffeine). Ureas are a carbonic acid derivative and may therefore be decarboxylated by heating. DMF is used as a solvent as well as a formylating agent, and its ammonia is released. Alkylation reactions of nitroimidazoles in apolar solvents are steered to the neighboring nitrogen atom by the nitro group. Treatment with HCl contracts six-membered pyrazine rings obtained from hydrazines and ketones to imidazole units. The nitrogen atoms of imidazoles and indole anions are also nucleophilic and the NH protons can easily be substituted.

barbiturates: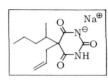

phenobarbital
(anticonvulsant,
hypnotic, sedative)

butalbital
(sedative)

secobarbital
sodium
(hypnotic)

pentobarbital
sodium
(hypnotic, sedative)

EtOOC Cl   H₂N
          C=O
H₂N EtO OEt   H₂N
   NH₂
O

(i) 0.5 h;
110°C

(ii) aq. KOH
2 min;
50°C

uric acid

(i) HCONH₂
3.5 h; Δ
−CO₂−NH₃

(ii) Me₂SO₄
(aq. NaOH)
pH 8−9
2 h; 35°C

caffeine
(central stimulant)

HNO₃/H₂SO₄

1 h; 140°C

O₂N

Br⌒OH

(C₆H₆); 3 d; Δ

O₂N

OH

metronidazole
(antiprotozoal, trichomonacide)

COOMe

MeO

N NH₂
H
O

(i) AcOH/H₂O
r.t.

(ii) HCl/EtOH
1 h; Δ
(iii) 6 M NaOH
3 h; r.t.

MeO

COOH

N
H

(i) BuᵗOH/DCC
(ZnCl₂/THF)

(ii) Cl—◯—COCl

(NaH/DMF)

(iii) 210°C; −⟍

MeO

COOH

N
O

Cl

indomethacin
(anti-inflammatory,
antipyretic, analgesic)

A related substance is famous for treatment of erectile dysfunction. It is called sildenafil and is available commercially under the name of Viagra™. It contains a caffeine-like pyrazolo[4,3-pyrimidine]-7-one heterocycle with pyrazole instead of imidazole and additional sulfonamide and piperidine substituents. Its activity depends on the inhibition of a phosphocyclase enzyme, namely phosphodiesterase-type 5. Sildenafil does not contain a chiral center, which could make its synthesis costly. It is a simple caffeine derivative. An early synthesis of the pyrazolo[4,3-d]pyrimidine-7-ones started with the condensation of a methylated acetoacetate and hydrazine. The resulting pyrazole was N-methylated at the nitrogen atom neighboring the electron withdrawing carbethoxy group. Nitration at C-3, reduction of the nitro group with SnCl₂ amidation, some standard condensation reactions and p-chlorosulfonation of the alkylphenolether yielded sildenafil or Viagra™ (Terrett, 1996). For a "one-step", solid state synthesis see Section 6.4.5.

Sildenafil (VIAGRA™)

Several other phosphodiesterase-type 5 inhibitors have been prepared and evaluated. One publication describes in detail the all-important test procedure to select active compounds. A direct electric assay of the erectile tissue of rabbits under a light microscope was applied (Rotella, 2000). Screening showed that nucleobases were activated as enzyme inhibitors by two additional benzene rings, one in the center, one at the periphery. Addition of a fluorophenyl substituent stabilized them against digestion and led to an orally active agent. Synthesis of the most active compound involved many classical reactions under carefully optimized conditions: dinitration of a substituted benzene ring, amidation of a carboxylate, nucleophilic substitution of chlorobenzene with benzylamide, NO₂ reduction to NH₂, cyclization to lactams first with formic then with benzoic acid, nucleophilic substitution of Br– by CH–, and the hydrolysis of a nitrile to an amide (Rotella, 2000).

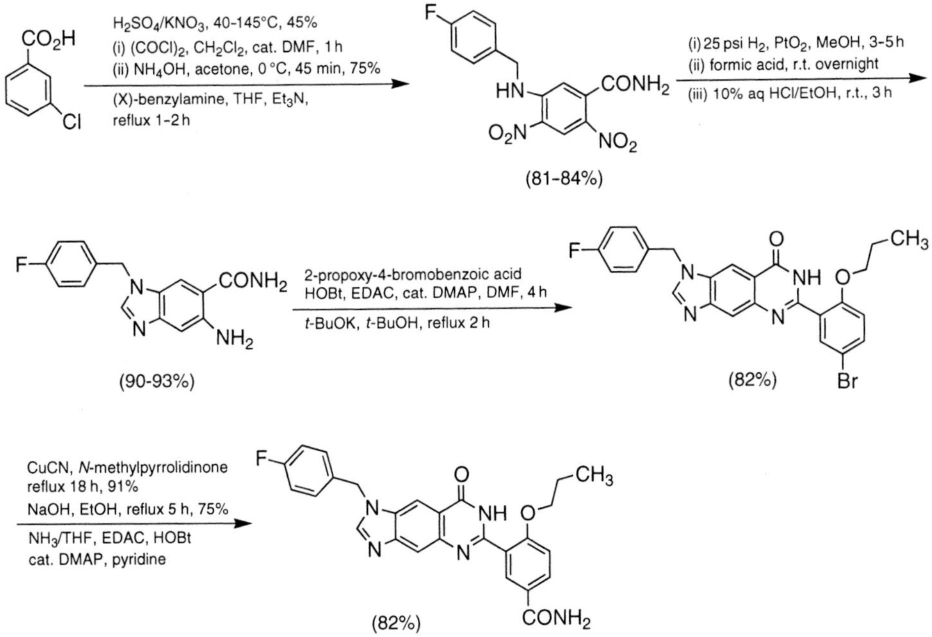

CO2H
Cl
H2SO4/KNO3, 40-145°C, 45%
(i) (COCl)2, CH2Cl2, cat. DMF, 1 h
(ii) NH4OH, acetone, 0 °C, 45 min, 75%
(X)-benzylamine, THF, Et3N,
reflux 1-2 h
F
HN
CONH2
O2N
NO2
(81-84%)
(i) 25 psi H2, PtO2, MeOH, 3-5 h
(ii) formic acid, r.t. overnight
(iii) 10% aq HCl/EtOH, r.t., 3 h
CONH2
NH2
(90-93%)
2-propoxy-4-bromobenzoic acid
HOBt, EDAC, cat. DMAP, DMF, 4 h
t-BuOK, t-BuOH, reflux 2 h
O
NH
CH3
Br
(82%)
CuCN, N-methylpyrrolidinone
reflux 18 h, 91%
NaOH, EtOH, reflux 5 h, 75%
NH3/THF, EDAC, HOBt
cat. DMAP, pyridine
(82%)
CONH2

The use of oximes as nucleophiles can be quite perplexing in view of the fact that nitrogen or oxygen may react. Alkylation of hydroxylamines can therefore be a very complex process which is largely dependent on the steric factors associated with the educts. Reproducible and predictable results are obtained in intramolecular reactions between oximes and electrophilic carbon atoms. Amides, halides, nitriles, and ketones have been used as electrophiles, and various heterocycles such as quinazoline *N*-oxide, benzodiazepines, and isoxazoles have been obtained in excellent yields under appropriate reaction conditions.

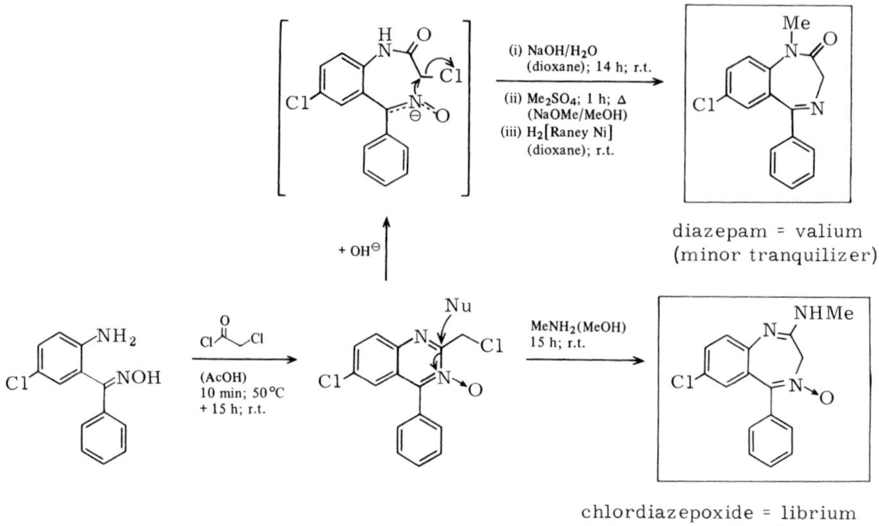

(i) NaOH/H$_2$O
(dioxane); 14 h; r.t.

(ii) Me$_2$SO$_4$; 1 h; Δ
(NaOMe/MeOH)

(iii) H$_2$[Raney Ni]
(dioxane); r.t.

diazepam = valium
(minor tranquilizer)

+ OH$^\ominus$

(AcOH)
10 min; 50 °C
+ 15 h; r.t.

MeNH$_2$(MeOH)
15 h; r.t.

chlordiazepoxide = librium
(minor tranquilizer)

(i)   SO$_2$Cl

HN$_{\diagdown Ac}$

[Et$_3$N]

(ii) aq. KOH

NH$_2$OH
[H$^\oplus$]

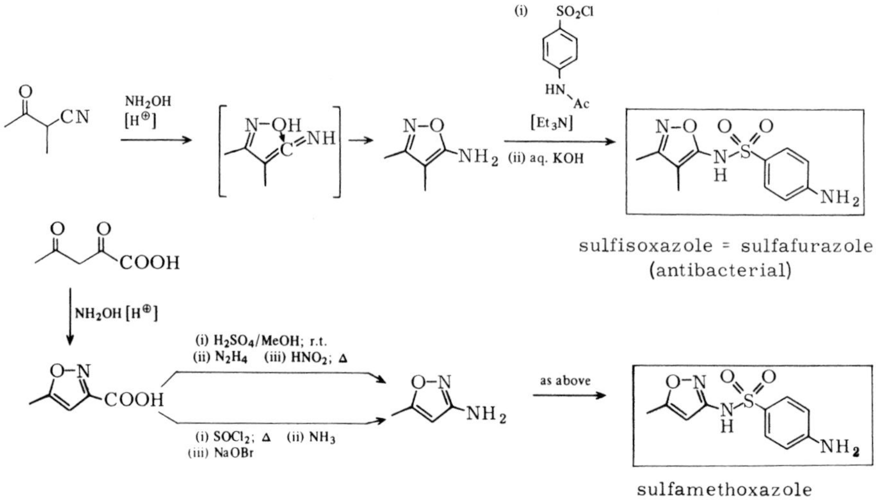

sulfisoxazole = sulfafurazole
(antibacterial)

NH$_2$OH [H$^\oplus$]

(i) H$_2$SO$_4$/MeOH; r.t.
(ii) N$_2$H$_4$ (iii) HNO$_2$; Δ

as above

(i) SOCl$_2$; Δ   (ii) NH$_3$
(iii) NaOBr

sulfamethoxazole
(antibacterial)

In the case of nitrated and carboxylated dinitrogen heterocycles we have already seen that the nucleophilicity of amine nitrogens is differentiated by their environment. There are many more examples. Triaminopyrimidine is easily accessible from malonitrile and guanidine. One gram of this versatile heterocycle costs less than a dollar from Aldrich. The corresponding triamine of 1,3,5-triazine is even three times less expensive (Melamin). In 2,4,5,6-tetraaminopyrimidine the most basic 5-amino group can be selectively converted to a Schiff base. It is *meta* to both pyrimidine nitrogens and does not form a tautomeric imine as do the *ortho-* and *para-*amino groups. This factor is the basis of the commercial synthesis of triamterene.

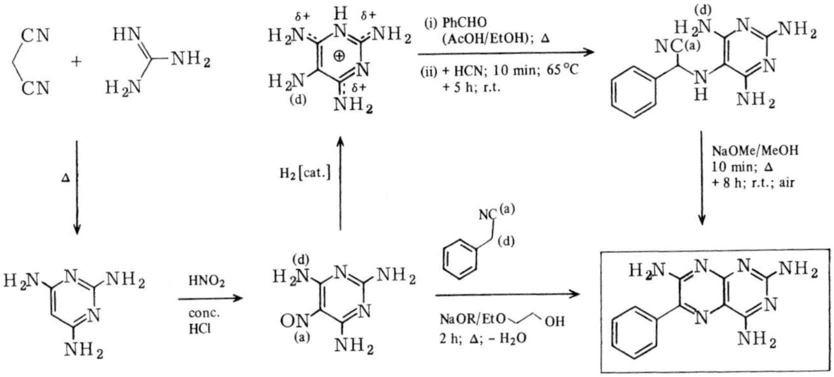

triamterene
(diuretic)

Other interesting regioselective reactions are carried out within the synthesis of nitrofurantoin. Benzaldehyde semicarbazone substitutes chlorine in chloroacetic ester with the most nucleophilic hydrazone nitrogen atom. Transamidation of the ester occurs with the diprotic outer nitrogen atom, the substitution of the chlorine atom occurs with the more delocalized electron pair of the second nitrogen atom. In the cyclization product the hydrazone is selectively hydrolyzed. As a hydrazine, its outer nitrogen now reacts exclusively with other carbonyl compounds. Neither of the two lactam nitrogen atoms is reactive.

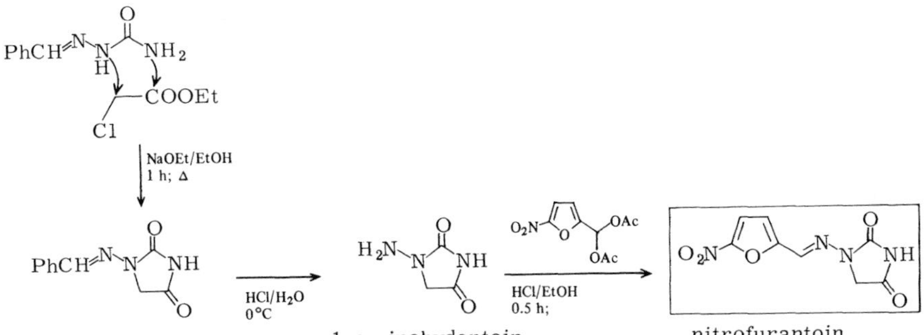

1-aminohydantoin

nitrofurantoin
(urinary antibacterial)

Bridged nitrogen heterocycles are also synthesized on a commercial scale. Synthesis of pentazocine consists of a reductive alkylation of a pyridinium ring, a remarkably selective and puzzling addition to the most hindered position. Synthesis of clidinium bromide implies an alkylation, catalytic hydrogenation of an electron-deficient pyridine ring and a Dieckmann cyclization (=intramolecular Claisen condensation). Both syntheses are applications of reactivity rules discussed in Section 4.5 (Alkaloids).

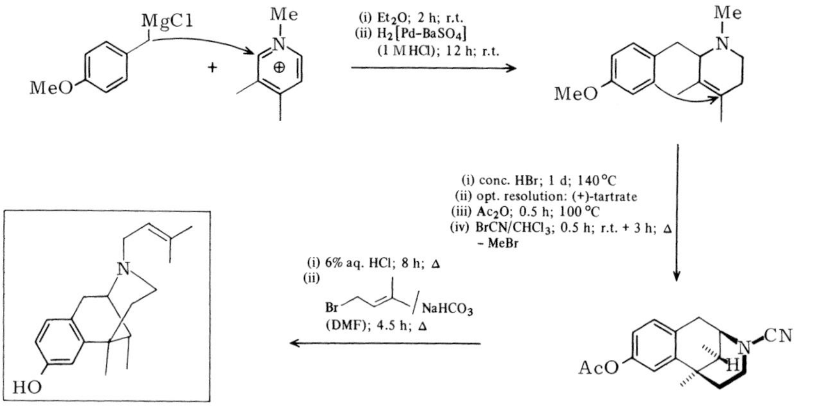
MgCl
Me
N
MeO
+
(i) Et2O; 2 h; r.t.
(ii) H2[Pd-BaSO4]
(1 M HCl); 12 h; r.t.
MeO
(i) conc. HBr; 1 d; 140 °C
(ii) opt. resolution: (+)-tartrate
(iii) Ac2O; 0.5 h; 100 °C
(iv) BrCN/CHCl3; 0.5 h; r.t. + 3 h; Δ
- MeBr
AcO
CN
H
(i) 6% aq. HCl; 8 h; Δ
(ii)
Br
NaHCO3
(DMF); 4.5 h; Δ
HO

(−)-pentazocine (analgesic)

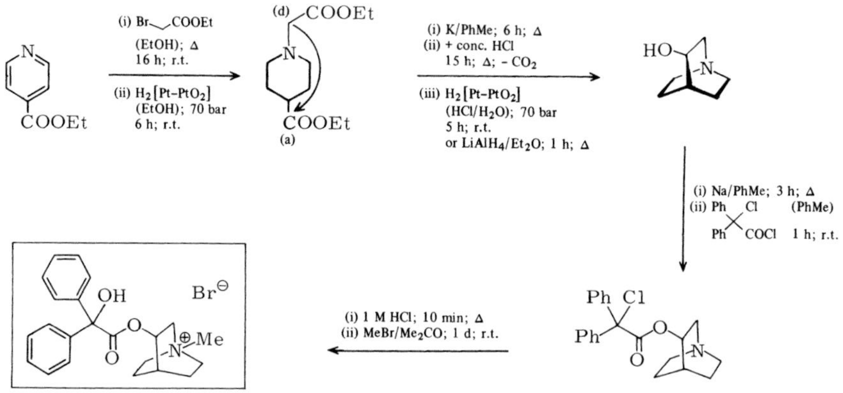
COOEt
(i) Br COOEt
(EtOH); Δ
16 h; r.t.
(ii) H2[Pt-PtO2]
(EtOH); 70 bar
6 h; r.t.
(d) COOEt
COOEt
(a)
(i) K/PhMe; 6 h; Δ
(ii) + conc. HCl
15 h; Δ; - CO2
(iii) H2[Pt-PtO2]
(HCl/H2O); 70 bar
5 h; r.t.
or LiAlH4/Et2O; 1 h; Δ
HO
(i) Na/PhMe; 3 h; Δ
(ii) Ph Cl (PhMe)
Ph COCl 1 h; r.t.
Ph Cl
Ph
(i) 1 M HCl; 10 min; Δ
(ii) MeBr/Me2CO; 1 d; r.t.
OH
Br⊖
⊕Me

clidinium bromide
(anticholinergic)

In the synthesis of commercial sulfur-heterocycles two interesting reactions are used: (i) diphenylamines may be connected by a sulfur bridge in the *ortho*-positions; (ii) the amino grouping of sulfonamides undergoes condensation reactions with neighboring imino- and amide groups.

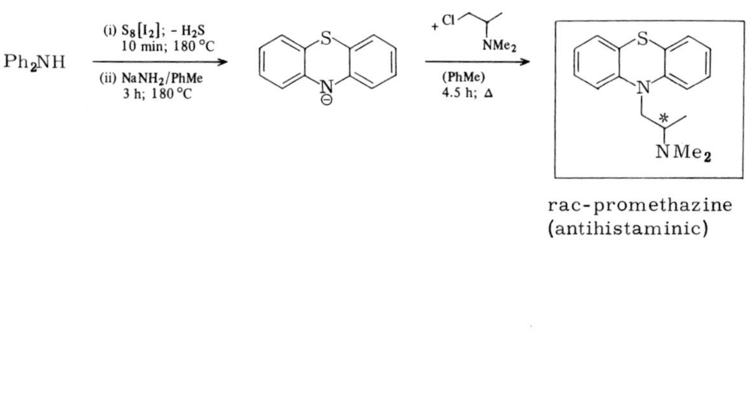
Ph2NH
(i) S8[I2]; - H2S
10 min; 180 °C
(ii) NaNH2/PhMe
3 h; 180 °C
S
N
+ Cl
NMe2
(PhMe)
4.5 h; Δ
NMe2

rac-promethazine
(antihistaminic)

prochlorperazine
(tranquilizer, antiemetic)

chlorothiazide
(diuretic, antihypertensive)

hydrochlorothiazide
(diuretic)

A large number of drugs contain functional or nonfunctional benzene derivatives, often with nitrogen-containing side-chains. Two special properties of this class of compounds are generally exploited: (i) both benzylic protons and nucleofuges (e.g. Br) can be substituted under relatively mild conditions; (ii) nucleophilic reactions at functional groups attached to a benzene nucleus are often highly regioselective, because the benzene nucleus is usually inert both to nucleophiles and to bases.

diphenhydramine
(antihistaminic)

(i) CH$_2$O/Me$_2$NH$_2^{\oplus}$ Cl$^{\ominus}$
(HCl/H$_2$O/EtOH)

(ii) opt. resolution:
(−)-dibenzoyl-
tartaric acid

(−)-(R)

(i) PhCH$_2$MgBr
(Et$_2$O); 3 h; r.t.

(ii)

(Py); 5 h; Δ

(+)-α-propoxyphene
(analgesic)

(i) NaNH$_2$/NH$_3$
1 h; −50 °C
+ Et$_2$O; 2 h; Δ

(ii) H$_2$SO$_4$/EtOH
1 d; Δ

EtOOC

(i) Et$_2$N⌒OH
(NaOR/PhMe)
Δ; − EtOH

(ii) ⌒SO$_3$H
+ ⌒SO$_3$H

−CH$_2$SO$_3^{\ominus}$]$_2$

caramiphene ethanedisulfonate
(anticholinergic, antitussive)

CHO

EtNO$_2$
[NaOEt]
(PhMe)

− H$_2$O

NO$_2$

(i) Fe
[FeCl$_3$]
conc. HCl

(ii) NH$_3$/HCN
(PrOH)
18 h; r.t.

H$_2$N CN

(i) opt. resolution:
(−)-camphor-
-10-sulfonate

(ii) conc. HCl
5 h; 130 °C; N$_2$
sealed tube

H$_3$N$^{\oplus}$ COO$^{\ominus}$

(−)-L-methyldopa
(antihypertensive)

A special problem arises in the preparation of secondary amines. These compounds are highly nucleophilic, and alkylation of an amine with alkyl halides cannot be expected to stop at any specific stage. Secondary amides, however, can be monoalkylated and hydrolyzed or reduced to secondary amines. In the elegant synthesis of phenylephrine, an intermediate β-hydroxy isocyanate (from a hydrazide and nitrous acid) cyclizes to give an oxazolidinone which is monomethylated. Treatment with strong acid cleaves the cyclic urethan.

CHO

(i) PhCH$_2$Cl
(K$_2$CO$_3$/Me$_2$CO)
10 h; Δ

(ii) Zn/Br⌒COOEt
(C$_6$H$_6$); 2 h; Δ

(iii) N$_2$H$_4$ · H$_2$O
(EtOH); 1 h; 50 °C

O⌒N$_2$H$_3$
HO

OBzl

(i) NaNO$_2$/
AcOH/H$_2$O
20 min; 0 °C

(ii) Δ; (C$_6$H$_6$)
− N$_2$

(iii) NaOMe
(MeOH)
evap.; r.t.

Na$^{\oplus}$

OBzl

(i) Me$_2$SO$_4$
(PhMe)
1 h; 100 °C

(ii) HCl/H$_2$O
5 h; 60 °C

(iii) optical
resolution:
(−)-tartrate

HN⌒Me
HO

OH

(−)-(R)-phenylephrine
(adrenergic)

Heterocycles containing the urea motif are of central importance to many drugs. Open-chain urea derivatives are also of interest. They may be obtained either from urea itself (barbiturates) or from amines and isocyanates. The latter are usually prepared from amines and phosgene with evolution of hydrogen chloride. Alkyl isocyanates are highly reactive in nucleophilic addition reactions. Even amides, e.g. sulfonamides, are nucleophilic enough to produce urea derivatives.

tolbutamide (hypoglycemic)

We conclude this section with short descriptions of partial syntheses or derivatizations of the ten top-selling drugs (Wang, 1999). The market for each one exceeds 1,5 billion dollars a year. Eight of them are mainly used by elderly people for the treatment of: cholesterol overproduction (No. 1), stomach trouble (Nos. 2 and 5), depression (Nos. 3, 9, and 10), high blood pressure (No. 4) and heart failure (No. 6). Most of these products are (i) quite expensive and (ii) used continuously over many years. Only two are equally useful for all age groups and are classic curing medicaments: No. 7 is an antiallergicum, No. 8 an antibioticum. We discuss their chemistry here, because they provide examples of special solutions to specific problems, which are connected with the economics of production or further development of already successful drugs

simvastin (No. 1)

omeprazole (No. 2)

fluoxetine (No. 3)

enalapril (No. 4)

ranitidine (No. 5)

amlodipine (No. 6)

loratadine (No. 7)

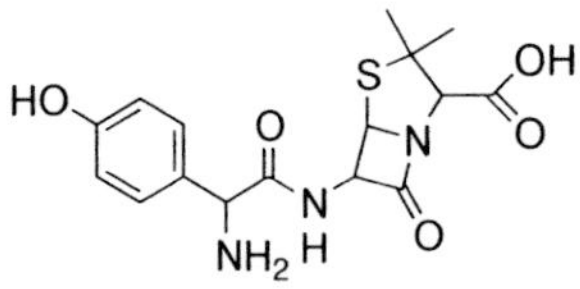

amoxicillin

(No. 8)

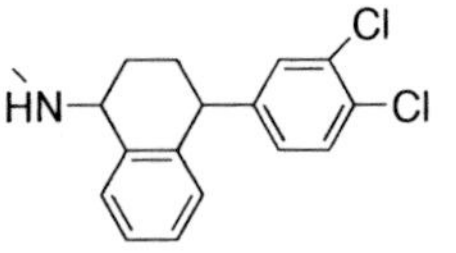

sertraline

(No. 9)

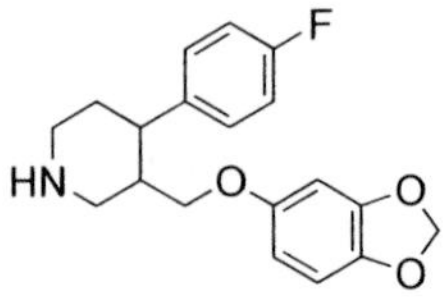

paroxetine

(No. 10)

No. 1: Simvastin is made from mevinolin, a natural product isolated from a soil fungus. It slows down the synthesis of cholesterol by obstructing a reductase. The only difference between simvastin and mevinolin is a second methyl group in an ester side-chain. This methyl group was initially introduced by de-esterification of the 2-methylbutyrate side-chain, protection of the 4-hydroxy group of the lactone, re-esterification with 2,2-dimethylbutyrate and deprotection of the 4-hydroxy group. The total yield of the conversion was 48%. The yield was increased to 87% by another four-step sequence using cyclopropylamine to protect the lactone, addition of a large excess of lithium pyrrolidide to block the OH-groups in the form of lithium imidates, and -methylation of the butyrate ester. Saponification of the cyclopropylamide and re-lactonization concluded the conversion (Thaper, 1999).

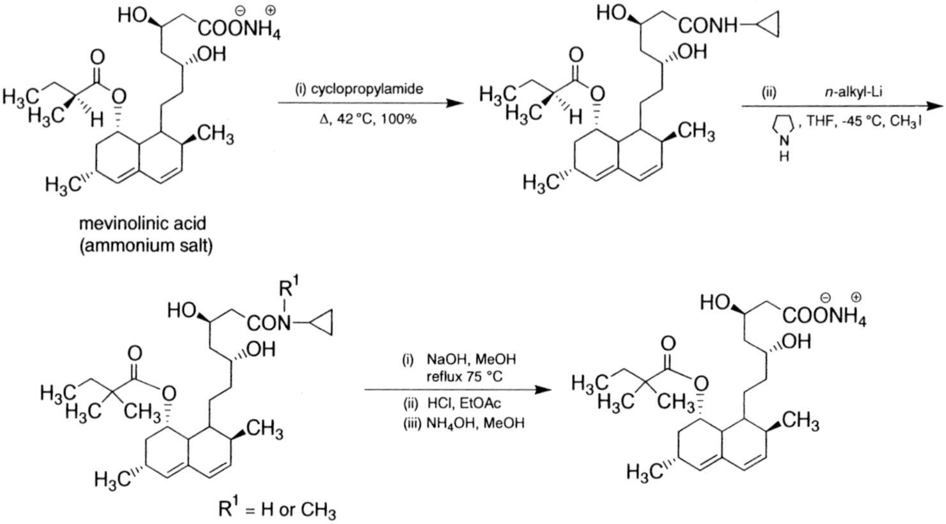

mevinolinic acid
(ammonium salt)

(i) cyclopropylamide

Δ, 42 °C, 100%

(ii)   *n*-alkyl-Li

, THF, -45 °C, CH₃I

(i)   NaOH, MeOH
      reflux 75 °C

(ii) HCl, EtOAc

(iii) NH₄OH, MeOH

R¹ = H or CH₃

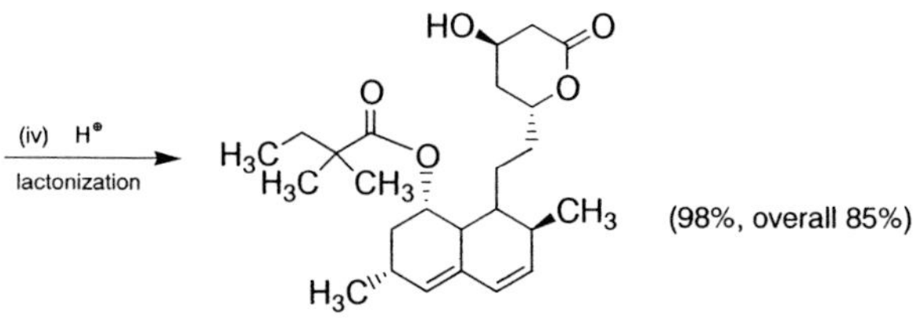

(iv) H$^\oplus$
lactonization

(98%, overall 85%)

No. 2: Omeprazole inhibits gastric acid secretion and blocks an ATPase. It is a synthetic 2-sulfinyl imidazole, which is made, for example, from substituted imidazoles (R_1–R_7) by treatment with *n*-BuLi, which removes the methane proton. The resulting anion then splits a disulfide. Mercapto-imidazoles are formed and the sulfur is then oxidized with peracids (Yamada, 1996).

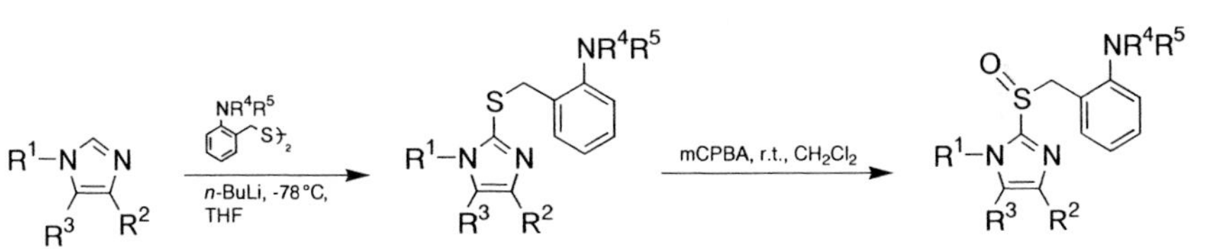

No. 3: Fluoxetine, an antidepressant which represses the uptake of serotonin, is made from methyl benzylamino-propiophenone (Mannich reaction of acetophenone and the amine) by reduction of the ketone, hydrogenation of the benzylamine and substitution with 4-trifluoromethylphenol (Wirth, 2000). The enantiomers are not separated.

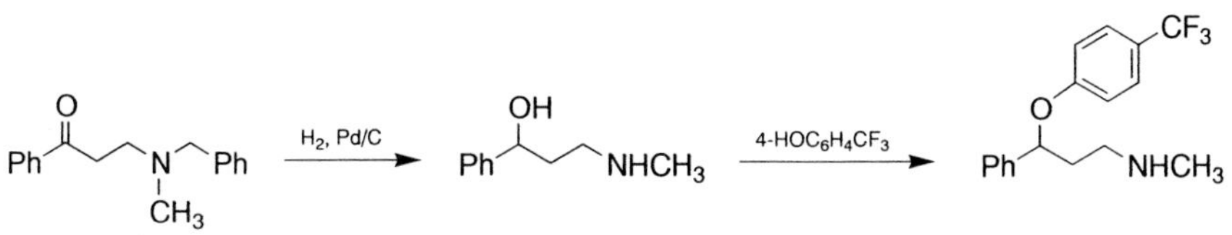

No. 4: Enalapril reduces blood pressure by blocking an enzyme. It is a dipeptide on the right side (Ala-Pro). The amino group of the L-alanine substitutes the bromide of an α-brominated ester (not shown).

No. 5: Ranitidine is another remedy for gastritis. It contains a furane ring and a peculiar ene-nitro group. It is obtained from dimethyl dithio-carbonic acid and nitromethane which produce 1,1-bismethylsulfanyl-2-nitroethylene. One of the sulfur atoms is subsequently substituted by two amines (Masereel, 1988; Gompper, 1967).

No. 6: Amlodipine blocks calcium channels in heart and brain neurons. Its 1,4-dihydropyridine ring is connected with a benzene ring on C-4. Optically active 1,4-dihydropyridines are obtained with asymmetric hydrogenation catalysts (Peri, 2000; Ashimori, 1991).

No. 7: Loratadine is an antihistamine containing a cycloheptane ring between two substituted benzene units. The dianion of the *tert*-butylamide given in the scheme below formed with 2 equivalents of *n*-BuLi at –40 °C, was selectively alkylated on carbon with 1 equivalent of methyl iodide. Adding an additional equivalent of *n*-BuLi followed by *m*-chlorobenzyl chloride, provided the *tert*-butyl amide. Dehydration to the nitrile occurred in refluxing POCl₃. It was treated with the Grignard reagent of 4-chloro-1-methyl piperidine, and acid hydrolysis yielded the piperidyl ketone. Friedel–Crafts cyclization using trifluoromethanesulfonic acid as catalyst gave the desired seven-membered ring (Piwinski, 1990; Kelly, 1998).

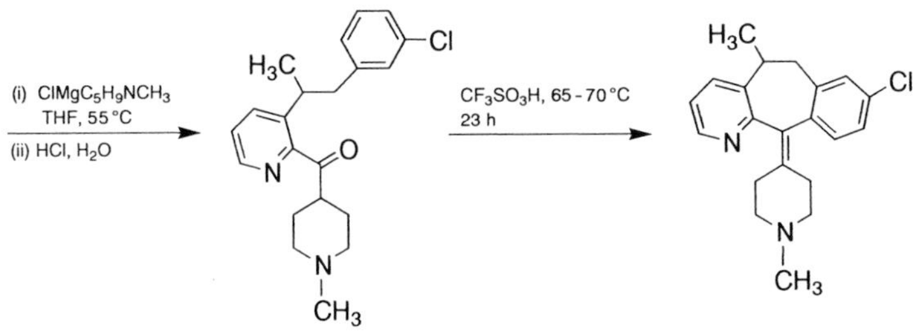
(i) ClMgC5H9NCH3
THF, 55 °C
(ii) HCl, H2O
H3C
Cl
O
N
N
CH3
CF3SO3H, 65 - 70 °C
23 h
H3C
Cl
N
N
CH3

No. 8: The antibiotic amoxicillin is accessible from fermentation processes. First penicillin G is made on a large scale and is then transesterified with an arylglycine in the presence of penicillin acylase (Bruggink, 1998). The enantiopure arylglycine can be obtained by asymmetric aminohydroxylation of styrene with a chloramine salt and Sharpless osmium catalysis (Reddy, 1998; see also Section 4.3).

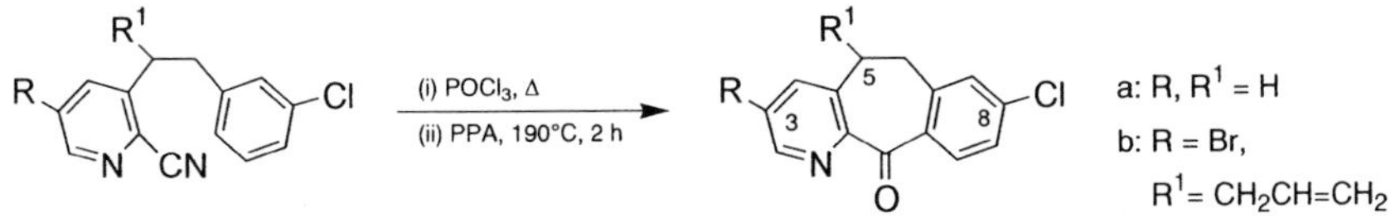
R1
R
Cl
N
CN
(i) POCl3, Δ
(ii) PPA, 190°C, 2 h
R1
R
5
3
8
Cl
N
O
a: R, R1 = H
b: R = Br,
R1= CH2CH=CH2

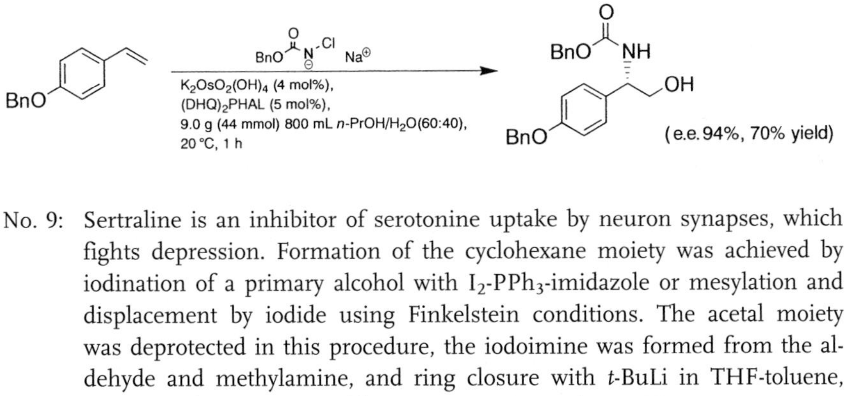
BnO
BnO
N
Cl
Na⊕
K2OsO2(OH)4 (4 mol%),
(DHQ)2PHAL (5 mol%),
9.0 g (44 mmol) 800 mL n-PrOH/H2O(60:40),
20 °C, 1 h
BnO
NH
OH
BnO
(e.e. 94%, 70% yield)

No. 9: Sertraline is an inhibitor of serotonine uptake by neuron synapses, which fights depression. Formation of the cyclohexane moiety was achieved by iodination of a primary alcohol with I_2-PPh_3-imidazole or mesylation and displacement by iodide using Finkelstein conditions. The acetal moiety was deprotected in this procedure, the iodoimine was formed from the aldehyde and methylamine, and ring closure with *t*-BuLi in THF-toluene, gave sertraline in 70% yield (Lautens, 1997; Davies, 1999).

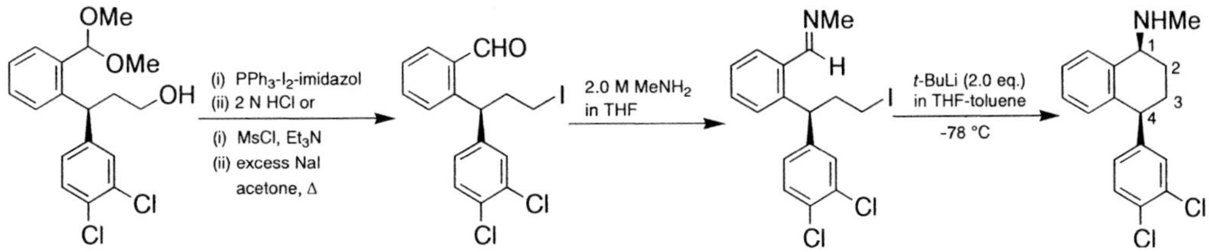
OMe
OMe
OH
Cl
Cl
(i) PPh3-I2-imidazol
(ii) 2 N HCl or
(i) MsCl, Et3N
(ii) excess NaI
acetone, Δ
CHO
I
Cl
Cl
2.0 M MeNH2
in THF
NMe
H
I
Cl
Cl
t-BuLi (2.0 eq.)
in THF-toluene
-78 °C
NHMe
1
2
3
4
Cl
Cl

No. 10: Paroxetine is the third antidepressant in the list. It also inhibits serotonin uptake. Its chiral piperidine ring is commercially made from amino acids (Bailey, 1998; Nadin, 1998). An enantioselective synthesis starts with the Michael addition of allylamine anions to a nitroalkene. The resulting ene-carbamate is cyclized to the lactam by removal of the protecting groups and hydrolysis of the enamine with HCl. Oxidation of the aldehyde to carboxylate with hypochlorite and esterification give an a,ω-nitroester. The nitro group was then reduced with hydrogen/Raney nickel and the a,ω-amino ester cyclized to the lactam which was reduced to the piperidine with LAH (Johnson, 2001).

4.3
Lactones (Oxetanones) and Lactams (Azetidinones)

Lactones and lactams are usually formed by intramolecular addition–elimination reactions of hydroxyl or amino groups to carboxylates. The pathway chosen by the amino nitrogen to the C=O double bond in addition reactions has been reconstructed from crystal structures. The intramolecular distance between the nucleophilic nitrogen atom and the electrophilic carbonyl carbon atom varies in three alkaloids, selected from the Cambridge data file from 2.56 Å (no binding) to 1.16 Å (covalent bond). The relative position of the two reactive groups, however, does not change. The final tetrahedral angle is kept at all distances from 3 to 1.5 Å. The weak N → CO interaction in protopin has the same geometrical ordering effect as the covalent bond in retusamin (Bürgi, 1973).

protopin   clivorin   retusamin

Baldwin (1976 B) introduced a set of rules for closure reactions of three- to seven-membered rings which are derived from arguments about "stereoelectronic control". Only cyclization reactions which involve the intramolecular attack of a nucleophile at an electron acceptor are considered. This may be tetrahedral (e.g. a bromide), trigonal (e.g. a carbonyl group as in the previous scheme or Michael-type C=C double bond) or digonal (an alkyne). The transition states which are first formed show the attacking nucleophile and the electron accepting group at bond angles of 180°, 109°, and 60° respectively. The chain that links the nucleophile and the electron accepting group may, however, restrict the relative motion of both reactants. The attainment of the transition state then requires distortion of bond angles and distances within the linking chain. In severe cases, the ring closure should be disfavored. Favored ring closures, on the other hand, are those in which the length and nature of the linking chain enables the terminal atoms to achieve the required trajectories to form the final ring bond. This generalization about selectivity in intramolecular reactions can be of help in synthesis design.

An example (Baldwin, 1976 C) is again given without referring to formal rules. A γ-hydroxy ester with a carbon–carbon double bond in conjugation with the ester group can undergo either lactonization or intramolecular Michael-type addition. The addition of the hydroxy group to the carbon–carbon double bond, however, is unlikely because the oxygen atom must approach the terminal carbon atom on a trajectory perpendicular to the plane of the C=C double bond (O–C–C angle ≈ 109°; Bürgi, 1973). The connecting chain of two methylene groups allows the correct distance between the reacting groups, but by no means the correct bond angles of the transition state. Lactonization, on the other hand, is without problems since the carbomethoxy group can be turned into a position which allows a transition-state-like trajectory of the hydroxyl group and the correct distance for bond formation. Experiments show that upon treatment with base only the expected lactone was formed, although the double bond is very susceptible to Michael-type additions with external nucleophilic reagents, e.g. methoxide.

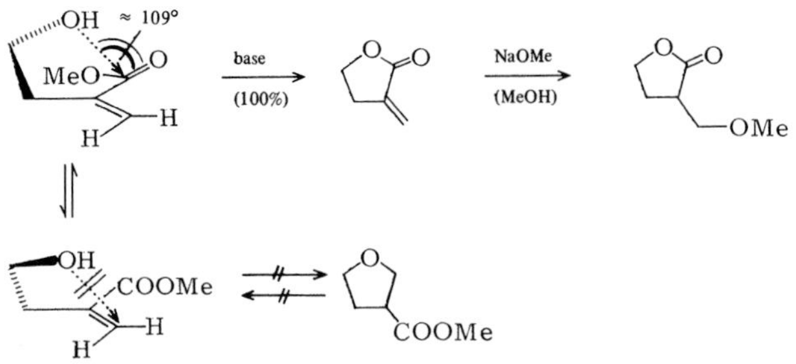

·······→ possible trajectory of the reacting groups

1,4-Disubstituted nitro compounds obtained from Michael additions of the d[1]-synthon nitromethane to electron-deficient carbon–carbon double bonds (a[3]-synthons) are useful precursors for lactams (Adams, 1979), because the CH–NO_2-group is transformed into an oxo group, if treated first with strong base and afterwards with strong acid (Nef reaction; W. E. Noland, 1955). Other examples of regio- and stereoselective Michael additions of d[1]-reagents are given below (Felner, 1967; Inhoffen, 1958, Gröbel, 1977 B).

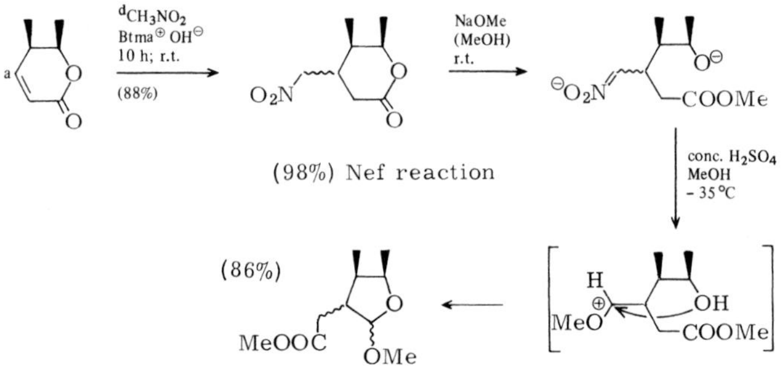

(98%) Nef reaction

(86%)

Nef reaction:

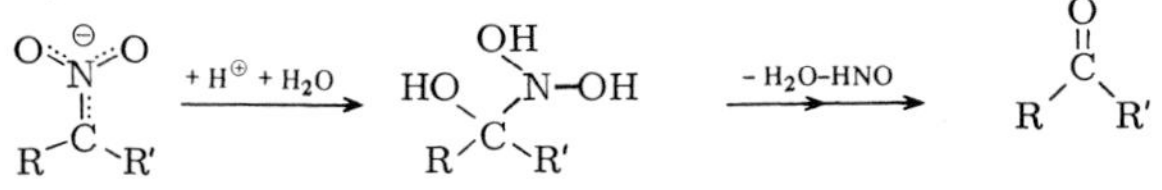

$$2\ HNO \longrightarrow N_2O + H_2O$$

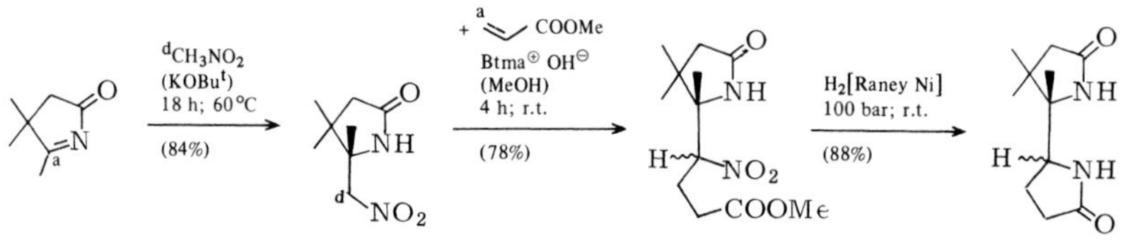

erythro : threo = 11 : 9

The addition of acetylides to oxiranes yields 3-alkyn-1-ols (Sondheimer, 1950; Adams, 1979; Carlson, 1974, 1975; Mori, 1976). Carboxylation with carbon dioxide or the use of propiolic acid then leads to lactones. The acetylene dianion and two a^1-synthons can also be used. 1,4-Diols with a carbon triple bond in between are formed from two carbonyl compounds (Jäger, 1977). The triple bond can be either converted to a *cis-* or *trans*-configurated double bond (Adams, 1979) or be hydrated to give a ketone. The regioselectivity of the addition of terminal alkynes to epoxides is improved, when the reagents prepared from the lithiated alkynes and either trifluoroborane of chlorodiethylaluminum are employed (Yamaguchi, 1983; Danishefsky, 1976). (Ethoxyethynyl)lithiumtrifluoroborane (1:1) is a convenient reagent for converting epoxides to γ-lactones (Nakatsuka, 1990; Danishefsky, 1976).

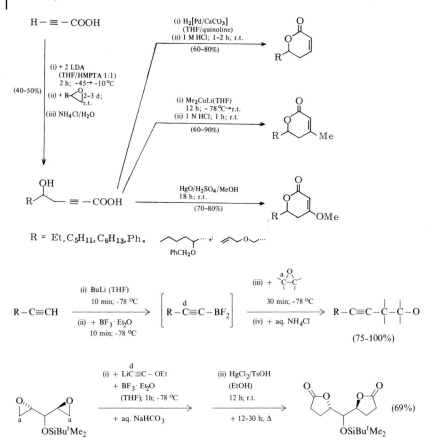

$R = Et, C_5H_{11}, C_6H_{13}, Ph,$ PhCH₂O

The active principle of the most commonly prescribed penicillin antibiotics is the β-lactam ring, which is under bond angle strain. These penicillins are called phenoxymethylpenicillin, ampicillin, and penicillin G.

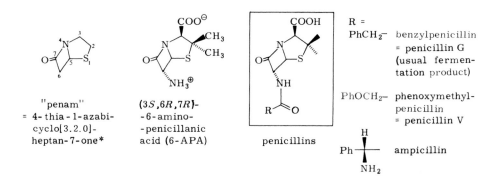

"penam"
= 4-thia-1-azabi-
cyclo[3.2.0]-
heptan-7-one*

(3S,6R,7R)-
-6-amino-
-penicillanic
acid (6-APA)

penicillins

R =
PhCH₂⁻ benzylpenicillin
= penicillin G
(usual fermen-
tation product)

PhOCH₂⁻ phenoxymethyl-
penicillin
= penicillin V

Ph—|— ampicillin
   NH₂

Under natural conditions various strains of *Penicillium* fungi produce either penicillin G or free 6-aminopenicillinic acid (= 6-APA). The techniques used to prepare analogs such as the ones given above have been: (i) fermentation in the presence of an excess of appropriate acids which may be incorporated as side-chains; (ii) chemical acylation of 6-APA with activated acid derivatives. 6-APA may be either obtained directly from special *Penicillium* strains or by hydrolysis of penicillin G with the aid of amidase enzymes. A major problem in the synthesis of different amides from 6-APA is the acid- and base-sensitivity of the β-lactam ring which is usually very unstable outside the pH range from 3 to 6. One synthesis of ampicillin applies the condensation of 6-APA with a mixed anhydride of N-protected phenylglycine. Catalytic hydrogenation removes the N-protecting group. Yields are low (≤30%; not shown). Another synthesis avoids the isolation of 6-APA and starts directly with penicillin G. Reaction with chloromethyl pivalate gives its pivaloyloxymethyl ester. This reacts with PCl$_5$ to an imidoyl chloride which may be solvolyzed with propanol. The acid chloride of (R)-phenylglycine is added to yield ampicillin pivaloyloxymethyl ester, or "pivampicillin", an ampicillin for oral application. After careful optimization of reaction conditions the cleavage of the β-lactam unit could be totally avoided (von Daehne, 1970; see also No. 8 of best-selling drugs).

pivampicillin

A similar cleavage of the amide linkage in natural cephalosporin C has been developed to synthesize 7-amino-cephalosporanic acid (=7-ACA). Treatment with nitrosyl chloride directly yields a cyclic imidic ester (presumably via an unstable diazonium salt) which can be hydrolyzed in the presence of weak acids. The carboxyl group need not be protected since it reacts only slowly with nitrosyl chloride

(Morin, 1962). Phosphorus pentachloride can also be used but requires low temperature. The carboxyl groups are protected by O-silylation (Fechtig, 1968) and deblocked upon acidic hydrolysis. Condensation with N-protected phenylglycine, e.g. by the mixed anhydride method, and removal of the protecting group by mild acid treatment after hydrogenolysis of the allylic 3'-acetoxy group yields the commercial antibiotic cephalexin (Spencer, 1966; Ryan, 1969).

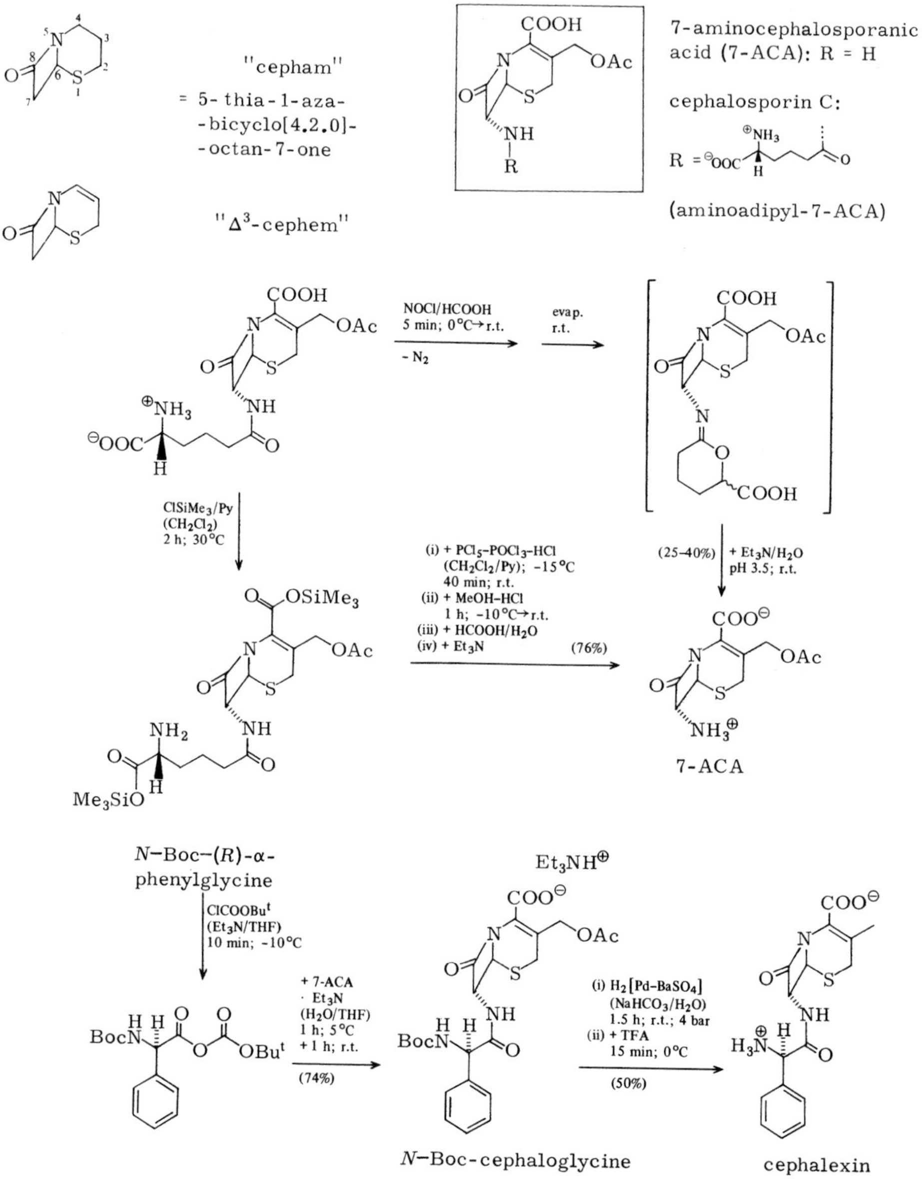

"cepham"

= 5- thia-1-aza-
-bicyclo[4.2.0]-
-octan-7-one

"Δ^3-cephem"

7-aminocephalosporanic
acid (7-ACA): R = H

cephalosporin C:

(aminoadipyl-7-ACA)

N−Boc−(R)−α−
phenylglycine

N−Boc−cephaloglycine

cephalexin

$n\text{-}C_5H_{11}CHO$ + (CH₂=CH-CO₂Et) $\xrightarrow[\text{THF, }t\text{-BuOH}]{\underset{0\,°C,\ 10\,h}{2\ SmI_2}}$ (lactone with $n\text{-}C_5H_{11}$) + diastereomer

70:30   (70%)

A very mild and efficient synthesis of N-substituted β-lactams uses the Mitsunobu reaction for the ring closure of seryl dipeptides protected at the terminal N as 4,5-diphenylosazol-1(3H)-one ("Ox") derivatives.

The synthesis of five-, six-, and seven-membered cyclic esters or amides uses intramolecular condensations under the same reaction conditions as described for intermolecular reactions. Yields are generally excellent. An example from the colchicine synthesis of E.E. van Tamelen (1961) is given below. The synthesis of macrocyclic lactones (macrolides) and lactams ($n>8$), however, which are of considerable biochemical and pharmacological interest, poses additional problems because of competing intermolecular polymerization reactions. Inconveniently high dilution, which would be necessary to circumvent this side-reaction, can be avoided if both the hydroxyl and the carboxyl groups are activated for mutual interaction before the actual condensation occurs. One way to achieve this is the use of a basic substituent bound to the active carboxyl derivative, which seizes and activates the remote hydroxyl group by hydrogen bonding. Successful derivatives of this kind are S-(2-pyridyl) and S-(2-imidazolyl) thioesters. If silver ions (AgClO$_4$, AgBF$_4$) are used to activate further the 2-pyridinethiol esters by complexation, cyclization occurs at room temperature in benzene solution. The best reagents to form the thioesters of 2-mercapto-pyridine and -imidazole with carboxylic acids are the corresponding disulfides together with triphenylphosphine (Nicolaou, 1977).

(i) DCC/Py; 1 d; r.t.
(ii) CH₂N₂(Et₂O/MeOH)
(iii) CC(Al₂O₃–Me₂CO/PE)

(41%)

Ph₃P/
PySSPy
(toluene)
5 h; r.t.; Ar

– Ph₃PO
– PySH

15 h addition
(high dilution)
Δ; (toluene); Ar
+10 h; Δ

$(CH_2)_{14}$ (80%)

(5%)

+ "trilide"   (1%)

(i) Ph₃P/PySSPy
(C₆H₆); 0.5 h; r.t.

(ii) slow addition to
AgClO₄/MeCN
3 h; 65 °C

(75%)

Some reactions proceed with enantiotopic group selectivity in the sense that a kinetic resolution is coupled to an initial asymmetric reaction. An example is the enzyme-catalyzed partial hydrolysis of achiral meso-diol diacetate esters to chiral, optically pure monoesters (Y.-F. Wang, 1984). The pro-S group of the diacetates is preferentially cleaved by pig pancreatic lipase. The other group is cleaved somewhat more slowly ($k_{fast}/k_{slow} = 15.6$). If one plots the contents of the di- and mono-acetates as a function of the diol content in the mixture, one finds that the "wrong" monoacetate disappears quickly, because the enzyme hydrolyzes its remaining "right" acetate group rapidly. If one waits until about 40% of diol is formed, the "right" monoacetate is essentially pure (e.e. $\approx 100\%$) and the diacetate has disappeared. The pure enantiomer of the monoacetate now only needs to be separated from the diol. Non-enzymatic reactions follow similar routes from achiral to optically pure chiral products if asymmetric reactions are coupled with kinetic resolution.

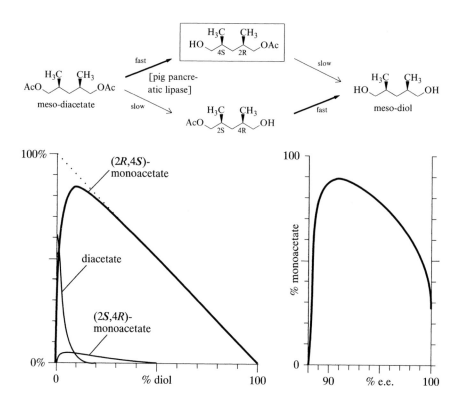

An example of a condensation reaction with diastereotopic group selectivity (see Appendix) is the lactonization of either of the two racemic, C-2-symmetrical 5-hydroxy-2,4,6,8-tetramethylnonanedioic acids. This reaction "desymmetrizes" a compound with two pairs of equivalent stereocenters (C-5 is achiral!) into a product with five different stereocenters. The trilithium salt is generated by reduction of the 2,4,6,8-tetramethyl-5-oxononanedioic acid with lithium in ammonia and then either acidified to pH $\approx$ 3 followed by rapid extraction of the lactones formed (kinetic control) or equilibrated at pH $\approx$ 1 to the lactone mixture (thermodynamic control). Depending on the steric interactions in the "chair-like" transition states and in the "half-chair" lactone products, either kinetic or thermodynamic control leads with high diastereoselectivity to the lactone with *trans*-configuration at C-5 and C-6 of the tetrahydro-2*H*-pyran-2-one ring (Hoye, 1984).

	R	R'	pH $\approx$ 3, rapid workup	pH $\approx$ 1, equilibration
	Me	H	55% trans	94% trans
	H	Me	99.7% trans	81% trans

pre-closure conformations
R" = CH$_2$–CHMe–COOH
$\rightarrow\leftarrow$ = repulsive forces

pre-cis   favorable for R = Me, R' = H   favorable for R = H, R' = Me   pre-trans

Metal complexes of synthetic macrocycles are of topical interest, since there is some hope that useful reactions found in nature (e.g. photolysis of water with sunlight; catalysis of hydrocarbon oxygenation; ion transport through membranes) can be reproduced with these easily accessible compounds. Various books (e.g. Melson, 1979; Voegtle, 1991; Cooper, 1992) provide detailed knowledge of this subject. We shall discuss some general problems in the synthesis of macrocycles together with specific examples for porphyrin-like tetraaza ligands and alkali-ion binding polyethers. The cyclization reactions discussed here either involve an intramolecular reaction or a donor group D with an acceptor group A or a cyclizing dimerization of two molecules with two terminal acceptors and two donors. A polymerization reaction will always compete with cyclization.

Intramolecular condensation reactions leading to macrolides (= macrocyclic lactones) or macrocyclic lactams can be favored by (i) intramolecular salt formation, (ii) high dilution techniques, (iii) the action of a template which forces the reacting ends together, and (iv) by alternating condensation and ring-opening hydrolysis reactions

Intramolecular salt formation and stiffness of the chain help in lactam formation catalyzed by N'-(dimethylaminopropyl)-N-ethylcarbodiimide methoiodide (EDCI) in aprotic, polar solvents (Cao, 2000).

Synthesis by high-dilution techniques requires slow admixture of reagents ($\approx$ 8–24 hrs) or very large volumes of solvents ($\approx$ 100 L mmol^{-1}). Fast reactions can also be carried out in suitable flow cells (J. L. Dye, 1973). High-dilution conditions have been used in dilactam formation from 1,8-diamino-3,6-dioxaoctane and 3,6-dioxaoctane-dioyl dichloride in benzene. The amide groups were reduced with lithium aluminum hydride to amines and a second cyclization with the same dichloride was then carried out. The new bicyclic compound was reduced with diborane. This ligand envelops metal ions completely and is therefore called a "cryptand" (Dietrich, 1969).

Another bicyclic compound with two condensed macrocycles was constructed as described above. Its secondary amino groups were deblocked by reductive detosylation and connected by a third cyclization with a diacid dichloride, again under high-dilution conditions. Upon reduction a tricyclic cryptand was obtained with four nitrogen bridgeheads, which could form a tetrahedral ligand field, and six octahedrally arranged ether bridges (Graf, 1975).

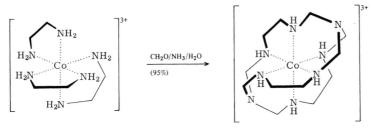

The expression *"template reaction"* indicates mostly a reaction in which a complexed metal ion holds reactive groups in the correct orientation to allow selective multi-step reactions. The template effect of the metal is twofold: (i) polymerization reactions are suppressed, since the local concentration of reactants around the metal ion is very high; (ii) multi-step reactions are possible, since the metal holds the reactants together. In the following one-step synthesis, eleven molecules (three ethylenediamine = "en", six formaldehyde, and two ammonia molecules) react with each other to form one single compound in a reported yield of 95%. It is obvious that such a reaction is dictated by the organizing power of the metal ion (Creaser, 1977).

clathrochelate, cryptate

Templating works best if the metal ion is bound solidly to the educt and even more tightly to the product. A common disadvantage of many template reactions is that it is often difficult to remove the metal ion. Such syntheses are therefore *in situ* syntheses of metal complexes, and can only occasionally be used for the synthesis of the metal-free ligands.

In 1,3-dicarbonyl compounds a special situation arises, because the prototropic ketoenols are formed. Both carbonyl groups are deactivated as acceptor synthons since neither the enol nor the α-β-conjugated ketone are very electrophilic. It is usually possible to condense one of the keto groups with amines. Acetylacetone and ethylenediamine will, for example, give bis(acetylacetone) ethylenediimine. The second keto group, however, will not react. If one wants to produce macrocyclic compounds by two nucleophilic reactions on the carbonyl groups, one has to activate them. In a successful synthesis of this kind acetylacetone was converted to its enamine and O-alkylated with Meerwein's reagent. Treatment with one equivalent of ethylenediamine and subsequently with two equivalents of sodium methoxide and a second equivalent of diamine gave the desired macrocycle in 35% yield. O-Alkylation made the carbonyl group more electrophilic, and the ketimino group is also more reactive toward nucleophiles than a keto group. The resulting macrocyclic ligand was then metallated with nickel(II) acetate. Hydride abstraction by the strongly electrophilic trityl cation and proton elimination resulted in the formation of carbon–carbon double bonds (Truex, 1972).

The "zip-reaction" (Kramer, 1978, 1979) leads to giant oligoamino macrocycles. Potassium 3(aminopropyl)amide = "KAPA" ("superbase") in 1,3-diaminopropane is used to deprotonate amines. The amide anions are highly nucleophilic and may, for example, be used to transamidate carboxylic amides. If N-(39-amino-

4,8,12,16,20,24,28,32,36-nonaazanonatriacontyl)dode canolactam is treated with KAPA, the amino groups may be deprotonated and react with the macrocyclic lactam. The most probable reaction is the intramolecular formation of the six-membered ring intermediate indicated in the scheme below. This intermediate opens spontaneously to produce the azalactam with seventeen atoms in the cycle. This reaction is repeated nine times in the presence of excess KAPA, and the 53-membered macrocycle is formed in reasonable yield.

(38%)

The reaction sequence is successful because reverse ring-contraction reactions are unlikely with polyanions.

4.4
Porphyrins, Chlorophyll a, and Corrins

Porphyrins and chlorophylls are the most widespread natural pigments. They are associated with the energy-converting processes of respiration and photosynthesis in living organisms, and the synthesis of specific porphyrin derivatives is often motivated by the desire to perform similar processes in the test tube. Metalloporphyrins are also redoxactive. Magnesium porphyrinates, for example, donate electrons and form cation radicals, tin(IV) porphyrinates pick up electrons to become anion radicals. Long-distance dimers are therefore optimized for charge separation (Skupin, 2001; Li, 2002). Furthermore the central ions are extremely electron-rich and may activate oxygen molecules or bind tightly to all kinds of nitrogen- or sulfur-containing ligands. A major factor for the popularity of synthetic metallo-porphyrin systems lies in the fact that visible absorption and fluorescence spectra are extremely sensitive to substitution patterns, central metal ions, and changes of the environment. It is for these reasons that we discuss the synthesis of porphyrins and largely omit other heterocycles of biological importance such as the flavins, nicotinamides, lipoic acid, etc. At present porphyrins seem to be promising materials for the construction of useful molecular machinery, possibly in connection with fullerenes (see Chapter 8). The chlorophylls behave similarly to porphyrins, but carry the light-absorption and electron-transfer processes to lower energies. The structurally and biosynthetically related corrins (e.g. vitamin B_{12}) catalyze alkylations and rearrangements of carbon skeletons via organocobalt intermediates.

If one heats acetone and pyrrole in the presence of catalytic amounts of acid, so-called "acetone pyrrole" is formed in over 80% yield. This colorless, macrocyclic compound contains four pyrrole units, which are connected by four dimethyl-methylene bridges. It is formed by electrophilic a-substitution of pyrrole by acetone, acid-catalyzed oligomerization, and spontaneous, non-template cyclization wherein four pyrrole units are combined. The reason for internal reaction instead of chain elongations with more acetone and pyrrole units is a purely statistical one: the intramolecular reaction is more probable (Mauzerall, 1960; Fuhrhop, 1974). No dilution technique is needed in this synthesis, presumably because the alkylated pyrrole unit at the chain end reacts faster than the unsubstituted pyrrole. Acetone pyrrole has recently been renamed calixphyrene and all kinds of expanded macrocycles have been synthesized, which bind anions in organic solvents (Bucher, 2001).

[HCl, MesOH, or TosOH]
acetone; 1–2 d; r. t.

"acetonepyrrole"
($\approx 90\%$)

Acetone pyrrole belongs to the class of tetrapyrrolic macrocycles in which the pyrrole units are connected by sp^3-hybridized carbon bridges. Such compounds are called porphyrinogens. Since C–C bonds are not cleaved easily, aromatization does not occur spontaneously and acetone pyrrole is stable. Similar porphyrinogens are formed if the α-aminomethyl-substituted pyrrole porphobilinogen is dissolved under anaerobic conditions in slightly acidic water. Since this pyrrole, which is the precursor of almost all tetrapyrrole pigments in nature, contains two different β-pyrrolic substituents a mixture of four porphyrinogen isomers is formed, which differ in the arrangement of the β-pyrrolic substituents. If this mixture is treated with an oxidant, e.g. oxygen, iodine, or a high-potential quinone, the methylene bridges are oxidized to form methane bridges, and red porphyrins are obtained in almost quantitative yield. The inner ring containing the methane bridges and nitrogen atoms is an 18π-electron system and therefore a stable aromatic system. Inner rings with four imine nitrogens (dehydroporphyrin) would only contain 16π-electrons and appear as a triplet state. Such a porphyrin has never been obtained so far. Since porphyrinogens are oxidized in air and tend to isomerize in protic solutions, they are usually prepared from the corresponding porphyrins by metal reduction (Zn/AcOH) or catalytic hydrogenation immediately before use, e.g. in biosynthetic studies. Pure uroporphyrinogen I can be obtained anaerobically from porphobilinogen by the action of a deaminating enzyme (uroporphyrinogen I synthetase, Smith, 1975; Battersby, 1979; Mauzerall, 1963).

porphobilinogen   uroporphyrinogens I–IV   uroporphyrins I–IV

*pH 0:   1 M HCl; 0.5 h; 98 °C ⟶ statistical ratio = 1:1:4:2;   (78%)
pH 7.6: phosphate buffer; 21 h; 60 °C ⟶ mainly I + III;   (55%)
pH 10: 0.001 M NaOH; 2 h; 98 °C ⟶ selectively I;   (70%)

A = –CH_2–COOH
P = –CH_2–CH_2–COOH

Because of the occurrence of isomers, the synthesis from pyrroles is only useful for porphyrins with eight identical β-pyrrolic and four identical methane bridge substituents. Important examples are the syntheses of chloroform-soluble β-octaethyl-porphyrin and *meso*-tetraphenylporphyrin which have been used in unnumerable studies on porphyrin reactivity (Smith, 1975). Porphyrins with four long *meso*-alkyl side-chains can be obtained by use of analogous reactions. These porphyrins have melting points below 100 °C and are readily soluble in petroleum ether. Sulfonation of olefinic double bonds leads to highly charged, water-soluble porphyrins (Fuhrhop, 1976).

β-Alkyl substituted meso-tetraphenylporphyrins with functional groups on the phenyl groups are particularly easy to obtain. The phenyl groups orient in a more or less perpendicular orientation to the porphyrin ring. This puts its ortho- and meta-substituents above and below the porphyrin plane. Substituents on one side may then be used to fixate the porphyrin on metal electrodes or colloidal particles, while substituents on the other side can interact with reagents in the bulk solvent or water phase.

This apparently extremely simple synthesis of symmetrical porphyrins from aldehydes and pyrrole has also been used to produce porphyrins with interesting stereochemical properties. *ortho*-Nitrobenzaldehyde yielded the corresponding tetrakis(nitrophenyl) porphyrin which was reduced to the tetraamine and converted into a tetraamide using pivaloyl chloride. Since the porphyrin macrocycle tends to retain a planar conformation, and because of strong steric interactions between β-pyrrolic and methane bridge substituents, the planes of the phenyl groups are forced into a position approximately perpendicular to the porphyrin plane. This minimizes interactions between β-pyrrolic H-atoms and phenyl substituents. If the phenyl substituents are large, e.g. pivaloylamide in the *ortho*-position, rotation of the phenyl ring is strongly impeded and rotational isomers are stable. The *all-cis* isomer is the famous "picket fence" porphyrin (Collman, 1975 B, 1977). Another, so-called "capped" porphyrin with substituents on one side of the porphyrin plane was obtained in 3% yield from a tetraaldehyde and pyrrole (Almog, 1975).

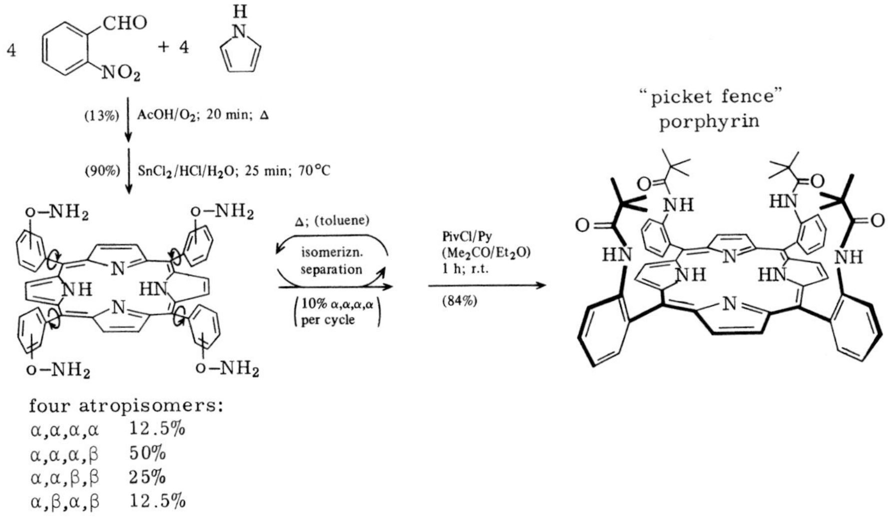

"picket fence" porphyrin

(13%) | AcOH/O₂; 20 min; Δ

(90%) | SnCl₂/HCl/H₂O; 25 min; 70 °C

o–NH₂   o–NH₂

Δ; (toluene)

isomerizn. separation

(10% α,α,α,α per cycle)

PivCl/Py (Me₂CO/Et₂O) 1 h; r.t.

(84%)

o–NH₂   o–NH₂

four atropisomers:
α,α,α,α   12.5%
α,α,α,β   50%
α,α,β,β   25%
α,β,α,β   12.5%

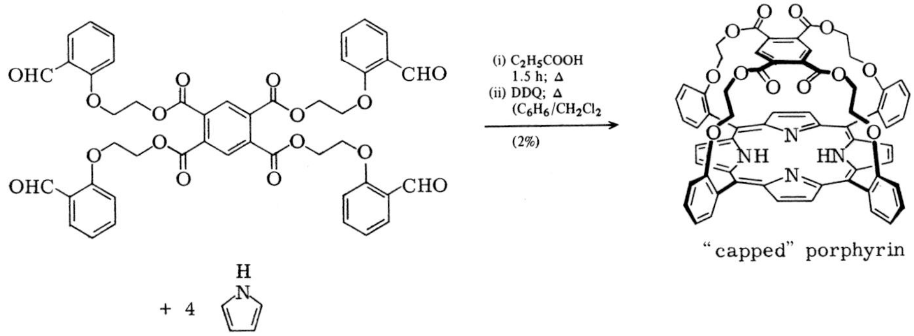

(i) C₂H₅COOH 1.5 h; Δ
(ii) DDQ; Δ (C₆H₆/CH₂Cl₂)

(2%)

"capped" porphyrin

+ 4

The precursor for substituted *meso*-tetraphenylporphyrins is a substituted benzaldehyde. The formyl group is often obtained by oxidation of a methyl or reduction of a carboxyl group or reduction of esters to the alcohol and its oxidation to the aldehyde. Three examples are given in the scheme below. One porphyrin carries two bromine groups and one *tert*-butyl group on each *meso*-phenyl substituent (Nakagawa, 2001), the other four ferrocene units at a long distance (Gryko, 2000). The iodobenzaldehyde can, of course, be modified by Heck reactions.

Double picket fence porphyrins were synthesized from 4,6-dichloro-pyrimidine-5-carbaldehyde, which was prepared using standard Vilsmeier conditions (DMF/POCl₃) for chloroformylation on 4,6-dihydroxypyrimidine. The chlorine functionalities on the pyrimidine ring are highly activated toward nucleophilic substitution, allowing the introduction of substituents at the aldehyde as well as the porphyrin stage. Functionalization of the two chlorine atoms of the aldehyde was then carried out under reflux in THF with several substituted phenolates in presence of K₂CO₃. This yielded the 4,6-disubstituted pyrimidine-5-carbaldehydes in excellent yields. These aldehydes were then reacted with pyrrole under conditions of high dilution (Smeets, 2000).

Conformationally constrained trimers and pentamers exhibit electronic spectra that help to verify models for energy transfer reactions, such as the exciton coupling theory (Kasha, 1965), and can lead to systems capable of useful charge separation.

= octaalkylporphyrin unit

The syntheses of these oligomers are surprisingly simple: HBr-catalyzed 1:1 condensations of linear tetrapyrrole precursors (10,23-dihydro-21H-bilines) with aromatic dialdehydes gave 5-(formylaryl)porphyrins in $\approx 70\%$ yield, which were used as key precursors for syntheses of di-, tri-, and pentamers. Porphyrin dimers were obtained by 1:2 condensation of the corresponding dimethyl acetals with a 5-unsubstituted ethyl pyrrole-2-carboxylate and reduction to give a 1,9-bis(hydroxymethyl)dipyrromethane group which was cyclized with a dipyrromethane in yields below 10% (Maruyama, 1989). The same authors later improved their procedures considerably and reported 30–60% yields of porphyrin trimers obtained by 2:2 condensations of 5-(formylaryl)porphyrins with dipyrromethanes (Nagata, 1990). Similar 1:1:2 condensations of a 5-(formylaryl)porphyrin and an aromatic dialdehyde monoacetal with a dipyrromethane, followed by acetal hydrolysis, giving mono(formylaryl) bis-porphyrins, and another 2:2 condensation with a dipyrromethane, furnished up to 10% overall yields of porphyrin pentamers (Nagata, 1990). The choice of acid catalyst and solvent – trichloroacetic acid and acetonitrile – was responsible for the high yields achieved in these syntheses. In a similar synthesis of a 1,8-anthrylene-linked porphyrin trimer using *p*-toluenesulfonic acid in methanol the yield amounted to only 5% (Abdal-muhdi, 1985).

The syntheses given are also useful for connecting porphyrins with other chromophores and reactive groups, e.g. quinines. If the reported yields are reproducible, large electron donor–acceptor "supramolecules" should become accessible on a large scale.

Monoethynyl porphyrin building blocks, accessible via mixed condensation of mesitaldehyde, 4-[2-(trimethylsilyl)ethynyl]benzaldehyde, and pyrrole, required lengthy purification. To facilitate separation of the desired monoethynyl porphyrin the more polar 2-hydroxyisopropyl unit was employed as the ethyne protecting group. A mixed-aldehyde condensation of mesitylaldehyde, *p*-acetylene benzaldehyde and pyrrole at a high concentration using BF$_3$-ethanol cocatalysis gave a mixture of porphyrins. The desired monoethynyl porphyrin containing only one OH-group was readily isolated by chromatography. Subsequent cleavage of the hydroxyisopropyl group afforded the desired porphyrin in gram quantities. The corresponding Zn or Mg chelates were obtained by reaction at room temperature either with Zn(OAc)$_2$ or with MgI$_2$ in the presence of DIEA (heterogeneous magnesium insertion conditions). A Glaser coupling reaction of an equimolar mixture of zinc- and magnesium porphyrinate was performed using the standard conditions plus Pd-mediation. As expected from this statistical reaction, three diphenyl-butadiyne(pbp)-linked porphyrins (Mg-pbp-Mg, Zn-pbp-Mg, and Zn-pbp-Zn) were formed. They were easily separated by chromatography owing to the significant polarity difference imparted by the respective metals, affording the desired dyad Zn-pbp-Mg in 46% yield. Mg porphyrins were demetallated by treatment with silica gel, while Zn porphyrins were not. Treatment of Zn-pbp-Mg with silica gel in CHCl$_3$ at room temperature caused selective demetallation of the magnesium porphyrin, affording the desired Zn-pbp-Fb (Fb=freebase) dyad in 84% yield. In this manner, the Mg chelate serves as (i) a polar site that facilitates separation of the hybrid product and (ii) a protective group that upon removal affords the linear heterodimer, which may again be metallated (Youngblood, 2002). Dynamic oligoporphyrin mixtures were established by Sanders using different metal–ligand interactions (Kim, 1999).

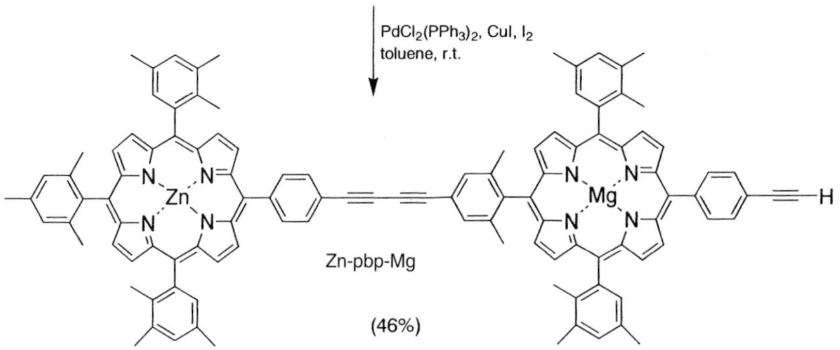

Zn-pbp-Mg

(46%)

In the [2+2] condensation reactions of a naphthyl-1,2-diketone with two dipyrro-methane molecules outlined in the scheme below, the overall yield was up to 26%. The molar ratio of the atropisomers was strongly dependent on the electronic and steric nature of the dipyrromethane starting materials as well as the reaction conditions (nature of the acid, dilution factors, and solvents) (Harmjanz, 2001).

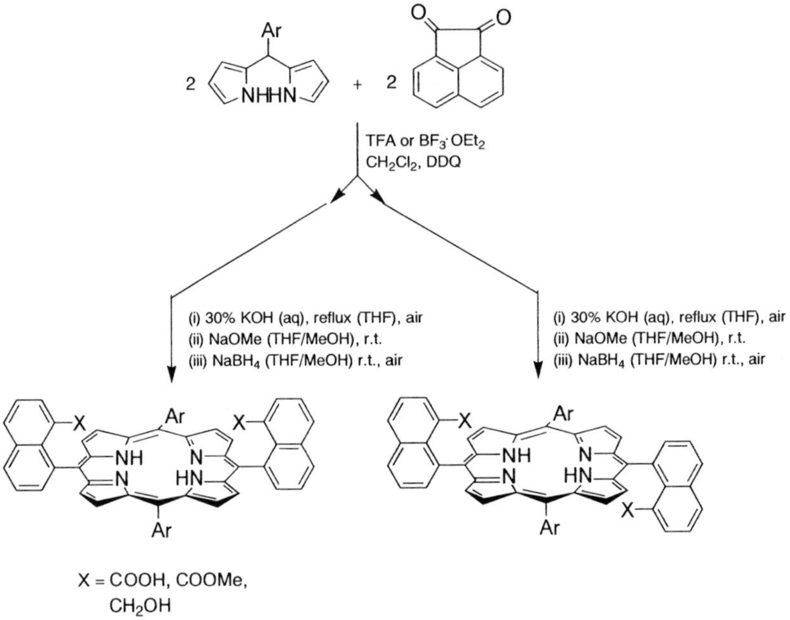

X = COOH, COOMe,
   CH₂OH

5,15-Diphenyl-10-bromo-porphyrin was converted to the Suzuki-borane and cross-coupled in DMF with a 15,20-dibromoporphyrin in the presence of a catalyt-ic amount (10 mol%) of Pd(PPh₃)₄ and 1.5 equivalents of Cs₂CO₃ at 80 °C under an inert atmosphere to give a trimer in 46% yield (Aratani, 2001).

Naturally occurring porphyrins are usually symmetrically substituted about the 15-methine bridge. These porphyrins can be synthesized by the condensation of two dipyrrolic intermediates. Typical dipyrrolic intermediates in current use are the dipyrromethanes and the dipyrromethenes. Both methods will shortly be described. This again is a highly specialized topic, discussed here mainly because general problems in the synthesis of complex aromatic molecules and the influence of π-electron distribution in the educts on reactivity can be exemplified.

Unsymmetrically substituted dipyrromethanes are obtained from a-unsubstituted pyrroles and a-(bromomethyl)pyrroles in hot acetic acid within a few minutes. These reaction conditions are relatively mild and the a-unsubstituted pyrrole may even bear an electron withdrawing carboxylic ester function. It is still sufficiently nucleophilic to substitute bromine or acetoxy groups on an a-pyrrolic methyl group. Hetero atoms in this position are extremely reactive leaving groups since the a-pyrrolylmethenium (= "azafulvenium") cation formed as an intermediate is highly resonance-stabilized.

X = NH$_2$, OH, OAc, OTos, halogen, etc.

Dipyrromethanes are inherently unstable toward "jumbling" or "redistribution" in the presence of acidic reagents. The dipyrromethanes are easily protonated at a-pyrrolic positions, and rearrangements occur through the relatively long-lived pyrrolylmethenium ions. This process is not only found with dipyrromethanes but is equally pronounced in porphyrinogens and open-chain tetrapyrrole pigments with methylene bridges (Jackson, 1973).

P = CH$_2$–CH$_2$–COOH
A = CH$_2$–COOH

A mild procedure which does not involve strong acids, has to be used in the synthesis of pure isomers of unsymmetrically substituted porphyrins from dipyrromethanes. The best procedure that has been applied, e.g. in unequivocal syntheses of uroporphyrins II, III, and IV, is the condensation of 5,5'-diformyldipyrromethanes with 5,5'-unsubstituted dipyrromethanes in a very dilute solution of hydriodic acid in acetic acid (Jackson, 1973). The electron withdrawing formal groups disfavor protonation of the pyrrole and therefore isomerization. The porphodimethene that is formed during short reaction times isomerizes only very slowly, since the pyrrole units are part of a dipyrromethene chromophore (see below). Furthermore, it can be oxidized immediately after its synthesis to give stable porphyrins.

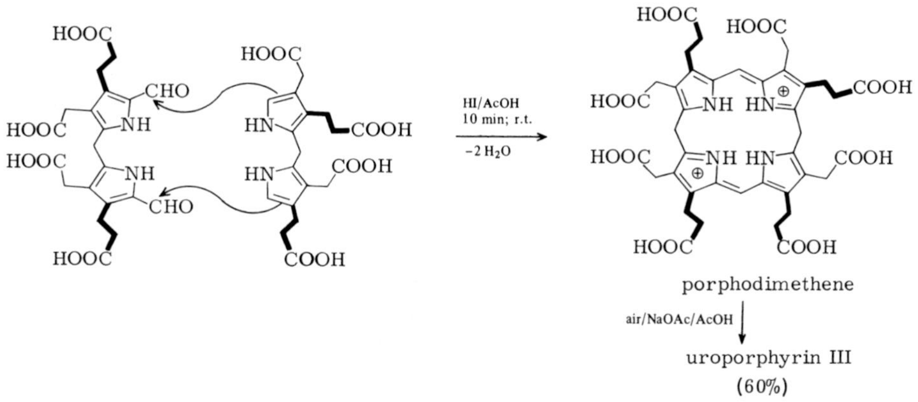

porphodimethene

uroporphyrin III
(60%)

Conjugated dipyrrolic pigments, the dipyrromethenes, are synthesized by acid-catalyzed condensation of an α-formyl pyrrole and an α-unsubstituted pyrrole. They are readily protonated and deprotonated, and are difficult to purify by chromatography.

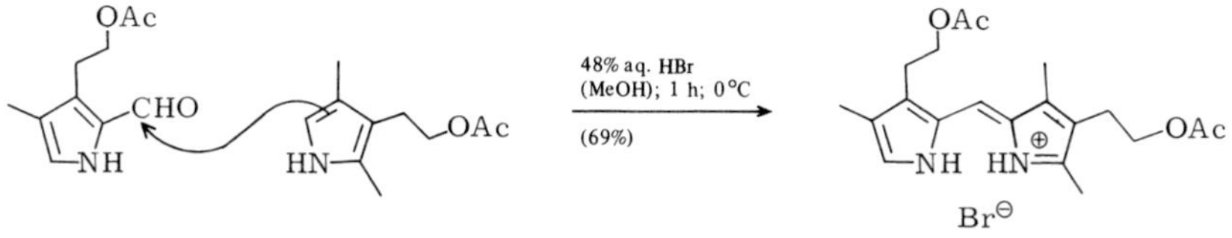

The pyridine-like nitrogen of the 2H-pyrrol-2-ylidene unit tends to withdraw electrons from the conjugated system and deactivates it in reactions with electrophiles. The acid-catalyzed condensations described above for pyrroles and dipyrromethanes therefore do not occur with dipyrromethenes. Vilsmeier formylation, for example, is only successful with pyrroles and dipyrromethanes, but not with dipyrromethenes.

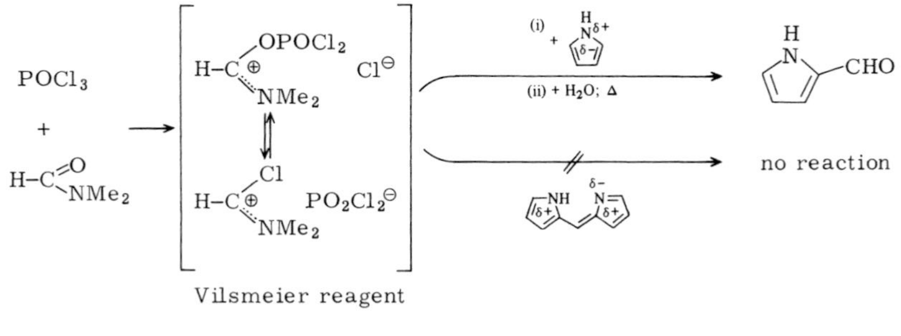

Vilsmeier reagent

α-Pyrrolic bromine substituents on dipyrromethenes, on the other hand, are reactive in nucleophilic substitutions. Under forcing conditions, i.e. fusion in a succinic acid melt, α-methyldipyrromethene cations can be deprotonated at the methyl group, and the resulting enamine develops some nucleophilic reactivity. This is the basis of Hans Fischer's classic porphyrin synthesis. The formation of isomer mixtures can only be avoided if one pyrromethene unit is symmetrically substituted about the methane carbon, or if the desired porphyrin is centrosymmetrical (Fischer, 1927, 1940; Treibs, 1971).

coproporphyrin I

R = H: 1 h; 195 °C (etioporphyrin III; 52%)
R = COOH: 2 h; 180 °C (mesoporphyrin; 29%)

This reaction sequence is much less prone to difficulties with isomerizations since the "pyridine-like" carbons of dipyrromethenes do not add protons. Yields are often low, however, since the intermediates do not survive the high temperatures. The more reactive, "faster" but "less reliable" system is certainly provided by the dipyrromethanes, in which the reactivity of the pyrrole units is comparable to activated benzene derivatives such as phenol or aniline. The situation is comparable with that found in peptide synthesis where the "slow" azide method gives cleaner products than "fast" DCC-promoted condensations.

With the catalysis of strong Lewis acids, such as tin(IV) chloride, dipyrromethenes may also be alkylated. A successful porphyrin synthesis involves 5-bromo-5′-bromomethyl and 5′-unsubstituted 5-methyl-dipyrromethenes. In the first alkylation step a tetrapyrrolic intermediate is formed which cyclizes to produce the porphyrin in DMSO in the presence of pyridine. This reaction sequence is useful for the synthesis of completely unsymmetrical porphyrins (Smith, 1975).

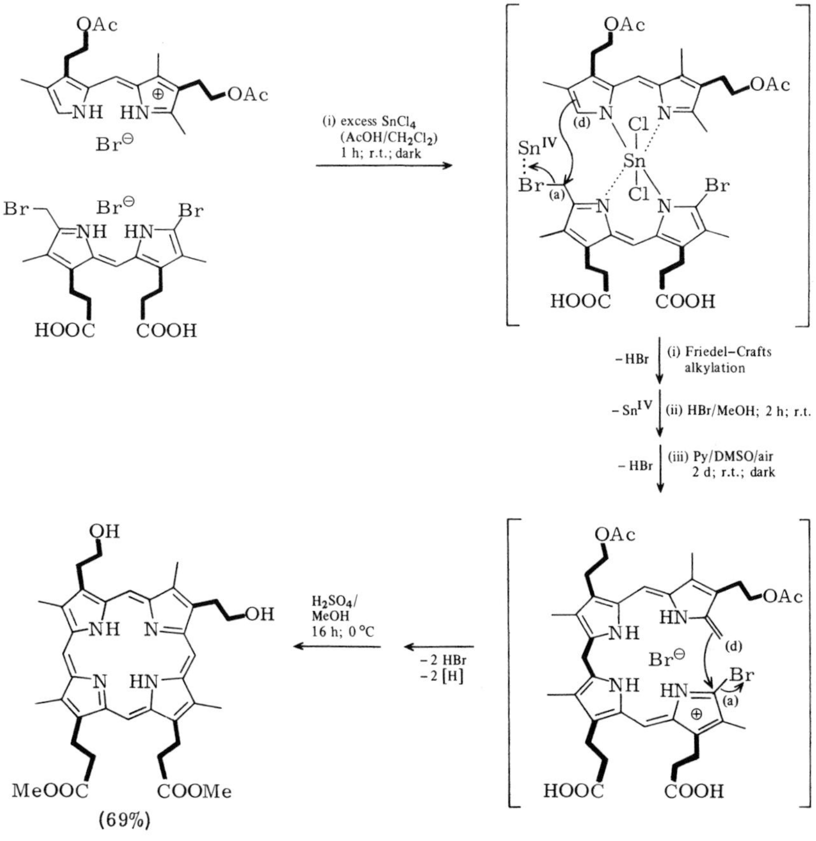
OAc
OAc
NH HN
Br⊖
Br
Br⊖
Br
NH HN
HOOC COOH
(i) excess $SnCl_4$
($AcOH/CH_2Cl_2$)
1 h; r.t.; dark
Sn^{IV}
Cl
Sn
Cl
(d)
(a)
- HBr
(i) Friedel–Crafts alkylation
- Sn^{IV}
(ii) HBr/MeOH; 2 h; r.t.
- HBr
(iii) Py/DMSO/air
2 d; r.t.; dark
OH
OH
NH N
N HN
MeOOC COOMe
(69%)
H_2SO_4/
MeOH
16 h; 0 °C
-2 HBr
-2 [H]

α-Siloxypyrroles are accessible from pyrrolenine lactams and silylchlorides. They condense with α-formyl pyrroles in the presence of $TiCl_4$ in essentially quantitative yield to give pyrromethenones. Selenylated ethyl groups proved to be ideal precursors for the introduction of vinyl groups, which are acid–base and redox sensitive. Peracid-induced selenoxide elimination set them free. Decarboxylation and treatment with trimethylorthoformate (TMOF) was used to introduce an α-formyl group for further condensation to yield open-chain tetrapyrrole bile pigments) (Jacobi, 2000).

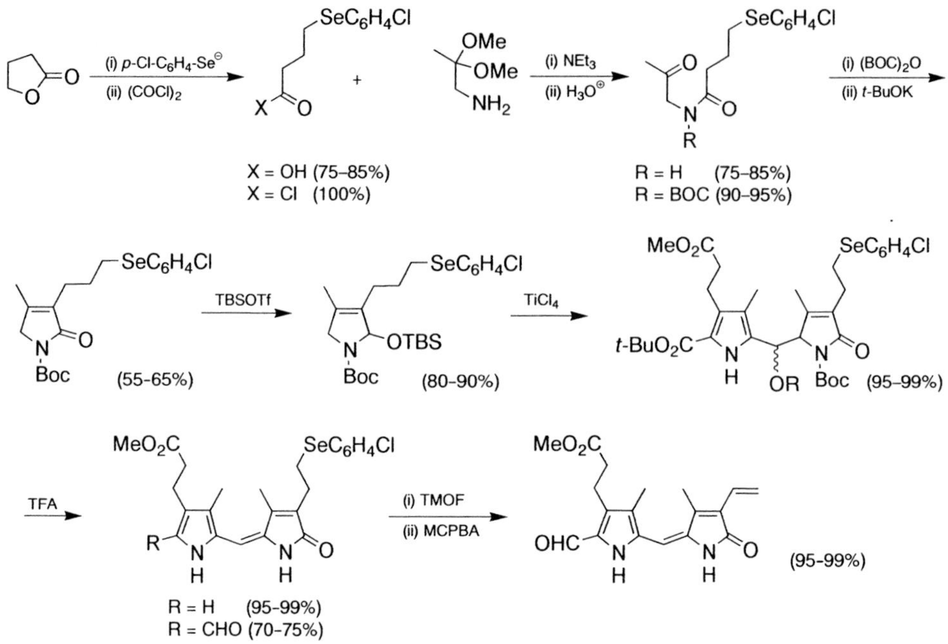

X = OH (75–85%)
X = Cl (100%)

R = H (75–85%)
R = BOC (90–95%)

(55–65%)

(80–90%)

OR Boc (95–99%)

R = H (95–99%)
R = CHO (70–75%)

(95–99%)

The unsymmetric 5-*tert*-butyl-10-methanol-diphenyl porphyrin A was obtained by a mixed condensation of the two aldehydes with the dihexyl-dipyrromethane shown in the scheme below, and chromatography. The symmetric diphenylporphyrin with two propionic ester side-chains above and below the porphyrin plane resulted from a similar condensation of the dipyrromethane with the *meta*-diphenolether-benzaldehyde. Mitsonobu condensation of four porphyrin-benzylalcohols with one porphyrin-tetracarboxylate then gave the pentamer with a central ruthenium metal ion (electron acceptor) and four peripheral zinc porphyrinates (electron donors; Mak, 2001).

Chlorophyll *a* (Vernon, 1966) contains an unsymmetrical porphyrin chromophore with two special features: the double bond between C-17 and C-18 is hydrogenated and carbon atoms 13 and 15 bear a carboxylated, "isocyclic" cyclopentanone ring E.

chlorophyll a : R = CH$_3$
chlorophyll b : R = CHO

Woodward's chlorophyll *a* synthesis started with the preparation of a porphyrin with an acetic acid side-chain on C-15 and a 13-methoxycarbonyl group, which helped to solve the major problem, namely the regioselective hydrogenation of ring D at C-17/C-18. This reaction should be simplified by repulsive interactions between the 15-methine bridge and the β-pyrrolic 13,17-substituents, which then distort the plane of the aromatic porphyrin ring. This "steric strain" could be released if one of the three carbon atoms 13, 15, or 17 became sp^3-hybridized. Its substituents would then be above and below the plane, and the chromophoric π-system could again be planar. Furthermore ring D should react much faster than ring C, because in the latter the double bond is deactivated by the carboxylic ester group. It turned out, however, and this is nowadays a very well-established fact in porphyrin chemistry, that it is always the methane whose bridges react much faster in all types of reactions than the aromatic pyrrole rings (Fuhrhop, 1974). Hydrogenation under various conditions always yielded the "phlorin" with a reduced carbon bridge but no "chlorin" with a reduced pyrrole ring. This failure led to a new concept on the same basis of "stereochemical strain release". Instead of the acetic acid, an acrylic acid side-chain was introduced at C-15. An acid-catalyzed intramolecular reaction between the olefinic double bond and the aromatic pyrrole unit D led to a methoxycarbonylated cyclopentene ring and a reduced ring D. The proton on C-18 is probably derived from an acid-catalyzed tautomerization of the acidic proton adjacent to the methoxycarbonyl group (Woodward, 1960, 1990).

aphlorin

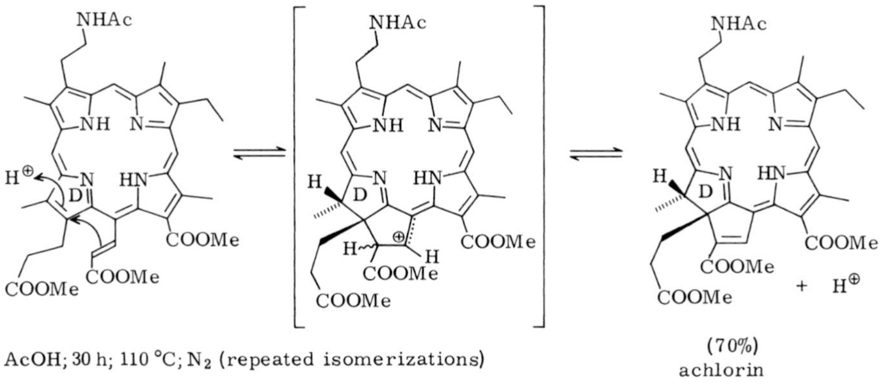

AcOH; 30 h; 110 °C; N₂ (repeated isomerizations)

(70%)
achlorin

The C=C double bond in the cyclopentene ring was then cleaved by photooxygenation. The methoxyl group on C-17 can, as a typical α-dicarbonyl system, be split off with strong base and replaced by a proton. Since this elimination occurs with retention of the most stable configuration of the cyclization equilibrium, the substituents at C-17 and C-18 are located trans to one another. The critical introduction of both hydrogens was thus achieved regio- and stereoselectively.

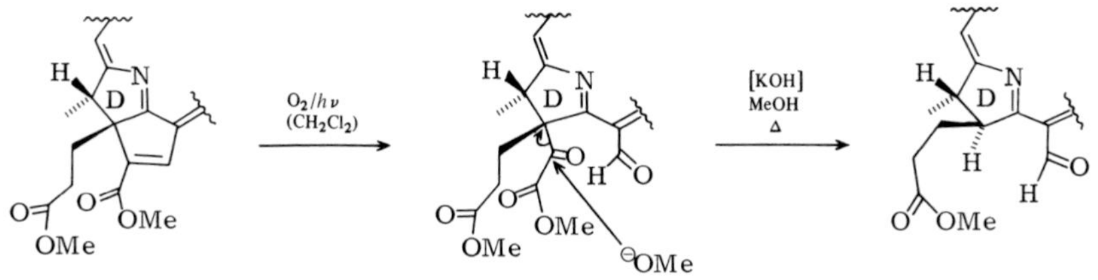

"Corrin" is the "porphyrinoid" chromophore of the vitamin B_{12} parent compound cobyrinic acid. Corrin itself has not yet been synthesized, but routes to cobyrinic acid and several other synthetic corrins have been described by Eschenmoser (1970, 1974) and Woodward (1967).

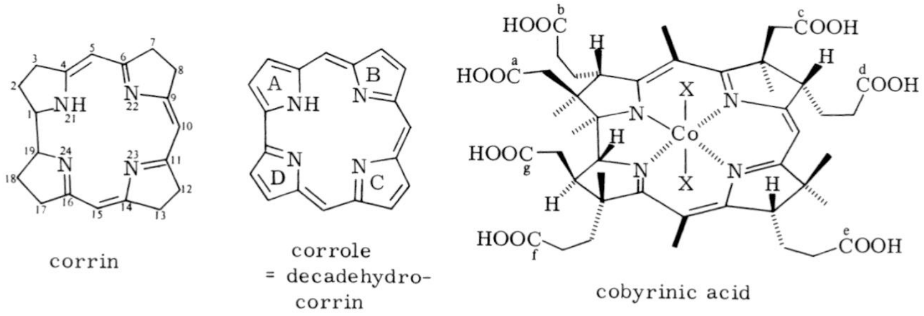

corrin

corrole
= decadehydro-
corrin

cobyrinic acid

It is conceivable that related ligands, e.g. dehydrocorrins, could be obtained from pyrrolic units using pathways similar to those used for porphyrins and could be hydrogenated to corrins. This has indeed been achieved (Dicker, 1971), but it is, of course, impossible to introduce the nine chiral centers of cobyrinic acid by such procedures.

The most straightforward synthesis of the corrin nucleus would be one in which the pyrroline rings are preformed acceptor and donor synthons similar to the α-substituted and α-unsubstituted pyrrole rings in porphyrin synthesis. The double-bonds of the corrin ring should be present in the condensation product since subsequent oxidation of a saturated system to the conjugated tetrazapolyene system of the corrins is difficult: corrins are easily destroyed by oxidants. The donor could be an enamine d^2-synthon in which the nitrogen atom belongs to the pyrrolidine ring and the double bond corresponds to an exocyclic methylene group, which would become the methane bridge. The choice of the a^1-acceptor synthon is more difficult. A good leaving group on a C=N-bond is provided, for example, by alkoxy groups. Since imidic esters can be synthesized from amides with Meerwein's trialkyloxonium reagents they were the first choice of Eschenmoser (1970). The condensation of imidic esters with enamines yields diimines connected via a methylene bridge. In order to facilitate isomerization to the desired a,β-unsaturated imines and the deprotonation of the enamine which raises its nucleophilicity, a nitrile group was introduced.

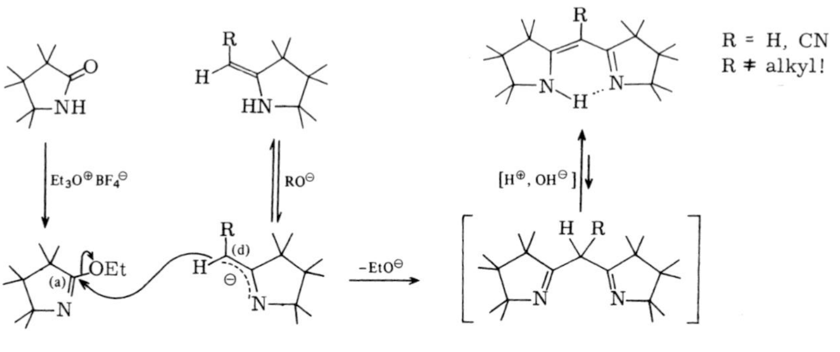

It then turned out that activation of enamine components could be achieved not only by deprotonation of the nitrogen atom but also by connecting it with certain metals, e.g. Ni(II), Pd(II), or Co(II), and subsequent treatment with base. The following scheme shows how the "eastern" and "western" parts of a corrin chromophore can be combined regioselectively. The western part has a more acidic enamine than the eastern part, whereas the imidic ester of the eastern part is more electrophilic.

Corrins bearing additional alkyl substituents next to the critical methylene reaction centers could not be synthesized by this approach. The intermolecular coupling step was therefore replaced by an intramolecular reaction. The lactam group of one condensation partner was converted into the thiolactim (via imidic ester; see second last reaction in the scheme below). The nucleophilic sulfur atom could be either condensed with a bromomethyl group adjacent to an imine or converted itself into an acceptor group: in the presence of hydrogen chloride it was dimerized oxidatively with dibenzoylperoxide to give the disulfide. This disulfide reacted regioselectively with nucleophilic enamides. The "sulfide bridge" was converted into the desired "methane bridge" by sulfide contraction using phosphite.

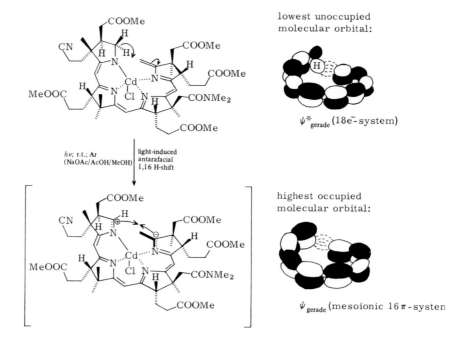

(i) MeHgOPr^i
(ii) Me$_3$O$^\oplus$ BF$_4^\ominus$
(iii) H$_2$S

(70%)

(i) KOBu^t
(ii) Δ; (EtO)$_3$P

4,5-secocorrin

The direct connection of rings A and D at C-1 cannot be achieved by enamine or sulfide couplings. This reaction has been carried out in almost quantitative yield by electrocyclic reactions of A/D-secocorrinoid metal complexes and constitutes a magnificent application of the Woodward–Hoffmann rules. First an *antarafacial* hydrogen shift from C-19 to C-1′ is induced by light (sigmatropic 18-electron rearrangement), and second, a *conrotatory* thermally allowed cyclization of the mesoionic 16 π-electron intermediate occurs. Only the 1,19-*trans*-isomer is formed (Eschenmoser, 1974; Pfaltz, 1977).

lowest unoccupied molecular orbital:

ψ^{*}_{gerade} (18e$^-$-system)

hv; r.t.; Ar
(NaOAc/AcOH/MeOH)

light-induced
antarafacial
1,16 H-shift

highest occupied molecular orbital:

ψ_{gerade} (mesoionic 16 π-system)

Carbon–oxygen bonds are formed by the Ullmann reaction (=coupling of aryl halides with copper) which has been varied in alkaloid chemistry to produce diaryl ethers instead of biaryls. This is achieved by the use of CuO in basic media (Kametani, 1969; Doskotch, 1971).

(S,S)-adiantifoline

The alkaloid dubamine contains a single bond between the two heteroarene units. This bond was formed in 79% yield by the generally valuable palladium-catalyzed coupling of an aryltrimethylstannane with an aryl triflate. The requisite stannane was prepared from 1,3-benzodioxol-5-yl triflate and hexamethyldistannane with the same palladium catalyst. The triflate ester was obtained from 2(1H)-quinolinone and trifluoromethanesulfonic anhydride (Echavarren, 1987). An earlier attempt to perform this aryl coupling by classical means gave a yield of only 1%.

(i): $(CF_3SO_2)_2O$/Py; 5 min; 0 °C + 1 d; r.t.
(ii): [Pd(PPh$_3$)$_4$]; LiCl/dioxane; 98 °C

Reactions that can be carried out by the oxidative coupling of radicals may also be initiated by irradiation with UV light. This procedure is especially useful if the educt contains olefinic double bonds since they are vulnerable to the oxidants used in the usual phenol coupling reactions. Photochemically excited benzene derivatives may even attack ester carbon atoms which is generally not observed with phenol radicals (Ninomiya, 1973; Yang, 1966).

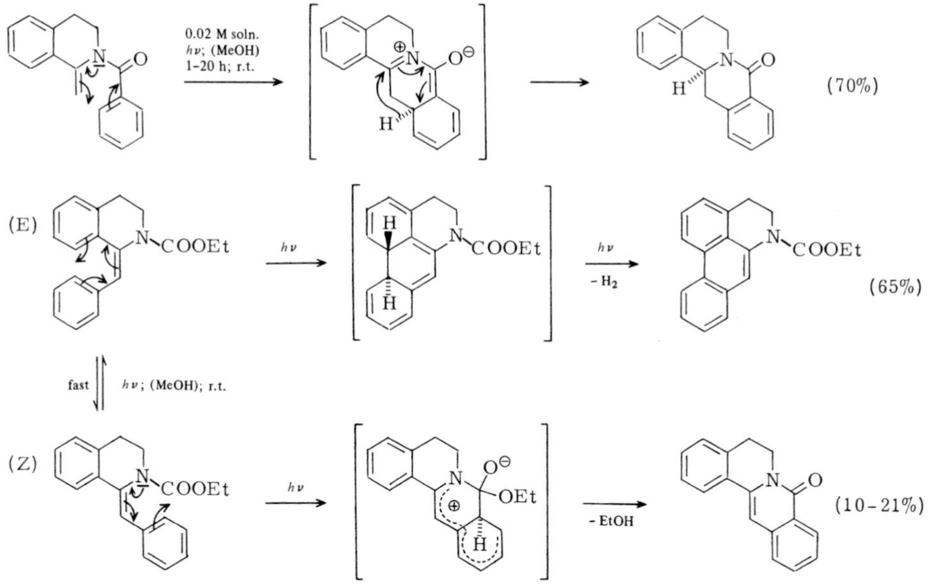

0.02 M soln.
hν; (MeOH)
1–20 h; r.t.

(70%)

(E)

hν

hν
– H₂

(65%)

N-COOEt

(65%)

fast | *hν*; (MeOH); r.t.

(Z)

N-COOEt

hν

N⊕
OEt
O⊖

– EtOH

(10–21%)

Indole derivatives are still synthesized by the classical method of E. Fischer. Its key step is an (intramolecular) [3,3]-sigmatropic rearrangement and it may therefore be used to produce sterically hindered products otherwise difficult to obtain. In the first example given below the (planar) lactam group causes a steric strain which prevents enolization of the bridgehead carbon neighboring the phenylhydrazone carbon. In the second example a nonplanar amine nitrogen does not have such effect and a highly hindered spirofused indolenine is formed besides the thermodynamically favored indole (Stork, 1963; Ban, 1965; Klioze, 1975).

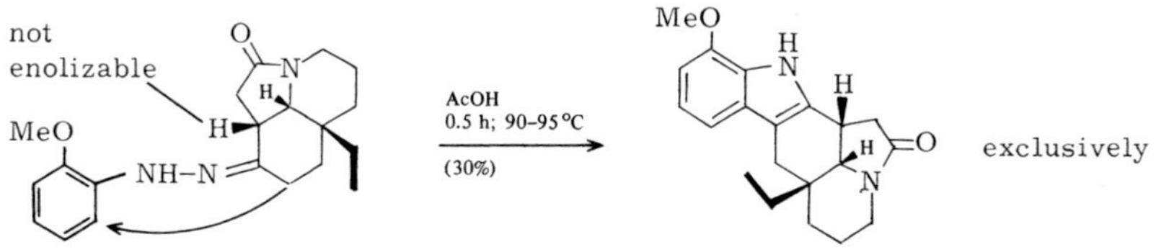

not
enolizable

MeO

NH–N

AcOH
0.5 h; 90–95 °C

(30%)

MeO

exclusively

In Diels–Alder reactions a nitroolefin may function as an electron-deficient ene component, or a 1,2-dihydropyridine derivative may be used as a diene component. Both types of reactant often yield cyclic amine precursors in a highly stereoselective manner (Hill, 1962; Büchi, 1965, 1966 A). The Diels–Alder reaction of *o*-quinodimethanes (from benzocyclobutenes) with nitrogen-substituted enes has also been applied to alkaloid synthesis (Kametani, 1972, 1973, 1974; Oppolzer, 1978 A).

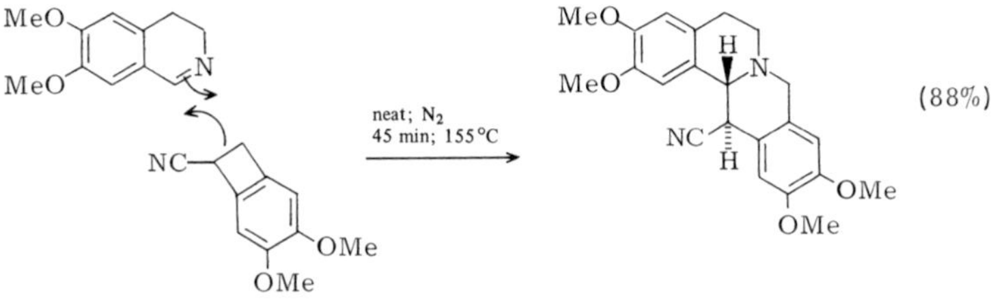

MeO / MeO ... N ... NC ...

neat; N₂
45 min; 155 °C

(88%)

NC ... OMe / OMe

Thermal and photochemical electrocyclic reactions are particularly useful in the synthesis of alkaloids (Oppolzer, 1973, 1978 B; Wiesner, 1968). A high degree of regio- and stereoselectivity can be reached, if cyclic olefin or enamine components are used in "ene" reactions or photochemical [2+2]cycloadditions.

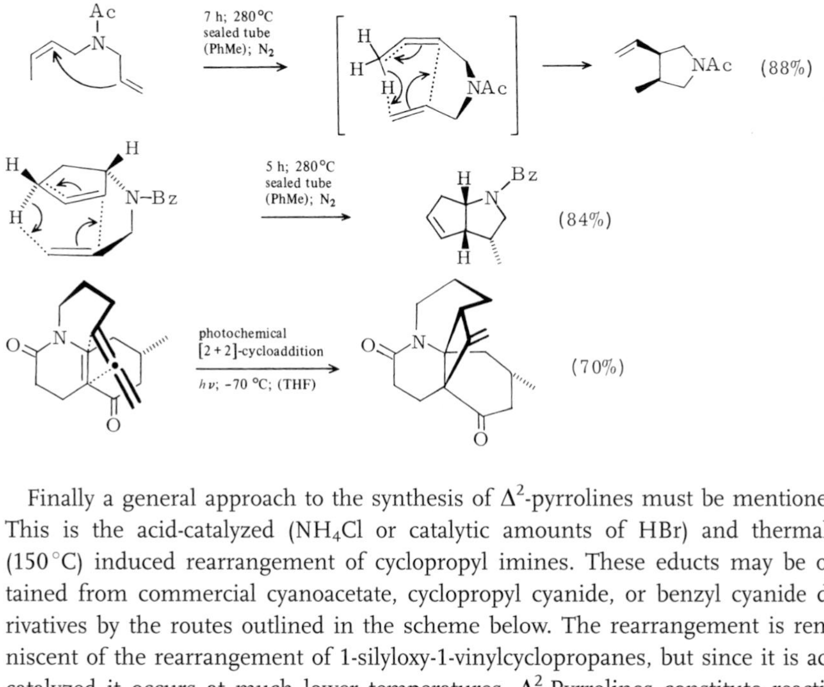

Ac / N

7 h; 280 °C
sealed tube
(PhMe); N₂

NAc

NAc (88%)

N–Bz

5 h; 280 °C
sealed tube
(PhMe); N₂

Bz / N

(84%)

O N

photochemical
[2+2]-cycloaddition

hν; –70 °C; (THF)

O N

(70%)

Finally a general approach to the synthesis of Δ²-pyrrolines must be mentioned. This is the acid-catalyzed (NH₄Cl or catalytic amounts of HBr) and thermally (150 °C) induced rearrangement of cyclopropyl imines. These educts may be obtained from commercial cyanoacetate, cyclopropyl cyanide, or benzyl cyanide derivatives by the routes outlined in the scheme below. The rearrangement is reminiscent of the rearrangement of 1-silyloxy-1-vinylcyclopropanes, but since it is acid catalyzed it occurs at much lower temperatures. Δ²-Pyrrolines constitute reactive enamines and may be used in further addition reactions such as the Robinson annelation with methyl vinyl ketone (Stevens, 1967, 1968, 1971).

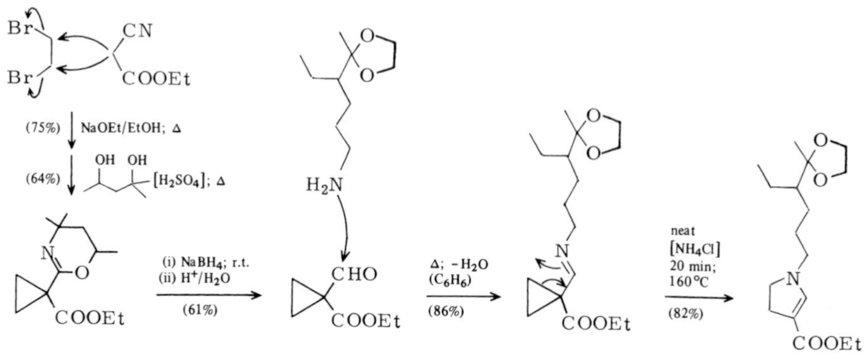

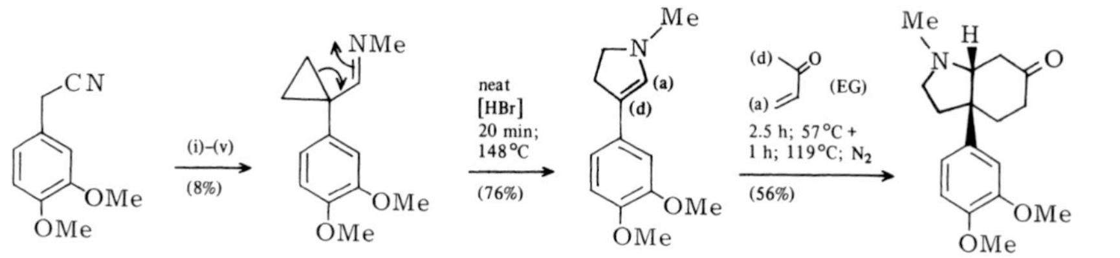

(i) BuLi/THF; 2 h; r.t.; N₂

(ii) Br∿∿Br (THF); 2 h; −78 °C → r.t. + 5 h; r.t. } (24%)

(iii) LiAlH₄/THF; 4 h; −78 °C → r.t. + 4 h; r.t.
(iv) + dil. aq. HCl } (38%)

(v) MeNH₂/MgSO₄(C₆H₆); 20; r.t.; −H₂O (91%)

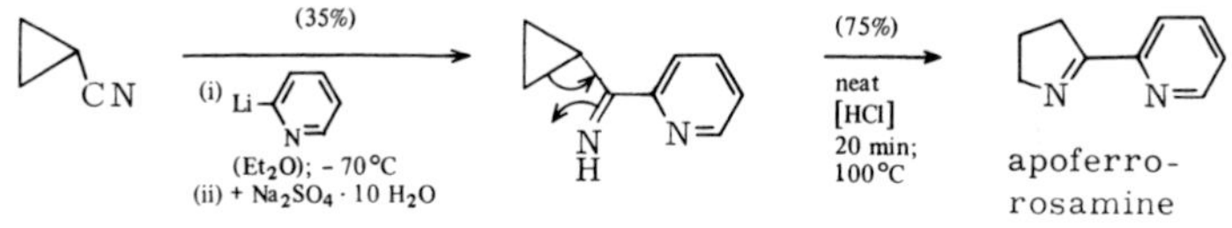

(±)-mesembrine

apoferro-
rosamine

Nowadays exotic marine alkaloids are the most prominent targets of total alkaloid syntheses. Some reviews of alkaloid syntheses discuss the application of more recent synthetic methods in the alkaloid field (Franklin, 1996; Hudlicky, 1996).

4.6
Self-Reproduction of Chirality

Proline, the only natural amino acid in which the chiral α-carbon atom lies in a ring, has been α-alkylated without loss of optical activity (Seebach, 1983). Proline was condensed with pivalaldehyde to give a pure bicyclic diastereomer with the bulky *tert*-butyl group on the *exo* face, i.e. *cis* to the angular hydrogen. LDA produces a chiral enolate which reacts with both alkylating reagents and carbonyl compounds attacking its *Re*-side, i.e. cis relating to the *tert*-butyl group. Thus this reaction sequence reproduces the original configuration with complete retention.

Furthermore, aldol-type additions occur with deastereoselectivities up to 95% i.e. probably due to lithium chelation control. The allylation of proline by this method has been applied in a complex alkaloid synthesis (Williams, 1990) containing a quaternary chiral center.

4.7

Monosaccharides and Protected Derivatives

Some monosaccharides and oligomeric carbohydrates are inexpensive and provide a great variety of functional and stereochemical features. The role of carbohydrates as low-cost starting materials ("feedstock") are briefly discussed in Chapter 10.

Enantiomerically pure tetroses, pentoses, and hexoses have been synthesized by a repetitive two-carbon homologization (Lee, 1982; Ko, 1983): (1) Wittig reaction of a protected hydroxy aldehyde with (triphenylphosphoranylidene) acetaldehyde, (2) reduction with sodium tetrahydroborate, (3) asymmetric epoxidation of the allyl alcohol, (4) nucleophilic epoxide opening with sodium hydroxide/benzenethiol, and (5) conversion of the (phenylthio)methyl group to an aldehyde group by oxidation and Pummerer rearrangement with (6a) or without (6b) inversion of the C-2 stereocenter. All diastereomeric L-hexoses have thus been synthesized.

Ph$_3$P=CH—CHO
(toluene); 1.5 d; 0 °C

R—CHO $\xrightarrow{(1)}$

$$R-\overset{\overset{H}{|}}{\underset{\overset{|}{H}}{C}}\overset{E}{=}C\diagdown_{CHO}$$

(**6a**) K$_2$CO$_3$/MeOH; 2 h; r.t. **or**
(**6b**) Bu$^i_2$AlH/toluene; 1.5 h; −78 °C

$$R-\overset{*}{CH}—\overset{*}{CH}-CHO \xrightarrow{(1)}$$
erythro
or threo

SPh
R—$\overset{*}{CH}$—$\overset{*}{CH}$—CH
OAc

(i) MCPBA/CH$_2$Cl$_2$; 1 h; −78 °C (**5**)
(ii) Ac$_2$O/NaOAc; 8 h; Δ

R = Ph$_2$CHO—CH$_2$—
or $\overset{O}{\underset{O-CH_2}{\diagup}}$CH— etc.

NaBH$_4$/MeOH
10 min; −40 °C

(**2**)

$$R-\overset{\overset{H}{|}}{\underset{\overset{|}{H}}{C}}\overset{E}{=}C\diagdown_{CH_2OH}$$

Bu^{t}OOH/Ti(OPr^i)$_4$
(+)- or (−)-diethyl tartrate
(CH$_2$Cl$_2$); 15 h; −20 °C
(**3**)

(**4**)

$$R-\overset{*}{CH}—\overset{*}{CH}-CH_2-SPh$$
erythro

(i) NaOH/PhSH;
(Bu^{t}OH); 3 h; 100 °C
(ii) OMe [H$^+$]; (CH$_2$Cl$_2$)
3 h; r.t.

threo

In all cases examined the (*E*)-isomers of the allylic alcohols reacted satisfactorily in the asymmetric epoxidation step, whereas the epoxidations of the (*Z*)-isomers were intolerably slow or nonstereoselective. The *erythro*-isomers obtained from the (*E*)-allylic alcohols may, however, be epimerized in 95% yield to the more stable *threo*-isomers by treatment of the acetonides with potassium carbonate. The competitive β-elimination is suppressed by the acetonide protecting group because it maintains orthogonality between the enolate π-system and the β-alkoxy group.

K$_2$CO$_3$
(MeOH)

erythro
(= cis)

threo
(= trans)
(> 95%)

no β-elimination

A soil bacterium, *Pseuodomonas putida*, degrades benzenes to substituted cyclohexadiene diols. These are versatile starting materials in the synthesis of a variety of pentoses and hexoses. The *cis*-diols are chemically convertible to both *trans*-derivatives. This allows simple cross-overs of diastereoselection (Hudlicky, 1996).

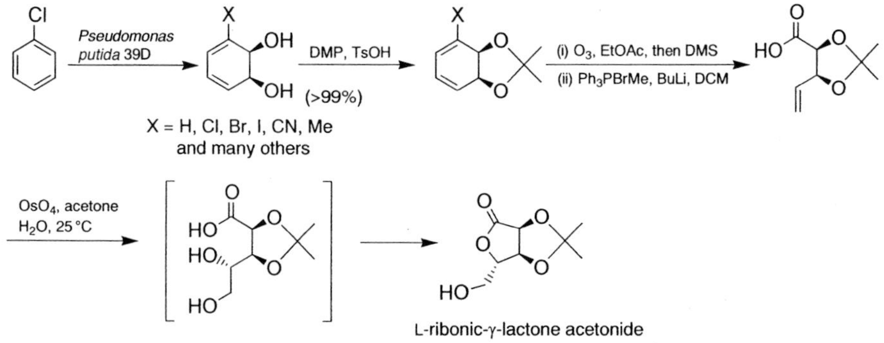
Cl
Pseudomonas putida 39D
X
OH
OH
DMP, TsOH
(>99%)
X = H, Cl, Br, I, CN, Me
and many others
(i) O_3, EtOAc, then DMS
(ii) Ph_3PBrMe, BuLi, DCM
HO
OsO_4, acetone
H_2O, 25 °C
L-ribonic-γ-lactone acetonide

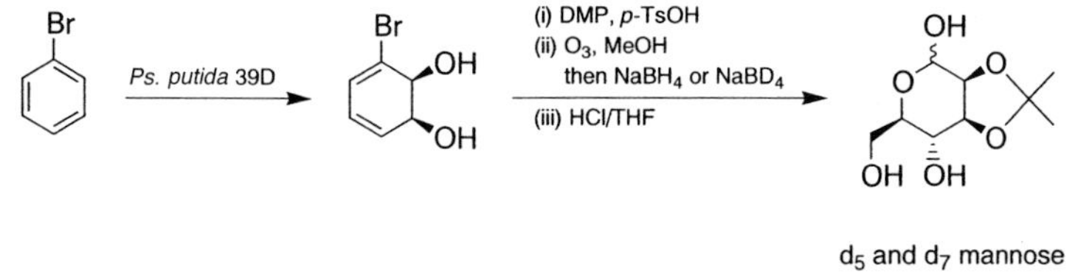
Br
Ps. putida 39D
OH
OH
(i) DMP, p-TsOH
(ii) O_3, MeOH
then $NaBH_4$ or $NaBD_4$
(iii) HCl/THF
d_5 and d_7 mannose

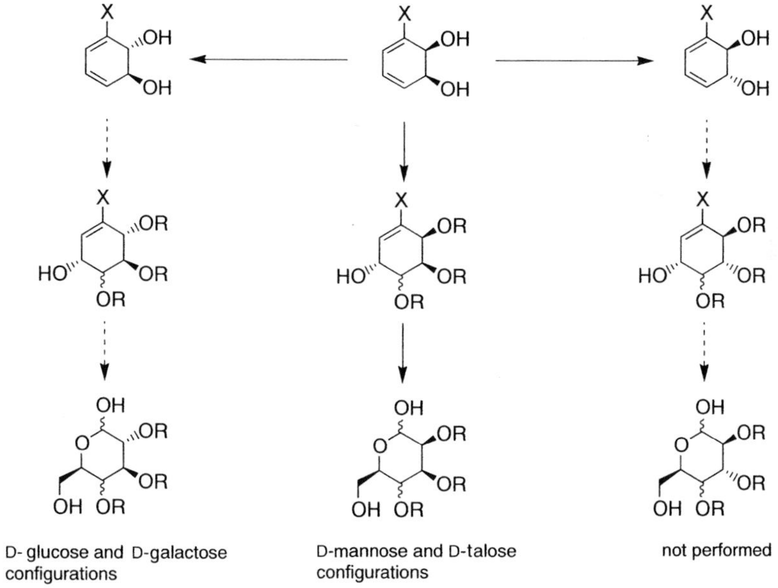
X
OH
OR
HO
D- glucose and D-galactose configurations
D-mannose and D-talose configurations
not performed

The hydroxyl groups of glucose (and, of course, other saccharides) must be regio- and stereo-selectively attacked and protected, if this most abundant natural carbon compound is to be used as starting material. We shall give a few selected examples to illustrate how this can be achieved (Haines, 1976; Lehmann, 1976; Hough, 1979).

The hemiacetal hydroxyl group on C-1 reacts selectively with alcohols in the presence of catalytic amounts of hydrochloric acid. The resulting acetal is called a glycoside or, in the case of glucose, a glucoside. These glycosides are relatively stable toward bases and nucleophiles. The hemiacetal, however, can also be activated selectively to nucleophiles. If glucose is peracetylated, only the C-1 acetate group is substituted by bromide anions. Usually only the more stable α-bromoglucose is formed. The corresponding β-chloro compound can be obtained by treatment of the α-acetate with hydrogen chloride and phosphorus trichloride. The primary alcohol group on C-6 has been converted to halides in high yields with carbon tetrahalides and triphenylphosphine. It is in general quite easy to protect or activate selectively C-1 or C-6 hydroxyl groups in aldohexoses (Lehmann, 1976, 1996; Györgydeák, 1998; Seeberger, 2001; Davis, 2002).

In aqueous solution glucose occurs only (>99%) as a six-membered ring hemiacetal (pyranose). With acetone and catalytic amounts of H_2SO_4 a five-membered ring derivative ("furanoside") is the major product. Heptanose forms have also been isolated from the reaction mixture (Stevens, 1972). A general rule in carbohydrate chemistry says that acetone tends to form five-membered 1,3-dioxolane rings ("isopropylidene derivatives") and reacts only with *cis*-1,2-diol groupings. Therefore, even though the most stable form of the hemiacetal of glucose itself is a pyranose, the reaction with acetone leads to a furanose derivative, because it provides two *cis*-diol groupings instead of one. In 1,2:5,6-di-*O*-isopropylidene-α-D-glucofuranose five of the six functional groups of glucose are protected. The free 3-hydroxyl group is a popular starting point in synthesis. It can be oxidized, e.g. with RuO_4 (Baker, 1976), tosylated, substituted, etc. Examples are the conversions of 3-keto groups of sugar derivatives to an α-hydroxy aldehyde (Paulsen, 1972, 1977) or to an acetic acid side-chain (Lourens, 1975). The exocyclic 1,3-dioxolane ring is much more vulnerable to acid hydrolysis than the ring connected with the acetal group. Partial deprotection of the side-chain is easily achieved by treatment with sulphuric acid.

When the 3-hydroxyl group has been converted to the desired functional group, e.g. to an amino group, the diacetonide is hydrolyzed, and the pyranose is recovered. The example synthesis of a steroid glycoside below indicates a typical pattern of protection, deprotection, and activation steps. The acetylated glycosidic hydroxyl group can be quantitatively exchanged with HBr, whereas the other acetyl groups remain intact (Meyer zu Reckendorf, 1971).

Ac₂O/Py; 12 h; r.t.
(ii) HBr/AcOH
(CH₂Cl₂); 3 h; r.t. (65%)

$$\text{CH}_2\text{OAc}\quad \text{AcO} \quad \text{TfacN} \quad \text{AcO Br}$$

(i) + strophantidin
[Hg(CN)₂/HgBr₂]
(C₆H₆); 8 h; Δ (70%)

(ii) NaOMe/MeOH; 1 h; r.t.
(iii) NH₃/MeOH; r.t. (33%)

CHO / CH₂OH / HO / H₂N / HO / OH

1,2-Dideoxy-1-enopyranoses ("glycals") are readily prepared from the corresponding glycosyl bromides by reduction with zinc in acetic acid (Fischer, 1914; Helferich, 1952). 2-Hydroxyglucal tetraacetate was obtained from tetra-*O*-acetylglucosyl bromide by a base-catalysed elimination of HBr (Lemieux, 1965) or via the corresponding iodides (Ferrier, 1966). In general, deoxy sugars can be synthesized from the corresponding tosylates or nitrobenzene-*p*-sulfonates by an iodide exchange followed by a reduction step. The addition of nitrosyl chloride to glycols gives 2-nitroso pyranosyl chlorides with high selectivity (Lemieux, 1968). The nitroso compounds may dimerize. Reduction with zinc in acetic acid yields 2-amino-2-deoxypyranoses.

Et₂NH/Bu₄N⊕ Br⊖ (MeCN); 20 min; r.t. (80%)
or: (i) NaI/Me₂CO; 15 min; r.t. (60%)
(ii) Et₂NH; 1 h; r.t.

2-acetoxy-3,4,6-tri-*O*-acetylglucal

(83%) Zn–Cu(NaOAc/H₂O) 1 h; –20 °C + 3 h; 0 °C

tri-*O*-acetyl-glucal

δ+ δ– + NOCl; –80 °C (CH₂Cl₂, Et₂O) (≈100%)

(i) Zn–Cu/AcOH (H₂O); 2 d; r.t.
(ii) Ac₂O; 1 h; r.t. (60%)

4 M aq. HCl 1 h; 100 °C (80%)

(60%) (i) H₂ [Pt]; (AcOH); r.t.
(ii) Ba(OH)₂/H₂O; Δ

dihydroglucal

dimer

glucosamine

Another common carbonyl compound for the protection of two sugar hydroxyl groups is benzaldehyde. It gives six-membered acetal rings (1,3-dioxanes) but no bridged bicyclic compounds. With glucose or methyl glucoside the 4,6-*O*-benzylidene derivatives can be obtained (Fletcher, 1963, Evans, 1980). Benzylidene protecting groups can be removed by catalytic hydrogenation.

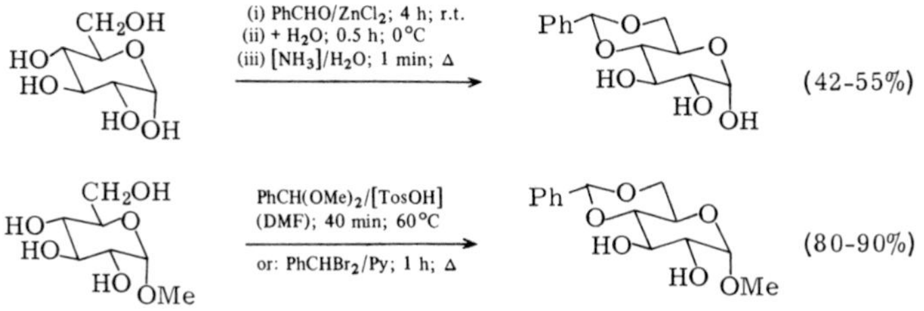

The benzylidene derivative above is used if both hydroxyl groups on C-2 and C-3 are needed in synthesis. This *trans*-2,3-diol can be converted to the sterically more hindered α-epoxide by tosylation of both hydroxy groups and subsequent treatment with base (Williams, 1970; Buchanan, 1976). An oxide anion is formed and displaces the sulfonyloxy group by a rearside attack. The oxirane may then be re-opened with nucleophiles, e.g. methyl lithium, and the less-hindered carbon atom will react selectively. In the sequence shown below, starting with an α-glucoside only the 2-methyl-2-deoxyaltrose is obtained (Hanessian, 1977).

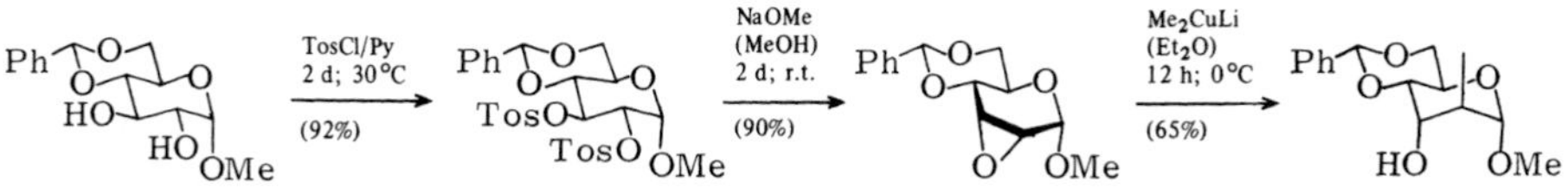

Glycosidic thiol groups can be introduced into glycosyl bromides by successive reactions with thiourea and aqueous sodium disulfite (Horton, 1963; Černý, 1961, 1963). Such thiols are excellent nucleophiles in weakly basic media and add to electrophilic double bonds, e.g. of maleic esters, to give Michael adducts in high yields. Several chiral amphiphiles have thus been prepared without any need for chromatography (Fuhrhop, 1986 A).

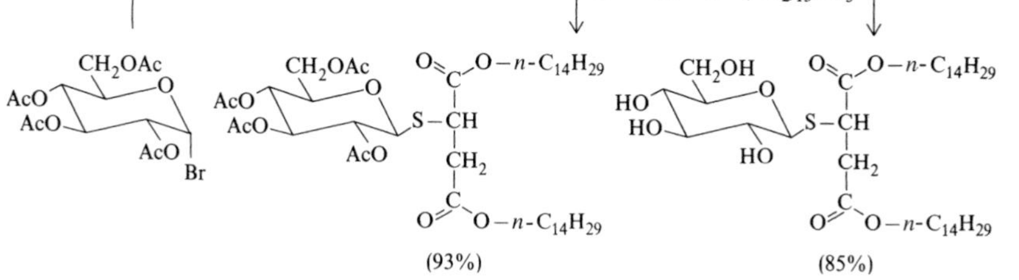

Carbohydrates are often used as chiral educts ("chirons") in synthesis. A book by S. Hanessian (1983) discusses this subject in detail. We give one example here, and one in Section 4.8.3. The synthesis of 11-oxaprostaglandins from D-glucose uses the typical reactions of glucofuranose diacetonide outlined above. Reduction of the hemiacetal group is achieved via a thioacetal. The carbon chains were introduced by Wittig reactions on the aldehyde groups, which are liberated stepwise by periodate oxidation and lactone reduction (Hanessian, 1979; Lourens, 1975).

11-oxa-
PGF$_{2\alpha}$

Oligosaccharide syntheses are discussed in Section 5.4.

5

Biopolymers and Dendrimers

5.1
Introduction

The most simple syntheses of complex molecules involve the joining of structural units in which all functional groups and asymmetric centers are preformed. This technique can usually only be applied to compounds in which these units are connected by C–X bonds rather than C–C. They are exemplified here by standard syntheses of oligonucleotides, peptides, and oligosaccharides, both in solution and on the surface of solid beads.

Spherical dendrimers are then discussed as one example of a nonbiological, uniform oligomer. The monodispersity of the oligomeric molecules described in this chapter always depends on controlled protection and deprotection of educts and intermediates in order to prevent uncontrolled growth.

5.2
Oligonucleotides

The fundamental problem of oligodeoxyribonucleotide synthesis is the efficient formation of the internucleotidic phosphodiester bond specifically between the C-3' and C-5' positions of two adjacent nucleosides. Any functional group (NH_2 of nucleic base; "the other" OH of deoxy-ribose; other phosphate groups) must be protected. In oligoribonucleotide synthesis the additional protection of the 2'-hydroxyl function is necessary. This group must be deblocked under conditions mild enough to avoid isomerization of the $3' \rightarrow 5'$ phosphodiester linkage to the unnatural $2' \rightarrow 5'$ arrangement. The following table summarizes protecting groups commonly used in nucleic acid synthesis (Narang, 1973; Köster, 1979).

Protecting groups are also used to solubilize synthetic intermediates in organic solvents, e.g. methylene chloride. Chromatography is then possible on a larger scale, since silica gel can be used as adsorbent. Six synthetic strategies have originally been developed (Köster, 1979):

Tab. 5.1 Protecting groups used for nucleotides

Functional group	Protecting group	Abbreviation	Reaction conditions for deprotection
5'-OH (primary)	Triphenylmethyl(= trityl)	Trit	ac. 80% AcOH; 100 °C
	Di-*p*-methoxytrityl	Dmtr	ac. 80% AcOH; r.t.
	Dimethylpropanoyl (= pivaloyl)	Piv	bas. Et$_4$N$^+$OH$^-$/MeOH; r.t.
3'-OH (secondary)	Acetyl	Ac	bas. 2 M NaOH; 0 °C
	β-Benzoylopropionyl	Bzpr	ac. N$_2$H$_4$/Py/AcOH
2'-OH (secondary)	Tetrahydropyranyl	Thp	ac. 0.01 M HCl; r.t.
NH$_2$ (nucleic bases A,C,G)	Acetyl	Ac	
	Bezoyl	Bz	bas. 1 M NaOH; r.t.
	p-Methoxybenzoyl (= anisoyl)	MrOBz	
Phophodiester	β-Cyanoethyl	Ce	bas. K$_2$CO$_3$
	β,β,β-Trichloroethyl	Tce	red. Zn-Cu/AcOH; r.t.
5'-Phosphate	2-(*p*-Tritylphenylthio)ethyl	Tpte	ox. + (i) NCs (→ sulfone)
			bas. (ii) NaOH; r.t.

1. the diester method
2. the triester method
3. the 1,3,2-dioxaphosphole method
4. the phosphite method
5. the solid-phase method
6. the combined chemical–enzymatic method.

In the diester method a deoxynucleoside-5'-monophosphate is condensed with the 3'-OH group of a deoxynucleotide to produce a 3',5'-phosphodiester. This is illustrated by a general method for dinucleotide synthesis developed by H.G. Khorana (K.L. Agarwal, 1976). One N-protected mononucleotide is condensed with an excess of 2-(*p*-tritylphenylthio)ethanol (Tpte-OH), and the 3'-OH group of the other N-protected mononucleotide is acetylated. Both components are then condensed with triisopropylbenzenesulfonyl chloride (TpsCl) in pyridine. Because of the hydrophobic trityl group, the protected dinucleotide product can be chromatographed on silica gel. The O-acetyl and N-protecting groups are removed with ammonia at 50 °C, and the Tpte group is first oxidized to the sulfone by N-chlorosuccinimide and then hydrolyzed with NaOH. Yields of isolated dinucleotides are around 70%.

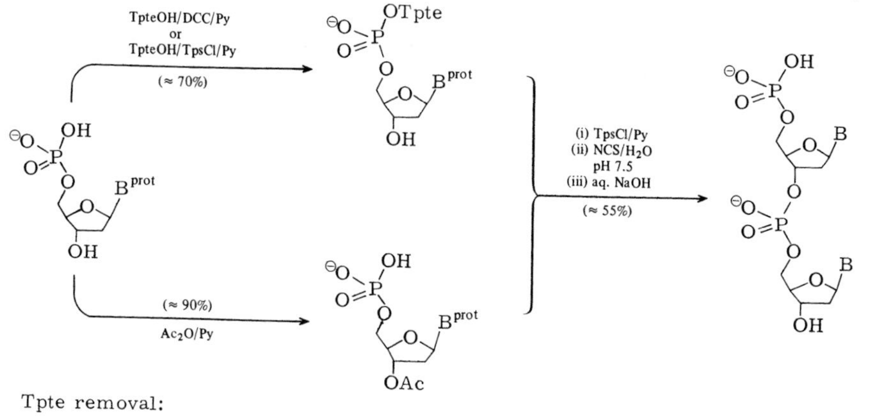

Tpte removal:

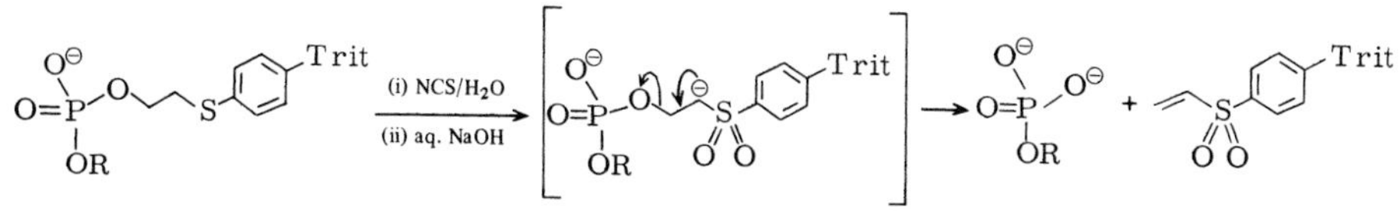

A major problem with the diester approach is the fact that one P–O⁻ group of the phosphate always remains free. Since this group is also activated by condensing agents, branched pyrophosphates or triester may also be formed. Cyclization may also occur when the protecting group of the 3′-OH group is removed before a third nucleotide is condensed to it. These side-reactions drastically reduce yields, especially if the condensing reagent is used repeatedly on a growing oligonucleotide chain with several phosphodiester groups.

The internucleotidic phosphodiester may be converted into a triester and thus be protected. In the usual "triester-method" a 3′-phosphodiester of a protected nucleoside is condensed with a nucleoside-3′-phosphate protected on all functional groups except for the primary 5′-OH group. The quantitative removal of several protecting groups (e.g. β,β,β-trichloroethyl, 4-chlorophenyl) from an oligomer, however, is a problem. It may be incomplete or be accompanied by isomerizations of the terminal phosphodiester groups, e.g. by formation of 3′,3′- or 5′,5′-phosphodiesters. Furthermore, long-chain oligomers of phosphotriesters are more difficult to separate by chromatography than are the corresponding diesters since diesters for different chain lengths are differentiated by different numbers of negative charges, whereas triesters are all electroneutral. The triester method therefore generally gives better yields but is more time-consuming; it is also more difficult to obtain pure products.

The main problem in the development of phosphotriester syntheses has been the lack of appropriate condensing agents. DCC cannot be used, because it will not activate phosphodiester functions. Triisopropylbenzenesulfonyl chloride (TpsCl) has been extensively applied, but gave low yields (10–20%) when condensations of products containing purine bases, especially guanine, were attempted. These low yields have been attributed to the liberation of hydrogen chloride, but trials with better or less innocuous leaving groups, e.g. azide, have also been unsuccessful. Only when triazoles were introduced by Narang (Katagiri, 1975) did the preparation of phosphotriesters become a general and high-yield procedure. Benzenesulfonyl chlorides are condensed with 1H-1,2,4-triazole in the presence of triethylamine in chloroform solution to form sulfonyltriazoles, e.g. MsT or pNbsT. These condensing agents will yield dinucleotides from protected mononucleotides, e.g. from mono(p-chlorophenyl)5'-O-(di-methoxytrityl)deoxythymidine-3'-phosphate (A) (1 mol equiv.) and thymidine-3'-phosphotriester (B) (1.2 mol equiv.) in 80% yield. Hexanucleotides have been made in 100 mg quantities by the triester method.

MsT: $R^2 = R^4 = R^6 = CH_3$
pNbsT: $R^2 = R^6 = H; R^4 = NO_2$

(i) coupling: MsT or pNbsT; (Py); 1-2d; r.t.
(ii) 5'-O-(di-p-methoxytrityl) removal: 80% aq. AcOH; 20 min.; r.t.
(iii) 2,2,2-trichloroethyl removal: Zn/AcOH/Py
(iv) 2-cyanoethyl and 4-chlorophenyl removal:
    0.1 M NaOH/H₂O/dioxane; 3-6 h; r.t.
(v) N-deblocking of nucleic bases B^{prot}: conc. aq. NH₃; 3 h; 50°C

The ideal phosphorylating reagents for phosphodiester syntheses should meet the following criteria:

- They should differentiate between primary and secondary alcohols.
- After the phosphorylation of a nucleoside the phosphorylating power of the reagent should be preserved or easily be restored to allow a second esterification.
- The reagent should carry an auxiliary group that prevents side-reactions of synthetic esters and can easily be removed when it is no longer needed.

In the 1,3,2-dioxaphosphole method a bis(2-butene-2,3-diyl) pyrophosphate is used as the condensing agent. It allows two successive esterifications of one phosphate group to be performed without additional activation. First a 5′-O-protected nucleoside is added in methylene chloride; in the second reaction an unprotected nucleoside can be used, since only the 5′-OH group is able to attack the cyclic enediol 3′-nucleosidyl phosphotriester. Protected dinucleoside triesters are obtained in 80% yield. Removals of protective groups, methoxytrityl by means of trifluoroacetic acid in methylene chloride and 1-methylacetonyl by aqueous triethylamine, also give about 80% yield (Ramirez, 1975, 1977).

(i) phosphorylation: pyrophosphate reagent/Et$_3$N (CH$_2$Cl$_2$; 5 h; 0 °C
(ii) coupling with unprotected deoxythymidine: Et$_3$N/DMF; 14 h; 0 °C + 2 h; r.t.
(iii) 5′-O-(p-methoxytrityl) removal: TFA/CH$_2$Cl$_2$; 20 min; 0 °C
(iv) acetoinyl removal: Et$_3$N/H$_2$O/MeCN; 2.5 h; 0 °C

The chemical basis for the phosphite triester approach is the observation, that dialkyl phosphorochloridites such as (C$_2$H$_5$O)$_2$PCl react very rapidly at the 3′-OH of nucleosides in pyridine even at low temperatures. In contrast, the reactions of analogous chloridates, e.g. (C$_2$H$_5$O)$_2$POCl, require several hours at room temperature. It was later found that phosphite esters can be oxidized quantitatively to the phosphates by using iodine in water and that clean condensation of phosphorochloridites with nucleosides can be achieved in THF at −78 °C. To develop this chemistry into a useful synthetic procedure it was necessary to establish which protecting groups are compatible with the highly reactive phosphorochloridites. It was found that *O*-trityl, methoxytrityl, acetyl, phenoxyacetyl, and benzoyl are stable.

In the condensation of 5′-*O*-(phenoxyacetyl)thymidine with 3′-*O*-(mono-*p*-methoxytrityl)thymidine, outlined in below, β,β,β-trichloroethyl phosphorodichloridite ($Cl_3CCH_2OPCl_2$) was used. The reaction time at −78 °C was only 5 min. Oxidation with iodine in water and chromatography on silica gel gave the desired dinucleoside phosphate triester. Cleavage of the methoxytrityl ether with aqueous acetic acid afforded the 3′-OH compound; similar treatment with ammonium hydroxide yielded the 5′-OH product. The synthetic sequence was repeated at the 5′-OH compound with more 5′-*O*(phenoxyacetyl)thymidine and phosphorodichloridite. This cycle yielded the trinucleotide derivative (69%), the next cycle gave the tetra-nucleotide derivative (75%). The removal of the trichloroethyl groups from the phosphate esters and of the mono-*p*-methoxytrityl group from 3′-oxygen to yield the trinucleotidyl-nucleoside dTpTpTpT was achieved in 70% yield by reduction with sodium-naphthalene in hexamethylphosphoric triamide (Letsinger, 1976).

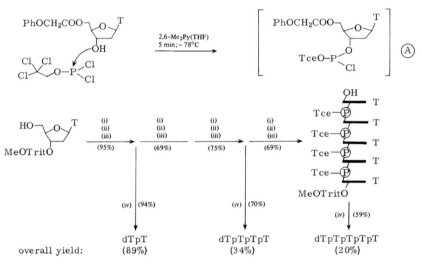

(i) coupling: Ⓐ /2,6-Me$_2$Py(THF); 15min; −78°C
(ii) phosphite triester oxidation: + I$_2$/H$_2$O; 5min; 0°C
(iii) 5′-*O*-phenoxyacetyl removal: conc. NH$_3$/H$_2$O/dioxane; 10min; r.t.
(iv) Tce and 3′-*O*-(*p*-methoxytrityl) removal: Na/ ⬡⬡ /HMPTA; 5min; r.t.

Phosphoramidites are a further improvement as stable phosphorylating agents, combined with tetrazole as activating agent (Caruthers, 1980; Matteucci, 1980, 1981; Beaucage, 1981). Methyl or 2-cyanoethyl protecting groups are used to prevent side-reactions of the phosphoramidite and to improve solubilities. The amino groups of adenosine and cytosine are protected by benzoyl, those of guanosine by isobutyryl groups. Thymidine is unreactive and does not need a protecting group.

Automated solid-phase DNA syntheses (see Section 5.3) usually start with a deoxynucleoside linked to the solid support through the 3'- or 5'-hydroxy group. A cross-linked poly(N,N-dimethylacrylamide) resin can be used, which swells to about ten times its dry bed volume in polar solvents (Atherton, 1975). It contains long spacers with terminal 4-(2-hydroxyethylthio)-phenyl groups for the reversible binding of the 5'-terminal deoxynucleotide (Gait, 1976). Other, more often used, non-swelling supports are silica or porous glass particles, e.g. about 0.15 mm in diameter with 50- or 100-nm pores, which have organic linkers attached to silanol groups on the surface. All free silanol groups are "capped" with chlorotrimethylsilane. The amino end groups of the linkers are connected to the 3'-O of the starting 5'-O-(4,4'-dimethoxytrityl) protected deoxynucleoside through a succinyl bridge by a stable amide and a base-labile ester bond.

In *Step 1* of each oligonucleotide-synthesis cycle the 5′-terminal 4,4′-dimethoxy-trityl protecting group is removed with trichloroacetic acid, and the support is washed with acetonitrile to prevent detritylation of the next incoming phosphoramidite. The 4,4′L-dimethoxytritylium cation produces a brilliant orange color (λ_{max}=498 nm) useful for monitoring the success of each cycle.

The acid treatment in each detritylation step may remove purines from deoxyriboses. Purine residues near the 3′-end will suffer the longest cumulative times of exposure to acid and therefore have the greatest chance for "depurination". Thus each acid treatment should be as brief as possible.

In *Step 2* of each synthesis cycle a protected deoxynucleoside 3′-phosphoramidite and tetrazole are added simultaneously. Tetrazole (pK_a=4.8) protonates the amidite nitrogen thus providing a very good leaving group. Nucleophilic attack by the free 5′-OH group yields the desired $5′ \rightarrow 3′$ internucleotidic linkage. 2-Cyanoethyl phosphoramidites react more slowly than the methyl-protected ones. In general, a tenfold excess of the phosphoramidite is needed to ensure nearly quantitative yields (>98%). At larger scales high coupling efficiencies can also be maintained with a fivefold excess.

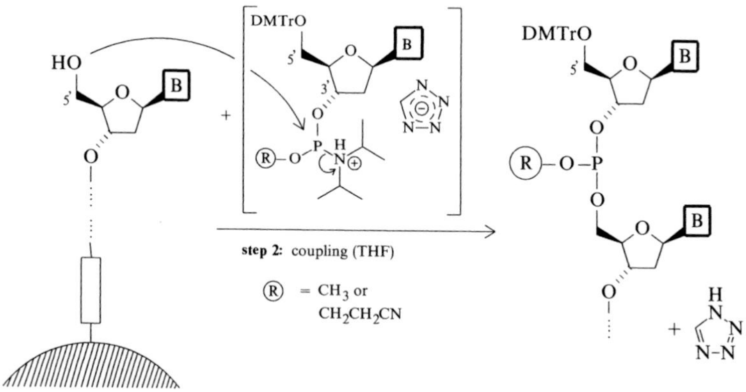

step 2: coupling (THF)

$\circledR$ = CH_3 or CH_2CH_2CN

Because coupling is not always quantitative, the nonreacted terminal deoxynucleoside must be excluded from the subsequent synthesis cycles. Otherwise deletion sequences will render the isolation of the pure final product difficult. Therefore a "capping" step (*Step* 3) follows, e.g. acetylation with acetic anhydride and N,N-dimethyl-4-pyridinamine (DMAP) in dioxane. "Capping" times should be as short as possible, especially with 2-cyanoethyl phosphite triesters, which are sensitive to bases such as DMAP.

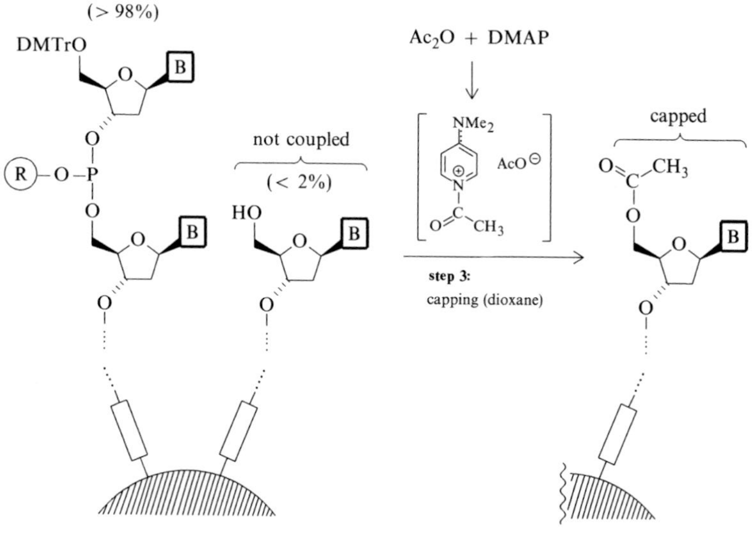

step 3: capping (dioxane)

The newly formed phosphite triester linkage is unstable to acids and bases and is immediately oxidized to a stable phosphate triester (*Step* 4). A solution of iodine, water, 2,6-dimethylpyridine, and tetrahydrofuran is commonly used. The oxidation is usually complete within 30 s.

One synthesis cycle is now complete. A new cycle can start with 5'-detritylation (*Step 1*). When the last cycle is complete, the terminal 5'-O-(4,4'-dimethoxytrityl) group of the desired oligonucleotide is either removed or left attached before the deprotection, liberation, and purification of the product.

In the *final steps* all protecting groups are removed, and the oligonucleotide is cleaved from the support. Methyl phosphate groups are cleaved selectively – without cleavage of other phosphate ester linkages – by benzenethiol/triethylamine in dioxane. 2-Cyanoethyl phosphate groups and the 3'-succinamic ester anchoring group are cleaved by concentrated aqueous ammonia. The benzoyl and isobutyryl N-protecting groups are then removed by warming the resulting ammonia-DNA solution at 55 °C for several hours. If the 5'-(4,4'-dimethoxytrityl) group has been left on, it is removed by 80% aqueous acetic acid, usually after HPLC.

The resulting oligonucleotide is often of surprising purity as judged by analytical HPLC or electrophoresis, and up to 30 mg of a deoxyeicosanucleotide (20-base DNA) can be routinely obtained. Nevertheless, small amounts of short sequences, resulting from "capping" and from base-catalyzed hydrolysis, must always be removed by quick gel filtration, repeated ethanol precipitation from water (desalting), reverse-phase HPLC, gel electrophoresis, and other standard methods.

The attraction for research chemists lies, of course, not in further perfection of the machine – which is best done systematically by the manufacturers – but in intelligent modifications of the target molecules.

Di- and trinucleotides may be used as units instead of the monomers. Such "convergent" synthetic strategy simplifies the purification of products, since they are differentiated by a much higher jump in molecular mass and functionality from the educts than in monomer additions, and it raises the yield. We can illustrate the latter effect with an imaginary sequence of seven synthetic steps, e.g. nucleotide condensations, where the yield is 80% in each step. In a converging seven-step synthesis an octanucleotide would be obtained in $0.8^3 \times 100 = 51\%$ yield, compared with a $0.8^7 \times 100 = 21\%$ yield in a linear synthesis. This applies, of course, to all multi-step syntheses.

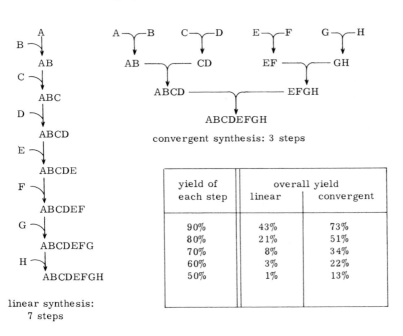

convergent synthesis: 3 steps

linear synthesis:
7 steps

yield of each step	overall yield	
	linear	convergent
90%	43%	73%
80%	21%	51%
70%	8%	34%
60%	3%	22%
50%	1%	13%

This "trick" can, of course, be applied in all multiple-step syntheses.

Synthetic oligonucleotides may be used as "primers" and be elongated stepwise with the aid of polynucleotide phosphorylase (PNPase) and nucleoside diphosphates.

$$d(TACG) + pp\text{-}dA \xrightarrow[{[Mn^{2+}]}]{[PNPase]^{*)}} d(TACGA) + P_i$$

* polynucleotide phosphorylase

More important is the enzymatic condensation of synthetic oligonucleotides by the "sticky end" approach. The term *"sticky ends"* indicates that oligonucleotides with at least four complementary bases at their ends (A=T and G≡C; the bonds indicate the number of hydrogen bridges between the paired bases) aggregate spontaneously and specifically in aqueous solution. One oligonucleotide molecule can bind two other oligonucleotides and bring them into close and well-defined contact with each other. An enzyme can then be used to condense the preformed aggregates to produce long, double-stranded DNA segments. Structural genes have already been synthesized from several oligonucleotides by simple mixing of all components and addition of DNA-ligase, which accepts all paired oligonucleotides as substrates (Itakura, 1977).

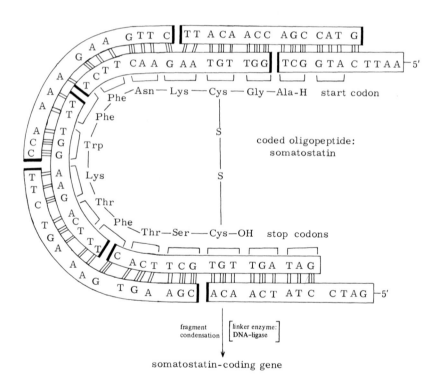

somatostatin-coding gene

The fastest synthesis of microgram quantities of deoxyribonucleic acid (DNA) is made possible by the *Polymerase Chain Reaction* (PCR; Arnheim, 1990; Mullis, 1990). Each PCR cycle begins by heat-denaturing a double-stranded DNA piece of any origin to give two single-stranded templates. Then the regions flanking the 3′-ends of the DNA and complementary DNA target sequences are annealed with site-specific primers. These primers, usually 20 nucleotides in length, need to be synthesized. The DNA-polymerase enzyme then catalyses the reaction with deoxynucleoside triphosphates to yield two double-stranded DNA pieces, each identical to the original piece, which serve as educts for the second synthesis cycle (denaturation, primer annealing, and polymerase reaction).

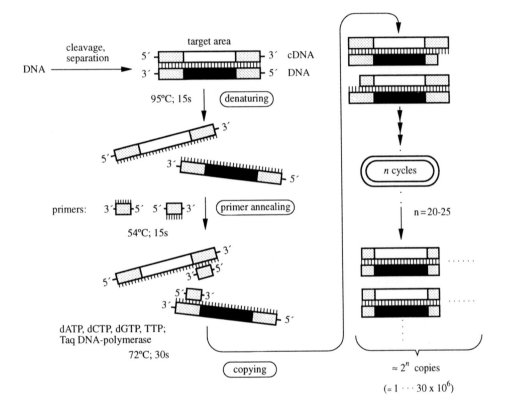

The reaction is usually carried out in a 0.1-mL volume in a 0.5-mL polypropylene tube. The reaction mixture contains picomoles (or less) of the target DNA, 20 nanomoles of each of the four deoxynucleoside triphosphates, 10 picomoles of each primer, and the appropriate salts and buffers. Denaturing occurs on heating to 95 °C for 15 s, another 15 s at 54 °C permit primer annealing, and a further 30 s at 72 °C are sufficient for the action of the thermostable DNA polymerase from *Thermus aquaticus* (Taq) archaebacteria. These three steps constitute one PCR cycle. Theoretically, n cycles produce 2^n times as much target DNA as before, e.g. 20 cycles result in a millionfold increase. PCR can generate several micrograms (approx. picomoles) of 1000-base pair targets representing less than one millionth of the genome of a higher organism from 1 µg of the total DNA. PCR can also be used to modify DNA sequences using primers differing at one or several positions from the target sequence. This is possible because PCR does not require perfect complementarity of a primer to the sequence flanking the target. Since all of the PCR products contain the primer sequence, an insertion or deletion can thus be incorporated into the product by modifying a primer. It is also possible to add new sequences to the 5'-ends of the primers. Modified or additional genetic information may thus be multiplied and transported.

The improvement of the technology of nucleic acid synthesis based on phosphoramidites is a matter of details nowadays, and these are published in data sheets for commercial apparatus. New strategies for the development of stronger binding, artificial "antisense strands" are needed. Three examples are given in the following for (i) rigidifying the furanose ring of the deoxyribose to produce locked DNA (=LNA), (ii) using peptide nucleic acids (=PNAs). Some of these DNA-analogues may inhibit the expression of specific genes and be used as therapeutic agent for the treatment of cancer and viral diseases.

The shape of a long DNA polymer strongly depends on the conformation of the furanose ring of deoxyribose. If that structural unit is immobilized, the binding between the single strands of double and triple helices becomes much stronger. Some bicyclic pentafuranose nucleoside monomers contain a rigidified ^{3}E conformation. In oligonucleotides they play the role of "locked nucleic acid" monomers and yield the most stable duplex-type nucleic acid system known. 2'-0,4'-C-linked bicyclo [2.2.1] nucleoside B was obtained from the nucleoside diol A by tosylation and treatment with sodium hydride. Subsequent deprotection–protection reactions, phosphoamidation, and oligomerization on a synthesizer yielded a locked nucleic acid oligomer (Wengel, 1999; Sørensen, 2002)).

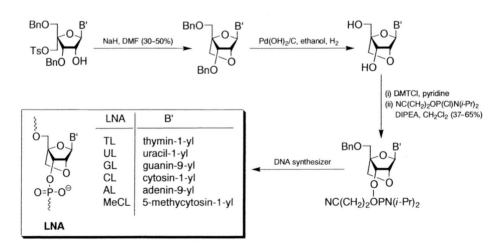

	LNA	B'
	TL	thymin-1-yl
	UL	uracil-1-yl
	GL	guanin-9-yl
	CL	cytosin-1-yl
	AL	adenin-9-yl
	MeCL	5-methycytosin-1-yl

LNA

Peptide nucleic acid polymers (PNAs) have uncharged internucleobase linkages, which do not produce any electrostatic repulsion to PNA and DNA/RNA targets. If the phosphate sugar backbone is replaced by (2-aminoethyl) glycine, the polymer is also achiral. It may easily be coupled with normal DNA or peptide synthesis in automated synthesizer machines. A synthetic scheme for assembly of PNA-peptide conjugates (Basu, 1997) is given in below. Section 5.3.1 provides an introduction to solid-state peptide synthesis.

Solid Phase Boc Peptide Nucleic Acid Synthesis

Deprotection and Cleavage: Anhydrous HF/Anisole, 0 °C, 45 min

H₂N-Gly-CCGCTTCCTTTC-C-Gly₄-Cys-Ser-Lys-Cys-CNH₂

PNA   peptide

If PNA conjugates become too insoluble or if their association becomes too slow, one may introduce some positively charged linkages. Guanidinium-linked nucleosides can, upon phosphitylation to an amidite, be incorporated into oligodeoxyribonucleotides (ODN) using standard automated DNA synthesis. The guanidinium linkages in the new dinucleosides remain protected with acetyl, allyl oxy carbonyl, or trichloroethyl carbonic acid groups and are deprotected at the end of the synthesis under mild conditions. The synthesis of the protected guanidinium dinucleosides involves coupling of the 5′-amino group of 5′-amino-5′-deoxythymidine with the isothiocyanates R–N=C=S of the corresponding protecting groups. Treatment with $HgCl_2$ yields a carbodiimide which is directly coupled with the 3′-protected 5′-thymidin-amine. The N-protected guanidine-coupled dinucleoside is formed and mild deprotection reactions release the cationic immonium group in almost quantitative yields (Barawkar, 1998). Nucleic bases are not attacked.

5.3
Peptides

In 1969 a series of short communications appeared in the *Journal of the American Chemical Society*, which announced the total syntheses of a few milligrams of complete or almost complete enzymes containing about 120 amino acid units (Gutte, 1969; Hirschmann, 1969). Their biological activities ranged from 1% to 13% of the natural enzyme. The papers were received by the editor between 25 November and 19 December 1968 and were printed in the copy of the journal that appeared on 15 January 1969. This is, to our knowledge, a record. The community of chemists obviously liked the idea "that a protein molecule with true enzymatic activity toward its natural substrate can be totally synthesized." The "Merrifield solid-phase synthesis" entered thousands of laboratories world-wide.

Seven years later, in a critical review on the synthesis of peptides, the following statement was made: *"Chemists in particular should respect the classical criteria of what constitutes synthesis of a natural product, i.e., that synthesis of a natural product has been achieved, when the physical, chemical and biological properties of the synthetic compound match those of the natural prototype. Unfortunately not a single one of the 'synthetic proteins' satisfies these criteria. It is frequently argued that these criteria are not applicable to more complex situations, but lowering standards of purity is not likely to advance the field. Presently available analytical methods cannot adequately detect inhomogeneity in a high molecular peptide that is produced by stepwise synthesis. Consequently, the synthetic method must be chosen so that the product can be purified and critically evaluated by the available analytical techniques . . . ."* (Finn, 1976). Both methodologies, sophisticated analysis and rise of yields, have been optimized within the last decades.

The following short descriptions of the steps involved in the synthesis of a tripeptide will demonstrate the complexity of the problem as well as the fragility of the (protected) amino acid units. In the later parts of this section we shall describe actual syntheses of well-defined oligopeptides by linear elongation reactions and of less well-defined polypeptides by fragment condensation.

5.3.1
Stages of Peptide Synthesis

One starts with individual amino acids or with peptides and tries to achieve the regioselective formation of a new amide bond. In its most general form such syntheses of peptides involve the following stages:

(i)   Preparation of a "carboxyl component" by blocking the amino group of an amino acid or a peptide with a group Y.

$$H_3\overset{\oplus}{N}-\underset{R^2}{\overset{\overset{\textstyle H}{|}}{C}}-COO^{\ominus} \quad \xrightarrow{N^{\alpha}\text{-protection}} \quad \boxed{Y}-\underset{R^2}{\overset{\overset{\textstyle H}{|}}{N}}-\underset{}{\overset{\overset{\textstyle H}{|}}{C}}-COOH$$

(ii)   Synthesis of an "amino component" by protecting the carboxyl group of another amino acid or peptide by a group Z.

$$\underset{\underset{R^1}{|}}{H_3\overset{\oplus}{N}-\overset{\overset{\displaystyle H}{|}}{C}-COO^{\ominus}} \quad \xrightarrow{\text{carboxyl}\atop\text{protection}} \quad \underset{\underset{R^1}{|}}{H_2N-\overset{\overset{\displaystyle H}{|}}{C}-\overset{\overset{\displaystyle O}{||}}{C}-O-\boxed{Z}}$$

(iii) Protection (= prot.) of all other functional groups that are reactive in condensations. The "best" reagents for side-chain protections are summarized in Tab. 5.2. The Fmoc group is highly stable under acidic and nucleophilic conditions. It allows activation of the carboxyl group as acid chloride (Carpino, 1973, 1996).

$$\boxed{Y}-\overset{\overset{\displaystyle H}{|}}{N}-\overset{\overset{\displaystyle H}{|}}{\underset{\underset{R^2}{|}}{C}}-COOH \quad \xrightarrow{\text{side-chain}\atop\text{protection}} \quad \boxed{Y}-\overset{\overset{\displaystyle H}{|}}{N}-\overset{\overset{\displaystyle H}{|}}{\underset{\underset{R^2}{|}\atop\boxed{prot}}{C}}-COOH$$

$$\underset{\underset{R^1}{|}}{H_2N-\overset{\overset{\displaystyle H}{|}}{C}-\overset{\overset{\displaystyle O}{||}}{C}-O-\boxed{Z}} \quad \xrightarrow{\text{side-chain}\atop\text{protection}} \quad \underset{\underset{R^1}{|}\atop\boxed{prot}}{H_2N-\overset{\overset{\displaystyle H}{|}}{C}-\overset{\overset{\displaystyle O}{||}}{C}-O-\boxed{Z}}$$

The phenolic hydroxyl group of tyrosine, the imidazole moiety of histidine, and the amide groups of asparagine and glutamine are often not protected in peptide synthesis, since it is usually unnecessary. The protection of the hydroxyl group in serine and threonine (O-acetylation or O-benzylation) is not needed in the azide condensation procedure but may become important when other activation methods are used.

(iv) Activation of the carboxyl component by a group X.

$$\underset{\boxed{prot}}{\boxed{Y}-\overset{\overset{\displaystyle H}{|}}{N}-\overset{\overset{\displaystyle H}{|}}{\underset{\underset{R^2}{|}}{C}}-COOH} \quad \xrightarrow{\text{carboxyl}\atop\text{activation}} \quad \underset{\boxed{prot}}{\boxed{Y}-\overset{\overset{\displaystyle H}{|}}{N}-\overset{\overset{\displaystyle H}{|}}{\underset{\underset{R^2}{|}}{C}}-\overset{\overset{\displaystyle O}{||}}{C}\sim\!\textcircled{X}}$$

(v) coupling of the activated carboxylic acid with the amino component to give a protected peptide. This is usually performed in an organic solvent in the presence of a slight excess of a base.

$$\underset{\boxed{prot}}{\boxed{Y}-\overset{\overset{\displaystyle H}{|}}{N}-\overset{\overset{\displaystyle H}{|}}{\underset{\underset{R^2}{|}}{C}}-\overset{\overset{\displaystyle O}{||}}{C}\sim\!\textcircled{X}} \; + \; \underset{\boxed{prot}}{H_2N-\overset{\overset{\displaystyle H}{|}}{\underset{\underset{R^1}{|}}{C}}-\overset{\overset{\displaystyle O}{||}}{C}-O-\boxed{Z}} \quad \xrightarrow[{-\,HX}]{\text{coupling}} \quad \underset{\boxed{prot}\quad\boxed{prot}}{\boxed{Y}-\overset{\overset{\displaystyle H}{|}}{N}-\overset{\overset{\displaystyle H}{|}}{\underset{\underset{R^2}{|}}{C}}-\overset{\overset{\displaystyle O}{||}}{C}-N-\overset{\overset{\displaystyle H}{|}}{\underset{\underset{R^1}{|}}{C}}-\overset{\overset{\displaystyle O}{||}}{C}-O-\boxed{Z}}$$

fully protected dipeptide

Tab. 5.2 Protection of amino acid side-chains

Functional group	Amino acid	Protected derivative	Reagent	Conditions for removal
$-NH_2$	lysine	N^{ω}-(diisopropylmethoxy-carbonyl	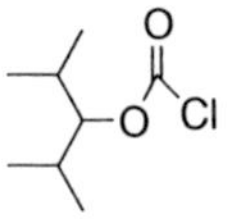	HF/anisole
		N^{ω}-benzyloxycarbonyl (=Cbz, Z) eta. X=H, Cl, NO_2	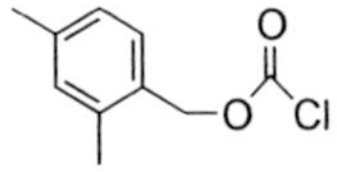	HBr/AcOH or H_2Pd
		N^{ω}-(9-fluorenylmethoxy-carbonyl)	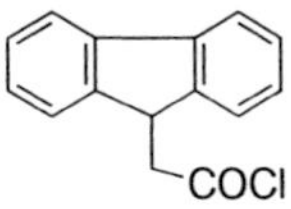 —COCl	morpholine
(guanidine structure shown) H N NH₂ NH	arginine	guanidinium cation N,N'-bis (adamantyloxycarbonyl) = Adoc N-nitro	PH < 10 HNO₃ / H₂SO₄	$H^{\oplus}$ H_2/Pd
$-SH$	cysteine	tripehnylmethyl (=trityl, Trit)	Ph_3CCL	$H^{\oplus}/Hg^{2\oplus}/I_2$
		acetamidomethyl (=Acm)	AcHN OH	$Hg^{2\oplus}/I_2$
		ethylcarbamoyl (=Ec)	EtNCO	$Hg^{2\oplus}$
$-S-CH_3$	methionine	sulfoxide	H_2O_2	$HSCH_2COOH$
$-COOH$	aspartic acide glutamic acid	benzyl and t-butyl esters or ethers, resp.	$PhCH_2Br$; /BF₃.Et₂O	HBr/TFA
(phenol structure) —OH	tyrosine serine			
$-OH$	threonine			

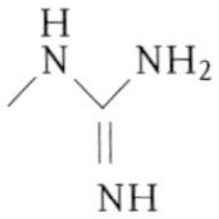

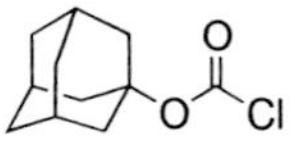

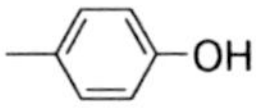

(vi) Removal of the blocking group Y.

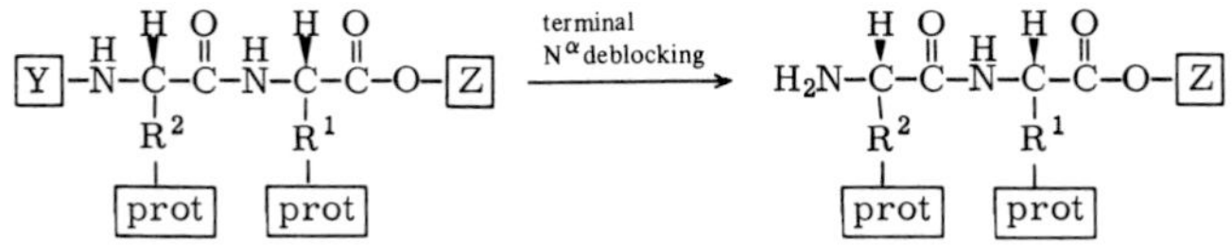

(vii) Repeated coupling with an activated acid.

$$Y-N-C-C\sim X + H_2N-C-C-N-C-C-O-Z \xrightarrow{-HX} Y-N-C-C-N-C-C-N-C-C-O-Z$$

fully protected tripeptide

(viii) Removal of protecting groups from side-chains R, amino, and carboxyl groups to give the free peptide.

$$Y-N-C-C-N-C-C-N-C-C-O-Z \xrightarrow[\text{deprotection}]{\text{complete}} H_3\overset{\oplus}{N}-C-C-N-C-C-N-C-COO^{\ominus}$$

tripeptide

Protected amino acids with either a free amino or a free carboxyl function can usually be prepared by proven methods or are even commercially available. Therefore stages (i)–(iii) may be considered as simple routine nowadays, although great care must be taken that the protected starting materials are pure enantiomers. The reactions that cause most trouble are in stages (iv), (v), and (vii). In these stages an activated carboxyl group is involved and the chiral centre adjacent to it is in peril of racemization. A typical reaction which causes epimerization is azlactone formation. With acids or bases these cyclization products may reversibly enolize and racemize. Direct racemization of amino acids has also been observed.

enantiomeric azlactones

Groups R, Y, and X affect the kinetics of these processes as do reaction conditions. The best X is azide, which gives little or no racemization. This has been rationalized with the assumption that the relatively electropositive azide group attracts the electronegative amide oxygen. This effect could stabilize the cyclic conformation given below, which does prohibit azlactone formation. Another explanation is that the azides are only moderately activated and therefore differentiate between the weakly nucleophilic oxygen and the strong amine-nucleophile.

R'O N H R

stable conformation of
N^{α}-acylated amino acid azides

With dicyclohexylcarbodiimide (DCC/EDCC) reagents, racemization is more pronounced in polar solvents such as DMF than in CH_2Cl_2, for example. An efficient method for reduction of racemization in coupling with DCC is to use additives such as *N*-hydroxysuccinimide or 1-hydroxybenzotriazole. A possible explanation for this effect of nucleophilic additives is that they compete with the amino component for the acyl group to form active esters, which in turn react without racemization. There are some other condensation agents (e.g. 2-ethyl-7-hydroxybenz[*d*]isoxazolium and 1-ethoxycarbonyl-2-ethoxy-1,2-dihydroquinoline) that have been found not to lead to significant racemization. They have, however, not been widely tested in peptide synthesis.

The protecting group Y of the amine is generally an alkoxycarbonyl derivative since their nucleophilicity is low. Benzyloxy- or *tert*-butoxycarbonyl derivatives usually do not undergo azlactone formation.

The best preventive measure against racemization in critical synthetic steps (e.g. fragment condensation) is to use glycine (which is achiral) or proline (no azlactone) as the activated carboxylic acid component. The next best choice is an aliphatic monoamino monocarboxylic acid, especially one with large alkyl substituents (valine, leucine). Aromatic amino acids (phenylalanine, tyrosine, tryptophan) and those having electronegative substituents in the β-position (serine, threonine, cysteine) are, on the other hand, most prone to racemization. Reaction conditions that inhibit azlactone formation and racemization are non-polar solvents, a minimum amount of base, and low temperature. If all precautions are taken, one still has to reckon with an average inversion of 1% per condensation reaction. This means, for example, that a synthetic hectapeptide contains only $0.99^{100} \times 100\% = 37\%$ of the fully correct diastereomer.

We now turn from the general problems of peptide synthesis to specific problems connected with three currently important procedures, namely:

- solid-phase peptide synthesis
- solution methods for peptide synthesis
- condensation of peptide fragments.

To illustrate the specific operations involved, the scheme below shows the first steps and the final detachment reaction of a peptide synthesis starting from the carboxyl terminal. *N*-Boc-glycine is attached to chloromethylated styrene polymer resin, cross-linked with 1% of divinylbenzene. This polymer swells in organic solvents but is completely insoluble. Polystyrene (PS) based resins have been classified as supports that retard reaction rates. An alternative popular support is the Tentagel® (TG) resin. It is based on a very small portion of cross-linked PS

grafted with long polyethyleneglycol (PEG) spacers containing 50–60 ethylene oxide units. Tentagel® has more "solution-like" properties, and reactions of solid-
phase peptide and nucleotide synthesis should be faster. The same should hold
true for all solid-phase-supported organic synthesis. Quantitative comparisons between PS and TG supports showed, however, that rate differences depended more
on the polarity of the reagents. Apolar reagents reacted faster on PS, polar reagents on TG. Both gel surfaces behave like another "solvent phase" rather than
as solids (Li, 1998). Among alternative supports for solid-phase synthesis are non-
swelling porous glass beads. They exhibit higher rates of mass transfer and higher reaction rates. Silanol groups on the surface are used to attach organic substituents. Treatment with HCl in acetic acid removes the *tert*-butoxycarbonyl (Boc)
group as isobutene and carbon dioxide. The resulting amine hydrochloride is neu-
tralized with triethylamine in DMF.

Then *N*-Boc-*O*-benzylserine is coupled to the free amino group with DCC. This
concludes one cycle (N^α-deprotection, neutralization, coupling) in solid-phase synthesis. All three steps can be driven to very high total yields ($\geq 99.5\%$) since excesses of Boc-amino acids and DCC (about fourfold) in CH_2Cl_2 can be used and
side-reactions that lead to soluble products do not lower the yield of condensation
product. One side-reaction in DCC-promoted condensations leads to N-acylated
ureas. These products will remain in solution and not react with the polymer-
bound amine. At the end of the reaction time, the polymer is filtered off and
washed. The times required for 99% completion of condensation vary from 5 min
for small amino acids to several hours for a bulky amino acid, e.g. Boc-Ile, with
other bulky amino acids on a resin. A new cycle can begin without any workup
problems (Merrifield, 1969; Erickson, 1976; Bodanszky, 1976; Pennington, 1997).

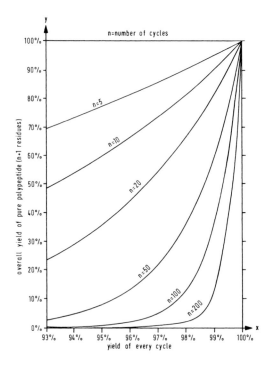

The purity of the peptide finally obtained depends critically on the yield of each cycle. It must be extraordinarily good to produce even moderately pure products (Lübke, 1975). If the average yield of amide formation in the synthesis of an unde-capeptide ($n=10$) is, for example, 98%, the product will already contain about 20% of different impurities which may be difficult to remove.

To keep reaction times low, one generally uses DCC as condensing agent. With easily racemizing amino acids, however, this procedure is not safe. The safe azide method, on the other hand, is too slow. Commercial 4-nitrophenyl esters are usually selected as simple alternatives (Bodanszky, 1976). 1,2,4-Triazole can serve as a bifunctional catalyst which accelerates the coupling rate of 4-nitrophenyl esters. Nevertheless, 50% condensation needs in the order of hours, so completion can rarely be expected in less than 24 h, and often 4-nitrophenyl esters do not couple quantitatively at all. On the other hand, these esters offer two more advantages over the DCC method: protection of side-chain hydroxyl groups is not required, and troublesome side-reactions such as the formation of nitriles by dehydration of asparagines or glutamine are not observed.

The N-to-C assembly of the peptide chain is unfavorable for the chemical synthesis of peptides on solid supports. This strategy can, in fact, be dismissed for the single reason that repeated activation of the carboxyl ends on the growing peptide chain would lead to a much higher percentage of racemization. Several other more practical disadvantages also tend to disfavor this approach, and acid activation on the polymer support is usually only used in one-step fragment condensations.

In each step of the usual C-to-N peptide synthesis the N-protecting group of the newly coupled amino acid must be selectively removed under conditions that leave all side-chain protecting groups of the peptide intact. The most common protecting groups of side-chains are all stable towards 50% trifluoroacetic acid in dichloromethane, and this reagent is most commonly used for N_α-deprotection. Only *tert*-butyl esters and carbamates (= Boc) are solvolyzed in this mixture. Peptide chains that fail to couple with the N-protected amino acid still bear the free amino group, while chains that do couple have a blocked α-amino group. If the resin is reacted irreversibly with a great excess of an acylating agent, these amino groups will be permanently terminal because chain growth is inhibited. Acetylation with acetic anhydride and triethylamine in DMF after each coupling step reduces the concentration of deletion peptides by a factor of at least ten. If 3-nitrophthalic anhydride is used, the "terminated" peptides bear an additional negative charge and are colored. They are easily separated by chromatography on an anion-exchange resin. Final detachment of the peptide bound by a benzyl ester link from the solid support usually occurs by acidolysis. If a side-chain-protected peptide is needed, trifluoroacetic acid is used. The usual reaction time at 25 °C is 60 min, although after 5 min a high percentage may already be set free. If the unprotected peptide is to be liberated, liquid HF for 1 h at 0 °C removes all side-chain-protecting groups and cleaves the anchoring bond in one step. For obvious reasons this method cannot be used with glass-bead supports. Anisole or ethyl methyl sulfide may be added as nucleophilic scavengers to protect tyrosine, tryptophan, histidine, and methionine residues from alkylation by carbonium ions (e.g. from Boc) produced during cleavage. It is usually advisable to go first for protected peptides, since they frequently crystallize whereas the free peptides rarely do so.

The problems of solid-phase synthesis are best exemplified by an early synthesis of cyclo-[-L-Val-D-Pro-D-Val-L-Pro-]$_3$ using the original experimental procedure of Merrifield himself, the inventor of the method (Gisin, 1972).

The linear peptide was synthesized first starting with L-proline at the C-terminal. Commercial polystyrene-co-1% divinylbenzene was treated with several solvents at 90–100 °C to transfer it to a swollen state and was then chloromethylated. The chloromethyl groups were converted into acetoxymethyl. Any remaining traces of chloromethyl groups were aminolyzed with boiling ethylamine, thus eliminating the chance of subsequent reactions of amines on the C-Cl bond. The resin was mixed at low temperature with *tert*-butoxycarbonyl-L-proline (=Boc-L-Pro) in methylene chloride and with carbonyldiimidazole as condensing agent. Reaction time was three days at room temperature. Hydrolysis of the resin indicated a substitution of approx. 130 mg (=0.6 mmol) of Boc-L-Pro per gram of resin. The starting material was now 4.5 g of Boc-L-prolyl resin.

The peptide chain was built up using the following cycle: deprotection with 50% trifluoroacetic acid in CH_2Cl_2, neutralization with diisopropylamine in CH_2Cl_2, coupling with a twofold excess of both DCC and Boc-D(or L)-Val-OH or Boc-D-(or L)-Pro-OH and again with 1 equiv. of each for 2 h. In order to block unreacted amino groups and also hydroxyl groups that may have formed by detachment of peptide from the resin, acetylation steps (Ac₂O/Py) were added after the tri-, tetra-, penta-, and nona-peptide stages. Afterwards the bound Boc-peptide was deblocked and the free amino groups were determined as picrates with picric acid. The total amount of amine was as follows: H-L-Pro-resin = 100%; tri-peptidyl resin 72%; pentapeptidyl resin 62%; heptapeptidyl resin 62%; nonapeptidyl resin 59%; dodecapeptidyl resin 60%. Thirty percent of educt was lost in the first two condensation steps (see below). The later condensation steps gave essentially quantitative yields.

The amino acid analysis of all peptide chains on the resins indicated a ratio of Pro:Val = 6.6:6.0 (calcd. 6:6). The peptides were then cleaved from the resin with 30% HBr in acetic acid and chromatographed on sephadex LH-20 in 0.001 M HCl, when 335 mg dodecapeptide was isolated. Hydrolysis followed by quantitative amino acid analysis gave a ratio of Pro:Val = 6.0:5.6 (calcd. 6:6). Cyclization in DMF with Woodward's reagent K (see scheme) yielded after purification 138 mg of needles of the desired cyclododecapeptide with one equiv. of acetic acid. The compound yielded a yellow adduct with potassium picrate, and here an analytically more acceptable ratio Pro:Val of 1.03:1.00 (calcd. 1:1) was found. The mass spectrum contained a molecular ion peak. No other spectral measurements (lack of ORD, NMR) have been reported. For a thirty-six step synthesis in which each step may cause side-reactions the characterization of the final produce should, of course, be more elaborate.

The cyclopeptide described above was tailored to form stable potassium complexes. It is one of the very few examples of complex peptide syntheses which did not lead to a natural compound.

The syntheses described met some difficulties. D-Valyl-L-prolyl resin was found to undergo intramolecular aminolysis during the coupling step with DCC. 70% of the dipeptide was cleaved from the polymer, and the diketopiperazine of D-valyl-L-proline was released into solution. The reaction was catalyzed by small amounts of acetic acid and inhibited by a higher concentration (protonation of amine). This side-reaction can be suppressed by adding the DCC prior to the carboxyl component. In this way, the carboxyl component is "consumed" immediately to form the DCC adduct and cannot catalyze the cyclization.

R–COOH + H–D-Val–L-Pro–O–[polymer]

cyclic dipeptide
(diketopiperazine)

Stepwise methods provide excellent routes to peptides up to the pentadecapep-tide range. Homogeneous peptides containing a hundred or more amino acid resi-dues can only be made from rigorously purified smaller protected peptides. In the fragment assembly technique peptides in the range of decapeptides are con-densed. The separation of unreacted small peptide educts from the peptide prod-uct of much higher molecular mass is without problems and rigorous proof of homogeneity can often be accomplished.

Problems in the condensation of oligopeptide blocks are the frequently observed low coupling yields of large fragments and the sparing solubility of large pro-tected peptides. The latter difficulty can be overcome by the use of DMSO or HMPTA as solvents. These highly polar solvents, however, favor racemization, and only peptides with carboxyl-terminal glycine or proline should be selected for the construction of complex peptides when DCC or similar condensing agents are used. Other N-protected peptides can only be coupled by time-consuming azide methods, and nonracemization has to be demonstrated after each fragment con-densation.

A stable "safety catch amide linker" (SCAL) was developed for such an assembly of large polypeptides on a solid support. All side-chains of the peptides had to be unprotected and mild thioester coupling and decoupling reactions were applied for support attachments. The carboxyl end of peptide 1 was bound to the amino group of SCAL, whose sulfoxide groups stabilize it and make it compatible with protection/deprotection as well as with ligation chemistries between peptides. After reduction of the aryl sulfoxide to the sulfide, however, SCAL becomes acid-labile and can be cleanly cleaved off with a mixture of thioanisol, $SiCl_4$, and TFA to give a C-terminal amide of proteins, e.g. the 71-mer MIP I, in >90% yield (Brik, 2000).

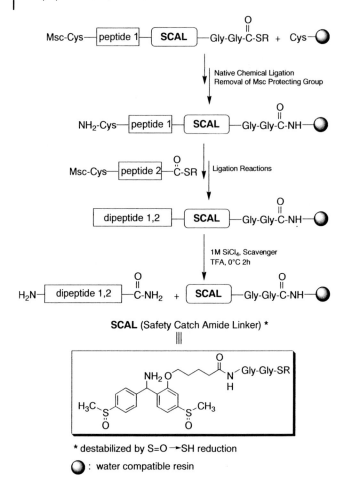

SCAL (Safety Catch Amide Linker) *

* destabilized by S=O →SH reduction

⬤ : water compatible resin

There is also a large choice of other "linkers" which may be introduced between the first substrate and the polymer. It may later be cleaved off by brief UV-irradiation, methylation with diazomethane, and similar rapid and efficient treatments (Balkenkohl, 1996).

The major disadvantage of solid-phase peptide synthesis is the fact that all the by-products attached to the resin can only be removed at the final stages of synthesis. Another problem is the relatively low local concentration of peptide which can be obtained on the polymer, and this limits the turnover of all other educts. Preparation of large quantities (>1 g) is therefore close to impossible. Thirdly, the racemization-safe methods for acid activation, e.g. with azides, are too mild (=slow) for solid-phase synthesis. For these reasons the convenient Merrifield procedures are quite generally used for syntheses of small peptides, whereas for larger polypeptides many research groups adhere to "classic" solution methods and purification after each condensation step (Finn, 1976).

The total synthesis of a protected heptatetracontapeptide (=47 amino acids) (Storey, 1972) may be used to compare the techniques and the results with those of solid-phase synthesis.

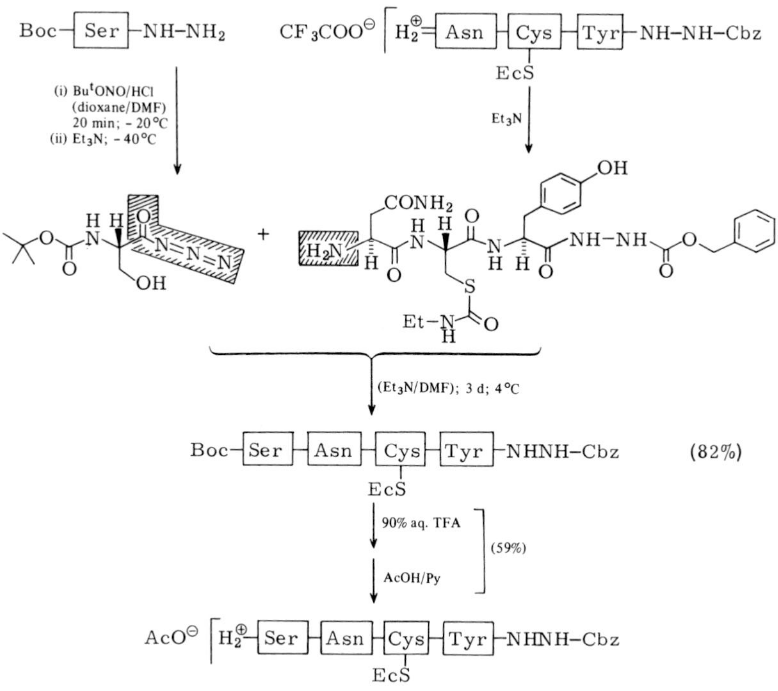

The comparison clearly demonstrates the main advantages of solution chemistry over the solid-phase approach:

- Most importantly, thin-layer chromatography can be used immediately to demonstrate homogeneity.
- Fewer protecting groups (Ser, Tyr unprotected) are sometimes needed.
- The azide method which gives only little racemization is complete within the reasonable time of 72 h at 4 °C.

On the other hand, a lot of material is lost in the chromatography and workup procedures ($\approx$ 40%), and the analytical data provided are not very convincing, either (except for TLC).

The analytical standards can, however, be raised considerably if ^{1}H-NMR and CD-spectroscopy are applied routinely in each step (Nutt, 1980). Fully synthetic oligopeptides of moderate size can thus be completely characterized.

An example of fragment condensation is taken from a convergent synthesis of bovine insulin (Zahn, 1970). The triacontapeptide B-chain of this hormone was built up from six protected oligopeptides. The following reaction sequence was used for every azide coupling cycle: (i) hydrazinolysis of the carboxyl-terminal methyl ester for 2 to 6 d, (ii) azide formation with isoamyl nitrite and HCl for 30 min at –20 °C, (iii) neutralization with triethylamine at –40 °C, and (iv) coupling with the amino component, first 12 h warming up from –40 °C to 0 °C, then 3 d at 0 °C. The thiol groups of cysteine residues were protected by formation of disulfide-linked dimers or polymers which were finally cleaved by "oxidative sulfitolysis". Large-scale peptide syntheses of pure materials have to rely on such procedures.

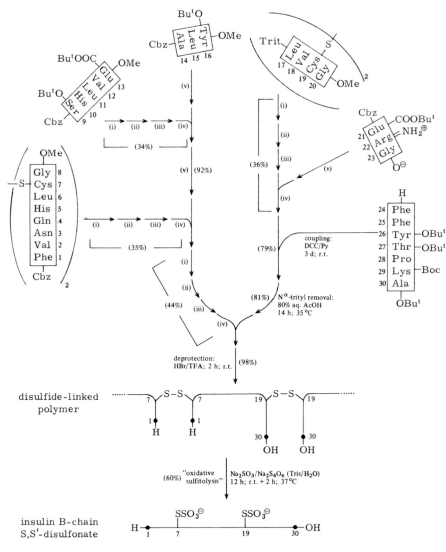

(i) hydrazinolysis:	$N_2H_4 \cdot H_2O/MeOH$; 2 - 6 d; r. t.
(ii) azide formation:	$Am^i ONO/HCl(THF/DMSO/DMF)$; 0.5 h; –20 °C
(iii) neutralization:	+ Et_3N; –40 °C
(iv) coupling:	12 h; –40→0 °C + 3 d; 0 °C
(v) Cbz hydrogenolysis:	$H_2[Pd]$; (AcOH, MeOH); 1-3 h; r. t.

Macrocyclic peptides and depsipeptides (= macrocyclic peptides with amide *and* ester linkages) are important natural compounds. They have been synthesized in low yield from open-chain precursors by DCC treatment at high dilution (Schröder, 1963; Shemyakin, 1961). More successful are solid-phase methods in which the linear precursor is attached through a labile ester bond (e.g. *o*-nitrophenyl) to a polymer.

First the protected oligopeptide is coupled with polymer-bound nitrophenol by DCC. N^α-Deblocking leads then to simultaneous cyclization and detachment of the product from the polymer (Fridkin, 1965). High dilution in liquid-phase cyclization is only necessary if the cyclization reaction is sterically hindered. Working at low temperatures and moderate dilution with moderately activated acid derivatives is the method of choice for the formation of macrocyclic lactams (Nutt, 1980).

The most reliable and efficient syntheses of specific proteins rely on living cells, especially "recombinant bacteria" (reviews: Robson, 1986; Itakura, 1980A, B, 1982; Gilbert, 1980; Glover, 1980; Williamson, 1979–1981; Sambrook, 2001). Organic chemists should therefore also be aware of the use of these "synthetic machines". Recombinant-DNA mutants, "recombinants", contain along with their own genes parts of foreign genes, e.g. DNA from human or animal cells. The combination of recombinant DNA techniques for placing and maintaining a new gene in dividing bacteria cells is called "gene cloning": a population of identical descendants, a clone, is formed. All cells will produce the foreign protein (e.g. a somatostatin, insulin, interferon, or growth hormone) and ideally release it into the culture medium. Today, even proteins that are easily accessible from natural sources are made on a commercial scale by recombinant synthesis, e.g. crystalline albumins (see Komatsu, 1999).

There are two convenient forms of genetic material that can be used as vehicles for introducing the new gene into the bacterium: a small circular DNA piece, called a plasmid, or a virus that grows in bacteria. The techniques described below apply to both plasmids and viruses.

The *first educt* (A) needed to build up recombinant bacteria which produce foreign proteins is a double-stranded DNA which encodes for the desired protein. This cannot be isolated directly from human or animal cells because it is a very small part of a large chromosome and often contains ten times more intervening DNA segments ("introns") which are not translated in protein synthesis. There are, however, messenger-RNA (mRNA) molecules present in the protein synthesis apparatus that carry just the required structural sequence plus the start and stop signals needed for the translation of the nucleic acid code into a protein structure. Such mRNAs can, for example, be isolated as electrophoretic spots from human pituitary gland tissue. Reverse transcriptase, an enzyme found in certain RNA viruses, is then used to copy the genetic information from the single-stranded mRNA into single-stranded DNA ("copy DNA", "cDNA"). The RNA template is removed, and a second complementary strand of DNA is attached with a DNA polymerase. After breaking the covalent bond between the two complementary strands with a nuclease, a double-stranded DNA is obtained, to which short terminal sequences of identical nucleotides, e.g. four cytosines, are attached with the aid of a terminal transferase to provide the sticky ends for further processing.

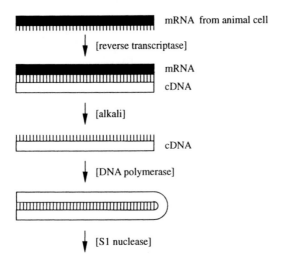

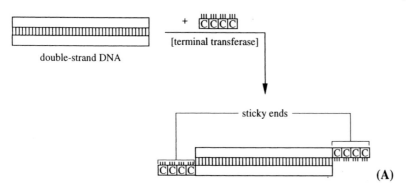

double-strand DNA

[terminal transferase]

sticky ends

(A)

As the *second educt* (B), the plasmid DNA with complementary sticky ends is prepared separately. In the first step the isolated plasmid DNA is cut open by a special type of enzyme called restriction endonuclease. It scans along the thread of DNA and recognizes short nucleotide sequences, e.g. CTGCAG, which are cleaved at a specific site, e.g. between A and G. Some 50 such enzymes are known and many are commercially available. The ends are then again extended with the aid of a terminal transferase by a short sequence of identical nucleotides complementary to the sticky ends of educt (A).

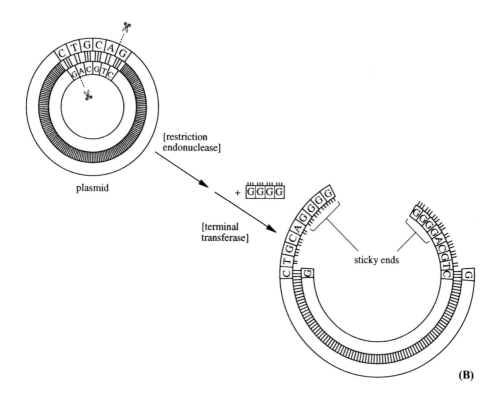

plasmid

[restriction endonuclease]

[terminal transferase]

sticky ends

(B)

The synthetic and plasmid DNAs are mixed and join their "sticky ends" spontaneously. They are covalently bound together by DNA ligases, when the resulting "hybrid plasmid" is inserted into bacterial cells. Dilute calcium chloride solutions render the bacterial membranes permeable and allow the passage of DNA into the cells.

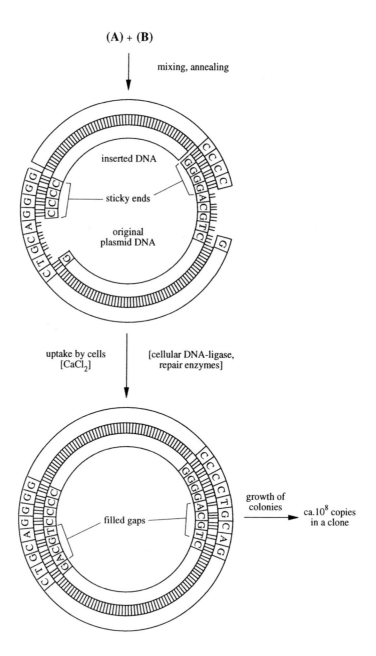

Their disadvantage lies mainly in their biological origin. Who guarantees the absence of trace amounts of carcinogenic or poisonous biomolecules, which are so abundant in microorganisms?

The yields of DNA intake by the cells are also very low. The few bacteria containing the hybrid plasmid need to be sorted out. One technique applies a plasmid which encodes for antibiotic-degrading enzymes, e.g. for penicillinase and tetracyclinase, and thus confers penicillin and tetracycline resistance if it is entrapped in a bacterium. The desired gene is now inserted within the penicillinase gene. The bacteria containing the hybrid plasmid will survive tetracycline treatment but will not be resistant to penicillin. Bacterial colonies containing the hybrid plasmids can thus be selected. A single colony contains about 100 million recombinant cells and from each single cell a new colony can be grown. Some of the selected clones will then produce small amounts of the desired protein sequence, e.g. of proinsulin ("expression of protein"). This will, however, be hidden in a larger penicillinase–proinsulin hybrid protein, which can be detected by antibodies directed against insulin or penicillinase. In order to make insulin, the hybrid protein has to be selectively digested by enzymes. Biologically active insulin was originally obtained in minute quantities (Gilbert, 1980), but similar processes have now become commercially successful. Human growth hormone, a protein with 191 amino acid residues, is nowadays produced in gram quantities.

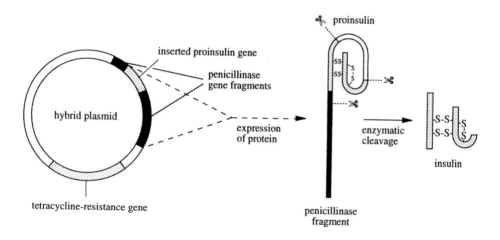

Plasmid biochemistry also allows modifications of given protein sequences (Robson, 1986). For exchanging a single amino acid ("single-site mutagenesis") a hybridized plasmid is isolated in the form of a circular single-stranded DNA. A synthetic oligodeoxyribonucleotide containing one deliberately mismatched base pair forms stable duplexes only at relatively low temperatures, say $10\,°C$, with this DNA. It acts as a primer for DNA polymerase to form a closed circular duplex after the action of a DNA ligase. After insertion of the plasmid into the host bacteria, about 10 to 40% will be "single-site mutants". These mutants can be identified by their ability to bind the original synthetic oligonucleotide not only at $10\,°C$ (in the case of partial mismatch) but also at $50\,°C$ (total match). Detailed procedures for site-directed mutagenesis are given by Zoller (1983) and Taylor (1985 A, B). Examples with high expression efficiency with respect to the total protein are rat liver cytochrome P-450$_d$ expressed in yeasts (Shimizu, 1988), human P-430$_F$ in plasmids (Domanski, 2001) and a bacterial cytochrome (Afolabi, 2001).

The synthesis of "partially modified retro-inverso" (PMRI) peptides, in which one –NH–CO– amide bond is rearranged to –CO–NH– has been largely evaluated. This modification results in a very strange –NH–CHR–NH–CO–CHR–CO– element within a peptide chain. Acidic CH–bonds between two carbonyl groups appear and the *gemini*-diamine may hydrolyze quickly and new hydrogen bonds form. Malonic acid is an important synthon in these syntheses, Hoffmann rearrangements are used to convert terminal amides to amines. Several other peptide bond surrogates are known (Fletcher, 1998). The condensation relations of N- or C-alkylated and a,β-dehydro amino acids as well as peptide macrocyclization have been reviewed (Humphrey, 1996, 1997).

parent peptide

one bond reversed

five bonds reversed

5.4
Di- and Oligosaccharide Syntheses

The central theme of carbohydrate chemistry is the construction of biologically active oligosaccharides and glycoproteins (Paulsen, 1990). Three closely related methods have proven to be reliable in early complex syntheses. They were all based on iterative Koenigs–Knorr reactions using halides, trichloracetimidates of thioglycosides, as glycosyl donors.

Halide derivatives may be fluorides, chlorides, or bromides. Fluorides are best prepared by the reaction of hydroxy groups with (diethylamino) sulfur trifluoride ("DAST"; M. Sharma, 1977) or of glycosyl thioethers with DAST/NBS (Nicolaou, 1990B). When the 2-hydroxy group of a monosaccharide reacts with DAST, quantitative and stereoselective rearrangements are observed (Nicolaou, 1986). This reaction may simultaneously introduce fluorine to C-1 and a new oxygen, sulfur, or nitrogen residue to C-2 with inversion of configuration. The other halides are usually only introduced at the glycosidic position, where treatment with hydrogen chloride or bromide is sufficient, and are easily exchanged for fluoride, e.g. with silver fluoride. Glycosyl fluorides react with free alcohol groups (e.g. of protected sugars, cholesterol) in the presence of tin(II) chloride and silver perchlorate or trifluoromethanesulfonate to give glycosides, with yields above 70% (Mukaiyama, 1981; Nicolaou, 1990 B). Another useful catalyst for the substitution of fluoride is dichlorobis(η^5-cyclopentadienyl)zirconium or -hafnium with silver perchlorate. It favors α-glycosidic linkages (Murakata, 1990). β-Glucosides and β-galactosides are obtained when a 2-O-acyl group blocks the α-position by formation of an intermediate cyclic acyloxonium cation (Nicolaou, 1990 B).

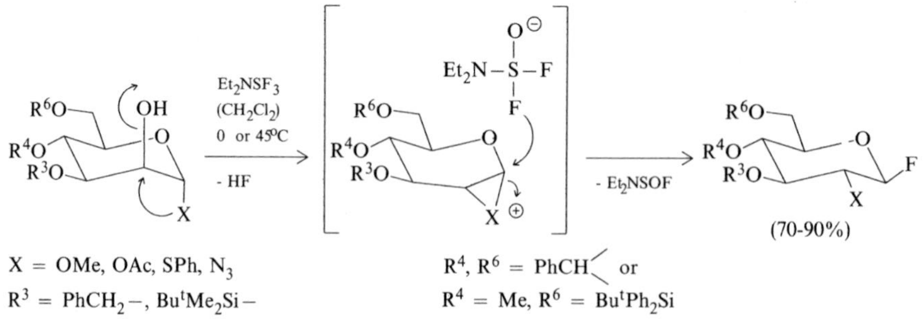

X = OMe, OAc, SPh, N_3

R^3 = $PhCH_2-$, Bu^tMe_2Si-

R^4, R^6 = $PhCH\diagdown$ or

R^4 = Me, R^6 = Bu^tPh_2Si

Trichloroacetonitrile reacts with glycosidic hydroxy groups of protected sugars to form glycosyl trichloroacetimidates (Schmidt, 1980, 1984, 1985, 1986; Wegmann, 1988). The imidate is substituted by alcohols in the presence of trimethylsilyl trifluoromethanesulfonate in ether to give high yields of glycosides. The silyl triflate catalyst favors the thermodynamically more stable product. More acidic catalysts such as boron trifluoride/diethyl ether are dangerous because they also cleave glycosidic linkages.

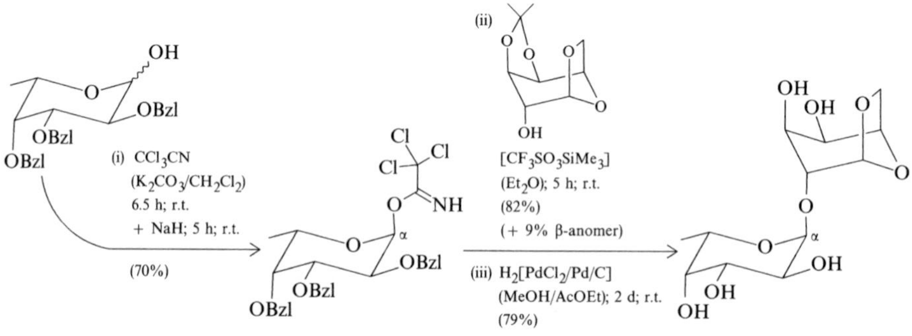

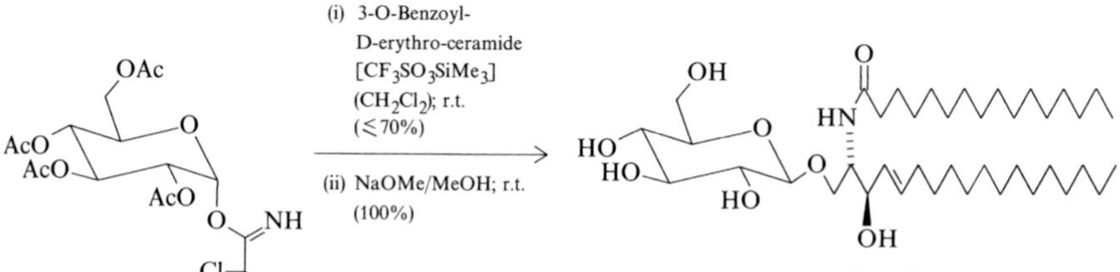

a cerebroside

The most versatile glycosylation reaction, however, is based on thioglycosides of sucrose (Ferrier, 1965; Oscarson, 2000). Thioglycosides are readily prepared, easily handled, and versatile. They are today the first choice in automated one-pot syntheses (Douglas, 1998, Zhang, 1999), because there is a broad spectrum of nonmetallic thiophiles, which activate the thiol as leaving groups. One possibility is oxidation by 1-benzenesulfinic piperidine (BSP) to a sulfoxide and subsequent triflation with trifluoromethanesulfonic anhydride. This treatment readily activates both armed and disarmed thioglycosides. It allows, for example, the double glycosylation of carbohydrate diols in high yield even with highly deactivated mannose derivatives (Crich, 2001).

Thioglycosides can be activated for glycosylation reactions with sulfur electrophiles, e.g. with dimethyl(methylthio)sulfonium triflate or with methanesulfenyl bromide and silver(I) to form reactive sulfonium intermediates (F. Dasgupta, 1988).

β-Mannosides are difficult to obtain since here a 2-O-acyl group blocks the β-position. 2-O-Benzyl-a-mannosyl bromides, however, give high yields of pure β-glycosides with a heterogeneous silver silicate catalyst preventing anomerization and S_N1 reaction of the bromide (H. Paulsen, 1981 B, C).

In contrast to the stepwise and highly stereoselective syntheses of hexoses and higher monosaccharides, carbohydrates $(CH_2O)_n$ have also been made by base-catalyzed oligomerization of formaldehyde in the presence of calcium hydroxide. A total yield of 55% of sugars was obtained comprising about 50% of hexoses. Glyceroaldehyde phosphate produces hexoses in the presence of sodium hydroxide alone and gives pentoses if formaldehyde is added (not shown).

In glycosylation reactions the component which contributes the anomeric carbon of the resulting glycoside is described as the glycosyl donor D. The most useful glycosyl donors are glycosyl halides, trichloroacetimidates or glycals. The corresponding glycosyl acceptor A furnishes the oxygen of the glycoside and consists usually of a free hydroxyl group in an otherwise fully protected carbohydrate. The following Scheme portrays, globally, the classical strategy of glycosylation, using first fully oxygenated pyranose donors and acceptors to produce the protected DA disaccharide and secondly the glycals, which have been activated by an electrophile E^+, e.g. I^+. Optimized yields with both methods usually do not exceed 60%, so solid-state synthesis of oligosaccharides is problematic (Wulff, 1974; Paulsen, 1990; Toshima, 1993; Schmidt, 1986).

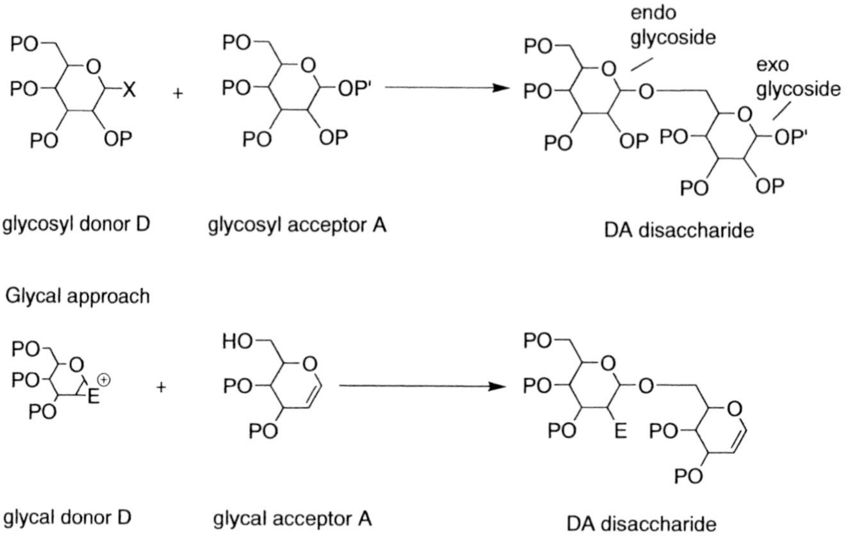

glycosyl donor D   glycosyl acceptor A   DA disaccharide

Glycal approach

glycal donor D   glycal acceptor A   DA disaccharide

P = protective group
X = activiting group, e.g. F, Cl, OC(NH)CCl$_3$
E = electrophile, e.g. O, I$^{\oplus}$

The most important factor in glycoside synthesis with respect to the glycosyl acceptor is steric hindrance. Two rules are of general significance:

- *Axial* hydroxyl groups usually react more slowly than *equatorial* hydroxyl groups.
- In the most important glucopyranosides, where hydroxyl groups in the ring and the CH$_2$OH group at C-5 are all *equatorial*, the order of reactivity is 6CH$_2$OH >> 3OH > 2OH > 4OH. Only the special situation at C-6 can, however, be utilized in regioselective synthesis. Partial protection of monosaccharides, which leaves only one OH-group free is usually a necessity and examples are described in Section 4.7.

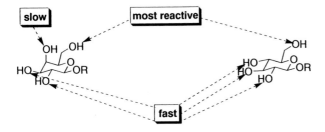

Apart from these rules each glycosidation has to be considered as an individual synthetic problem and the literature should be consulted extensively. Success depends upon the ability to form glycosidic bonds stereoselectively and to apply efficient protecting group strategies.

The common Koenigs–Knorr synthesis of glycosides consists of the reaction of an appropriate glycosyl halide with the alcohol in the presence of an acid acceptor such as silver oxide or silver triflate. Glycosyl halides are prepared by reaction of the corresponding hydrogen halide with a per-O-acyl-aldose or a per-O-benzyl-aldose, in which only the 1-O-substituent is replaced by halogen. Under acidic conditions the α-anomer always predominates because of the powerful anomeric effect. The relative reactivities increase in the order F << Cl < Br << I. Chlorides and bromides have been used most extensively, but fluorides offer the advantage of being stable in silica gel chromatography. A neighboring *trans*-acetoxyl group at C-2 participates in the halide elimination and leads to a cyclic 1,2-acyloxonium cation. Acetoxyl groups are therefore considered as participating groups, whereas benzylether groups do not form cyclic intermediates and are called nonparticipating. Participating groups on C-2 invariably lead to high yields of 1,2-*trans* glycosides (Lehmann, 1976, 1996).

1,2-*cis*-Glycosides are difficult to synthesize for steric reasons: there is little space on the surface of a pyran ring to accommodate a protected OH-group and an incoming saccharide at two neighboring carbon atoms with ease, cyclization of neighboring substituents with neighboring groups is therefore favored. *cis*-Glyco-

sides are, however, accessible from β-D-glucopyranosides and α-D-galactopyrano-
sides provided a non-participating group is present at C-2, i.e. benzyloxy or acetonide. The β-D-glycopyranosyl halide affords the α-glycoside when reacted with an
alcohol, silver perchlorate and *sym*-collidine in ether. The α-anomer of the glyco-
pyranosyl bromide is first converted by tetraethyl ammonium bromide to the β-
anomer in dimethyl formamide in the presence of a bulky base and then the alcohol and catalysts are added (not shown). Another common route to 1,2-*cis*-glyco-
sides applies solid catalysts, to which the α-halide is adsorbed first.

If the usual Koenigs–Knorr syntheses using chlorides and bromides are not
stereoselective enough, it is advisable to use stereochemically pure fluorides or trichloro-acetimidates as educts (Schmidt, 1986; Toshima, 1993). The use of glycosyl
fluorides needs a fluorophilic $SnCl_2$-$AgClO_4$ activator and leads preferentially to
retention of configuration. Trichloro-acetimidates are made of the otherwise protected half-acetal of a carbohydrate and trichloroacetonitrile. Glycosidation in presence of trifluoromethanefulgonate leads again preferentially to glycosides with retention of configuration. Treatment of the α-glycosides with base may, however,
convert them to the β-diastereomer, provided there is a participating substituent
on C-2.

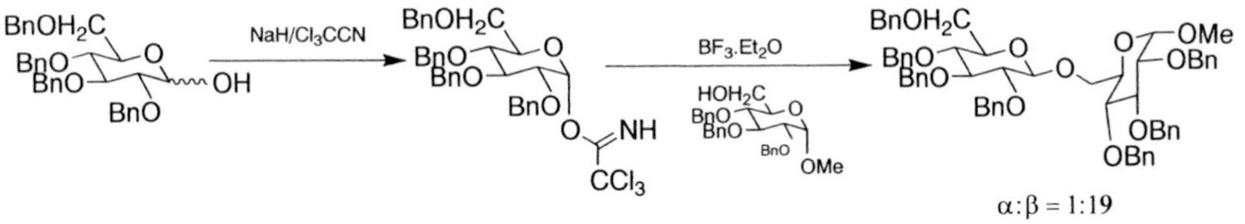
NaH/Cl3CCN
BF3.Et2O
α:β = 1:19

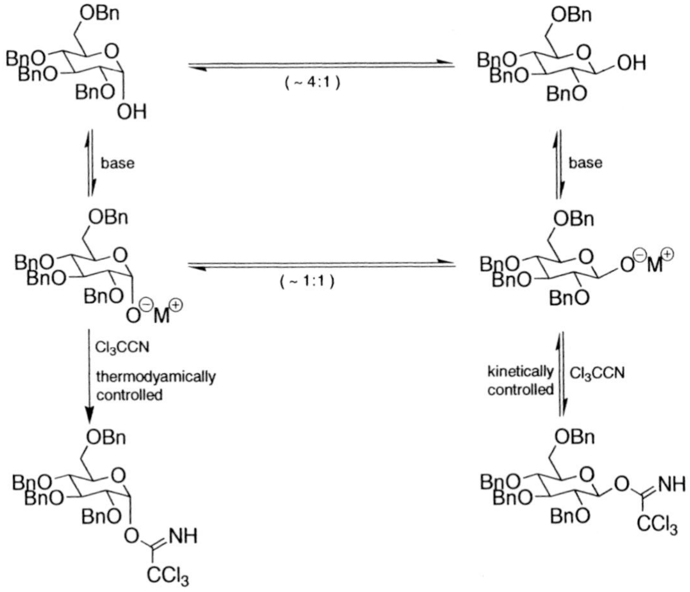
(~ 4:1)
base
base
(~ 1:1)
Cl3CCN
thermodyamically controlled
kinetically controlled
Cl3CCN

Another glycosylation method uses 4-pentenyl glycosides with different protecting groups as donors and acceptors. 4-Pentenyl glycosides with acyloxy groups at C-2 are much less reactive than glycosyl donors having a benzylether group at the same position. The activated 2-acyloxy donor is called a "disarmed" sugar, the activated benzylether an "armed sugar. Carbohydrates with the same leaving group, in particular *O*-pentenyl, *S*-pentenyl or *S*-phenyl, were combined in armed–disarmed pairs upon oxidation of the armed glycoside mixed with the unreactive disarmed partner, which only reacts as an alcohol component. Glycoside yields between 60 and 80% are typically achieved (Toshima, 1993).

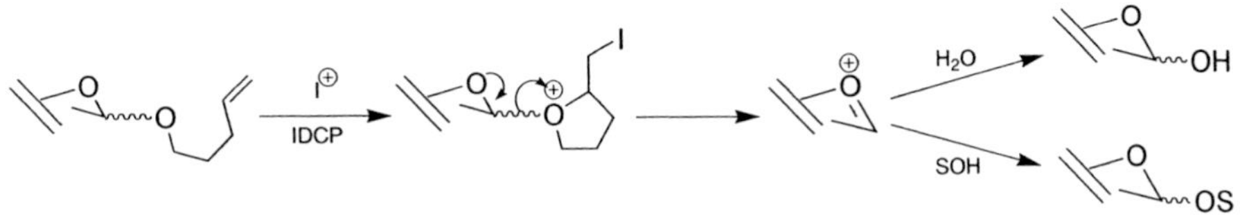
IDCP
H2O
SOH

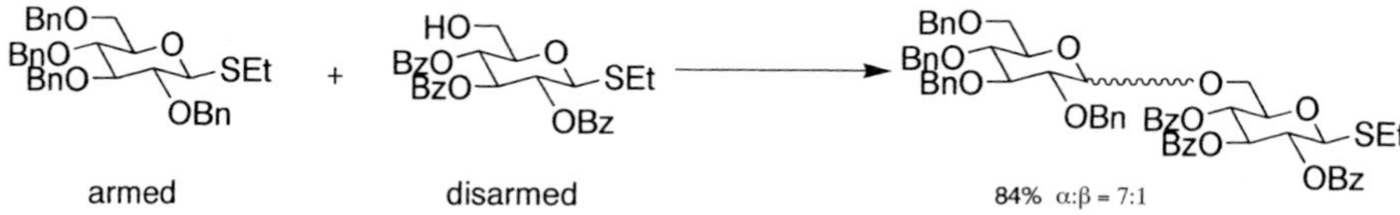

armed + disarmed → 84% α:β = 7:1

Oligosaccharides are usually assembled by iterated Koenigs–Knorr and related reactions described above: a glycosyl halide or related glycosyl electrophile is condensed with a partially protected second sugar unit. The crucial glycosidic bond is thus obtained by alkylation of an alcohol group. In complex oligosaccharide syntheses phenyl thioglycosides are preferably used as universal building blocks (Garegg, 1997), because they are stable under most reaction conditions applied in carbohydrate condensation reactions. They are activated by N-iodosuccinimide (NIS) under acidic conditions (CF_3COOH; TfOH). If two thioglycosides compete in a reversible reaction for the iodonium oxidant, only one may collapse to the corresponding transient oxocarbenium ion, which then rapidly and irreversibly reacts with the acceptor molecule to give a disaccharide. The second glycosyl donor D2 is not activated and does only react after D1 has disappeared. Side-chains can be introduced with appropriate reactive disaccharide synthons (Zhang, 1999).

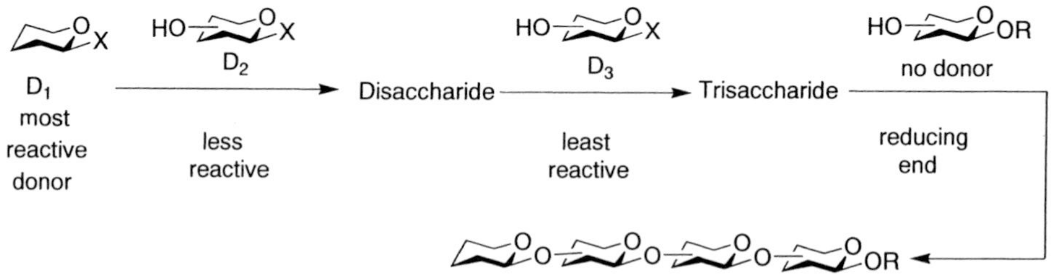

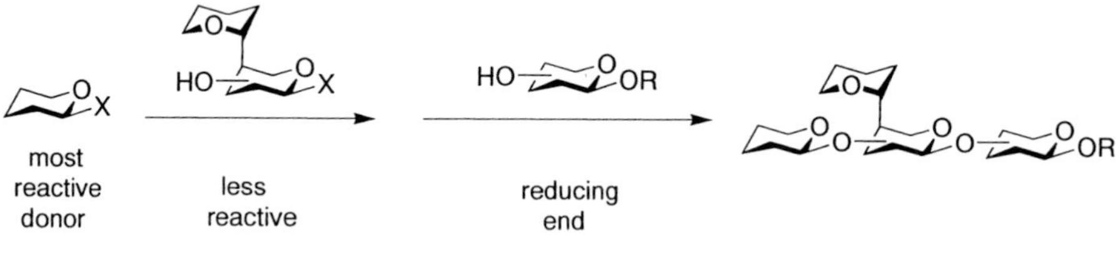

How can one now activate or de-activate a glycosyl donor in order to allow linear assembly in one-pot reactions? In the following scheme we give a simplified model, which has been quantified for many monosaccharide derivatives by quantum mechanical calculations. Galactose-tetraacetate is more active in the formation of the oxocarbenium ion than the glucose-analog, because the favorable 1,3-interaction of the basic acetyl oxygen atom stabilizes the positive charge on oxygen. The phthalimido group on C-2 of glucose does the same for the positive charge on the carbenium carbon even more effectively. One may also assume that the basic oxygen atoms pull the cationic oxidants toward the sulfide atom of the thiophenol group, the preferred leaving group in glycosylations. The relative reactivities of the 50 thioglycoside donors were thus tested, first by calculation and then by experiment (Zhang, 1999). The least reactive donor in the library, namely the tetraacetate of mannose **1**, was assigned the unit reactivity of 1.0. The most reactive thiophenol group was the one in tetrabenzylated galactose **9**. It reacted 7.2×10^4 times faster than the mannose tetraacetate. Compound **10** with a free hydroxyl group on C3 was even faster. Other protected monosaccharide thiophenols are in between these extremes.

disaccharide product

| 1 (1.0) | 2 (1.3) | 3 (1.7) | 4 (2.7) | 5 (5.7) |

| 6 (57.3) | 7 (185.4) | 8 (731.4) | 9 (17000) | 10 (20000) |

One-pot oligosaccharide synthesis can thus be achieved in the following way. The most reactive donor A will be mixed first with a promoter and monoalcohol B of a less reactive donor. AB will be formed. AB is then again activated and mixed with the least reactive glycosyl donor C, again containing one free OH group. If the reactivity of C is at least five times smaller than that of B then C will only add to AB to form ABC and not to another molecule of C to give the undesired CC. In the final step of a one-pot tetrasaccharide synthesis one could than add a monoalcohol D, which has no active acetal group at all. ABCD will be formed, because DD is impossible. Any protected tetrasaccharide, for example, can thus be produced by selecting four derivatives with reactivities far enough apart. A computer program (Optimer) not only stored the reactivity parameters under given conditions (solvent, promoter), but also predicted yields and proposed procedures for deprotonation. A closer look at the scheme above suggests, for example, that the synthesis of the pentasaccharide 1-5-7-8-10 should be possible in a one-pot reaction, because the glycosylation rates are quite different.

Two active glycoside donors have also been condensed with one acceptor containing two free OH-groups, and branched trisaccharides have thus been obtained (Crich, 2001). In all cases the deactivation power of 2-benzoate (Bz), 2-benzyl (Bn) and 2-chloroacetyl (ClAc) groups has to be determined individually for each mono- and disaccharide. The qualitative sequence is always ClAc > Bz > Bn, but the degree of the deactivating influence is very dependent upon the structure of the rest of the molecule (Ferrier, 1973; Oscarson, 2000).

C-Glycosides are stable against hydrolysis and enzymatic degradation. They are potential inhibitors of enzymes. Several classical methods have been reviewed (Levy, 1995; Postema, 1995, 2000). Modern approaches favor protected glycols in the form of their epoxides (Rainier, 2001) or enones (Ramnauth, 2001). Nucleophiles add to the epoxides at C-3, rhodium catalysts attach carbon substituents at C-2 in Suzuki-type reactions.

Nu: Grignards, cuprates, alkyl lithiums, organostannanes

a-O-Linked glycopeptides are best obtained by the so-called "cassette-approach", in which the terminal carbohydrate unit already bears an in-place serine or threonine unit. The coupling of a 3-OH group to a suitable donor unit, e.g. a thioacetal, is usually much more reliable than stereoselective glycosylation with amino acids. *a*-Threonine or *a*-serine glycopeptides are best obtained from either the pure *β*-anomer or the epimer mixture of trichloroacetimidates and protected amino acids (Schwarz, 1999).

Cassette approach:

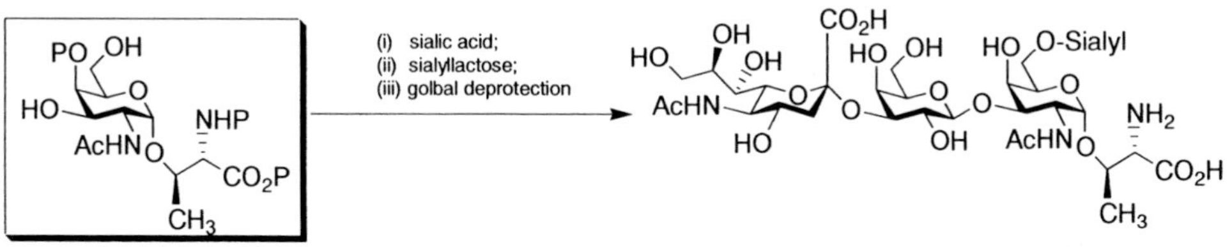

P = Protecting group

Hydrophobic amino- or hydroxymethyl polystyrene or hydrophilic Tentagel are used as polymer supports for glycopeptide soild-state syntheses (Herzner, 2000). The SiCl-activated acetonide given below reacted directly with benzyl alcohol groups on the polystyrene beads. Conversion of the glycal into the epoxide with dimethyl dioxirane (DMDO) and $ZnCl_2$-catalyzed glycosylation yielded a 1,3-disaccharide on the bead. Iodination, azide addition, and sulfonamidation gave the 1-azido-2-sulfonamide. The anthracene sulfonamide at C-2 was cleaved by 1,3-propanedithiol and the amide reduced to amine by the same reagent. One-step coupling of the disaccharide amine with the pentapeptide by 1-isobutyloxycarbonyl-2-isobutyloxy-1,2-dihydroquinoline (IIDQ) gave the resin-bound glycopeptide, which was readily disengaged with HF/pyridine (yield: 44%; Savin, 1999).

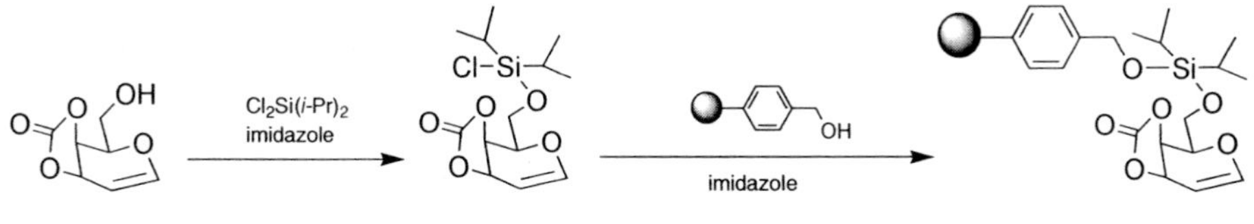

Tetraallyl derivatives of glucose derivatives and β-cyclodextrin open the way to polyfunctional side-chains and dendrimers. They are easily accessible from free OH-groups of carbohydrates and allyl bromide (Boysen, 1999; Dubber, 2000; Fulton, 2000).

5.5
Dendrimers

Spherical polymers shaped like the head of a tree are called dendrimers (dendri=tree-like) and can also be synthesized as a single compound. The purity or monodispersity of these polymers results from iterative syntheses using conditions where reactions are driven to completion, side-reactions are avoided and polymer growth is limited by the finally closed spherical structure ("de Gennes dense packing"; de Gennes, 1982; Grayson, 2001). Defined branches of the molecular tree tops are obtained by protection–activation series of the monomers as described in the previous sections. The most popular approach to dendrimers is convergent. One trivalent monomer is coupled with two trivalent, diprotected monomers. The heptamer may then be dimerized with a disubstituted alkyne, which upon cobalt-catalyzed [2+2+2]-cyclotrimerization yields a hexasubstituted benzene core dendrimer (Hecht, 1999).

Coupling of a hydrophobic and a hydrophilic half to a half-protected pare-bis-phenol gave a perfectly half-water and half-toluene soluble particle (Hawker, 1991, 1993).

Several other convergent dendrimers with all kinds of cores and peripheries have been synthesized. Purification and analytical data are often very impressive. Uni-dispersity was achieved in many cases and proven by NMR analysis (Schlueter, 2000). Functionalized dendrimers may be as uniform as globular proteins and it may be possible also to introduce well-defined gaps into their surface to mimic enzyme clefts.

6
Combinatorial Mixtures and Selection

6.1
Introduction

New pharmacological screening assays are capable of testing a large series of substances simultaneously in a remarkably short time. These new analytical procedures require the combinatorial synthesis of libraries of compounds in micro- or nanogram quantities, which are sufficient for screening with biological receptors. Classical methods of synthesis are too slow to provide the large number of compounds that can be tested, but automated solid-state synthesis on beads, as described in the previous chapter, provides the necessary methodology.

Two different strategies are applied to prepare large series of structurally similar compounds:

- In the multiple parallel synthesis of individual compounds, one substance is prepared per reaction vessel. This simple procedure helps to optimize bead structures and allows aconvergent syntheses, namely the coupling of large blocks.
- In the multiple parallel synthesis of compound libraries, on the other hand, several structurally similar compounds are prepared in each vessel, but only one compound per solid support bead (one bead, one compound). This divergent synthesis gives a large number of compounds in small quantities. Convergent syntheses cannot be carried out in parallel.

The use of these procedures is not limited to drug companies. Labeled receptor proteins have become commercially available and tests of catalytic activity or binding of metal ions carried out in microliter vessels make it possible to investigate the efficiency of substances in each chemical lab with minute quantities. Furthermore mass spectrometry fluorescence tests and simple colorimetric methods on the beads of Merrifield synthesis become very sensitive under the microscope. Even NMR spectra can be measured. It therefore makes sense to produce mixtures of hundreds of compounds in nanogram quantities and to search for the most useful one in the chaos, provided that all the beads or vessels are labeled, so that the compounds they carry can be assigned to individual beads.

Since the nano-analytical procedures are often more important than the synthetic methodology, for each combinatorial synthesis one has to optimize methods which

- give collections of compounds, in which each component appears in comparable quantities,
- label each bead or vessel containing the different reaction products, and
- identify compounds, which merit synthesis on a large scale.

The analogy of compound mixtures with a library is correct with respect to their inherent information content. It gives, however, a false impression about selection methods. Compounds are not selected mentally or by a register of structures, but by a chemical interaction with fluorescing receptor molecules, colorizing metal ions, or activated molecules in catalytic cycles.

In Section 6.2 we discuss methods that produce compound mixtures in which each possible compound appears in approximately the same quantity. The book in the library must have a few pages at least, in order to be detectable. Only quantity guarantees that the most efficient possible compound is present and can be detected. In Section 6.3 we describe a few chemical and non-chemical "codes" for beads, vessels, or locations on a chip containing a particular product and in Section 6.4 we discuss reactions and concepts that have led to potentially useful libraries of compounds and to the corresponding screening methods.

6.2
The Split Method and Multicomponent Reactions

The first and most established procedure of combinatorial synthesis is called the split method. It produces a number, n, of compounds in n vessels, in comparable yields. In order to obtain this result, reagents are applied in large excess and reaction times are adjusted for each vessel to the individual reaction rates of the educts. For example, if solid-state synthesis is used (see Section 5.3.1) and $n=3$, each bead must be loaded with similar amounts of the first reactant A^1, A^2, or A^3. The loaded beads are then mixed and divided into three portions. Each portion is reacted with a large excess of B^1, B^2, or B^3. The reaction time depends on the reactivity of the B-components. It should be long enough to convert the load of all beads to AB-compounds. The first portion of the beads then carries all $A^{1,2,3}$–B^1, the second all $A^{1,2,3}$–B^2, and the third all $A^{1,2,3}$–B^3 combinations (see Fig. 6.1). The beads are again mixed and divided into three portions and combined with three C compounds, which now form the terminal in a library of $3 \times 3 \times 3 = 27$ ABC-compounds. The method can be generalized for any number of A, B or C components. The same technique can be applied using containers instead of beads (see Section 6.3), and the method can be extended to any value of n.

Uncontrolled multicomponent reactions or one-pot syntheses are hardly useful. For example, cubanetetracarbamoyl chloride reacts with 19 amines (mostly partially protected amino acids) to yield more than 11,000 different tetraamides in so-

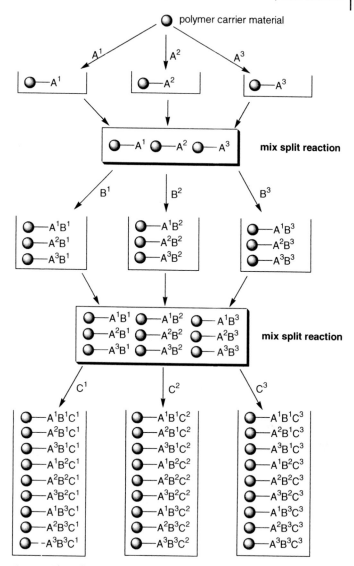

Fig. 6.1 The split method for the synthesis of ABC compounds.

lution. The functional groups are equally distributed on the surface of a spherical molecule and each of these compounds may bind selectively into an enzyme cleft. Analysis of the binding process is, however, close to impossible as long as the compounds are not isolated in separable containers (Cheng, 1996). The parallel three-component synthesis of individual compounds (approach 1) has been exemplified by Boger (1998). Stepwise amidation of iminodiacetic anhydride with three different amines and simple purification by extraction at different pH-values produced hundreds of triamides in milligram quantities.

A few multicomponent reactions between components proceed in a given order. The best-known example is the Ugi four-component-condensation (4CC). A carboxylic acid R^1COOH, a primary amine R^2NH_2, an aldehyde R^3CHO, and an isocyanide R^4NC react in a one-pot manner to afford an N-substituted acyl amino amide, introducing four R-groups in the order R^1, R^2, R^3, and R^4. The sequence is determined by the velocity of the reactions, namely (i) formation of a Schiff base, (ii) formation of an acyl immonium salt, (iii) a rearrangement to an amide. If cyclohexen-1-isonitrile is used as final component, only one compound will be isolated at first, but it may then be transformed into a large variety of derivatives by "postcondensation modifications" (Keating, 1996).

Multiple parallel syntheses of libraries (approach 2) will be discussed in Section 6.4. The fourth realistic possibility for combinatorial synthesis consists of the creation of dynamic combinatorial libraries (DCL). Here the individual members of the library are constantly interconverting in reversible reactions and its composition is under thermodynamic control. Introduction of a template molecule will stabilize and remove from the library those members that happen to bind to the template. The most active compounds may thus be collected in high yield. So far, research on DCLs is only at the proof-of-principle stage, and they are not yet practical. Self-assembled host capsules forming around spherical guests (Hof, 2000; Hiraoka, 1999) and cyclic disulfide oligomers for metal binding (Otto, 2000) are promising examples. A metal ion, which binds and stabilizes one receptor in the library, will also increase the concentration of this receptor in the mixture at the expense of other members. This has been demonstrated with a diamide made of Phe-Pro-benzoic acid, which bears an aldehyde group on one end and a hydrazone on the other. Acid-catalyzed cyclization of this trimer generates a DCL of macrocycles, but in the presence of lithium iodide the trimer of trimers is formed almost exclusively.

Two receptors for two different guests from a single library have also been amplified simultaneously. Small differences in host–guest binding were found to translate into useful enrichments (Otto, 2002).

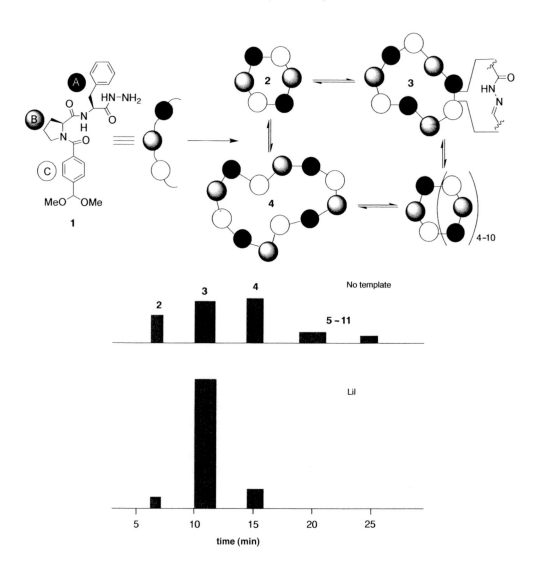

6.3
Coding

A bead in solid state synthesis carries less than a picomole of a compound, so 10^3 beads thus carry less than a nanomole. Receptor tests are mostly carried out in solution, which makes efficient detachment reactions of the products mandatory.

One possibility of coding is to bind a label-compound to the beads after each synthetic step. Each label must then survive the detachment reaction of the synthetic target molecule and it must be possible to release all tag molecules without any decomposition. If these conditions are fulfilled one may analyze the tag mixture by mass spectrometry. The procedure is, however, costly and unreliable.

Sophisticated binary coding procedures, which can be evaluated with a computer, have therefore been developed. We give three examples.

(i) Halogenated arenes are available in large numbers and in pure form because they are used as standards in environmental analysis. They are coupled to the beads via alkoxy spacers of different chain lengths. Aryloxy alcohols, for example, are decoupled by oxidation with Ce(IV) salts and silylated. Automated mass spectrometry then only checks, whether the tag molecule is present (logical digit 1) or absent (digit 0). The printed binary number decodes the sequence of reaction steps performed on the given bead (Still, 1996; Balkenkohl, 1996). The target molecule X is released photochemically or by other selective methods.

R = H, $n = 3 - 12$
R = Cl $n = 4 - 6$

(ii) Milligram combinatorial syntheses are carried out in microreactor vessels labeled by a radiofrequency or optical encoding (see Fig. 6.2). They may be constructed as closed tubes, which contain solutions as well as diaphragms with porous side panels. Automated loaders fill the microreactors with beads, the reactants for the entrapped beads diffuse through the panels, and the codes are read and registered automatically by a computer (NanoKan System; Nicolaou, 2000). About 10^4 vessels can be handled routinely. If one cannot afford the expensive computer machinery, combinatorial syntheses in hand-labeled tea bags follow the same principle.

(iii) Photolithography is used to activate a smooth glass or oxidized silicon wafer surface covered with photolabile protecting groups, e.g. UV-sensitive polymers or o-nitrobenzyl groups. After irradiation of a masked surface leaving only a 1-μm stripe free, the substrate, e.g. amino acid 1, can be fixed by a dark reaction on this stripe containing free hydroxy, carboxy, or amino groups. Moving the slit of the mask by 1 μm then allows fixation of another substrate, e.g. amino acid 2. This procedure can be repeated for, say, 10 different amino acids so that one obtains

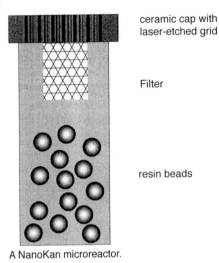

ceramic cap with
laser-etched grid

Filter

resin beads

Fig. 6.2 A NanoKan microreactor.

A NanoKan microreactor.

ten lines containing amino acids 1–10, all protected with a photolabile group. UV-irradiation through a 1 µm slit now lying perpendicular to stripes 1–10 then removes the photolabile protecting group from the first 1 µm of stripes 1–10 and solid-phase synthetic procedures are used to make a dipeptide there with amino acid 11, again containing a terminal photolabile protecting group. The process is repeated on each stripe with a known amino acid. Ten stripes in one direction with ten different amino acids 1–10 and ten perpendicular stripes with different amino acids 11–20 thus yield 100 different dipeptides in 1-µm² spots of identified and known position. Six reaction cycles would yield one hundred different identified hexapeptides. The wafer is then plunged into an enzyme solution, which cuts off a terminal acetyl group from a "recognized" hexapeptide. Fluorescing substrates are used to detect the most active peptides directly under the fluorescence microscope (Baum, 1994).

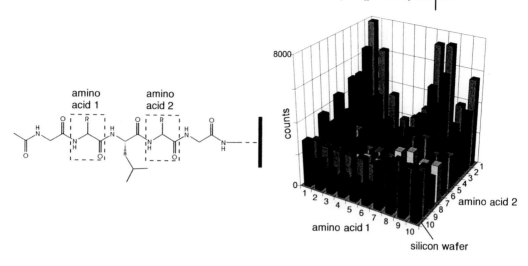

Simple chemical reactions can also be followed on microchips, where electric fields control the flow. Sequential delivery of reagents can be programmed to perform combinatorial syntheses in microreactors (glass chips) of 90-µm depth. Product formation is monitored by absorbance at 488 µm using an Ar-ion laser source (Salimi-Moosavi, 1997).

Pulse field gradient (PFG)^{1}H-NMR spectroscopy allows structural determination of individual components in mixtures which have much lower diffusion coefficients than others. Low molecular weight educts dissolved in $CDCl_3$ (10 mM) do not give any signals in the PFG spectrum, but all components that are part of a molecular complex produce a spectrum. The massive simplification of the spectrum of combinatorial mixtures promises many applications in supramolecular and combinatorial chemistry (Lin, 1997).

6.4
Libraries for Different Purposes

We now give a few examples of successful combinatorial syntheses using the standard split method on bead surfaces or in microliter vessels. The most efficient products were selected

- as optimal catalysts,
- as selective ligands for metal ions and
- as inhibitors for enzymes.

6.4.1
Ligands for Metal Ions

Azo dye chelators can be used for optical detection of multiple metal ions in sensor arrays. Signal intensity and broadness changes, fluorescence, and visual color changes can be used to differentiate metal ions in solution. A library of azo dyes was formed by coupling of azotized aromatic amines with phenol derivatives. Six different types of metal chelates were envisioned, with one or two phenol groups adjacent to the diazo bridge. Each reaction mixture containing the product of the azo coupling was then incubated with a series of solutions each containing a heavy metal ion. The coupling product of the diphenylamine with an azidine bis-phenol component proved to show very large differences in the absorption spectra of complexes with different metal ions. It was thus selected as an efficient analytical tool (Szurdoki, 2000).

Combinatorial mixtures of Schiff bases made of 2-formyl-pyridine and various aniline derivatives proved to be active in the selective extraction of zinc or cadmium ions from water into chloroform solution (Epstein, 2001). The ^{19}F-shifts for the fluorobenzene labels were organized into a group of 16 tags that had chemical shift ranges from −25 to +5 ppm. The fluorobenzene is chemically inert and was used in combinatorial syntheses using a variety of cyclic anhydrides and amines to form 1,4-dicarbonyl and amine ligands as well as a ^{19}F-tag. Binding experiments with different transition metals in acetonitrile solution were performed, the NMR-signal of the fluorine tag was detectable on the solid beads and metal binding showed up in visible spectra (Pirrung, 2002).

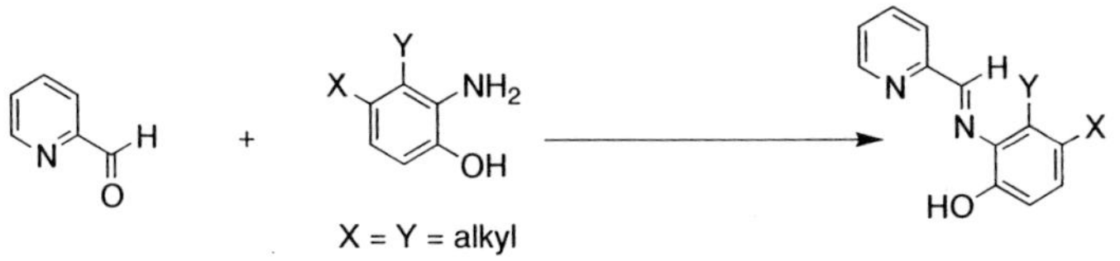
X = Y = alkyl

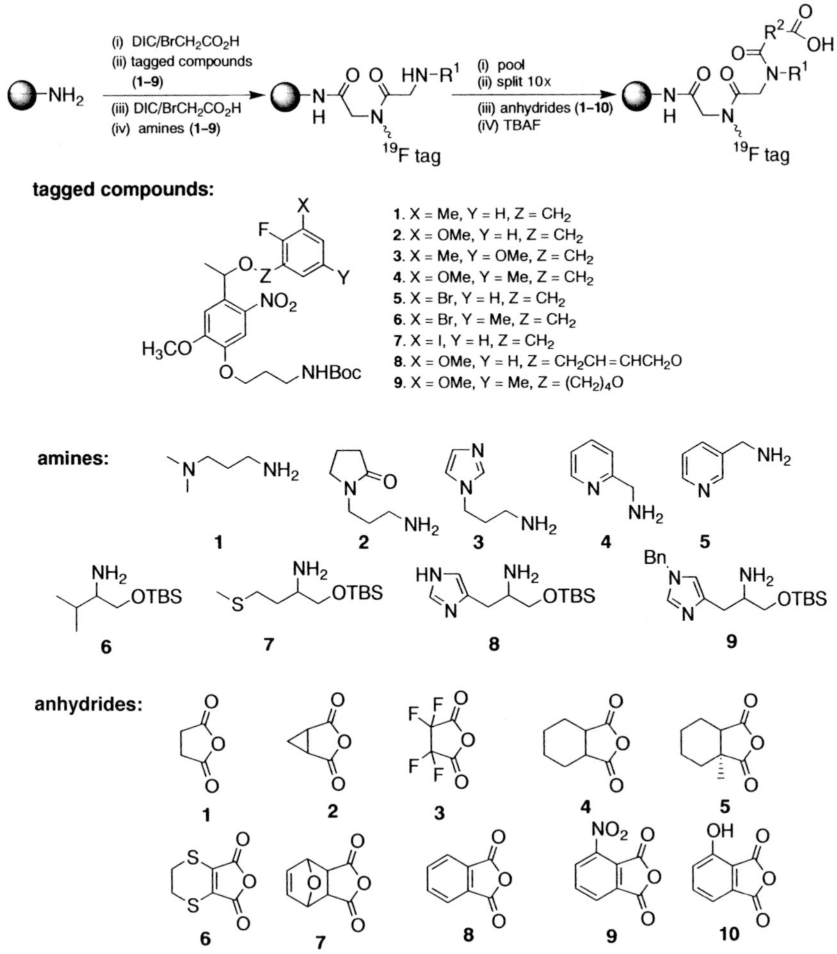
(i) DIC/BrCH2CO2H
(ii) tagged compounds (1–9)
(iii) DIC/BrCH2CO2H
(iv) amines (1–9)
(i) pool
(ii) split 10x
(iii) anhydrides (1–10)
(iV) TBAF
19F tag
tagged compounds:
1. X = Me, Y = H, Z = CH2
2. X = OMe, Y = H, Z = CH2
3. X = Me, Y = OMe, Z = CH2
4. X = OMe, Y = Me, Z = CH2
5. X = Br, Y = H, Z = CH2
6. X = Br, Y = Me, Z = CH2
7. X = I, Y = H, Z = CH2
8. X = OMe, Y = H, Z = CH2CH=CHCH2O
9. X = OMe, Y = Me, Z = (CH2)4O
amines:
anhydrides:

6.4.2
Catalysts

Capillary array electrophoresis (CAE) coupled with microreactors (100 nL) is most useful to combine variations of catalysts with those of reaction mixtures. A palladium-catalyzed annelation of an indole Schiff base with a disubstituted (Me, Ph) alkyne gave two isomeric γ-carbolines. They readily separated from each other and from the indole educt on 50-cm long capillary columns and the concentrations were determined quantitatively. Side-products were also detected. The reaction was catalyzed by eight different palladium complexes and eleven different bases in separate microreactors. Eighty-eight different combinations have thus been tested. The total yields of both products with respect to catalysts and bases are given below. Inorganic bases proved to be more effective than organic bases, where yields were low and side products appeared. None of the test conditions gave complete regioselectivity for one of the two carboline products. The most selective conditions gave low yields. The content of the microreactors was also discarded from the capillaries after different periods of time in order to determine the kinetics of the reactions (Zhang, 2000). Multiplexed CAE thus appears to be an ideal method for establishing the optimal reaction conditions for any reaction.

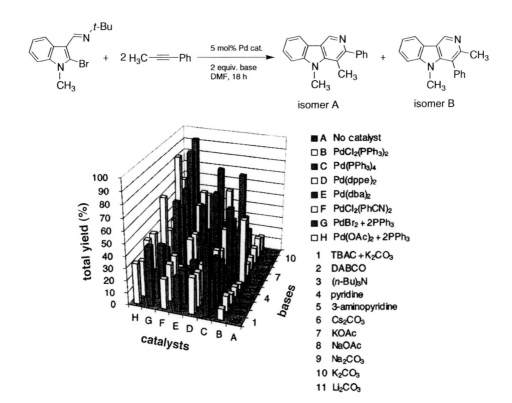

Combinatorial chemistry has also been applied to the development of homogeneous catalysts, where a wide variety of parallel assays is needed. The easiest test involves catalytic reactions of dyes. D–CH=CH–A and D–N=CH–A dyes were used, where A is a pyridinium electron acceptor and D a ferrocene electron donor. Hydrosilylation of cyclooctene with Ph_2Sih_2 was applied as catalytic reaction. Solutions of possible catalysts (3×10^{-4} M in THF) were tested and the bleaching of the dyes monitored. Twelve catalysts for the reduction of four dyes were evaluated rapidly and one new catalyst was discovered. None of the catalysts differentiated the reduction of C=C from that of C=N bonds (Cooper, 1998).

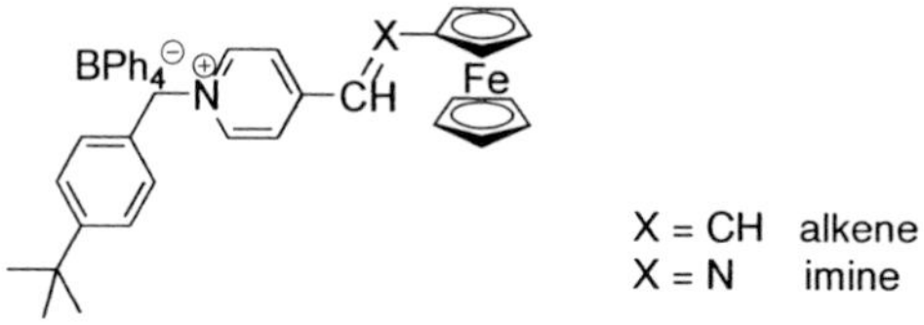

X = CH   alkene
X = N   imine

Results of hydrosilation catalyst screens using dyes

compound	time for dye bleaching (min)	
	alkene	imine
	t	t
1 [Ir(cod)(PPh$_3$)$_2$]BF$_4$	0:05	0:05
2 [Rh(cod)(PPh$_3$)$_2$]PF$_6$	0:03	0:02
3 [Rh(nbd)(PPh$_3$)$_2$]PF$_6$	0:04	0:15
4 RhCl(PPh$_3$)$_3$	0:03	0:03
5 [Rh(octanoate)$_2$]$_2$	NR	NR
6 RuCl$_2$(PPh$_3$)$_3$	0:10	0:10
7 NiCl$_2$(PPh$_3$)$_2$	NR	NR
8 [Ni(tss)]$_2$	NR	NR
9 Cp$_2$ZrClH	NR	NR
10 [Pd(Ar$_2$PC$_6$H$_4$CH$_2$)OAc]$_2$	0:01	0:01
11 [(nbd)Rh(triphos)]SbF$_6$	NR	NR
12 PtCl$_2$(NH$_3$)$_2$	insoluble	insoluble

NR = no reaction

Catalysts for the stereoselective hydrocyanation of imines (Strecker reaction) have been optimized to 91% e.e. by combinatorial synthesis of tridentate Schiff bases without metal ions (Sigman, 1998). One catalyst is shown below. It was the most successful in the combinatorial contest.

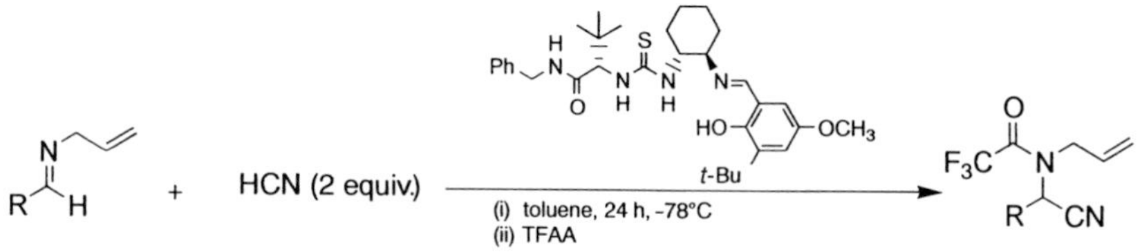

The scope of the Sharpless asymmetric dihydroxylation was studied by solid-state chemistry on Tentagel S-OH and by solid-state high-resolution magic angle spinning NMR. On-bead measurements allowed the determination of enantiomeric excess (e.e.) using Mosher's acid (Riedl, 1998).

Determination of enantiomeric excess:
on-bead by HRMAS-NMR of the bis-Mosher ester
or
by chiral HPLC after reductive cleavage and derivatization

AD-mix

Palladium-mediated combinatorial syntheses and epoxidations on solid phases have been reviewed. Heck, Suzuki and Stille reactions were performed with yields of >80% (Wendeborn, 2000).

A combinatorial phage-Fab library with 2×10^8 members was produced in order to find a catalyst for primary amide bond hydrolysis at pH 7.5. Such a catalyst was indeed identified by "panning", i.e. selective enrichment of phage-Fab members. It bound to terminal boronic acid haptens which were present in the form of a hydrated tetrahedral anion, and thus constituted a transition-state analog for primary amide bond hydrolysis. The isolated Fab fragment hydrolyzed a dipeptide to 50% at pH 9.5 in 4 h, and did not accept the corresponding methyl ester as a substrate (Gao, 1998).

carrier protein-spacer- HN

hapten

(inhibitor has Boc on amino group
instead of carrier)

Lewis Base

+

virus-produced
phage - Fab
library

carrier protein-spacer-HN

selected antibodies,with high affinity to
tetrahedral anionic boronate;they
selectively hydrolyze carboxyamides at the
same site where boronate was situated

6.4.3
Proteins

Convergent syntheses of multicomponent libraries are only possible in the solution phase, where large monomers can be combined to dimers. In the following we introduce "do it all at once" work executed by Boger et al. Synthesis, work-up and biological analysis were all combined and optimized at the same time.

Iminodiacetic acid diamides (see page 362 f.) were synthesized in separate vessels by amination of the anhydride with amines A^1–A^{10} and by amination of the resulting carboxylate with amines B^1–B^{10}. This yielded 100 different compounds in 100 vessels. The secondary amino group in the center was then disconnected from its protecting group and acylated with another iminodiacetic molecule carrying also a vinyl substituent with alkyl linkers of four different chain length. 20,200 Diamides containing all permutations of A^1–A^{10}, B^1–B^{10} and C^1–C^4 were obtained $(n = (100 \times 101/2) \times 4)$. The vinyl group at the end of the alkyl spacer was then dimerized by a metathesis reaction using $RuCl_2$ [(PCy$_3$)$_2$=CHPh] (Grubbs, 1995). The central double bond was then hydrogenated to yield 114×10^6 compounds $(n = (100 \times 101/2) \times 9 + \{(100 \times 101/2)[(100 + 101/2) - 1]/2\} \times 9)$ in the form of several compound libraries (Boger, 1998).

Convergent synthesis with multiplication of diversity (solution phase only)

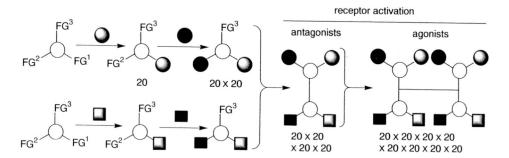

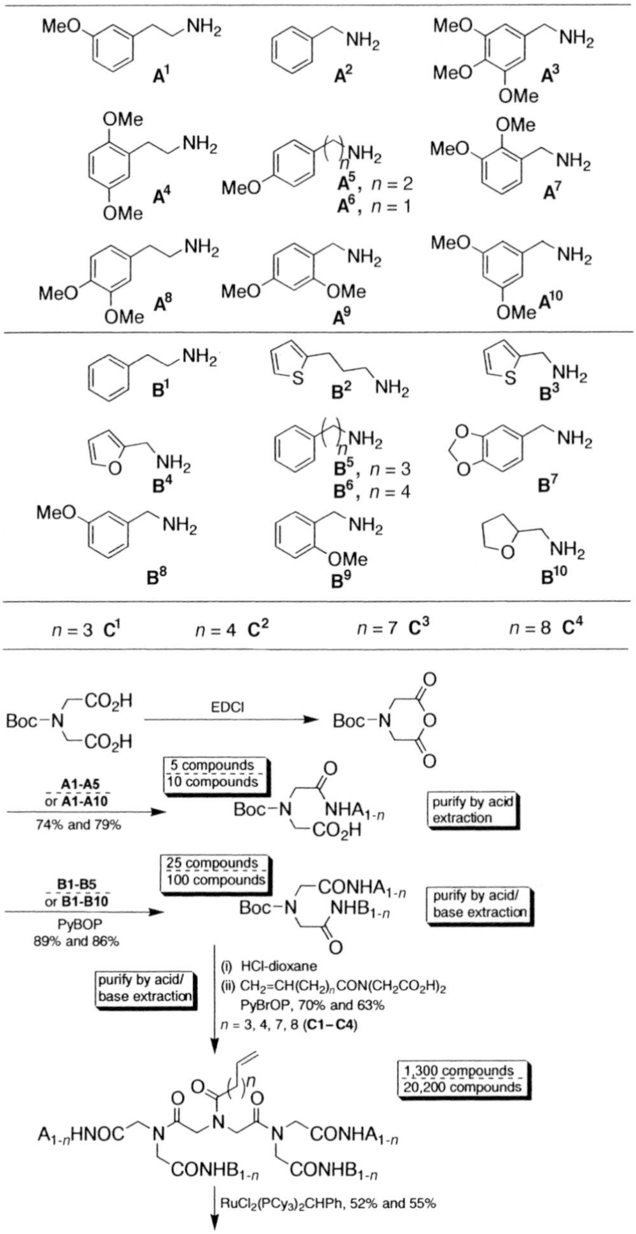

A¹ A² A³ (MeO groups)

A⁴ **A⁵**, n = 2 **A⁶**, n = 1 A⁷

A⁸ A⁹ A¹⁰

B¹ B² B³

B⁴ **B⁵**, n = 3 **B⁶**, n = 4 B⁷

B⁸ B⁹ B¹⁰

n = 3 **C¹**   n = 4 **C²**   n = 7 **C³**   n = 8 **C⁴**

Boc–N(CH₂CO₂H)₂ →(EDCI)→ Boc–N(anhydride)

A1-A5 or **A1-A10**
74% and 79%

5 compounds / 10 compounds

Boc–N(CH₂CONHA$_{1-n}$)(CH₂CO₂H)

purify by acid extraction

B1-B5 or **B1-B10**
PyBOP
89% and 86%

25 compounds / 100 compounds

Boc–N(CH₂CONHA$_{1-n}$)(CH₂CONHB$_{1-n}$)

purify by acid/base extraction

purify by acid/base extraction

(i) HCl-dioxane
(ii) CH₂=CH(CH₂)$_n$CON(CH₂CO₂H)₂
PyBrOP, 70% and 63%
n = 3, 4, 7, 8 (**C1–C4**)

1,300 compounds / 20,200 compounds

A$_{1-n}$HNOC... N(...CONHB$_{1-n}$)...N(CH₂CONHA$_{1-n}$)(CONHB$_{1-n}$)

RuCl₂(PCy₃)₂CHPh, 52% and 55%

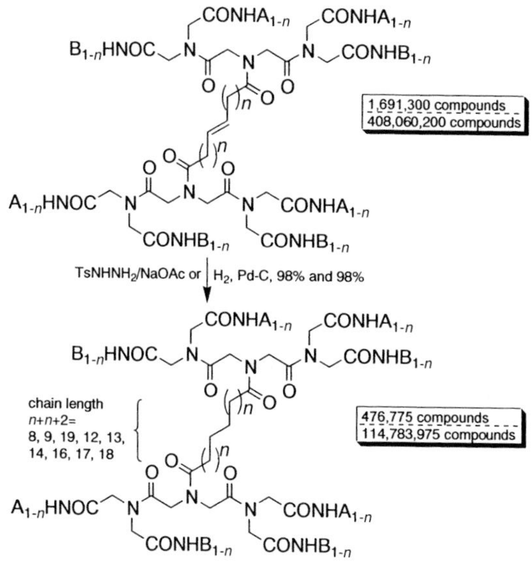

1,691,300 compounds
408,060,200 compounds

TsNHNH$_2$/NaOAc or | H$_2$, Pd-C, 98% and 98%

chain length
n+n+2=
8, 9, 19, 12, 13,
14, 16, 17, 18

476,775 compounds
114,783,975 compounds

From scanning deconvolution

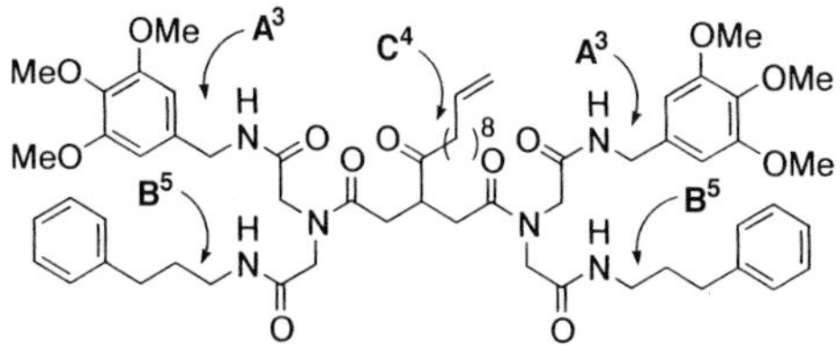

A^3B^5-C^4-A^3B^5

From deletion deconvolution

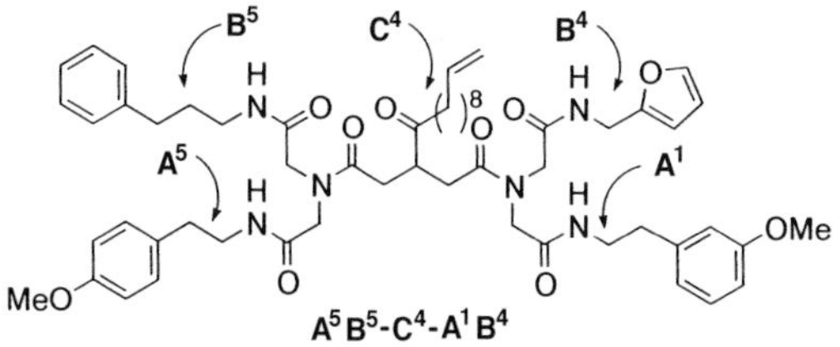

A^5B^5-C^4-A^1B^4

The library had to be deconvoluted twice, first by "positional scanning", then by "deletion". The library was therefore divided into non-overlapping subsets. The subsets were tested separately, and the one with the greatest activity was identified. In sublibraries where only one of the A components was present, i.e. A^1 or A^2 or A^3 or A^4 or A^5, it was found that the A^3 and A^5 libraries were more active than the others (scan A). The same procedure was executed with the B and C libraries. In the present case it was found that out of the A^1–A^5, B^1–B^5, C^1–C^4 sublibraries those containing A^3, A^4, C^4, B^1, and B^5 were most active. The second protocol called "deletion" was then applied synthetically and produced 14 libraries where only one of the fourteen components A^1–A^5, B^1–B^5, and C^1–C^4 was left out. Otherwise the 10^5 compound libraries were complete. This time the loss of activity was measured. The omission of A^1, A^5, B^4, B^5, or C^4 had the largest negative effects. Typically, the deletion synthesis mixtures contain what the scanning mixtures do not contain, and the combination of both sublibrary collections reconstitutes the full mixtures. Scanning showed strongest cytotoxicity for the A^3 and B^5 sublibraries, while deletion gave the least toxic libraries when B^5 or C^4 were missing. Only B^5 and C^4 featured prominently in both libraries. A few pentaamides were synthesized following the "scanning" and "deletion" sublibrary leads and it was indeed found that the two most toxic compounds contained B^5–C^4 combinations (printed in bold in Tab. 6.1).

The tests were executed as follows: The compound mixture was dissolved in DMSO and brought to concentrations of 0.1, 1, 10 and 100 mg mL^{-1}. In the test, medium containing 3000 cells of L-1210 mouse lymphocytic leukemia cells, which had been incubated for 24 h on 96-well plates, were incubated for another 72 h with the test compounds. The number of cells per well were then plotted as a function of the concentration of the test agent. The dose that reduced the cell count to 50% (= IC50) was thus determined. Alternatively the hydrolysis of *p*-nitrophenyl phosphate by phosphatases in the cells was measured by colorimetry (Boger, 1999, supporting information; Boger, 1988). High activity in "scanning" and large loss of activity in "deletion" indicated bioactive amides.

The development of the complementary identification protocols of "scanning" and "deletion" was absolutely necessary to differentiate the activity of the dimers obtained by metathesis. The biological tests became an integral part of the synthesis. Without such microbiological methodology in the laboratory the compound libraries would have been useless, because no other analysis makes sense.

Tab. 6.1 Cytotoxic activity of compounds identified by deconvolution.

From scanning deconvolution		From deletion deconvolution	
Compound	*IC$_{50}$ (µg/mL)*	*Compound*	*IC$_{50}$ (µg/mL)*
A^3B^5–C^4–A^3B^5	0.6	A^5B^5–C^4–A^5B^5	3.2
A^3B^1–C^4–A^3B^1	22	A^1B^5–C^4–A^1B^5	6.2
A^3B^1–C^4–A^3B^1	40	A^5B^4–C^4–A^5B^4	11
A^4B^5–C^4–A^4B^5	3.2	A^1B^4–C^4–A^1B^4	10
A^4B^1–C^4–A^4B^1	7.6	A^5B^5–C^4–A^5B^4	5.5
A^4B^2–C^4–A^4B^2	18	A^1B^5–C^4–A^1B^4	11
		A^5B^5–C^4–A^1B^5	4.9
		A^5B^4–C^4–A^1B^4	4.4
		A^5B^5–C^4–A^1B^4	0.8
		A^1B^5–C^4–A^5B^4	12

See previous scheme for molecular structures, and text for the applied test

The glycopeptide antibiotic vancomycin is active against Gram-positive bacteria and widely used for the treatment of serious infections. The emergence of global resistance to vancomycin has serious clinical consequences. It is effected by replacement of the terminal D-Ala in L-Lys-D-Ala-D-Ala wall proteins by D-lactate. The presence of a terminal ester bond to the lactate hydroxyl group instead of an amide bond causes a 1000-fold reduction in binding. The identification of synthetic receptors that bind to L-Lys-D-Ala-D-Lac could provide a useful strategy for the development of a vancomycin substitute. A receptor was searched for and found by combinatorial synthesis. The library did not consist of molecules as complex as vancomycin itself, but contained only variations of partial structures. The usual scheme, where a peptide is optimized to bind to a known receptor, was thus inverted, and relatively simple peptides were found, which bound the depsipeptide L-Lys-D-Ala-D-Lac with a fluorophore on the lysine end much tighter than vancomycin itself (Xu, 1999). These peptides cannot only replace vancomycin but they can also be modified easily if resistance occurs again.

The right-hand carboxylate binding pocket of vancomycin was retained, the left-hand side was replaced with a variable tripeptide. Chlorine and sugar residues were left out. The biaryl ether linkage was formed by oxidation of phenols with $Tl(NO_3)_3$. The tripeptide cyclization precursor was thus prepared in the ten gram scale by standard peptide chemistry, oxidative cyclization followed by zinc metal reduction and removal of the alkyl ester with tributyltin hydride and dehydrohalogenation with nickel(II) acetate and borohydride. A library of about 40,000 theoretical members with 34 amino acid inputs was prepared with variations at R^1, R^2, R^3. For bead assays the fluorophore-labeled nitrobenzodioxazole-ligands were used. The ester is analogous to the Lac-terminal amide, the amide to the D-Ala terminal (Xu, 1999).

Vancomycin:

Synthetic receptor:

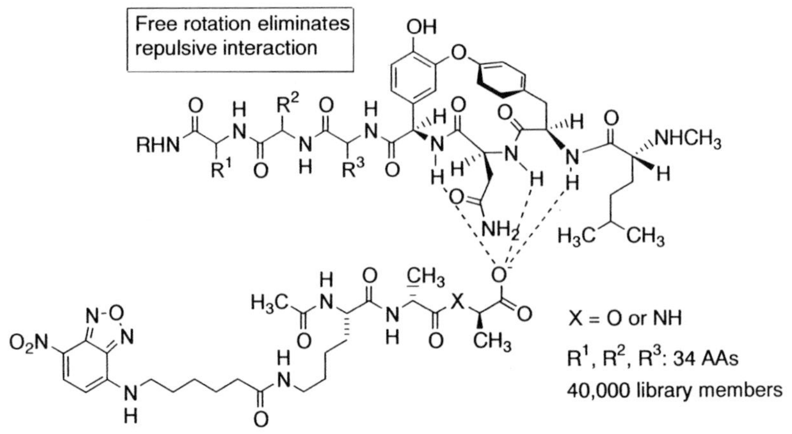

Sarcodicytin from soft corals disturbs the tubulene–microtubuline interplay in tumor cells, resulting in tumor cell death. A combinatorial library based on this natural product has been prepared by attachment of synthetic compounds, which had served as intermediates in its total synthesis, to a solid support. The fragments of sarcodicytin were then modified in a combinatorial manner on beads and detached again. The modification of an allylic alcohol to an acroleïn and acrylic ester derivatives are given below. All modified substances were then tested as inhibitors of the tubuline–microtubuline interactions (Nicolaou, 1998).

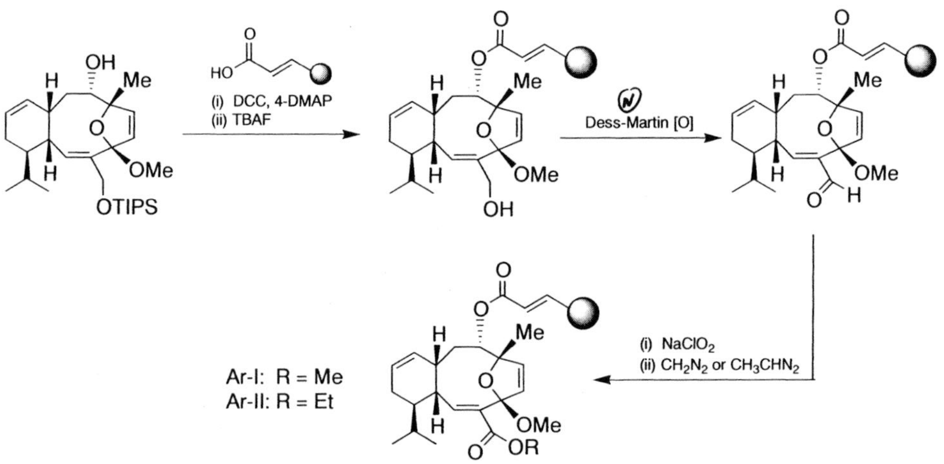

Cytotoxicity data of sarcodictyins and related compounds

compound	% tubulin polymerization	inhibition of carcinoma cell growth, IC_{50} (nM)		
		1A9	1A9PTX10	1A9PTX22
1 Taxol	65	2	50	40
2 epothilone A	73	2	19	4
3 epothilone B	97	0.04	0.035	0.04
4 Ar-I	18	430	1800	>2000
5 Ar-II	4	>2000	800	385

A glycopeptide library was linked to the solid support via a photolabile linker and released from the beads by a MALDI-TOF-MS laser. High-quality spectra were obtained from a simple bead. Analysis of the mass spectrum identified labels and the glycopeptides, which reacted with fluorescent labeled proteins. A glycosyl amino acid building block used in glycopeptide Merrifield synthesis is shown below, together with its mass spectrum without assignment of the peaks (see St. Hilaire, 1998).

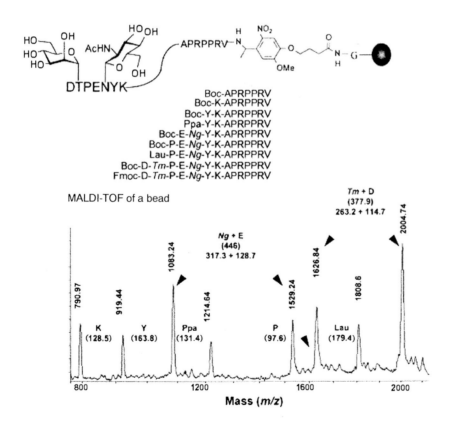

Boc-APRPPRV
Boc-K-APRPPRV
Boc-Y-K-APRPPRV
Ppa-Y-K-APRPPRV
Boc-E-*Ng*-Y-K-APRPPRV
Boc-P-E-*Ng*-Y-K-APRPPRV
Lau-P-E-*Ng*-Y-K-APRPPRV
Boc-D-*Tm*-P-E-*Ng*-Y-K-APRPPRV
Fmoc-D-*Tm*-P-E-*Ng*-Y-K-APRPPRV

An example of a combinatorial synthesis in tea bags as well as in the solid phase has been carried out with oligoureas. They were obtained from N-protected *α*-amino-*ω*-isocyanates and screened by a competitive radioimmunoassay with monoclonal antibodies. The results were the same with both synthetic methods (Burgess, 1997).

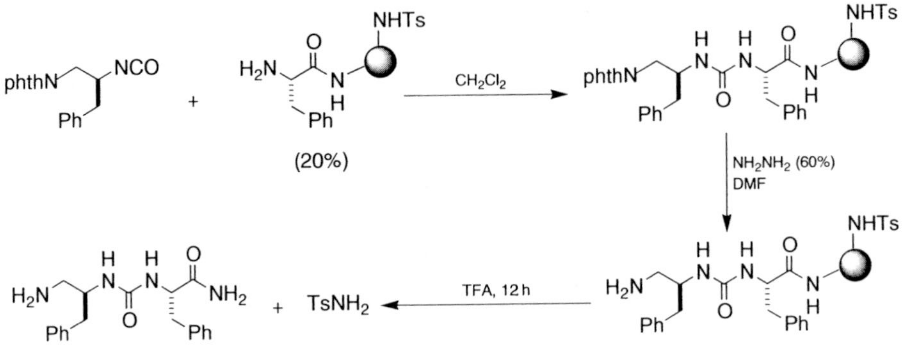

6.4.4
Nucleic Acids

Aminoglycoside antibiotics recognize RNA specifically and interfere with protein biosynthesis. A *gluco*-configured 1,3-hydroxamine library has been synthesized by reductive amination at C-1 and acylation at C-2 (Wong, 1998). Binding to RNA was assayed by angle change measurements of surface plasma resonance on a sensor chip (Hendrix, 1997).

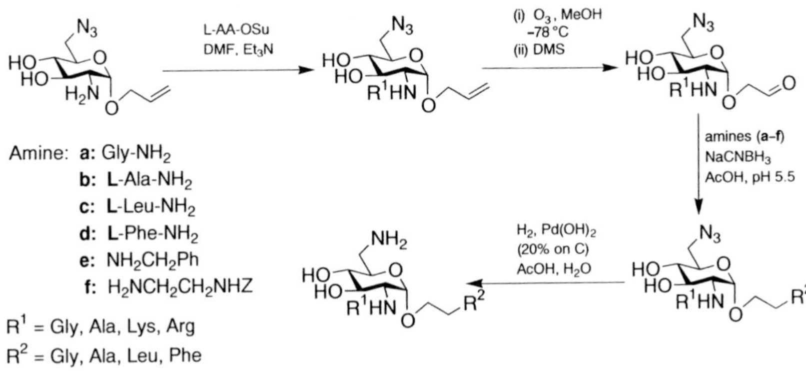

Amine: **a**: Gly-NH$_2$
  b: L-Ala-NH$_2$
  c: L-Leu-NH$_2$
  d: L-Phe-NH$_2$
  e: NH$_2$CH$_2$Ph
  f: H$_2$NCH$_2$CH$_2$NHZ

R^1 = Gly, Ala, Lys, Arg
R^2 = Gly, Ala, Leu, Phe

Stability Constants

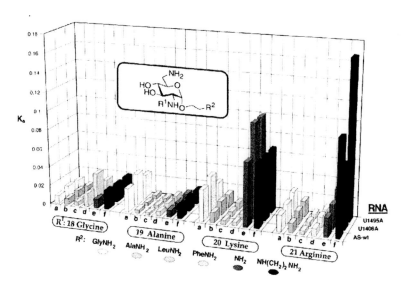

Standard peptide chemistry (see Section 5.3) and workup have been used to synthesize 2640 distamycin A analogs in solution to 95% purity by acid/base precipitation–dissolution cycles. Binding to different ethidium bromide-stained DNAs was detected on several 96-well plates after addition of strongly binding distamycin analogs by the loss of ethidium fluorescence. Such fluorescence decay screenings of 2640 distamycin analogues on DNA homopolymers or specific hairpin oligonucleotides revealed compounds that were 1000 times more potent than natural distamycin A (Boger, 2000, 2002). Follow-up cytotoxic assays then indicated first the most potent libraries made of tricyclic diamides and, after deconvolution, resynthesis determined the most potent single compound.

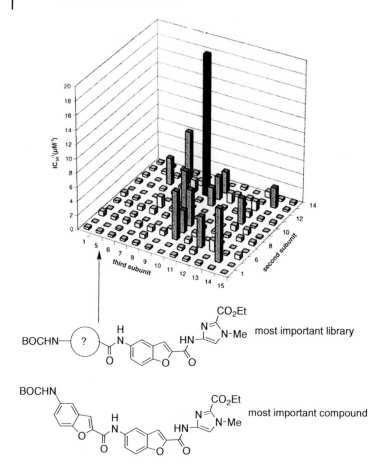

most important library

most important compound

6.4.5
Heterocycles

With respect to drugs, combinatorial syntheses should involve a small number of steps, rely on starting materials with a diverse selection of substituents, and lead to interesting cyclic structures (Marx, 1997). A large variety of heterocycles is made from di- or tripeptides on solid surfaces (see Section 5.3.1), but peptides themselves have strict limitations on their later development as pharmaceuticals owing to their poor bioavailability and proneness to enzymatic degradation. The focus of combinatorial drug chemistry has therefore shifted from peptide libraries to libraries of heterocycles (Houghton, 1999).

Combinatorial syntheses starting from polymer-bound amino acids have been elaborated extensively. 1,4-Benzodiazepines are obtained with amines, pyrrolidines by cycloaddition of Schiff bases (in the form of azomethine ylides) (Marx, 1997), and β-lactams from Schiff bases and acid chlorides (Balkenkohl, 1996).

An example of the solid-state synthesis of a single compound, which can easily be converted to a combinatorial procedure, is the one-pot preparation of Sildenafil by Ley et al. (Baxendale, 2000). An efficient polymer-supported coupling reagent, derived from 1-hydroxybenzotriazole (=HOBt), was applied for amidations. The procedure for amide synthesis involves formation of a triazole-activated ester by condensation of the N–OH group with an acid in the presence of bromo-tripyrrolidinophosphonium-hexafluorophosphate (=PyBrOP). Addition of an amine liberates the desired amide into the solution (Pop, 1997). The same amidation procedure was then applied to couple and cyclize the acidic part of Sildenafil with its pyrazole amine part in quantitative yield. Sildenafil was then released into the solution and unreacted pyrrole amine reacted with additional polymer beads containing isocyanate groups and filtered off. It was said that the synthesis only suffered from inconvenient properties of polymer beads. *"They picked up water and static electricity, were difficult to weigh out and decomposed mechanically, when they were stirred."* In other words: no problem at all with solid-state synthesis of single compounds on a 10–100 mg scale.

An application of Ugi's four component reaction on a polystyrene bead leads to a molecule in which a furane unit and a Michael double bond are close to each other. Heating then leads to a Diels–Alder cyclization. Allylation of two amide nitrogens and a ring opening–closing reaction gave to two additional lactam rings (Lee, 2000; Schreiber, 2000). Other reactions in a different order would lead to other scaffolds. The major problem with such complicated products lies in the analysis and screening of the libraries.

A major dilemma of library synthesis is aimlessness. This may be overcome by equilibrium mixtures (see Section 6.2) or by *in situ* click chemistry. Sharpless "clicked together" two oligocycles in the presence of acetylcholinesterase providing a rigid lipophilic cleft. Both units contained either acetylene or azide substituents, connected by spacers of different length. Triazole formation was slow in solution, but accelerated by adsorption of the educts by the enzyme. This acceleration was most important for one out of eight alkyne and one out of eight azide educts thus leading to only one single compound. This product turned out to be the strongest noncovalent inhibitor yet identified for the enzyme. "Screening" not only selected the most potent inhibitor, but produced it. The "click" approach is so far very limited, but extremely promising. Let the protein select what it wants and just provide efficient educts! (Sharpless, 2001; Borman, 2002).

6.4.6
Carbocycles

Prostaglandin combinatorial syntheses have been based on three components, namely a cyclopentene-3-on-5-ol, a tributyltin alkene, and a β-alkyne triflate. The cyclopentenol is bound reversibly to the solid support, the tin compound is added by Michael addition and the triflate by nucleophilic substitution. The cleavage of the THP ether with HF/THF also removed the TBS protecting group. PGE_2 methyl ester was obtained in 37% yield and TLC showed only one major spot (Chen, 1997).

Combinatorial synthesis of vitamin D_3 libraries was based on two coupling reactions:

- connection of ring A with the CD-unit by a Wittig reaction and
- build-up of the side-chain by alkylation with Grignard reagents. Both classical reactions have been performed on polymer support in high yield and without apparent complications (Doi, 1999).

6.4.7
Chiral Compounds

β,γ-Unsaturated α-amino acids have been synthesized in one step by a Mannich-type reaction using an amine, an α-keto acid, and a boronic acid. The boronic acids were obtained from alkynes in geometrically pure form. They tolerate air and water and react with amines and glyoxylic or pyruvic acid to give the corresponding α-amino acids after deprotection of readily cleavable amines. Bromo-substituted derivatives form bromoalkenyl amino acids, asymmetric precursors with phenyl substituents gave pure enantiomers (Petasis, 1997).

β-Amino alcohols were obtained in a one-step three-component reaction from an ene or arene organoboronic acid, aminodiphenylmethane and an α-hydroxy aldehyde. The *anti* diastereomer is formed to more than 99% d.e. When pure enantiomers of the α-hydroxy aldehyde are used, single enantiomers are obtained with more than 99% e.e. The experimental procedure only involves mixing the three components together and stirring for 12–48 h at ambient temperatures. Water and oxygen may be present. The benzylamino groups are removed by hydrogenolysis. Conversion to BOC derivatives, diol splitting with periodate, and oxidation with RuCl$_3$ yield α-amino acids (Petasis, 1998).

Enzymatic syntheses of β-hydroxy-α-amino acids (=threonine analogs) using D-

and L-aldolases start from an aldehyde and glycine. High stereoselectivity is occasionally observed (Kimura, 1997).

7
Concepts in Nanometer-Sized Skeletons and Modules

7.1
Introduction

Nanometer skeletons are molecular assemblies that allow the positioning and ordering of molecules at typical distances of 1–3 nm. The most interesting ordering of this kind concerns a photoactive electron donor–acceptor pair, which allows charge separation and ultimately water splitting with the energy of visible sun light. Nanometer modules are parts of signal receivers, mechanical machines, or computers, which can be altered or exchanged without disrupting the functioning of the system. Typical examples are rotors, valves, switches or programs.

The distance between functional parts in biological reaction systems such as chloroplasts or mitochondria is in the order of 1.0 nm. Semiconductors or light-emitting elements in computers or displays are usually a few micrometers apart. One major synthetic aim of modern organic chemistry is to close the size gap, which exists between micrometer assemblies, which are observable under the light microscope, and nanometer assemblies, which are of molecular dimensions. One likes to produce molecular assemblies in which the reaction partners are separated by about one or two nanometers. In charge separation modules they are then close enough to allow rapid electron tunneling, but far enough apart to prevent fast recombination (Li, 2002).

An organic chemist can learn most from Nature. Detailed knowledge of biological molecules and processes is a prerequisite for this learning process. Most profit is earned from simplified systems, which fulfill just one task perfectly. This approach has been valuable for the development of synthetic reagents (e.g. metal complex and organyl catalysts instead of metalloproteins) and of industrial or academic target molecules (e.g. nylon instead of protein fibers, crown-type ligands instead of natural ion carriers). At the present time, the details of the module function and arrangement of cell membranes, enzymes, and nucleic acid assemblies are elucidated by electron microscopy, crystal structure analysis, molecular biology, and modern spectroscopic methods. Attempts are then made to understand the construction principles and to apply them in syntheses of molecular systems of "supramolecules" which are held together by weak intermolecular forces. Sometimes it turns out that with simple, repetitive reaction sequences and well-planned or luckily discovered self-assembly processes, molecular skeletons and chemical machines of astounding complexity may be synthesized on often surprisingly large scales. Alternatively, organic chemistry may be combined with biochemical expertise and computer-aided molecular modeling in order to suppress enzyme action and nucleic acid replication.

A second motivation comes from the desire to miniaturize the modules of known mechanical machines, electronic signal receivers and transformers, and computers to molecular size. This is a dream of many would-be nanoengineers, but has also already been realized in modules of modest complexity by organic chemists.

7.2
Nucleic Acids

The unique combination of organic chemistry and molecular biology which is nowadays applied in nucleic acid synthesis and analysis led to a great interest in nucleic acid bioorganic chemistry. We give a few examples of synthetic endeavours in this field.

7.2.1
DNA with a Site-Specifically, Covalently Bound Mutagen

Some epoxides and halomethyl derivatives of aromatic polycycles are known to bind to DNA *in vivo* to cause mutations and thus carcinogenic effects. To study these mutagens in plasmids and bacteriophages, site-specifically modified oligodeoxynucleotides were synthesized (Casale, 1990). Anthracene was chosen as a simple mutagen model, and a 13-base DNA single strand complementary to the sequence of position 6183 to 6195 of a bacteriophage DNA as genetic material. The tridecanucleotide was prepared by the usual automated phosphoramidite procedure and a 9-anthracenyl-methyl residue was attached to N-2 of the eighth nucleotide, guanosine 6190. It is, however, known that the N-2 position of guanosine cannot be alkylated to any significant extent. Alkylations at positions N-7, N-3, N-1, or sometimes O-6 are always predominant. But a tricyclic imidazol[1,2-*a*]purine precursor (Kasai, 1976) reacted at the desired position with 9-(chloromethyl)anthracene in 80% yield. The C=C double bond in the auxiliary imidazole ring was then brominated with NBS (Boryski, 1984, 1985) and hydrolyzed with aqueous ammonia to give the desired N^2-substituted guanosine. Conventional protection at O-5′ of the deoxyribose with 9-chloro-9-phenyl-9*H*-xanthene ("pixyl chloride", PxCl; Chattopadhyaya, 1978) and activation at O-3′ with 2-cyanoethyl *N,N*-diisopropylphosphoramidochloridite followed to give the nucleotidylation reagent for incorporation into the oligonucleotide by automated chemical synthesis (see Sections 5.2.1 and 5.3.1).

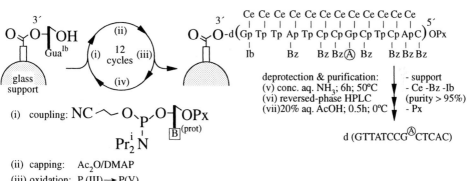

(i)   CH₂Cl
K₂CO₃/DMF
6 h; r.t.

(ii) NBS; aq NaOAc/AcOH
(DMSO); 15 min; r.t.
+ aq. NH₃; 15 min; r.t.

(iii) PxCl/Py; 2 h; r.t.

(iv) Pr⁺₂NP(Cl)OCH₂CH₂CN
(EtNPr⁺₂/CH₂Cl₂)
80 min; r.t.

(80%)   (26%)   (52%)   (53%)

$$\left(\quad = \quad \frac{Px}{(CeO)(Pr_2^iN)P} dG^{\textcircled{A}} \right)$$

Poor coupling yields were observed even with a 30-fold excess of the 2-*N*-(9-anthracenylmethyl)guanosine reagent. An increase of reaction time from the few minutes sufficient for the coupling of the four common 2'-deoxynucleoside 3'-phosphoramidites up to 60 min, gave coupling yields of 65–70%. This is far below the approximately 98% obtained in routine synthesis steps, but is acceptable if it occurs in only one step of an oligomer synthesis. After detachment from the solid support, removal of the 2-cyanoethyl groups and chromatography, about 1 mg of the 5'-*O*-pixyl tridecanucleotide was routinely obtained from 1 μmol of the polymer-bound first nucleoside. The pixyl group was finally removed with 80% acetic acid.

```
                              Ce Ce Ce Ce Ce Ce Ce Ce Ce Ce Ce Ce
   3'                    3'    |  |  |  |  |  |  |  |  |  |  |  |    5'
O—O—OH          (ii)   O—O-d(Gp Tp Tp Ap Tp Cp Cp Gp Cp Tp Cp ApC) OPx
    Gua^Ib  (i)  12  (iii)     |           |     |  |         |  |  |
        glass   cycles         Ib          Bz   Bz Bz(A) Bz  Bz Bz Bz
        support       (iv)
```

(i) coupling: NC~O–P(OPx)
                  Pr⁺₂N   B(prot)

deprotection & purification: - support
(v) conc. aq. NH₃; 6h; 50°C   - Ce -Bz -Ib
(vi) reversed-phase HPLC   (purity > 95%)
(vii)20% aq. AcOH; 0.5h; 0°C - Px

d (GTTATCCG^(Ⓐ)CTCAC)

(ii) capping: Ac₂O/DMAP
(iii) oxidation: P (III)→ P(V)
(iv) O-5' deprotection (- Px):
    2% Cl₂CHCOOH/DCE; 90 s

The unmodified and complementary oligonucleotides were also synthesized, in order to detect thermodynamic and spectroscopic differences between the double helices. Circular dichroism spectra revealed that the covalently bound anthracene does not stack in the centre of the DNA double helix. Mutagenic activity by intercalative binding of the anthracene residue is thus unlikely. Only *in vitro* and *in vivo* replication experiments with site-specifically modified bacteriophage DNA can tell more about the type and intensity of mutagenic effects of this site-specific modification.

7.2.2
Cationic Oligonucleotides

With the objective of gaining further understanding and control of reversible association processes of oligonucleotides, analogs carrying positive charges along the
backbone were synthesized (Letsinger, 1988). Methyl phosphoramidite chemistry
was used in generating phosphodiester links. Phosphonic diester links were generated and oxidatively coupled (Froehler, 1986) with appropriate diamines to obtain the cationic N-(2-aminoethyl)phosphoramidate links. Tetrachloromethane is a
recommended oxidant for this reaction. Complexes of cationic oligonucleotides
with their complementary anionic counterparts turned out to be more stable than
the natural, fully anionic DNA double strands. Non-complementary oligonucleotides with consequent improper base pairing, however, did not form stable complexes in either case. Steric fitting of bases and hydrogen bonds within the double
helix is thus more important than optimum ion pairing.

$$d(T\overset{\oplus}{-}T\overset{\oplus}{-}C\overset{\oplus}{-}T\overset{\oplus}{-}G\overset{\ominus}{-}A\overset{\oplus}{-}A\overset{\oplus}{-}A\overset{\oplus}{-}A\overset{\oplus}{-}T)$$ **1** cationic

$$d(T-T-C-T-G-A-A-A-A-T)$$ **2** normal (anionic)

$$d(A-T-T-T-T-C-A-G-A-A-T-T-G-G-G-)$$ **3**

complementary sequence

	binding
1+2	none
1+3	enhanced
2+3	normal

7.2.3
A Synthetic Scissor for Sequence-Specific DNA Cleavage

The recognition of specific nucleic acid sequences is important in the regulation
of many biological processes. The antibiotic distamycin binds to the minor groove
of the DNA double helix with a strong preference for A–T-rich regions. The sequence specificity presumably results from hydrogen bonding between the amide
NH groups of distamycin and O-2 of thymine and N-3 of adenine. The stepwise
"peptide" synthesis of distamycin (Bialer, 1978; Grehn, 1983, 1986; Lown, 1985)
can be modified by attachment of EDTA-type chelands at the terminal amide and
insertion of oligoethylene glycol-type chelands in the centre of the oligomer chain

(Griffin, 1987). The binding of the central amide groups to DNA in the hypothetical molecular complex shown below enforces a cyclic conformation of the oligoethylene glycol unit. Experimentally, the iron ion bound to the terminal EDTA group was found to cleave nucleic acids only in the presence of oligoethylene glycol-fitting metal ions, e.g. Ba^{2+} or Sr^{2+}. Several other reagents have been synthesized by Dervan and co-workers from DNA-binding natural products, e.g. from distamycin (Baker, 1989) and proteins (Sluka, 1990), and have been used for site-specific cleavages (affinity cleaving) as well as for protection of nucleic acids.

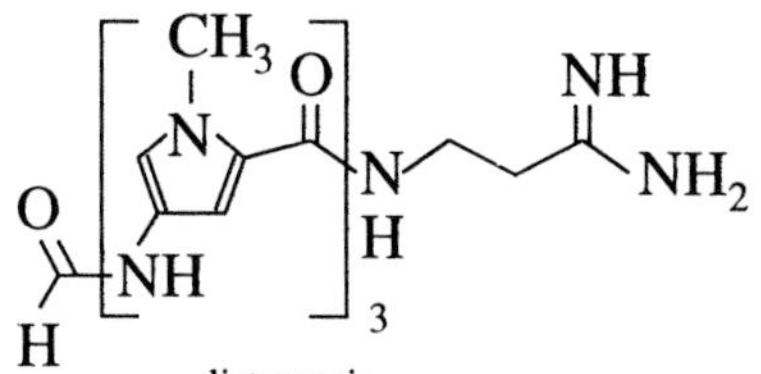

distamycin

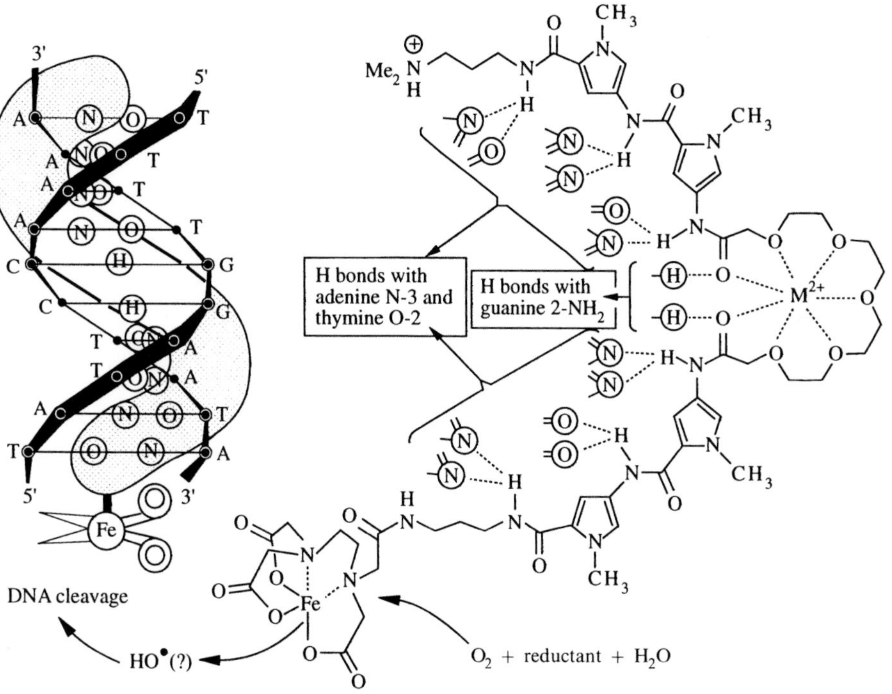

H bonds with adenine N-3 and thymine O-2

H bonds with guanine 2-NH_2

DNA cleavage

$HO^{\bullet}(?)$

O_2 + reductant + H_2O

7.2.4
Hexose Oligonucleotides

The furanose rings of the deoxyribose units of DNA are conformationally labile. All flexible forms of cyclopentane and related rings are of nearly constant strain and pseudorotations take place by a fast wave-like motion around the ring. The flexibility of the furanose rings (Levitt, 1978) is presumably responsible for the partial unravelling of the DNA double helix in biological processes.

Eschenmoser (1991, 1992) asked: what happens if the flexible furanose ring is replaced by a more rigid pyranose ring of similar overall geometry? He chose 2,3-dideoxy-*β*-D-*erythro*-hexopyranose as a model and linked the corresponding "2′*a*-homo-deoxynucleosides" by 4,6-phosphodiester bonds to give "homo-DNA" using the automated procedures described in Section 4.1. In contrast to the inherently highly helical DNA single strands, homo-DNA single strands are nearly linear as predicted by conformational analysis. The 6′-CH$_2$ group and the phosphodiester link allow the bending of homo-DNA to form antiparallel double helices which are much less twisted than DNA double helices. As expected, the pyranose rings of homo-DNA are very rigid, and homo-DNA pairs more tightly than DNA. Not expected was the tight pairing of identical purine homonucleotides, i.e. homo-(A···A) and homo-(G···G). Homo-DNA and DNA form no mixed double helices.

homo-G homo-G-pair

homo-A homo-A pair

Syntheses of sterically modified biopolymers can clearly yield insights into the presuppositions and possibilities of the biological self-organization processes of biopolymers which go far beyond the general thermodynamic and kinetic descriptions of natural systems.

7.2.5
DNA Helicates with a Central Copper(I) Ion Wire

Copper(I) tends towards a tetrahedral coordination geometry in complexes. With 2,2'-bipyridine as a chelate ligand a distorted tetrahedral coordination with almost orthogonal ligands results. 2,2'-Bipyridine oligomers with flexible 6,6'-links therefore form double helices with two 2,2'-bipyridine units per copper(I) ion (Lehn, 1987, 1988). Lehn (1990; Koert, 1990) has also prepared such "helicates" with nucleosides, e.g. thymidine, covalently attached to suitable spacers to obtain water-soluble double helix complexes, so-called "inverted DNA", with internal positive charges and external nucleic bases. Cooperative effects lead preferentially to two identical strands in these helicates when copper(I) ions are added to a mixture of two different homooligomers.

7.3
Kemp's Acid: Enzyme-Cleft and Self-Replication Models

Biological catalysts – enzymes – are usually proteins. The development of new protein syntheses is nowadays dominated by "genetic protein engineering" (see Section 5.3.2). Bioorganic approaches towards novel catalytically active structures and replicating systems try to manage without biopolymers.

In this area, functional groups which converge in a molecular cleft in a rigid molecule are of central interest. 1,3,5-Trimethyl-1,3,5-cyclohexanetricarboxylic acid was first prepared by Kemp (1981) by oxidative degradation of 3,5,7-trimethyl-1-adamantanol. This acid is a surprising example of a cyclohexane ring with three *syn*-oriented axial carboxylic acid groups. This unlikely conformation is stabilized by the singular structural feature that steric interaction of the methyl groups at the same carbon atoms would be even more repulsive were they axial. Furthermore, attractive forces are likely to hold the polar functional groups together. Kemp's acid is easily converted to monoimides by heating with amines.

7.3.1
Enzyme-Cleft Models with Convergent Functional Groups

Rebek, Jr. (1987) first developed a new synthesis of Kemp's acid and then extensively explored its application in model studies. The synthesis involves the straightforward hydrogenation (Steitz, 1968), esterification and methylation of inexpensive 1,3,5-benzenetricarboxylic acid (trimesic acid; approx. $ 30/100 g). The methylation of the trimethyl ester with dimethyl sulfate, mediated by lithium diisopropylamide (Shiner, 1981), produced mainly the desired *all-cis*-1,3,5-trimethyl isomer, which was saponified to give Kemp's acid.

Heating Kemp's acid with appropriate aromatic diamines yields bis-imides with two convergently oriented carboxylic acid groups on the edges of a hydrophobic pocket (Rebek, Jr., 1987; Tjivikua, 1990 B).

7.3.2
A Synthetic Self-Replicating System

5'-Amino-5'-deoxy-2',3'-O-isopropylideneadenosine was acylated at N-5' with an activated derivative of the 6-carboxy-2-naphthyl ester of Kemp's acid amide. The resulting molecule possesses self-complementary binding sites, the key feature of replicating molecules that act as templates for their own reproduction. The dimer of this molecule is, however, not very stable ($K = 630$ L mol^{-1}). When the two initially mentioned educts are added, a small proportion of the ternary complex is also formed and undergoes a fast, template-catalyzed amidation to yield the binary product complex. The complex dissociates, and each product molecule binds a further two educt molecules and so on. Autocatalysis is thus observed together with self-replication of the product (Tjivikua, 1990A; Nowick, 1991).

7.4

Covalent and Non-Covalent Porphyrin Assemblies

An important type of chemical module was inspired by the crystal structure determination of bacterial photosynthetic reaction centers (Deisenhofer, 1984), where two magnesium porphyrins (the "special pair") and one metal-free porphyrin are bound in a fixed array. Irradiation of this arrangement with visible light induces an extremely rapid charge separation. Several bis-porphyrins with rigid aromatic spacers in which the components are also fixed, porphyrins and bis-porphyrins with redox-active groups, and other model structures have been synthesized (e.g. Sessler, 1986, 1987, 1988 A, B).

Porphyrin monomers have a strong tendency to form stacked dimers in solutions. The most stable dimers in water were obtained with negatively charged *meso*-substituents and positively charged *β*-substituents. Binding constants are in the order of 10^7 M^{-1}. Long-distance dimers for charge separation were realized in 2 nm gaps of rigid membranes on solid electrodes or colloidal particles. Porphyrin 1 was deposited flat on the surface of the electrode or particle and surrounded by a rigid monolayer of diamido bola-amphiphiles. Gaps with a width of 2 nm were thus formed in the monolayer. They contained C=C double bonds, which were then aminated. In a water phase a ring of positive charges was thus formed within the gaps and could be used to bind a second porphyrin. A long-distance heterodimer for charge separation was thus established (Fudickar, 2000; Skupin, 2001).

Another possibility was the attachment of porphyrins to negatively charged gold particles by positively charged viologen substituents. They were separated by a long alkyl spacer from the porphyrin, which floated in water or a solvent. Upon flashing with visible light, electrons were transported from the porphyrin to the viologen molecules and could be trapped by the gold particles. If they were adsorbed to the surface of a gold electrode, a photocurrent became detectable, as light energy was converted to electrical energy (Shipway, 2000).

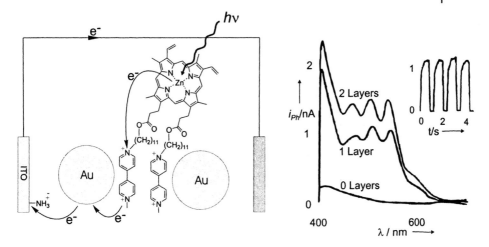

If the porphyrins possess side-chains with hydrophilic, hydrogen-bonding head groups, they assemble to form long fibers in aqueous suspension (Fuhrhop, 1992, 1994, 2000). The scheme below models the dimeric unit of such electronegative pyrrole rings pack over the electronic void in the porphyrin's center (Hunter, 1990). The properties of non-covalent oligomers forming by spontaneous self-assembly are often found to be similar to those of certain covalent synthetic oligomers, thus providing helpful information on structures and reactivities.

Fullerene acts as an electron acceptor with a remarkably small reorganization energy of electron transfer. This results from its π-electron delocalization over a three-dimensional surface, together with the rigid, confined structure of an aromatic π sphere. Conventional condensation chemistry coupled a 10,20-bis-anilino porphyrin with a ferrocene-carboxylate. Schiff base formation between a glycine–fullerene derivative gave the Fe-zinc porphyrin-porphyrin-fullerene tetramer. The lifetime of its charge-separated state ($Fe^{3+}-C_{60}$ radical anion) was extremely long (0.4 s!; Imahori, 2001).

Fc - ZnP - H$_2$P - C$_{60}$

7.5
Molecular Assemblies of Amphiphiles

In biological systems molecular assemblies connected by noncovalent interactions are as common as biopolymers. Examples are protein and DNA helices, enzyme-substrate and multi-enzyme complexes, bilayer lipid membranes (BLMs), and aggregates of biopolymers forming various aqueous gels, e.g. the eye lens. About 50% of the organic substances in humans are accounted for by the membrane structures of cells, which constitute the medium for the vast majority of biochemical reactions. Evidently organic synthesis should also develop tools to mimic the structure and properties of biopolymer, biomembrane, and gel structures in aqueous media.

A review by Ringsdorf (Ahlers, 1990) on biomembrane models and books by Voegtle (1991) and Fuhrhop (1994, 2000) on "supramolecular chemistry" are recommended for further studies in this area.

7.5.1
Vesicle Membranes

A typical biomembrane consists largely of amphiphilic lipids with small hydrophilic head groups and long hydrophobic fatty acid tails. These amphiphiles are insoluble in water ($< 10^{-10}$ mol L^{-1}) and capable of self-organization into ultrathin bilayer lipid membranes (BLMs). Until 1977 only natural lipids, in particular phospholipids such as lecithins, were believed to form spherical and related vesicular membrane structures. Intricate interactions of the head groups were supposed to be necessary for the self-organization of several tens of thousands of amphiphilic molecules. Kunitake (1977), however, demonstrated that almost any water-insoluble amphiphile can self-organize to stable vesicles either spontaneously or, more rapidly, on ultrasonication or heating. His most simple and therefore particularly striking example was commercial dimethyldioctadecylammonium chloride ("DODAC"). Hundreds of vesicle-forming amphiphiles have been synthesized since then (Fuhrhop, 1994, 2000).

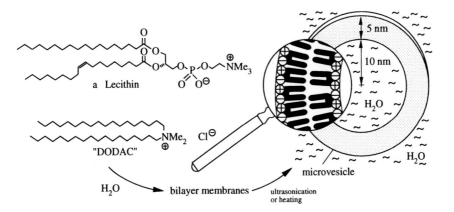

Artificial vesicles usually show perfect spherical or ellipsoidal symmetry. The bilayer symmetry, however, may be broken, since the convex outer surface and the concave inner surface are easily differentiated, e.g. by the low accessibility of the interior space. The bilayer asymmetry is important for the efficient separation of redox-active systems, e.g. of the components of the respiratory assembly or of photosynthetic systems. Fully asymmetric vesicle membranes have been obtained from so-called bola-amphiphiles consisting of a long hydrophobic core with hydrophilic head groups on both ends. (A bola is a South American throwing weapon. Its simplest form consists of two metal or stone balls fixed to the ends of a cord.)

A 36-membered macrocyclic tetraester was prepared in two steps from maleic anhydride and 1,12-dodecanediol. Michael addition of mercaptosuccinic acid converted the isopropanol-soluble macrocycle into an insoluble amphiphile with one "large" succinate head group. The precipitating monoaddition product was redissolved in a more polar solvent and subjected to a second Michael addition with aqueous hydrogen sulfite to give a "small" sulfonate head group at the other end of the complete bola-amphiphile. On ultrasonication in water, such bola-amphiphiles form vesicles of a uniform outer diameter of 30 nm. The thickness of the monolayer membrane is 2 nm, and the ratio of the outer to the inner vesicle surface area is thus about 4:3. This size difference can enforce a totally unsymmetrical arrangement of the head groups in the vesicle membrane. In the case given, all of the succinate head groups (>99%) are located on the outer surface, but all sulfonate head groups on the inner surface. Such asymmetry is typical of vesicular structures and may be useful in charge separation processes across membranes (Fuhrhop, 1983, 1986 A, B).

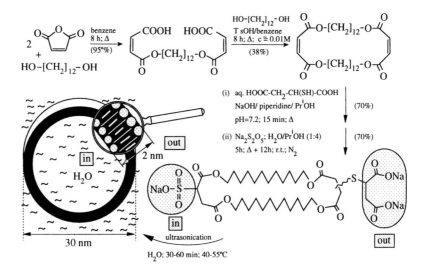

7.5.2
Helical Micellar Fibers

The self-organization of amphiphiles with chiral head groups often results in complex helical superstructures. The easily accessible *N*-octyl-D-gluconamide, for example, dissolves in hot water and forms double and quadruple helical micellar fibers with ratios of length to diameter of more than 10^4:1. The diameter of the single twisted helical cylinders, 3.6 nm, corresponds to the length of two amphiphile molecules. The hydrophobic effect and the helical chains of amide hydrogen bonds cause the formation of these cylindrical micelles, and the chiral carbohydrate head groups enforce helical twists and determine the right-handed helicity of the chiral superstructures (Fuhrhop, 1987, 1988, 1990 A, B).

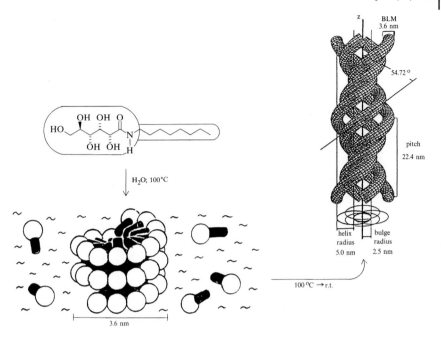

The racemic mixture *N*-octyl-DL-gluconamide does not form micellar fibers, but long flat sheets precipitate on cooling the micellar solution. The properties of the aggregates described, and related ones, suggest that chiral surfaces are a prerequisite for the formation of ultrathin fibers with high helical curvature and large surfaces in aqueous suspensions and gels ("chiral bilayer effect"; Fuhrhop, 1987) as found in biological systems. Many different structures are formed when other chiral hexonamide diastereomers, homologous alkyl derivatives, and other solvents are applied (Fuhrhop, 1988, 1990 A), e.g. rods, platelets, twisted ribbons, tubes, rolled up sheets, and clusters of globular micelles. When a rigid 2,4-hexadiyne unit is inserted into the alkyl chain of *N*-octyl-D-gluconamide, tubular aggregates are formed instead of bulgy multiple helices (Fuhrhop, 1991). The size, shape, and symmetry of micrometer-sized molecular assemblies can be lyophilized and stored indefinitely. On shaking with water, stable suspensions of the tubules are restored (Fuhrhop, 1994, 2000).

Covalent micelles ("dendrimers") are introduced in Section 5.5.

7.6
Capsules

The hollow interior of dodecahedrane and other organic cage compounds described in Section 7.7 is much too small to envelop atoms, ions, or molecules. Tight closed-shell macromolecules have been obtained from vesicles by several research groups, by polymerization of amphiphiles possessing double or triple bonds with the membrane or at the head groups. Smaller, but well-defined, closed-shell containers have been obtained by two other methods described below, namely by directed synthesis and by formation of closed-shell all-carbon molecules in graphite vapor.

An uneventful coupling of two hemispherical "cavitand" molecules – a tera-methanethiol and a tetrakis(chloromethyl)precursor – yielded Cram's (1988) "car-cerand". It entraps small molecules such as THF or DMF, cesium or chloride ions, or argon atoms as permanently "imprisoned guests". Only water molecules are small enough to pass through the two small pores of this molecular prison.

+ carcerates with entrapped
THF, DMF, $Cs^{\oplus}$, $Cl^{\ominus}$, or Ar

Self-assembling capsules show a binding behavior that none of the individual components display alone (Conn, 1997). Calixarene-based cyanuric acid derivatives, for example, form a box, if held together by a barbiturate "glue" in chloroform solution. This solvent is noncompetitive for hydrogen bonds (Vreekamp, 1996).

Cyclic and bent derivatives of urea, e.g. glycouril and its derivatives, form hydrogen-bonded capsules even in polar solvents such as DMF. They entrap hydrophobic guests, such as methane, ethane, and cyclohexane (Branda, 1995).

A widely applied method for capsule building applies metal ions as joints (Conn, 1997). A spectacular example is the adamantane-shaped palladium assembly given below.

a: Ar = none; b: Ar = —⟨◯⟩— ; c: Ar = —⟨◯⟩—⟨◯⟩—

7.7
Fullerenes and Carbon Rods

Fullerenes form by self-assembly of carbon atoms on evaporation of graphite in an inert gas atmosphere (synonyms: (foot)ballene, buckminsterfullerene ("BF"), carbosoccer, soccer(ball)ene, spherene. Buckminster Fuller was the inventor of self-supporting polygon frameworks, e.g. of pavilions). Up to 20% of [5,6]fullerene-C_{60}-I_h (the proposed bridged-fused ring system and von-Baeyer names and numberings vary) a regular C_{60}-polyhedron consisting of 12 pentagons and 20 hexagons, and ellipsoidal higher fullerenes can be isolated from the condensed carbon matter and used as starting materials for a wide range of interesting new compounds (Krätschmer, 1990 A, B; Diederich, 1991, 1992; Smalley, 1992). Fullerenes are also isolable from ordinary sooting flames on the gram scale (Howard, 1991). Fullerenes with metal atoms trapped inside have been prepared by laser vaporization of graphite-metal oxide composites (Weaver, 1992).

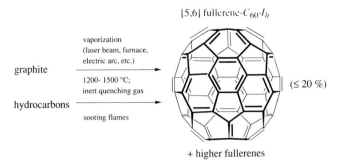

[5,6] fullerene-C_{60}-I_h

graphite $\xrightarrow{\substack{\text{vaporization} \\ \text{(laser beam, furnace,} \\ \text{electric arc, etc.)} \\ \text{1200- 1500 °C;} \\ \text{inert quenching gas}}}$

hydrocarbons $\xrightarrow{\text{sooting flames}}$

($\leq 20\,\%$)

+ higher fullerenes

Similar procedures also lead to carbon rods and tubules. They can be oxidized at the rims and then attached to positively charged surfaces (Liu, 1999). Carbon rods are favored candidates for the function of molecular wires in computer devices of molecular size.

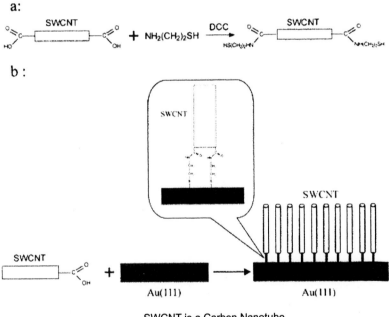

a:

b:

SWCNT is a Carbon Nanotube

Fullerene spheres and rods can be considered as a molecular full stop to organic synthesis: highly complex and possibly very useful molecules are formed by self-organization of carbon atoms in the vapor phase. Synthetic chemists may, however, manipulate new molecular materials to integrate them into functional devices.

7.8

Rotaxanes and Catenanes

Catenanes consist of two or more interlocked rings. They are synthesized in high yield by binding of an electron-deficient guest (e.g. bipyridine) to a macrocyclic ring containing electron-rich parts (e.g. tetrathiafulvalene on p. 409 or dioxynaphthalene below). The bipyridine rings may then be connected by bis-alkylation in order to produce the interlocked catenane (Raymo, 1999).

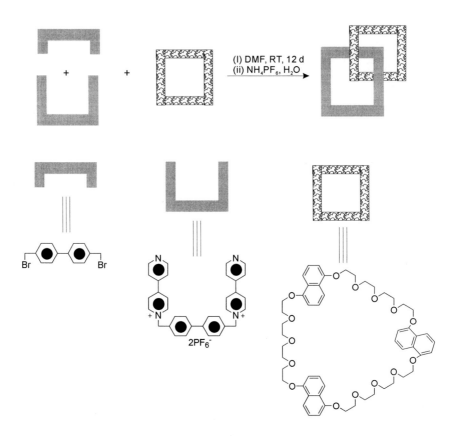

Upon oxidation of the thiafulvalene cycle to a cation or dication (350 and 720 mV), the viologen cycle is repulsed and moves to the naphthalene part of the interlocking partner. This mechanical movement is the basis of a molecular-based device on silicon wafers. The catenane is integrated into a Langmuir-Blodgett monolayer and top electrodes are deposited on them through a shadow mask using electron-beam deposition. A series of voltage pulses moves the system stepwise from the bottom current (thiafulvalenes are positioned to the right and left) to the top current (thiafulvalenes above and below). Change of the voltage recovers the original state through a hysteresis loop (Figure 7.1). A solid-state molecular switch has thus been established, which exhibits a "react" current difference between "open" (–1.0 V) and "closed" (+1.8 V) states of an approximate factor of 3 (Pease, 2001).

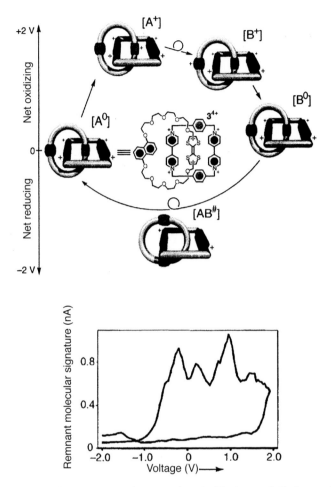

Fig. 7.1 A catenane monolayer was deposited by Langmuir-Blodgett techniques onto parallel polysilicon wires etched into a smooth silicon thin film. Top metal electrodes were then deposited through a shadow mask using electron-beam vaporization. The device thus becomes a switchable conductor. A series of voltage pulses moves the device from the bottom to the top of a current loop and then back again. After each voltage pulse, current through the device is read at a low voltage. Only monolayers of the catenane with two stable orientations of the rings produces such loops. The thiofulvalene-rotation-switch was stable over many cycles over a two-month period before showing signs of failure (from A. R. Pease, J. O. Jeppesen, J. F. Stoddart, Y. Luo, C. P. Collier, J. R. Heath, 2001, Acc. Chem. Res. 34, 433).

In rotaxanes one or more rings encircle a dumbbell-shaped component. In a pseudo-rotaxane at least one of the stoppers at the dumbbell ends is absent, so that the ring(s) can slip off. Synthesis depends on the same type of complex formation as with the catenanes (Raymo, 1999).

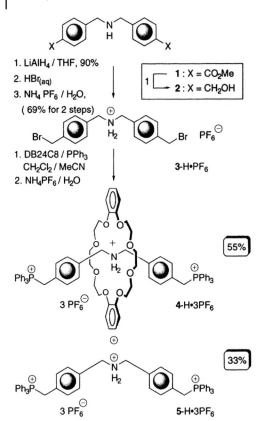

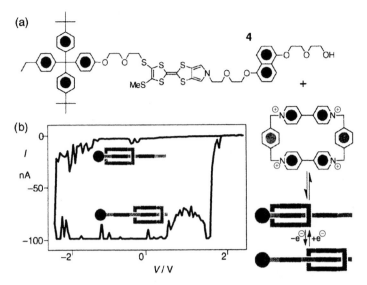

Fig. 7.3 The room temperature solid-state switching signature of a bistable pseudorotaxane. The signal amplitude is much larger than that of the catenane switch of Figure 7.1 (100 nA instead of 1 nA) and the slopes are steeper. A superior switch was obtained with the same chemical motifs in a different arrangement (Pease et al., see Figure 7.1).

8
The Internet and the Feedstock

8.1
Introduction

Since the early 1990s the internet has provided a very rapid and simple information system for chemists. The classical chemical library and its search tools are needed for long-term conservation of chemical knowledge as well as for systematic historical work. Short-term information comes, however, almost exclusively from the net. *Chemical Abstracts*, Beilstein and Houben-Weyl, to name just a few information sources, are being replaced by the electronic Chemical Abstract Services (in particular SciFinder Scholar) and similar services. Searches of the chemical journals of libraries are replaced by American and European Net services. The ne plus ultra is the "anywhere in the article" search category in the ACS journal address. Catalogues of chemical suppliers can also be found on the net. It is the quick access to complex information which makes the electronic medium unbeatable. A brief review of practical hints is given in the next few pages.

Another aspect of synthesis is: where do all the commercial compounds come from? The answer is: mainly from mineral oil. However, this is only expected to last until about 2040 or so. Young chemists should be thinking about how to replace petrol by renewable carbohydrate sources. The subject of chemical feedstocks will be discussed briefly in this chapter.

8.2
Starting Materials

The first requirement for a synthesis is the knowledge and purchase of *commercial starting materials*, reagents, and appropriate solvents. In the early 1990s this would have meant a thorough search in several catalogues, e.g. from Aldrich, Fluka, Merck, Sigma, Alpha, Roth and Strem. This had to be repeated every year as new catalogues were issued, and quite often the most appropriate compounds for a given purpose were not discovered because they appeared under an unknown name and contained unexpected substituents, which made formula indices useless.

Tab. 8.1 Commercial quinones as obtained from an internet search under SigmaAldrich.com

1	Quinonene Oxidoreductase 2 (NQO2) human
2	1,4-Benzoquinone, 98%
3	1,8-Dihydroxy-2,4,5,7-tetranitroanthra-9,10-quinone
4	2,6-Bis((4-methyl-1-piperazinyl)sulfonyl)anthra-9,10-quinone
5	2-(3-Chlorophenyl)benzo-1,4-quinone
6	1-(4-nitrophenyl)benzo-1,4-quinone
7	6-Anilinoquinoline-5,8-quinone
8	Cellobiose quinone oxidoreductase
9	Cyssor I, 2-methyl-N^1-benzenesulfonyl-N^4-(bromoacetyl)quinonediimide
10,11	Rat or human Quinone reductase
12,13	Methoxatin, 4,5-dihydro-4,5-dioxo-1H,pyrrolo[2,3-f]quinone-2,7,9-tricarboxylate Phenyl-p-quinone

Nowadays one does better to look in the electronic catalogue of the same suppliers on the internet. If you work, for example, on non-covalent assemblies in water, you may need a water-soluble quinone for the synkinesis of functional molecular assemblies. In order to search for useful compounds, for example a quinone, you look first under SigmaAldrich.com and print "quinone" into the "product name" box. The quinone search provides mostly trivial compounds, but 2,6-bis((4-methyl-1-piperazinyl)sulfonyl) anthra-9,10-quinone and Methoxatin come as surprises and sound promising.

2,6-bis-((4-methyl-piperazinyl)-1-sulfonyl)-anthra-9,10-quinone

Methoxatin = 4,5-dihydro-4,5-dioxo-1H,pyrrolo[2,3-f]quinone-2,7,9-tricarboxylate

A similar search for "amines", however, ends in disaster. "Over 150 documents" are indicated as a warning and only a small selection of 30 amines is given. A more specific search may help. I may, for example, look for carbohydrates with amino groups. Neither "amino carbohydrate" nor "amino saccharide" produces any result. I try a related compound which I know, namely Kanamycin, and 20 entries are produced. No systematic name of Kanamycin is, however, given in the Sigma catalogue. This is the reason why this important amino pseudo-trisaccharide cannot be found under "amino saccharide". Glucosamine, on the other hand, is found in 75 documents, but again Kanamycin is not among them, although it is a 6-amino-glucopyranose derivative. The conclusion is that you first have to become familiar with the Sigma-Aldrich and other chemical catalogues in order to know what you can expect. The catalogues provide many systematic names, but "amino", "glucose", or "hydroxy" derivatives are not systematized, because there are too many entries. Even the entry "pyrrole" gives you hundreds of compounds, including the quinone methoxatin mentioned above. The structure of "Kanamycin" remains, however, a black hole, and Karakoline, Karanjin, and Katacine, which you find on the same page of the Sigma catalogue, are also given no structural formula or systematic name. You have to use the SciFinder Scholar (=chemical abstracts service CAS; see next section) in order to know what you can buy from the supplier. The Katacine structure remains obscure even in the Chemical Abstract Service ("no structure diagram available"). Here you can buy a chemical from a supplier even though nobody seems to know the structure.

Katacine also appears in the Sigma catalogue(€ 220/100mg). A reference is provided, which is difficult to obtain. No structure could be found in SciFinder.

These search operations take a few minutes. If you print out the results you can then decide which compounds you want to buy.

Commercial reagents can also be scanned. If you look for a reducing agent for a ketone you may search for hydrides and you will quickly find 81 of them. Here you discover, for example, the surprising fact that DIBAH can be purchased in dichloromethane solution.

The electronic catalogues also contain several special listings of reagents and auxiliaries, which you must know as a synthetic chemist. Several "Oxazolidone" chiral auxiliaries for Evan's enantioselective aldol synthesis can be bought and helpful literature procedures are indicated in the Aldrich catalogue. Hundreds of chiral ligands are listed there too and open the way to the immediate application of Heck, Suzuki, Stille, Brown, and Sharpless reactions.

Tab. 8.2 Commercial hydrides from the SigmaAldrich catalogue on the internet

1	(Methylthio)methyltriphenylphosphonium chloride – potassium hydride mixture
2	2-Pocolyltriphenylphosphonium chloride-potassium hydride mixture
3	*R*-Alpine-Hydride® (lithium *B*-isopinocampheyl-9-borabicyclo[3.3.1]nonyl hydride
4	*S*-Alpine-Hydride® (lithium *B*-isopinocampheyl-9-borabicyclo[3.3.1]nonyl hydride
5	Bis(cyclopentadienyl)tungsten chloride hydride
6,7	Bis(cyclopentadienyl)zirconium chloride hydride, Schwartz' reagent
8–15	Calcium hydride
16	Carbonylhydridotris(triphenylphosphine)rhodium(I), 97%
17	Cyclopentadienylmolybdenum tricarbonyl hydride
18	Cyclopentadienyltungsten tricarbonyl hydride, 98%
19	Deuterium hydride, 96 mmol% HD
20	Diethylaluminum hydride
21–37	Diisobutylaluminum hydride, DIBAH, invarious solvents, e.g. dichloromethane
38	Lithium 9-BBn hydride, 1.0 *M* solution in tetrahydrofuran
39–55	Lithium aluminum hydride, LAH
56–59	Lithium hydride
60–61	Lithium tri-*tert*-butoxyaluminum hydride
62	Lithium triethylborohydride solution, Super-Hydride®, purum, lithium triethylborohydride, 1.0 *M* in tetrahydrofuran solution
63	Magnesium hydride, 90%
64–67	Poly(dimethylsiloxane), hydride terminated
68–69	Potassium hydride
70	Red-al®, 65+wt.% solution of sodium bis(2-methoxyethoxy)aluminum hydride in toluene
71	Sodium aluminum hydride, tech.
72	Sodium borohydride
73–79	Sodium hydride (60% suspension in oil)
80	LS-Selectride® (lithium trisiamylborohydride, 1.0 *M* solution in tetrahydrofuran)
81	N-Selectride® (sodium tri-*sec*-butylborohydride, 1.0 *M* solution in tetrahydrofuran)

8.3
Literature Search

Starting materials and reagents are one thing, knowing how to use them efficiently another. The scientific journals are a unique source for recipes. The best recipe is, first, the one you trust. Read carefully and compare before you start. The second aspect is the reported yield, which is given in percent and quantity. Both are equally important. A yield of 95% calculated from the isolation of 1.2 mg of an oil is not as good as an 80% yield of 100 mg of crystalline material. Compare workup and elemental analysis. Never trust mass spectra with respect to purity or structural proof, be cautious with NMR spectra as criteria for purity.

8.3.1
Science Finder Scholar (Chemical Abstract Service, CAS)

The most general source of scientific papers and chemical substances is Scifinder Scholar (Ridley, 2002), to which you have to subscribe. The following items can be selected on the screen

(1) Chemical Substance or Reaction
Subtitles are:
(a) Draw chemical structures. There are 407 references given for the following example:

(b) Type name of compound. Our example Methoxatin produces 716 references. The search within these references can be specified by selecting one of the following topics: toxicity, analysis, biological, combinatorial, crystal structure, preparation, properties, and spectra.
(c) Give a molecular formula. Notation of $C_{17}H_{19}N$, for example, produces 859 structural formulae and corresponding references.

(2) Research Topic
You describe your topic using a phrase: e.g. Heck reaction of porphyrins This will give you 18 references which include title, authors, and abstract. The search can be refined by a more specific chemical structure, product yield, number of steps, and classification of reaction conditions. If you ask, for example, for a yield above 90% only 12 references remain. As always with SciFinder: your chance of finding a specific reference is in the order of 50% only. The oldest references are from 1975. Many papers, even in the most prominent journals are not found. One accessible recent reference discussing "Heck reactions on porphyrins" will, however, help you immensely. This paper will contain the citations which the SciFinder did not find.

(3) Author's name
This will give you a quick overview over some of the author's papers. Fuhrhop, for example, provides 108 references, which is about 50% of the existing papers. This is a typical number for SciFinder and not impressive. It comprises, however, all of the chemical, biological, and medical papers dealing with substances, and is therefore the most general and fastest path finder.

(4) Document identifier
This is used to find patents and chemical abstracts when you know the patent number. Not useful.

(5) Company's name
You'll find amazing and surprising research topics, which are investigated in your immediate environment, if you look for "company". Freie Universitaet Berlin, for example, produced a list of 23,852 papers ordered by date, the latest coming first. The recent papers provided me with a whole new view on my campus called Dahlem, where I have now worked for 25 years. "Company name" is a very social address.

(6) Browse table of contents
The table of contents of 1741 journals is given. You get the newest issue if you click one of them and you can go back to the very first issue. An abstract is provided for each paper as well as the reference list.

8.3.2
American Chemical Society (ACS)

The ACS has developed a spectacular chemistry information system on the Internet (*http://pubs.acs.org*) based on its journals. There is an archive version, which will comprise all journals which the society ever published, and the standard version, which covers only the last six years. Both are available on subscription and are worth every cent you pay.

The first option on the ACS page is "search the journals." A click opens the Journals Search page for Web Editions Subscribers. The section gives five options: Title, Author, Abstract, Title or Abstract, and Anywhere in the Article. Each of these options can be combined with the others in three different lines connected by "and" or "or".

Transformation of isomaltulose into products with industrial "application profiles"

The other major mass bioproducts are fats. About 80% of them go directly into the food industry, and most of the remainder is converted either to detergents or to methyl esters of fatty acids. The latter are used as "biodiesel" (Haas, 2001). Conversion to feedstock chemicals outside the detergent and fuel markets is negligibly small.

By the mid-twenty-first century this type of chemistry should have replaced petrol-based chemistry as a source of feedstock. Mass products from farmland should be considered seriously as alternatives to mineral oil as a source of feedstock chemicals, and their utilization in industry requires synthetic chemists. Molecular biology and biotechnology as carried out in fermenters will always remain expensive and will presumably concentrate on pharmaceuticals and possibly some foodstuffs.

9
Retrosynthetic Analysis, Sketches of Syntheses, Tandem Reactions and Green Chemistry

9.1
Introduction

So far we have described single steps in syntheses and some pathways leading to heterocycles, biopolymers, and dendrimers. Extended voyages to target molecules located far from commercial starting materials have not yet been described. We shall first exemplify some short synthetic excursions into the environment for beginners and then sketch plans of a few more demanding hiking tours. These apply tools that only recently became available. For detailed discussions of a large number of classical and modern total syntheses the reader should look into the book by Sörensens and Nicolaous (1996). We avoid the military term "strategy" here, which points more in the direction of decomposition than that of synthesis, and replace it by "concept" or "sketching".

9.2
Elementary Retrosynthetic Analysis

The sketch of a synthetic plan starts, of course, with the structural formula of a target molecule. The target molecule is methodically broken apart in such a way that reassembling the pieces can be done by known or conceivable reactions. This analytical reverse-synthetic procedure is called retrosynthesis or antithesis (Corey, 1967A, 1971; Warren, 1978). The structural changes in the antithetic direction are named transforms, whereas the same operations in the synthetic direction are reactions. A double-lined arrow ($\Rightarrow$) will be used to indicate the direction associated with a transform in contrast to a single arrow in the direction of a synthetic reaction. Since the structural units within a molecule, which are combined by a synthetic operation, are called synthons, one can also say that antithesis is the analytical process by which a molecule is converted into synthons and into their equivalent reagents. The flow of events (a computer language term) within antithesis may be regarded as:

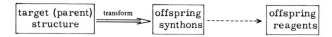

Each transform should lead to reagents that are more easily accessible than the target molecule. In the subsequent steps of antithesis the reagents are defined as new target molecules, and the transform procedure is repeated until the reagents needed are identical with commercially available starting materials.

The following terms are used to specify the type of transform in the subsequent discussion of antitheses:

Antithetical disconnections are thought to divide the target molecule at a C–C bond into an electron acceptor synthon (a) and an electron donor synthon (d), into two electroneutral radical synthons (r·), or into two electroneutral nonradical synthons (e). The latter may undergo electrocyclic reactions and may be stable reagents, whereas other synthons are unstable carbanions, carbocations, or radicals. The disconnection site in the structural formula of the target molecule may be indicated by a wavy line and the symbols (a, d, r·, e) for the respective synthons. If the target molecule contains only one functional group, it is generally split by an antithetical "one-group" disconnection of the α- or β-carbon atom. This corresponds to a synthetic alkylation reaction. If the target molecule contains two proximate functional groups (1,2- to 1,6-difunctional), a "two-group" disconnection between the functional groups may be applied, which usually produces two carbonyl-derived synthons. The synthons (often a carbanion, an alkene, and a carbonyl compound) may be written down before the structures of the proposed equivalent reagents or may be left out. Disconnection operations are indicated by a double-lined arrow. Between "formal" synthons and their equivalent reagents dashed arrows may be introduced and reaction conditions proposed.

Transform type	Target molecule	Synthons	Reagents and reaction conditions
one-group *disconnection*	![OH target]	$\overset{(a)}{\oplus}$ OH—H + $C_2H_5^{\ominus}$ (d) retro-Grignard transform	CH_3CHO + C_2H_5MgBr (i) 0°C (THF) (ii) NH_4Cl/H_2O
two-group *disconnection* *(heterolytic)*	![=O, OH target]	$\ominus$—O + $\overset{\oplus}{H}$—OH retro-aldol type transform	OLi + CH_3CHO (i) −78°C → r.t. (THF) (ii) NH_4Cl/H_2O

Transform type	Target molecule Synthons	Reagents and reaction conditions
two-group disconnection (homolytic)	retro-acyloin transform	(i) Na/Me$_3$SiCl (toluene); Δ (ii) H$_2$O
electrocyclic disconnection	retro-Diels-Alder transform	$\left(\begin{array}{l}\text{synthons} \\ \equiv \text{reagents}\end{array}\right)$ (C$_6$H$_6$); Δ [hydroquinone]

Antithetical connections (the reverse of synthetic cleavages) and rearrangements are indicated by a "con" or "rearr" on the double-lined arrow. Here it is always practical to draw the reagents immediately instead of synthons. A plausible reaction mechanism may, of course, always be indicated.

Transform type	Target molecule	Reagents and reaction conditions
connection	retro-ozonolytic transform	O$_3$/Me$_2$S (CH$_2$Cl$_2$); -78°C
rearrangement	retro-Beckmann transform	H$_2$SO$_4$; Δ

Antithetical functional group interconversions (FGI), additions (FGA), removals (FGR) are symbolized by the given abbreviations on the double-lined arrow.

Transform type	Target molecule	Reagents and reaction conditions
functional group interconversion (FGI)		CrO_3/H_2SO_4 (acetone)
		$HgCl_2(CH_3CN)$
		$HgCl_2$ (aq. H_2SO_4)
functional group addition (FGA)		$PhNH_2; \Delta$
		$H_2[Pd-C]$ (EtOH)
functional group removal (FGR)		(i) LDA(THF) $-25°C$ (ii) O_2; $-25°C$ (iii) $I^{\ominus}/H_2O$

Tab. 9.1 Terms and symbols in antithetical schemes.

Terms	Symbols	
disconnection (of C—C bonds)	(a) C $\not\mid$ (d) C (r) C $\not\mid$ (r) C (e) C $\not\mid$ (e) C	
connection (of C—C bonds)	con	
rearrangement	rearr	
functional group interconversion	FGI	
functional group addition	FGA	
functional group removal	FGR	

Tab. 9.1 (cont.)

Terms	Symbols
transform	all of the operations above taken collectively
synthon	structural formula of molecular fragment with (a), (d), (r), or (e)
(equivalent) reagent	structural formula of a reactive molecule, which produces the synthon under the reaction conditions. It is often, although incorrectly, called a synthon. Only in electrocyclic transforms both terms may denote the same.

In Section 9.2 we shall concentrate on the stereochemistry of target molecules, although functional group chemistry will, of course, remain the basis of all synthetic operations. In this section we shall analyze synthetic functional group chemistry in two ways:

(i) Systematic variations of functional groups (FGI, FGA, FGR) in the target molecule to produce "alternative target molecules", which may be easier to synthesize and which can be converted into the target molecule by a known or conceivable reaction.

(ii) Systematic generation of synthons from the target molecule and all "alternative target molecules" by disconnections, connections, and rearrangements and formulation of the corresponding reagents (=educts, intermediates). The reagents should be more easily available from simple starting materials than the target molecule.

Both items are repeated until the intermediates are reasonably priced commercial starting materials. The resulting antithetic schemes are then evaluated and the most promising ones can be converted into synthetic plans and investigated in the laboratory.

9.2.1
Open-chain Compounds

The systematic application of both antithetic steps will now be exemplified with the admittedly trivial synthesis of 3-methylbutanal (isovaleraldehyde). Functional group operations would yield the following alternative target molecules:

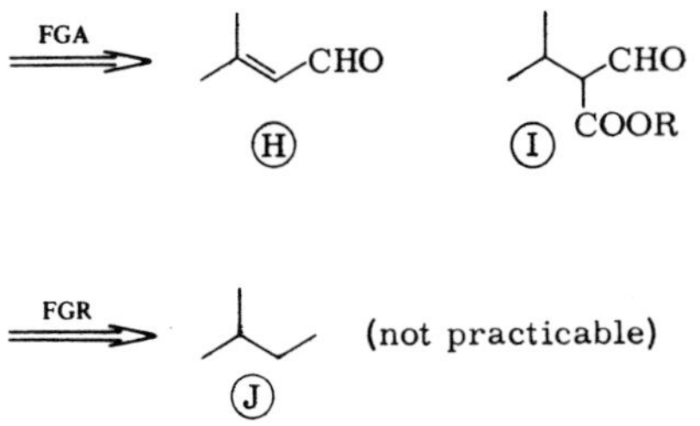

The following reactions could be used to convert compounds (B)–(I) into the target molecule (A). Doubtful or difficult reactions are indicated by a question mark.

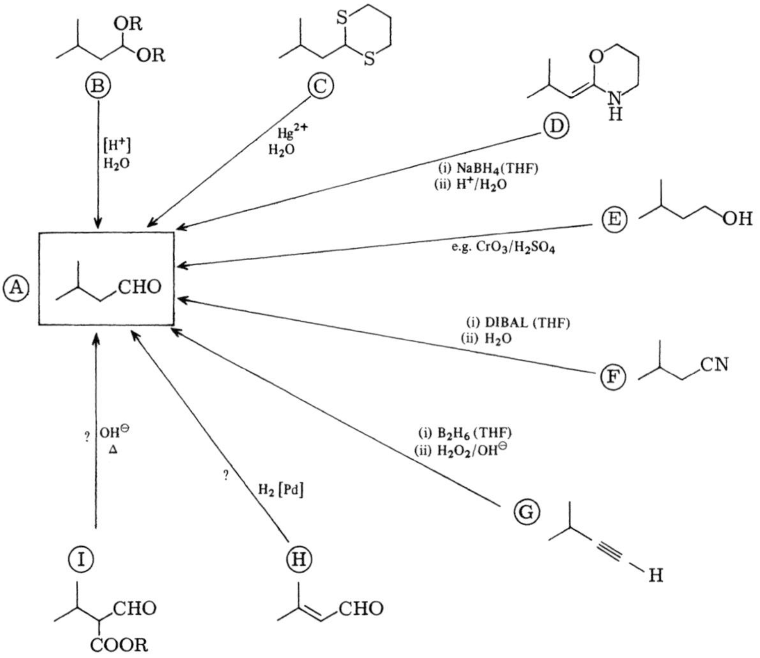

Since (A) does not contain any other functional group in addition to the formyl group, we may predict that suitable reaction conditions could be found for all conversions into (A). Many other "alternative" target molecules can, of course, be formulated. The reduction of (H), for example, may require the introduction of a protecting group, e.g. acetal formation. The industrial synthesis of (A) is based upon the oxidation of (E) since 3-methylbutanol (isoamyl alcohol) is a cheap distillation product from alcoholic fermentation ("fusel oils"). The second step of our simple antithetic analysis – systematic disconnection – will now be exemplified with all target molecules of the scheme above. For the sake of brevity we shall omit the synthons and indicate only the reagents and reaction conditions.

(A)

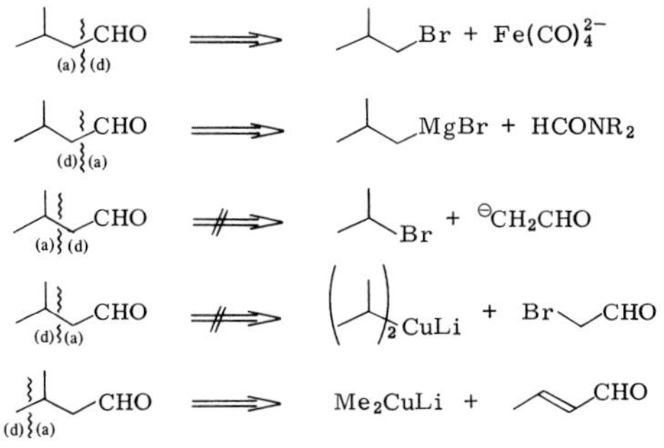

    CHO ⟹ Br + Fe(CO)$_4^{2-}$ (i) CO/Ph$_3$P(THF); low T
(a)⎰(d)   (ii) AcOH

    CHO ⟹ MgBr + HCONR$_2$ (i) (THF); low T
(d)⎰(a)   (ii) H$^+$/H$_2$O

    CHO ⇏ Br + $^\ominus$CH$_2$CHO
(a)⎰(d)

    CHO ⇏ ( )$_2$CuLi + Br⌒CHO
(d)⎰(a)

    CHO ⟹ Me$_2$CuLi + ⌒⌒CHO   (i) (THF); low T
(d)⎰(a)   (ii) NH$_4$Cl/H$_2$O

} protection of the formyl group would be necessary.

(B)

    OR / OR ⟹ Li + HC(OR)$_3$ (THF); −LiOR
(d)⎰(a)

(C)

    (dithiane) ⟹ Br + Li(dithiane)   (THF); −LiBr
(a)⎰(d)

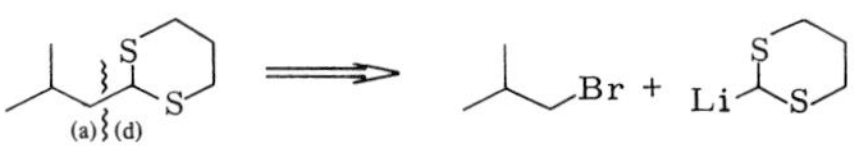

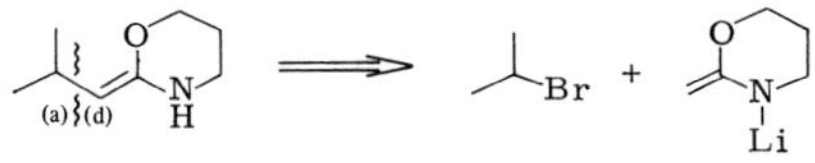

(D)

      ⟹ Br +   (i) (THF); low T
(a)⎰(d) H   N / Li   (ii) H$_2$O

(E)

    ⌒OH ⟹ MgBr + CH$_2$O   (Et$_2$O); low T
(d)⎰(a)

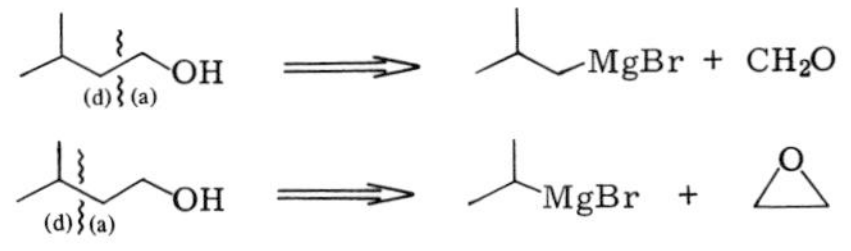

    ⌒OH ⟹ MgBr + (epoxide)   (Et$_2$O); low T
(d)⎰(a)

(F)

    CN ⟹ Br + CN$^\ominus$ KCN(MeCN)
(a)⎰(d)

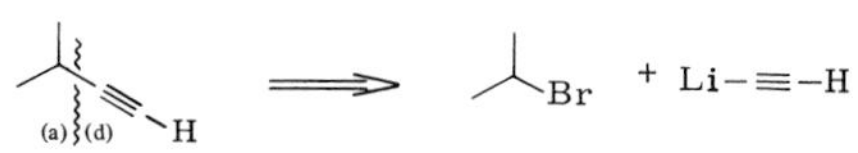

(G)

    ≡—H ⟹ Br + Li−≡−H   (TMEDA/THF)
(a)⎰(d)

(H)

    CHO ⟹ O + ⌒N—R / Li   (i) (THF); low T
(a)⎰(d)   (II) H$^+$/H$_2$O; Δ

(I)

(d)⎰(a)
    CHO ⇏ RO⌒OLi + HCOOR'
    COOR

(a)⎰(d)
    CHO ⇏ Br + RO—O·Li
    COOR

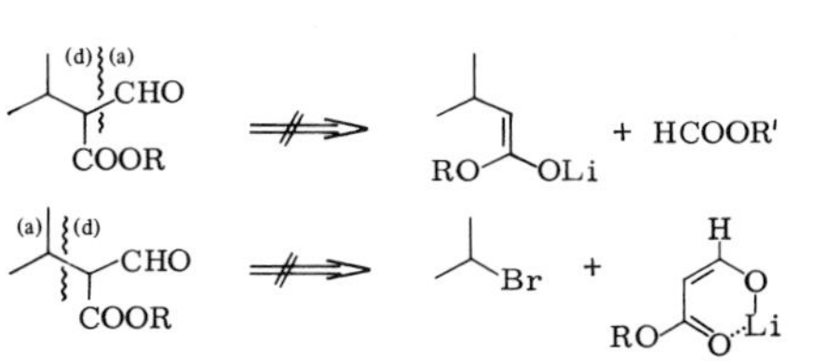

} unprotected formyl groups would undergo side-reactions

It is clearly evident from the extremely simple example of the synthesis of iso-valeraldehyde that a fully systematic approach to antithesis is not very useful. Chemists are not interested in encyclopaedic catalogues of synthetic routes. We shall now discuss a few simple examples, where the availability and price of starting materials are considered. This restriction generally reduces long lists of "alternative target molecules" and "precursors" to a few proposals.

2-Ethylpentanoic acid is our next target molecule. No commercial compound of acceptable price with all seven carbon atoms and a functional group in the right order could be found. The usual α- and β-disconnections of monofunctional target molecules yield either secondary halides (disconnections 1) or primary halides (disconnections 2 and 3) as precursors. Reagents with terminal functional groups are usually much more abundant and inexpensive than chemicals with branched substituents. Butanoic and pentanoic acid esters and nitriles and ethyl and propyl bromides are all fairly low-cost reagents. Alkylations with these reagents, however, all need highly purified, aprotic solvents and very strong bases as auxiliary reagents. The pathway via dialkylmalonic ester is less demanding. It could be carried out in ethanol and would probably be the reaction of choice for a low-cost preparation in large quantity.

The target molecule above contains a chiral center. An enantioselective synthesis can therefore be developed. We will use this opportunity to summarize our knowledge of enantioselective reactions. They are either alkylations of carbanions or addition reactions to C=C or C=O double bonds:

Tab. 9.2 Some enantioselective reactions to produce chiral monofunctional products.

No	Educts	Chiral products
1		
2	+ RLi (CuI)	
3	+ H$_2$ (Noyori)	
4	+ H$^\ominus$	

Since our target molecule is acyclic and monofunctional the obvious solutions to our problem are stereoselective alkylations and hydrogenations (see scheme above) e.g.

Now we turn to syntheses of simple, non-commercial or very expensive difunctional open-chain compounds. We have chosen one or two examples each of 1,2- up to 1,6-difunctional target molecules.

Our first example is 2-hydroxy-2-methyl-3-octanone. 3-Octanone can be purchased, but it would be difficult to differentiate the two activated methylene groups in alkylation and oxidation reactions. Typical syntheses of acyloins are based upon addition of terminal alkynes to ketones (disconnection 1). For syntheses of unsymmetrical 1,2-difunctional compounds it is often advisable to look also for reactive starting materials, which already contain the right substitution pattern. In the present case it turns out that 3-hydroxy-3-methyl-2-butanone is an inexpensive commercial product. This molecule requires disconnection 3. Another practical synthesis starts with acetone cyanohydrin and pentylmagnesium bromide (disconnection 2). Many 2,2-difunctional compounds are accessible via oxidation of C–C multiple bonds. In this case the target molecule may be obtained by simple permanganate oxidation of 2-methyl-2-octene, which may be synthesized by a Wittig reaction (disconnection 1).

We now consider another 1,2-difunctional target molecule, butyl glyoxylate. This simple compound is quite expensive in the pure state because it is rather unstable. It is best prepared freshly in solution and used immediately for further reactions (Wolf, 1963). Obvious FGI-precursors are dibutyl oxalate and butal glycolate. Aldehydes, however, can also be prepared very efficiently by cleavage of olefins or glycols. Our list of starting materials contains three symmetrical diacids, which would be ideally suited for this purpose, namely fumaric, maleic, and tartaric acids. Cleavages of olefins and glycols are such common synthetic procedures, that they would probably be preferable to partial reduction or oxidation procedures, which are often difficult to control.

The general syntheses of alkenes (Wittig), dienes (Heck, Suzuki, etc.) and 1,2-di-hydroxy compounds (alkene oxidation) are not repeated here. But there is an important "chiral pool" for chiral 1,2-disubstituted compounds, namely *a*-amino acids.

A simple example of an optically active target molecule is the triol given below (Ravid, 1977; Valentine, 1978). The amino acid resembling this molecule is glutamic acid, of which the expensive (*R*)- and the low-cost (*S*)-enantiomers can both be purchased. The primary alcohol groups are transformed to the carboxyl groups and the chiral secondary alcohol group into the amino group by FGI. An obvious reaction for the synthetic conversion of the amine into the alcohol is treatment with nitrous acid. The stereoselectivity of this reaction varies, but retention of configuration can often be achieved to a high degree.

Another example is a chiral olefinic alcohol, which is disconnected at the double bond by a *retro*-Wittig transform. In the resulting 4-hydroxypentanal we again recognize glutamic acid, if methods are available to convert regio- and stereoselectively:

(1) the *γ*-carboxyl group into the aldehyde group,
(2) the *a*-carboxyl group into a methyl group, and
(3) the amino group into the hydroxyl group with retention of configuration.

Conversion (3) has already been mentioned in the example above. For conversions (1) and (2) we need a synthetically useful differentiation of the *a*- and *γ*-carboxyl groups. This is quite simple, because only the *γ*-carboxyl group can react with the α-hydroxyl group to form a lactone. The remaining free α-carboxyl group could be converted into the chloride, then selectively reduced to a hydroxymethyl group and converted via the tosylate into an iodomethyl group, which is finally reduced to yield the desired methyl group. The lactone can be converted into the cyclic hemiacetal by DIBAL (Mori, 1975).

1,3-Difunctional products often come from aldo-type reactions. The following example, 2-(hydroxymethyl)-2-methylbutanal needs no further comment (Nerdel, 1968).

This target molecule again contains a chiral center and we inspect Tab. 9.3 for help.

Only reaction 1 provides a direct pathway to this chiral molecule: the intermediate 2-methyl-butanal may be silylated and reacted with formaldehyde in the presence of the boronated tartaric ester described in Chapter 1. The enantiomeric excess may, however, be low.

Tab. 9.3 Some enantioselective reactions that produce difunctional products.

No	Educts		Chiral products
1	R^1—CH=CR^2(OSiMe$_3$)	+ R^{3}CHO	R^1—CO—CHR^2—CHR^3(OSiMe$_3$)
2	R—CHO	+ (cyclohexyl ketone OSiBu^{t}Me$_2$)	R—CH(OH)—CHR'—COOH
3	(CH$_2$=CH—CH$_2$—Br)	+ RCHO	R—CH(OH)—CH$_2$—CH=CH$_2$
4	R—CH=CH—R'	+ Bu^{t}OOH	R—(epoxide)—R'
5	R—CH=CH—R'	+ OsO$_4$	R—CH(OH)—CH(OH)—R'

In stereoselective antitheses of chiral open-chain molecules, transformations into cyclic precursors should be tried. The *erythro*-configured acetylenic alcohol given below, for example, is disconnected into an acetylene monoanion and a symmetrical oxirane (Adams, 1979). Since nucleophilic substitution occurs with inversion of configuration this oxirane must be *trans*-configured; its precursor is commercially available *trans*-2-butene.

(erythro)   + Li—≡—H   (trans)

2,4-Nonanedione, another 1,3-difunctional target molecule, may be obtained from the reaction of hexanoyl chloride with acetonide anion (disconnection 1). The 2,4-dioxo substitution pattern, however, is already present in inexpensive, symmetrical acetylacetone (2,4-pentanedione). Disconnection 2 would therefore offer a tempting alternative. A problem arises because of the acidity of protons at C-3 of acetylacetone. This, however, would probably not be a serious obstacle if one produced the dianion with strong base, since the strongly basic terminal carbanion would be a much more reactive nucleophile than the central one (Hampton, 1973).

Our example of a 1,4-difunctional target molecule is 6,6-dimethyl-2,5-heptane-dione (Hankinson, 1972). It should be synthesized by conventional combination of a d^2- and an a^3-synthon (disconnection 1) or of a d^1- and an a^3-synthon (disconnection 2). Disconnection 3 produces more problems. The keto group of 4-oxo-pentanoyl chloride (levulinic acid chloride) would have to be protected, before a re-action with t-butylzinc or -copper reagents or with $Na_2Fe(CO)_4$ and t-butyl halide could be attempted.

1,5-Difunctional target molecules are generally easily disconnected in a *retro*-Michael-type transform. As an example we have chosen a simple symmetrical molecule, namely 4-(4-methoxyphenyl)-2,6-heptanedione. Only *p*-anisaldehyde and two acetone equivalents are needed as starting materials. The antithesis scheme given below is self-explanatory. The aldol condensation product must be synthesized first and then reacted under controlled conditions with a second enolate (e.g. a silyl enolate plus TiCl$_4$ or a lithium enolate), enamine (Pfau, 1979), or, best, acetoacetic ester anion as acetone equivalents.

Methyl 6-hydroxy-3-methylhexanoate is our 1,6-difunctional target molecule. Obvious precursors are cyclohexene and cyclohexadiene derivatives (Section 1.14). Another possible starting material, namely citronellal, originates from the "magic box" of readily available natural products (Overberger, 1967, 1968; Corey, 1968 D; Clark, 1976).

The 1,6-difunctional hydroxyketone given below contains an octyl chain at the keto group and two chiral centers at C-2 and C-3 (Magnusson, 1977). In the first step of the antithesis of this molecule it is best to disconnect the octyl chain and to transform the chiral residue into a cyclic synthon simultaneously. Since we know that ketones can be produced from acid derivatives by alkylation, an obvious precursor would be a seven-membered lactone ring, which is opened in synthesis by octyl anion at low temperature. The lactone in turn can be transformed into *cis*-2,3-dimethylcyclohexanone, which is available by FGI from (2,3-*cis*)-2,3-dimethylcyclohexanol. The latter can be separated from the commercial *cis–trans* mixture, e.g. by distillation or chromatography.

We end this section on synthesis sketches of open-chain molecules with an example of a trifunctional target molecule. This does not introduce any fundamentally new problem. In antithetic analysis one simply chooses an appropriate difunctional starting material, which may be further disconnected into monofunctional starting materials.

Diethyl 3-oxoheptanedioate, for example, is clearly derived from glutaryl and acetic acid synthons (e.g. acetoacetic ester; Guha, 1973; disconnection 1). Disconnection 2 leads to acrylic and acetoacetic esters as reagents, but the reaction with acrylic ester would inevitably yield by-products from aldol-type side-reactions.

From the above discussion it should be obvious that antithetical analyses of difunctional molecules are simply reversals of the synthesis schemes already described in Chapter 1.

9.2.2
Mono- and Bicyclic Target Molecules

The conformational stability of substituted cyclic compounds can be used in synthesis to take advantage of neighboring group effects. The directing effect of ring substituents is the basis of most regio and stereoselective syntheses. In this section we shall again only analyze target molecules that can be obtained in a maximum of four steps from commercial, inexpensive, starting materials. The examples have been chosen mostly from the recent literature, and references are given for the actual syntheses. We shall restrict ourselves to a *retro-synthetic* analysis.

We begin with two examples of combined electronic and steric effects in benzene-derived target molecules. These are generally obtained from starting materials already containing a benzene ring. An obvious precursor for our next target molecule, methyl 2-formyl-4,6-dimethoxybenzoate, would be 3,5-dimethoxyphthalic acid (A) (Brockmann, 1957). One might predict that the less-hindered carboxyl group could be reduced selectively, e.g. by a bulky hydride such as DIBAL. However, neither diacid (A) nor alternative derivatives are commercially available. Monoacids (B) and (C) can be purchased. Acid (B) is not a suitable substrate for the introduction of a carbon substituent at C-6, since all three substituents would direct an incoming electrophile towards C-3 and C-5. Substrate (C) is more promising since the electron-donating methoxy groups activate all unsubstituted positions (C-2, C-4, C-6) and since its symmetry equalizes positions C-2 and C-6. If the monoformylation of (C) occurred statistically at positions 2, 4, and 6, about 67% of the product would be the desired isomer.

Our second aromatic target molecule is a trisubstituted tetralin derivative with two vicinal *trans*-oriented substituents at the reduced ring (platyphyllide; Bohlmann, 1979). Its antithetic analysis is dictated by the requirement to transform it into the only reasonable starting material 1-naphthoic acid (A). Naphthalenes can be reduced to give tetralins, which in turn can be oxidized at the benzylic methylene groups. If a metal salt is used as oxidant, one might hope that a carboxylate or amide group would bind the metal ion, which would then preferentially attack the neighboring, sterically hindered *peri*-position. Thus molecule (B) should be available. The alternative target molecule (C) could now be obtained from (B) by aldol addition with acetone. We need, however, the enolate of the tetralone compound and a reaction that leads stereoselectively to the trans product. The actual synthesis has been carried out by a Lewis acid-catalyzed aldol addition with the trimethylsilyl enolate of the tetralone. Reduction of the ketol with LiAlH$_4$ produces the *trans*-diol because the tertiary alcohol group binds the hydride reagent below the ring plane. The final steps are re-oxidation of the primary alcohol group and dehydration.

Our first non-aromatic carbocyclic target molecule contains two condensed five-membered rings, an angular methyl group, and an α,β-unsaturated ketone moiety (Welch, 1979). The first disconnection is clearly of the retro-aldol type. The resulting diketone poses no problem since 2-methylcyclopentanone can be purchased and selective formation of the thermodynamically more stable, more substituted enolate is possible. The a2-synthon could be bromoacetone, but its keto group must be protected, e.g. as a ketal. Nucleophilic substitution of this bulky bromide by the highly hindered enolate, however, could be difficult. Inexpensive 2,3-dibromopropene is a good alternative. The allylic bromine should be substituted much faster than the vinylic one. Conversion of the vinylic bromide into the ketone is catalyzed by acids and mercury(II) ions.

The most common stereoselective syntheses involve the formation and cleavage of cyclopentane and cyclohexane derivatives or their unsaturated analogues. The target molecule (*all-cis*)-2-methyl-1,4-cyclohexanediol has all of its substituents on the same side of the ring. Such a compound can be obtained by catalytic hydrogenation of a planar cyclic precursor. Methyl-1,4-benzoquinone is an ideal choice (*p*-toluquinone; Makazaki, 1966).

The following example of an antithesis of a complicated polycyclic and chiral (although racemic) compound, which can be transformed into commercially available starting materials in a few steps, provides convincing evidence for the power of the *retro*-Diels–Alder transform. Its application can result simultaneously in a decrease of the number of rings and chiral centres, in disconnection of a molecule into two stable fragments, and finally in simplification of functionality patterns. The target molecule (Danishefsky, 1979) is first transformed into the more symmetrical dicarboxylic acid (A) by FGA. Synthetic decarboxylation of this vinylogous β-keto acid should occur readily in mildly basic media. The cyclohexenone system must then be transformed into a substituted cyclohexane derivative to prepare for the anticipated *retro*-Diels–Alder disconnection. The choice of (B) and the corresponding disubstituted butadiene is dictated by the (commercial) availability of 4-methoxy-3-buten-2-one, which can be converted into the silyl enol ether. The next transforms are FGI of the cyclic anhydride, removal of the isopropylidene group, and FGI of the *cis*-glycol. As a synthetic reaction osmium tetroxide-catalyzed *cis*-dihydroxylation of the C=C double bonds is advisable since oxidation should occur at the less hindered *exo*-side. The glycol would be ketalized with acetone and acidic catalysts. A final *retro*-Diels–Alder transform yields the readily available acetylenedicarboxylic ester and cyclopentadiene.

Antithesis of the cyclopentene derivative given below with its three chiral centres is another example of the utility of the *retro*-Diels–Alder transform, which is used here for simultaneous stereospecific disconnection of two ring substituents. The acetic acid and hydroxyl groups are cis to each other and can therefore be transformed into a lactone ring system, which can be produced from a bicyclic ketone precursor by Baeyer-Villiger oxidation. The cyclopentadiene derivative from the following *retro*-Diels–Alder-type disconnection can be synthesized easily from cyclopentadienyl anion and chloromethyl methyl ether. One should, however, choose the mildest possible reaction conditions for the alkylation of cyclopentadiene (low temperature) and for the Diels–Alder reaction (acid catalysis), since substituted cyclopentadienes isomerize quickly. The "ene" reagent could not be ketene itself but "protected" equivalents such as 1,1-dichloroethene or 2-chloropropenonitrile. The latter would probably be the reagent of choice because of its activating, electron-withdrawing cyano group. The third chiral centre could be expected to be formed predominantly with the correct relative stereochemistry since addition of the "ene" should occur preferably on the less hindered side of the cyclopentadiene ring (Corey, 1969).

Antithetic analysis of another chiral cyclopentane derivative (Cavill, 1967) is simple, if one remembers that cyclopentanecarboxylic acids are easily obtained by ring contraction from cyclohexanones in a Favorskii rearrangement and can then be modified by FGI in the *a*-position. Otherwise it would be quite complex. The obvious precursor is low-cost pulegone. It has the right number and arrangement of carbon atoms and contains only one enolizable carbon suitable as electron donor in the Favorskii rearrangement. The electron acceptor for this ring contraction must be C-4, which is, however, an electron-rich olefinic carbon atom. An "umpolung" is obtained if the double bond is epoxidized. This epoxide is useful since its opening yields the desired hydroxyl group at C-8. The only remaining problem is the stereoselectivity of the epoxidation of pulegone. The directing steric influence of the remote methyl group at the chiral center A-1 upon peroxide attack can be expected to be quite small. Thus a nearly equimolar mixture of *cis*- and *trans*-epoxidized products must be separated, e.g. by distillation or chromatography.

9.2.3
Bridged Polycyclic Molecules

Topological analysis is the first task in antitheses of complex molecular structures having sets of interconnecting bridges. Both bridged and fused oligocyclic precursors have to be defined, from which the desired skeleton can be produced by formation of one or two connecting bonds. We shall exemplify the analytical procedure with a simple bridged carbocyclic and two heterocyclic target molecules.

6-Methyltricyclo[4.4.0.0^{2,7}]decan-3-one is our first target molecule. We designate all carbon atoms, which belong to more than one ring, by a dot. First, disconnections of bonds between these "common atoms" are examined, which lead to the decalones (A) and (B) in our example. Synthesis of the target molecule from (A) would be relatively simple, since C-2 is already activated by the keto group and therefore liable to act as a donor group. The enolate would readily substitute an appropriate leaving group, e.g. tosylate, at C-7. (B) would give many more problems, e.g. synthesis of a suitable trifunctional *cis*-decalone derivative, prevention of *cis–trans* isomerization, and protection of the keto group.

Next we consider disconnections of bonds ending in only one "common atom". These yield bicyclo[3.1.1]heptane derivatives with a carbon substituent at a methylene bridge. Compounds of type (C) could be made from cyclohexanone derivatives, but again this synthetic route would be much more difficult than via (A). For precursors of types (D) and (E) even lengthier syntheses would certainly be required. Therefore we settled on precursor (A) and called the 2,7-bond (or the equivalent 1,2-bond) our "strategic" *bond* since in the development of a synthetic strategy we would first have to consider how to establish this bond. Since precursors of type (B) and (C) should also be synthesizable, we designate the 1,6(=6,7-) and 1,10-(=7,8-)bonds as our "reserve" strategic bonds if we meet problems with our first choice (A). The retrosynthetic analysis of (A) given below, however, does not show any serious difficulties (Heathcock, 1966, 1967; Ramachandran, 1961).

In the preceding example we did not consider cycloaddition reactions since these would not offer any suitable alternative synthetic pathway. The bicyclic iso-quinuclidine derivative given below (Büchi, 1965, 1966 A) contains only un-strained six-membered rings, and the *retro*-Diels–Alder transform is obviously the furthest-reaching simplification and the fastest antithetical route to commercial starting materials. Both bridgehead atoms can be introduced in one step.

The two-bond disconnection (*retro*-cycloaddition) approach also often works very well if the target molecule contains three-, four-, or five-membered rings (see Chapters 2 and 4). The following tricyclic aziridine can be transformed by one step into a monocyclic amine (Nagate, 1968). In synthesis one would have to convert the amine into a nitrene, which would add spontaneously to a C=C double bond in the vicinity.

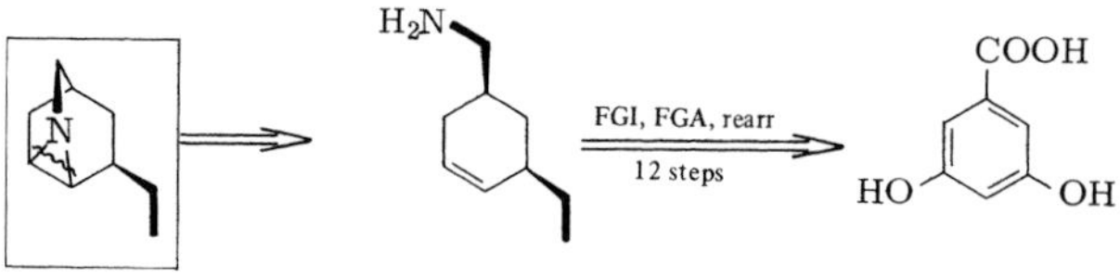

9.2.4
Summary of Antithetical Analysis of Simple Molecules

We have considered the following structural features affecting the choice of antithetical steps starting from simple target molecules:

(i) the arrangement of functionality and interconversion of functionality ("alternative target molecules")
(ii) the presence of special substructures close to commercial starting materials
(iii) stereochemical configuration
(iv) symmetry in (alternative) target molecules
(v) the presence of "strategic" bonds in polycyclic structures

The same considerations are fundamental in the analysis of more complex molecules, which we shall discuss in the Chapter 10, with the exception that in item (ii) the "special substructures" are generally related to more complex, noncommercial precursors, which are known from the literature.

9.3
Learning from Research Papers

The best textbooks for the advanced student of chemistry are the scientific journals for synthesis: the *Journal of the American Chemical Society* is recommended. The experimental parts either in the Journal itself or in the additional material published on the internet are superb. On my desk I have Vol. 103, No. 25 (1981). The following seven compounds are discussed in this particular issue (page numbers in brackets).

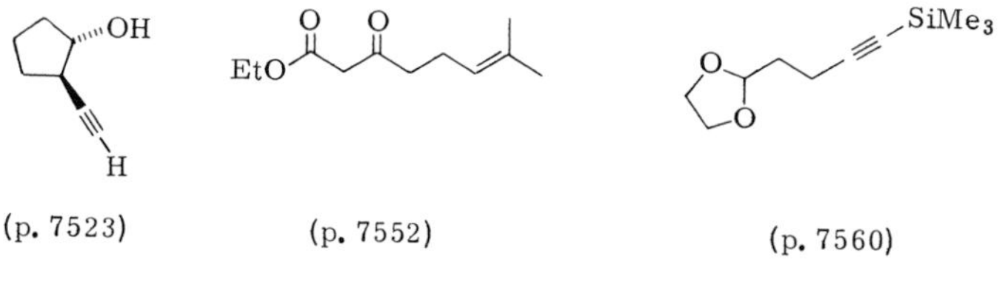

(p. 7523)   (p. 7552)   (p. 7560)

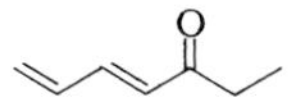

(p. 7573) 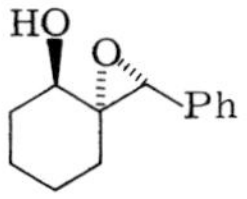   (p. 7663) 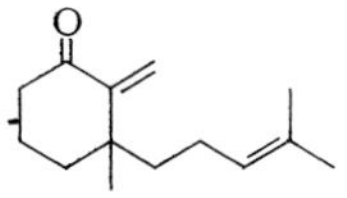   (p. 7668)   (from

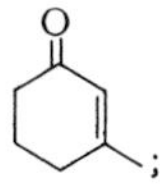

enolate quenching
with CH₂O)

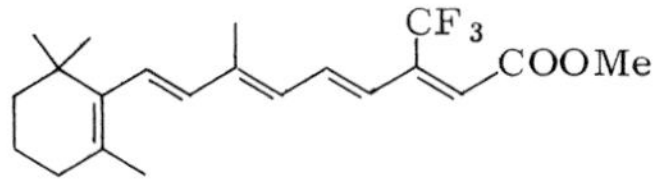

(from β-ionone)

(p. 7642)

These compounds can be made from the starting materials listed in this chapter in a few steps. Try to suggest starting materials and propose synthetic procedures of your own! Then compare these with the procedures given in the Journal. If you regularly make up your own problems from scientific journals, work through them seriously, and slowly get to more complicated target molecules, you cannot fail to learn a lot about solving synthetic problems in a realistic manner!

About ten years later in *J. Am. Chem. Soc.*, Vol. 113, No. 25 (1991) and Vol. 114, No. 21 (1992) we still find simple target molecules. Stereoselective reactions, mostly alkylations, are now prevalent.

1991

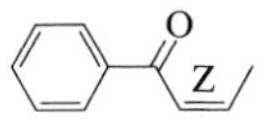

(p. 9589; Stille reaction) 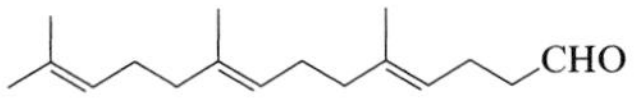   (p. 9674, from squalene)

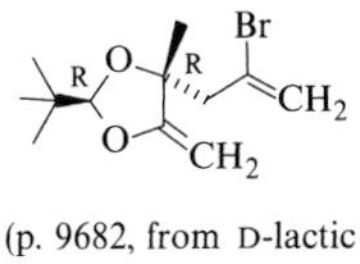

(p. 9682, from D-lactic acid) 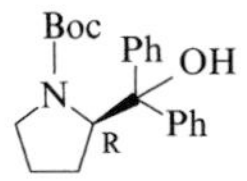   (p. 9709, asymmetric deprotonation)

1992

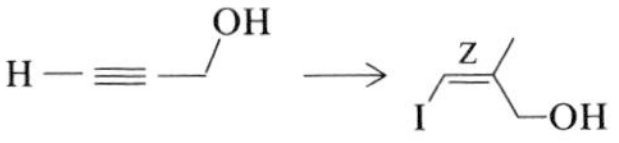

(p.8009) 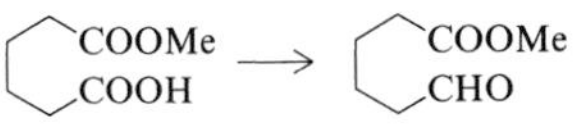   (p.8161)

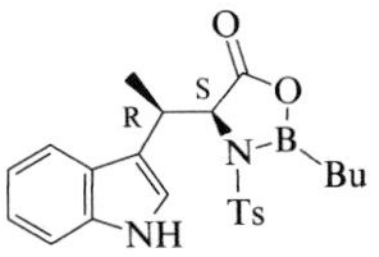

(p. 8290)

In 2002 one we need to look not only at an individual JACS issue, but we need to check on the internet for the whole period of January 2001–July 2002 and include *Org. Lett.* and *J. Org. Chem.* Altogether, 269 articles are found that display "total synthesis" in the title. The names of all the target compounds sound strange: pumiliotoxin, pyridovericin, herbarumin, rhainalam, etc. If familiar names occur (prostaglandin, retinal), the synthesis has to do with special aspects, such as urinary degradation products or special binding to proteins. There are very few target molecules whose names sound familiar. This will very likely also happen to you if you get involved in a synthetic project. The structure of your target molecule will be known from NMR studies, and X-ray structures but virtually nothing will be known about the chemistry. Worse – even if you have reached your goal, your compound will soon be forgotten. Nearly all modern target compounds, as first isolated from "marine sponges" or other exotic species, "attract considerable attention because of their potent antitumor activity" and also "exhibit potent antimalarial and antituberculosis activity", but they never make it beyond structural elucidation and synthesis of a few milligrams. Do not let this discourage you from working in organic synthesis, but, rather, let us have a look at new popular reactions within these syntheses.

A few typical retrosynthetic cleavages of large molecules conclude this chapter. We looked for target molecules of total syntheses on the internet under pubs.acs.org. Altogether, 279 documents were indicated for "all journals" by title and authors in the period from January 2001 to August 2002. Selecting "pdf" conjured the whole papers onto the screen, including plenty of "additional material" with experimental data. A few of those which produced sensible retrosynthetic analyses of compounds were selected. We reproduce the analyses of the authors. The syntheses themselves are detailed in the cited papers and the corresponding synthetic steps described in this book.

A potent cancerostaticum and fungicide was named Ratjadone after Cala Ratjada, the favourite swimming and diving place of Germans on Mallorca island. The chemists among them isolated Ratjadone from soil bacteria and synthesized it (Bhatt, 2001). The molecule contains a chain of ten carbon atoms with one *trans,trans*-diene and one *cis,trans*-diene unit each, as well as two terminal perhydro-pyran units with and without a double bond. The *trans,trans*-diene was, of course, disconnected by a retro-Heck coupling, the *cis,trans*-diene was traced back to two consecutive Wittig reactions. The right-hand cyclohexene-type pyran ring should be made, of course, by a hetero-Diels–Alder reaction of an acrolein derivative. The left-hand tetrahydropyran was retro-opened to a δ-hydroxy epoxide. With the retrosynthetic steps carefully selected, synthesis was very successful.

Ratjadone

The synthesis of fostriecin, another compound with antitumor activity, was retro-synthesized in a very similar manner. There were, however, two distinct differences. The central chiral center was introduced in a very clever way with D-glutamic acid, a member of the chiral pool, and the chirality at C-11 was obtained with a Sharpless osmylation. A Still–Gennari coupling, step 3, which introduces the phosphonate group for the Wittig reaction, after oxidation, leaves the carbonyl group intact (Boger, 2001).

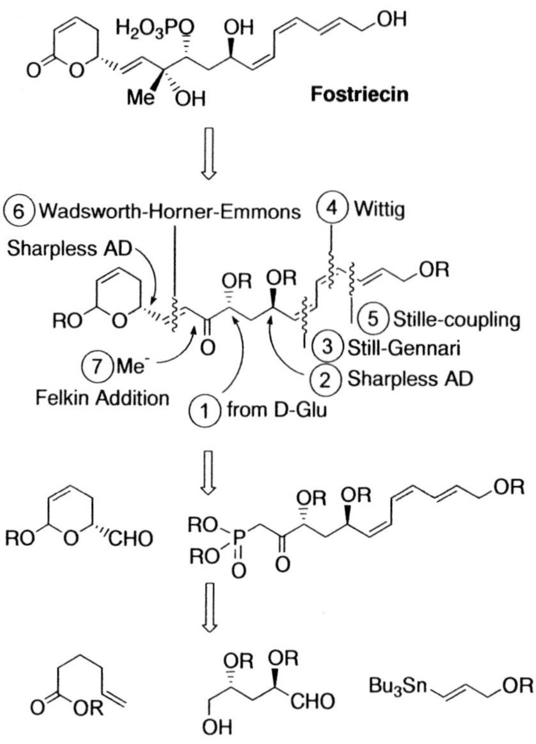

Fostriecin

⬇

⑥ Wadsworth-Horner-Emmons ④ Wittig

Sharpless AD

⑤ Stille-coupling
③ Still-Gennari
② Sharpless AD

⑦ Me⁻
Felkin Addition ① from D-Glu

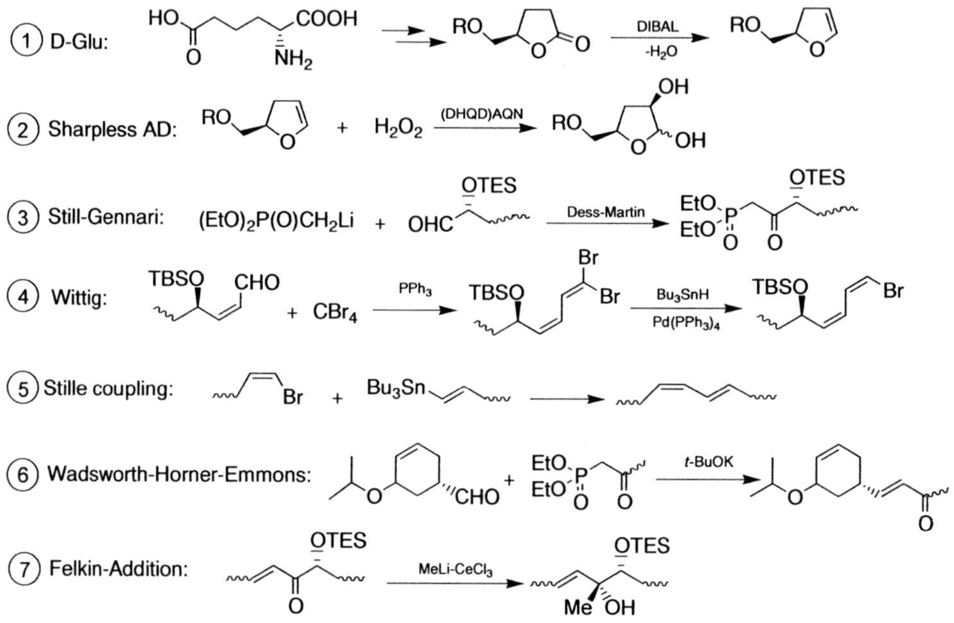

⬇

⬇

① D-Glu:

② Sharpless AD:

③ Still-Gennari: $(EtO)_2P(O)CH_2Li$ + OHC

④ Wittig: + CBr_4

⑤ Stille coupling: Br + Bu_3Sn

⑥ Wadsworth-Horner-Emmons:

⑦ Felkin-Addition:

Frondosin B isolated from a sponge is anti-inflammatory and contains four condensed rings, namely phenol, a furane, a cycloheptene, and a cyclohexene. Cyclohexene comes, of course, again from a Diels–Alder reaction, cycloheptene from an intramolecular Friedel-Crafts acylation, furane from a base- and palladium-catalyzed intermolecular addition of a phenolate to an alkyne. The alkyne was thought to come from a Sonogashira coupling. The rest is Sharpless and phenol chemistry (Inoue, 2001).

Frondosin B

Our example for a heterocycle is an alkaloid called batzelladine F. It was isolated from a Jamaican sponge, may be useful to treat autoimmune responses, and
inhibits protein–protein interactions. It contains two tricyclic guanidine derivatives
and a branched octanoic acid chain as connecting link. At first the right-hand guanidine was disconnected to give a guanidine hemi-aminal and a chiral alcohol in
the side-chain, which could be substituted stereoselectively. The left part was then
attached via a β-ketoester carbanion. Therefore the OH-group in the side-chain of
the left-hand triazaacennaphthalene was coupled with acetoacetate and its triaza-
tricycle disconnected to an allyl-protected β-ketoester and a guanidine aminal. The
acrylic ester obtained had to be decarboxylated and hydrogenated. The guanidinium bicycle was accessible from a 1,3-diamine and CbZ-protected carbonimi-
dothioate (Franklin, 1999; Cohen, 2001).

Batzelladine F

Batzelladine is a bolaamphiphile carrying similar heterocycles on both ends. The molecule is, however, not symmetrical, which would make it much less soluble and much more prone to crystallization. Unsymmetry is, in addition to asymmetry or chirality, a very important property of biologically active compounds. Symmetrical compounds are much easier to synthesize, but usually inactive. This principle, which has been pointed out by Winterfeldt (Drögemüller, 1998; see also La Cour, 1998), makes the life of drug-producing chemists much harder than they would like. Nature loves not only asymmetry (=chirality) but also unsymmetrical molecular skeletons.

The final example for a recent retrosynthesis originates from the idea to find a suitable "lock" for a given "key", the reverse scenario of the familiar combinatorial synthesis of a library of ligands and its testing with receptors. The receptor library wanted consisted of tweezers with three different flexible arms hold together by a hinge, which would align them. A triamino heterocycle was needed which could (i) be selectively protected three times and (ii) be deprotected selectively. A triazacyclophane was chosen as hinge and its synthesis planned. The primary amino end groups of the initial open-chain triamine could be selectively sulfonated and trifluoroacetylated. Acetylation of the remaining secondary amine with alkyl chloroformate gave the desired triply protected building block, which cyclized with 3,5-bis(bromomethyl)-benzoic acid methyl ester. The triflate was then removed together with the methanol of the ester and reprotected with Fmoc-*N*-hydroxysuccinimide. Deprotection was possible in the order Fmoc, ONBS (thiolysis) and Aloc (Pd-catalyzed alkyl transfer to anilinium *p*-toluenesulfinate). Split-mix syntheses of three libraries of peptides on each arm could then be performed on solid beads, which were tested for ion binding (Opatz, 2002).

9.4
Tandem Reactions

Tandem reactions form several covalent bonds in one sequence without isolating the intermediates. Such reaction sequences are also called "domino" or "cascade" reactions, but "tandem" is used most frequently. The ACS search program produces 507 "tandem" titles published since 1996, as compared to 115 using "cascade" and 34 using "domino". "Multistep reaction" or "one-pot sequence" are more appropriate descriptions of the procedure.

Most tandem reactions are used to produce complex cyclic or polycyclic architectures and some of them have been described earlier in the book, e.g. Johnson's and Corey's steroid syntheses (p. 125 ff., 131), the zip reaction (p. 268 f.), the Robinson annelation (p. 113 f.) and the radical bis-cyclization (p. 3, 72 f.). A book by Ho (1992) and a review by Tietze (1996) systematize all kinds of reaction types that have been combined in tandem reactions. Another review (Winkler, 1996) specializes in the most popular Diels–Alder cycloadditions.

In the following, we provide a few examples which show that tandem-synthetic double or triple steps can, indeed, be used in all kinds of reaction sequences.

Some sterically hindered, SnCl₄-catalyzed hetero-Diels–Alder cyclizations of α-unsaturated ketoesters with alkene alcohols do not occur intramolecularly. Large substituents on the ketoester prevent the formation of medium-sized rings and the first reaction is a linear dimerization combined with the formation of one dihydropyran unit. The second reaction then gives a second dihydropyran and produces a macrocyclic oligo-ether with good yield (Bear, 2001).

The catalytic effect of Lewis acids on Diels-Alder reactions is so strong, that it is even possible to overcome Bredt's rule. Double bonds to bridge atoms become accessible in cyclohexene units (Nicolaou, 2002).

Acetylacetate reacts with zinc methylene iodide (Furukawa reagent) first to give a zinc enolate, and then to add its methylene group to the enolate's double bond. Aldehydes then decompose the cyclopropane formed and undergo a Reformatsky addition. A chain extension–aldol addition tandem is thus realized. In a related three-step reaction a cyclohexanone derivative underwent zinc enolate formation and Michael addition in one step. Acetals were then reactive enough to decompose the enolate and form a second CC bond stereoselectively the presence of chiral phosphines (Alexakis, 2001; Lai, 2001).

The coupling of a cyanide Michael addition to propargylic acid and Dieckmann cyclization with a neighboring benzyl ester provides an example for the synthesis of a highly functional arene derivative (Deville, 2002).

9.5
Green Chemistry

There are U.S. and European Green Chemistry Programs, which try to establish environmentally benign synthetic procedures (Anastas, 2002). The approach is not so much scientific as educational with respect to good chemical behavior. The foundational pillar is catalysis in synthetic systems. Energy requirements, waste, and the number of separation steps are all minimized by increased selectivity of the reactions catalyzed. Heck-, Sharpless- and Noyori-type reactions are successful endeavors in this direction and are already applied as often as possible both in the laboratory and in industry. So synthetic chemists have a good conscience here.

Another approach is to replace solvents by water or by supercritical fluids, in particular CO_2. Not only has the latter been achieved in the dry cleaning industry, but it is found in general that CO_2 can replace chlorinated solvents (Hitzler, 1997). Water is so far the principal solvent of choice in noncovalent synthesis (=synkinesis, see Chapter 6). Many stereoselective organic reactions leading to covalent bonds have, however, also been carried out in water using micellar media, solubilizing auxiliaries, in particular carbohydrates or pyridine-derivatives, or THF and acetonitrile as additive solvents (Lindström, 2002).

Replacement of soluble Lewis acids by mesoporous solids containing bound sulfonates or aluminum chloride should also become common practice (Clark, 2002). The solids can be filtered off and usually reactivated and recycled. This helps to prevent waste. Aqueous gels have also been used to bind contrasting reagents, such as oxidants and hydrogenating systems, or acids and bases in one reaction volume (Gelman, 2000)

Finally, and most typical for green chemistry, educts should preferably come from renewable sources, in particular glucose.

Furthermore syntheses should be atom-efficient, and reagents as simple as possible. Catalysed reactions are preferable.

Household and large-scale industrial chemicals, e.g. chelators, should always be biodegradable, as should the intermediates in their synthesis. Boger's iminodiacetic acids are good examples, because they only use succinic acid derivatives (see p. 373 ff).

In all cases the purchase price should not be the only criterion to apply to the choice of a chemical for a given purpose. Waste and renewal problems should also be considered, and nobody can judge the problems involved here better than chemists. This means, then, that only a well-trained, independent, and responsible chemist should determine ecological policy, not the "alert" politicians, who tend to overreact to local junk scandals, listen to computer-model apocalyptic theories, and give no help. None of the solutions to the ecological problems of the last century came from politicians, including those of the green parties. It was always the active chemists who changed products and habits as problems arose. It seems clear that, within a few generations, we shall have to change our food stuff and energy basis from petrol to glucose and hydrogen. Synthetic chemists will be heavily involved in this necessary revolution.

10
Seventy-Seven Conclusions in Alphabetical Order

10.1
Introduction

The large number of subjects which are dealt with in this book on organic synthesis are far too complex to end it with a simple conclusion. We provide 77 summarizing remarks, which try to highlight important points in simple words and without structures.

10.2
Conclusions in Alphabetical Order

Alcohols – They are the result of Grignard-type syntheses, see also chiral alcohols.

Aldehydes – They can be obtained by hydration of terminal alkynes or by oxidation of primary alcohols: Moffat-Pfitzner, Dess-Martin, silver oxide.

Alkenes – They are bisfunctional compounds after reaction with oxidants.

Alkynes – They can be hydrated to form aldehydes or ketones, their radicals undergo efficient C–C coupling.

Allyl groups – They open the way to stereoselective reactions in acyclic compounds and to the synthesis of complex monomers for polymerization.

America – Modern organic synthesis needs a large choice of starting materials, sophisticated reagents, and detailed recipes. Therefore the synthetic chemist anywhere in the world depends on co-operation with America, its chemical suppliers and Chemical Society. Commercial catalysts, chiral ligands, and organometals come from America; the "anywhere in the article" or "choose a scientific subject" searches are only possible in America. Up to now, these services have been available to the whole world at a reasonable price. It is the American attitude "to serve science" which makes the life of chemists so pleasant. Europeans and Asians should adopt the same attitude in order to preserve this situation.

Behavior – Educated behavior is called green chemistry in organic synthesis. Nasty behavior may be more successful. But: do not bother your "fellow citizen with chemistry" as Linus Pauling would have warned you.

Biopolymers – Proteins, nucleotides, and polysaccharides are or will be all synthesized in automated machines using commercial monomers. Enzymes help to multiply polynucleotide quantities and to use them as templates for protein synthesis. These target molecules belong in biochemical laboratories, where ability in chemical synthesis is traditionally restricted to reading the manufacturer's instructions for synthesis machines and chromatography equipment. The chemist's task is to provide new polymer backbones and optimized monomers (Chapter 5).

C=C-double bonds – They can be converted to two neighboring chiral centers (Sharpless oxidation with peroxide and tartrate or OsO_4 and chiral amides. Terminal double bonds or arenes can be coupled to dienes with palladium catalysts (Heck, Suzuki, Stille).

Carbenes – Two-valent carbon with a free electron pair may be formed by a,a'-dehydrohalogenation or by ruthenation of halides. They form cyclopropane with C=C double bonds or undergo metathesis.

Carbohydrates – Carbohydrates, in particular glucose and its polymers, provide the most important renewable chemical feedstock.

Carbon – The magic of the carbon atom lies in its eight oxidation states from -4 in methane which is buried deep in the reducing earth crust, to $+4$ in carbon dioxide, which floats next to molecular oxygen in the atmosphere. No other element can so easily be oxidized, reduced, hydrated, dehydrated, and metallated and forms such interesting covalent compounds.

Catalysis – It used to be only important in oxidations (Sharpless) and reductions (heterogeneous catalysis with Pd, Wilkinson, Noyori), but appeared in synthesis with the discovery of Heck and metathesis reactions. It also helps to avoid the need for low-temperatures during enantioselective reactions, high-temperatures to overcome unfavorable transition states, and strong reductants or oxidants for unlikely reactions. Pd(II) and Pd(0) are very good, Rh and Ru come next.

Chiral alcohols – They can be obtained from ketones by application of a variety of chiral reductants.

Chirality – Determines the helicity and curvature of biopolymers, which then assemble to form gels in water. Admixture of enantiomeric helices leads to the precipitation of sheets (chiral bilayer effect). Without chirality the molecules of living organisms would crystallize or dissolve, not form metastable gels. In protein sheets chirality allows the fixation of three binding sites at an exact distance and orientation. These three binding sites may react with three substituents of a chiral carbon center. The fourth substituent of the latter then points into either a hydrophobic or a hydrophilic medium. Specific recognition is thus achieved between receptor proteins and chiral drugs.

***cis*-Selectivity** – *cis*-Configured C=C bonds are synthesized by carefully controlled Wittig reactions or obtained by catalytic hydrogenation of alkynes.

Compound libraries – Together with synkinesis of noncovalent molecular assemblies, these are the most promising as well as the most questionable endeavour of modern chemistry. It is easy to make, register, and store mixtures of tens to millions of different molecules, either bound to beads (one bead – one compound) or loaded into microvessels. So far, very few "good molecules" have come out of compound libraries, whereas real libraries retain only good books. There is some hope that selection and enrichment of the most potent compounds can be enforced during the synthesis of combinatorial libraries (multicomponent reaction series, dynamic libraries, clicking together. Do not produce libraries of compounds if you do not have a really selective test procedure. If your test cannot differentiate between ten library compounds, do not produce ten thousand. Rational libraries are small ones, which are only used to optimize the reaction conditions of a catalyzed reaction, or larger ones if you have a well-defined labelled receptor molecule for your compounds.

Copper – This metal is the soft analogue of magnesium in organometals and is used in Michael additions and as a redox-active catalyst for alkyne coupling reactions.

Covalent and noncovalent molecular assemblies – Covalent assemblies of several organic molecules, usually prepared in quantities of a few milligrams only, are mostly of modest solubility in organic solvents and insoluble in water. Their advantage is the perfect definition of intermolecular bond angles and lengths. Noncovalent molecular assemblies are accessible in gram to kilogram quantities and they often form stable gels in water. Their disadvantage is the statistical arrangement of fibers.

Cyclic oligomer structures – Covalent bonds (porphyrins, calixarenes) or hydrogen bonds (*a*-helix, DNA, starch) tend to favor the formation of four-membered rings. The tetramer is the first oligomer for which end-to-end interaction becomes possible without angle strain if the monomers are not bent. Intramolecular interactions are then favored by entropy over intermolecular interaction.

Cyclohexane and pyran – *β*-D-Glucose with five equatorial substituents is the most abundant stable conformer on earth. Equatorial, cyclohexane-chair-like transition states are formed whenever possible. The most popular route to cyclohexene is the Diels–Alder cycloaddition.

Cyclopentanes – Cyclopentane-derivatives and tetrahydrofurans are extremely flexible (pseudorotation), do not crystallize easily and are highly soluble. Several individual synthetic pathways, but no general method, lead to five-membered rings.

Cyclopropanes – Cyclopropanes are accessible from alkenes and diazomethane or carbenes. Hydrogenation leads to the formation of a methyl group, cyclopropanated allylalcohols give *γ*-alkenylbromides with HBr. Resemble olefins in their reactivity.

Dendrimers – Spherical dendrimers are the only synthetic polymers of limited growth. Quantitative branching within each generation leads to a single compound made from up to hundreds of monomers.

Distance – The distance between molecules of different reactivity must be carefully controlled in order to construct functional modules for energy conversion or molecular switching. The distance between bulky substituents in a single molecule determines the enantioselectivity of addition reactions.

DNA – It seems to be useful only in molecular biology. The overwhelming stability of the CG hydrogen bridge in water may, however, also be taken as molecular mortar in all kinds of molecular and colloidal assemblies (Section 7.2).

Do everything at the same time – Organic synthesis covers several areas of research, which so far have hardly been beneficial to human society. Noncovalent and combinatorial syntheses are modern examples. Progress in artificial photosynthesis, nanometer modules, and compound libraries that contain useful drugs, are envisaged in myriads of publications, but the commercial output is negligibly small. Natural scientists at universities should work together widely, in order to advance new methodologies significantly. Boger, for example, not only produced complex heterocycle libraries, but explored their medical action and then optimized the libraries to find optimal compounds (p. 373 ff and 381 f). Stoddart not only synthesized rotaxanes and named them "molecular switches", but, together with physicists and hardware engineers, developed working molecular computer modules. Supramolecular architecture does not mean anything without TEM and/or AFM images (p. 401 ff). The aim should not be to sell drugs or molecular machines but to provide realistic guidelines for academic synthesis and synkinesis research.

Epoxides – Epoxides (= oxiranes) behave like aldehydes and ketones, but yield different addition products. Opening with Grignard reagents produces a-alkyl-alcohols, alkinyl anions give γ-alkinyl alcohols, and acetylene carboxylates yield 1,5-hydrocycarboxylic acids with a triple bond in between. All kinds of $1,n$-disubstitution patterns can be obtained from C=C double bonds after cyclopropanation or epoxidation.

Esters – They are formed from acids under neutral conditions by mild dehydrating agents.

Examples – Interesting synthetic reactions are usually exemplified by specific examples. Most of the examples will either be too simple and trivial or too exotic and complicated for readers. This is caused by the unavoidable fact that the example may not belong to their particular chemical world. Examples need to be transferred by the readers into their own field of interest.

Fluoride – Fluoride anions make tin nonvolatile, act in phase transfer reagents to push salts into organic solvents, and cleave Si–C bonds.

Functional side-chains – Allylation of aldol-type educts (β-hydroxy or β-keto carbonyl) and subsequent ozonolysis produces a versatile aldehyde group in the side-chain.

Heterocycles – Heterocycles containing nitrogen and oxygen provide the best electron donors and acceptors in redox chains and the best proton donors and acceptors in hydrogen bond chains. Phenol and aniline derivatives are of similar versatility. Molecular machines and architectures, poisons, and drugs are mostly based on the interaction of such molecules (Chapters 4, 6, and 7).

Hydride – It is a hard reducing agent and reacts with hard carbonyl double bonds. DIBAL is an exception. It attacks CC bonds and leaves aldehyde derivatives intact.

Hydrocarbons – Carbon–hydrogen bonds of primary, secondary, and tertiary hydrocarbons are oxidized with chromic acid at relative rates of $1:100:10,000$.

Hydrogen – Can be removed as a proton from activated hydrocarbons and added as an H_2 molecule or hydride to C=C and C=O double bonds. Hydrogen gas is a soft reducing agent and reacts with soft alkene and alkyne multiple bonds, but not with hard C=O bonds.

Ketones – They can be made from carboxylates by using Weinreb amides.

Light – It triggers movements around C=C double bonds. It also forms excited states (leading to *cis–trans* isomerization or rearrangements) from organic molecules in the ground state. Excited states are both strong reductants *and* strong oxidants (=electron donors and acceptors).

Magnesium – Strong reductant for C–Hal bonds. Produces strongly polarized, hard metal organic compounds.

Metathesis – The word indicates change of position (from Greek via late Latin) and is thus quite abstract. It means exchange of two alkylidene groups from one molecule to another in the presence of catalysts. In synthesis it is usually applied either in cyclizations of dienes or for the rearrangement of enes.

Nature – Macroscopic analogies to nature are seldom helpful in the construction of machinery. Nature does not use the wheel, nor do our airplanes have flapping wings. On the molecular scale, however, thinking in analogies has been valuable. For example, nylon is a simplified protein, mimicked intermediates in the biosynthesis of human molecules are powerful drugs, and water seems to be the best medium for the construction of complex molecular architecture in nature as well as in the laboratory.

Nitriles – They are accessible from aldehydes via dehydration of an oxime or from ketones with TosMic.

Nitrogen – Nitrogen reagents and compounds come as strong nucleophiles (N_3^-), bases (NH_2^-, NH_3), electrophiles (NO_2^+, ArN_2^+), oxidants (HNO_3, NO_2), converters of light to mechanical movement (azo dyes) and metal ligand (porphyrins, EDTA). FGIs in nature are dominated by proton and electron shifts in Schiff bases, and the molecular assembly of proteins by secondary amide hydrogen bonds. The Gries reaction of aniline with nitrous acid to yield diazonium compounds and, subsequently, azo dyes marks the start of the chemical industry, e.g. IG Farben in Germany. The versatility of this element following carbon in the periodic system is unsurpassed.

OH-Groups –They are best substituted by halides or amines using the Mitsonobu reaction with phosphine-diazenedicarboxylate.

Oligosaccharides – Glycosyl halides, sulfides, trichloroacetimidates, or glycols act as glycosyl donors; the corresponding glycosyl acceptor furnishes a free OH-group for oligosaccharide synthesis.

Oxazolidones – Sound complicated, but they are useful chiral donor synthons and they are commercially available or can be made from amino acids and phosgene as pure enantiomers.

Oxidation – Means going upwards on the ladder of carbon oxidation states: alkanes alkenes arenes alkynes alcohols aldehydes carboxylic acids carbonic acids. Sharpless, Collins, and ozonolysis reactions are the most important oxidations in synthesis. Oxidation makes molecules more stable to air and more soluble in water.

Oxygen – The slow reaction of the triplet state of molecular oxygen with diamagnetic organic compounds is the most useful prohibition on earth. It allows the co-existence of a strong volatile oxidant in air and fluid reducing biological cells on the surface of earth. The tendency of oxygen atoms to pull electrons from carbon in the double bond of carbonyl compounds constitutes the most beneficial polarization in organic molecules.

Ozone – It cleaves C=C double bonds.

Palladium – The perfect electron shuttle and removable connecting link between carbon atoms and hydrogen.

Phosphorus – Useful nucleophile in organic synthesis (Wittig reaction) with the ability to form five covalent bonds. C–P bonds are often spontaneously replaced by C–O bonds.

Radicals – The only intermolecular radical reactions of synthetic use are the Glaser and Cadiot–Chodkiewiez couplings of alkynes. Otherwise they are usually applied in intramolecular cyclizations and SmI_2 is an unsurpassed initiator.

Reduction – Going downwards on the ladder of carbon oxidation states (see Oxidation). Most important is the hydrogenation of C=C (palladium catalysts) and C=O (hydrides, Noyori). Reduction makes molecules more stable to treatment with acids, bases, nucleophiles, and electrophiles. The molecules often become more hydrophobic.

Regioselectivity – Forget about Markovnikov and Sayzev in synthetic sketches. Hydroboration and Hofmann elimination are better.

Reversible reactions – They are used in synkinesis, enzymatic condensation reactions, recognition processes, and dynamic combinatorial synthesis.

Ring closure – Functional group inversion (e.g. lactonizations) and synthetic (e.g. aldol-type) cyclizations depend on bis-functional chains or enes and dienes (e.g. Diels–Alder).

Ring contraction – Ring contraction is done routinely by expulsion of CO_2, CO, N_2, S, and SO_2 or by Favorskii rearrangement of a cyclic ketone or by oxidation of cycloalkenes with Fe(III) and acid treatment.

Ring expansion – Ring expansion is less common than contraction. It is easy with angle-strained cyclopropane and cyclobutane systems. Fragmentation of fixed rings, the zip reaction, and ozonolysis of cycloalkenes followed by aldol condensation-type cyclizations provide possible pathways to ring enlargement.

Samarium – Samarium iodide with partly occupied f-orbitals produces more stable organometallic radicals than transition metals with d-electrons only. C-Hal reduction and intermolecular cyclization chains are preferred applications of these long-lived, soft radicals.

Selectivity – Selectivity requires recognition. A base finds the most acidic proton in a compound and steers the regioselectivity of nucleophilic substitutions and addition reactivity. A drug recognizes clefts on the surface of proteins. A protein recognizes a fitting drug on a bead, which can then be identified under the light microscope, if it carries a fluorescent dye. Stereoselectivity depends on steric repulsion between substituents on chiral centers in cyclic transition states.

Selectride – It reduces lactones to hemiacetals.

Silicon – Silylethers provide the most versatile protecting groups for alcohols. They are cleaved by potassium carbonate, F^-, and acetic acid.

Sodium – Sodium metal reduces C=C (Birch) and C=O (acyloin) double bonds to radicals.

Steric hindrance – Bulky neighbors prevent addition reactions to C=O and C=C double bonds. Hindrance is overcome by catalysts, e.g. HBT in acylation, Lewis acids in aldol, Diels–Alder, and Michael additions. Radical dimerization also overcomes hindrance very well.

Steric overpopulation – Overlap between bulky substituents in a plane may change the character of an aromatic molecule to that of a polyene.

Steroid synthesis – It is dominated by derivatization of 1,4-androstadiene-3,17-dione as isolated from microorganisms on a large scale.

Strategy – It is an old-fashioned name for the synthesis plan. Use "concept" or "sketch" instead.

Sulfur – Sulfur atoms are very good nucleophiles. They can be reversibly oxidized to S-radicals and disulfides and expelled to yield C–C bonds. A sulfur atom is a powerful nucleophile, even as a partner of an electron withdrawing carbonyl group on the same carbon atom.

Synkinesis – The preparation of molecular assemblies by well-defined noncovalent bonds, mainly directed hydrogen bonds and non-directed van der Waals and electrostatic interactions. Synkinetic reactions are always fully reversible (Chapter 7).

Synthesis – The preparation of large molecules from smaller ones by the formation of carbon-carbon bonds. The Wittig, Heck, and Diels–Alder reactions are all the methods a beginner needs to know. Most compounds can be made by these reactions if they are followed by intelligent FGIs.

Tandem – A good example of a bad symbolic image. What has a two-seated bicycle or a two-horse-in-a-row carriage to do with the formation of several bonds in a sequence? The alternative expressions "domino" and "cascade" are no better. A row of falling dominoes or the rules of the game have nothing to do with the construction of architecture. A cascade falling over rocks is a waterfall and makes no sense either. Chemists are usually bad poets, but draw many great structures. The terms "multistep reaction" or "one-pot sequence" may be acceptable.

Tartaric acid – It is useful wherever chiral selectivity is asked for. Aldol reactions and Sharpless oxidation are good examples of successful applications of Pasteur's discovery.

Thoughts and bandwagons –Thoughts are the product of your own mental activity and lead to experiments with uncertain outcomes. Jumping on a bandwagon saves you from thinking – it also supports and neutralizes the mental activity of other successful chemists. More importantly, it is helpful in the organization of conferences as well as in collecting grant money. Thoughts lead to scientific freedom for scientists, but only grant money lets them survive. A good compromise is a life on the coachman's seat.

Transition states – The appropriate positions of electron donating and accepting groups in energetically accessible transition states determine the major pathways and stereochemical outcome of organic reactions.

Vinylbromides and -iodides – They are best obtained from alkynes or aldehydes. They are not only most useful in Pd-catalyzed reactions (Heck, Suzuki, Stille, etc.), but also add to epoxides.

Water – Its surface tension dictates the ultra-thinness of molecular bilayers and the formation of well-defined minimal surfaces in chiral molecular assemblies and transition states. Organometals and complexes of Pd(II) and similar metal ions of high electronegativity survive in water, and synthetic reactions become feasible with such reagents. Several redox reactions, the Michael addition, and the Gries reaction also take place in water.

Workup – Workup is as important as the execution of a chemical reaction in the daily life of a chemist. Whenever possible, use efficient extraction procedures.

11
References

AALBERSBERG, W.G.; BARKOVICH, A.J.; FUNK, R.L.; HILLARD, R.L., ILL; VOLLHARDT, K.P.C. **1975**, J. Am. Chem. Soc. *97*, 5601

ABDALMUHDI, L.; CHANG, C.K. **1985**, J. Org. Chem. *50*, 411

ABE, Y.; HARUKAWA, T.; ISHIKAWA, H.; MIKI, T.; SUMI, M.; TOGA, T. **1956**, J. Am. Chem. Soc. *78*, 1422

ACHESON, R.M. **1976 B**, *An Introduction to the Chemistry of Heterocyclic Compounds,* Wiley: New York London – Sydney

Acheson, R.M.; Paglietti, G. **1976 A**, J. Chem. Perkin Trans. 1, 45

ACHMAD, S.A.; CAVILL, G.W.K. **1963, 1965**, Aust. J. Chem. *16*, 858; *18*, 1989

ADAM, W.; ERDEN, I. **1978**, Angew. Chem. *90*, 223 (Int. Ed. Engl. *17*, 210)

ADAMS, M.A.; DUGGAN, A.J.; SMOLANOFF, J.; MEINWALD, J. **1979**, J. Am. Chem. Soc. *101*, 5364

ADICKES, H.W.; POLITZER, I.R.; MEYERS, A.I. **1969**, J. Am. Chem. Soc. *91*, 2155

AFOLABI, P.R.; MOHAMMED, F.; AMARATUNGA, K.; MAJEKODUNMI, O.; DALES, S.L.; GILL, R.; THOMPSON, D.; COOPER, J.B.; WOOD, S.P.; GOODWIN, P.M.; ANTHONY, C. **2001**, Biochemistry *40*, 9799

AGARWAL, K.L.; BERLIN, Y.A.; FRITZ, H.J.; GAIT, M.J.; KLEID, D.G.; LEES, R.G.; NORRIS, K.E.; RAMAMOORTHY, B.; KHORANA, H.G. **1976**, J. Am. Chem. Soc. *98*, 1065

AGER, D.J.; PRAKASH, I.; SCHAAD, D.R. **1996**, Chem. Rev. *96*, 835

AHARONOWITZ, Y.; COHEN, G. **1981**, Sci. Am. *245*, 140

AHLERS, M.; MÜLLER, W.; REICHERT, A.; RINGSDORF, H.; VENZMER, J. **1990**, Angew. Chem. *102*, 1310 (Int. Ed. Engl. *29*, 1269)

AKHREM, A.A.; RESHETOVA, I.G.; TITOV, YU.A. **1972**, *Birch Reduction of Aromatic Compounds,* Plenum: New York

ALEXAKIS, A.; TREVITT, G.P.; BERNARDINELLE, G. **2001**, J. Am. Chem. Soc. *123*, 4358

Allen, W.S.; Bernstein, S.; Littell, R. **1954**, J. Am. Chem. Soc. *76*, 6116

ALMOG, J.; BALDWIN, J.E.; DYER, R.L.; PETERS, M. **1975**, J. Am. Chem. Soc. *97*, 226, 227

AMBROISE, A.; WAGNER, R.W.; RAO, P.D.; RIGGS, J.A.; HASCOAT, P.; DIERS, J.R.; SETH, J.; LAMMI, R.K.; BOCIAN, D.F.; HOLTEN, D.; LINDSEY, J.S. **2001**, Chem. Mater. *13*, 1223

AMES, D.E.; GOODBURN, T.G.; JEVANS, A.W.; McGHIE, J.F. **1968**, J. Chem. Soc. C *1968*, 268

ANANCHENKO, S.N.; LIMANOV, V.YE.; LEONOV, V.N.; RZHEZNIKOV, V.N.; TORGOV, I.V. **1962**, Tetrahedron *18*, 1355

ANASTAS, P.T.; KIRCHHOFF, M.M. **2002**, Acc. Chem. Res. *35*, 686

ANDO, K.; GREEN, N.S.; LI, Y.; HOUK, K.N. **1999**, J. Am. Chem. Soc. *121*, 5334

ANDO, S.; MINOR, K.P.; OVERMAN, L.E. **1997**, J. Org. Chem. *62*, 6379

ANDRUS, M.B.; LEPORE, S.D. **1997**, J. Am. Chem. Soc. *119*, 2327

ANGST, C.; KAJIWARA, M.; ZASS, E.; ESCHENMOSER, A. **1980**, Angew. Chem. *92*, 139 (Int. Ed. Engl. *19*, 140)

ANTONSSON, T.; JACOBSSON, U.; MOBERG, C.; RÄKÖS, L. **1989**, J. Org. Chem. *54*, 1191

ARATANI, N.; OSUKA, A. **2001**, Org. Lett. *3*, 4213

ARDUENGO, A.J., III; MORAN, J.R.; RODRIGUEZ-PARADA, J.; WARD, M.D. **1990**, J. Am. Chem. Soc. *112*, 6153

ARIGONI, D.; VASELLA, A.; SHARPLESS, K.B.; JENSEN, H.P. **1973**, J. Am. Chem. Soc. *95*, 7917

ARNDT, D. **1975**, in: HOUBEN-WEYL, *Methoden der Organischen Chemie*, Vol. IV/1 b, p. 505 ff., Thieme: Stuttgart

ARNHEIM, X.; LEVENSON, C.H. **1990**, Chem. Eng. News *68*, 36

ARNOLD, L.A.; NAASZ, R.; MINNAARD, A.J.; FERINGA, B.L. **2001** J. Am. Chem. Soc. *123*, 5841

ARREDONDO, V.M.; TIAN, S.; McDONALD, F.E., MARKS, T.J. **1999**, J. Am. Chem. Soc. *121*, 3633

ARRIETA, A.; COSSIO, F.P.; LECEA, B. **2001**, J. Org. Chem. *6*, 6178

ARZOUMANIAN, H.; METZGER, J. **1971**, Synthesis *1971*, 527

ASAKAWA, M.; ASHTON, P.R.; BALLARDINI, R.; BALZANI, V.; BĚLOHRADSKÝ, M.; GANDOLFI, M.T.; KOCIAN, O.; PRODI, L.; RAYMO, F.M.; STODDART, J.F.; VENTURI, M. **1997**, J. Am. Chem. Soc. *119*, 302

ASAKAWA, M.; DEHAEN, W.; L'ABBÉ, G.; MENZER, S.; NOUWAN, J.; RAYMO, F.M.; STODDART, J.F.; WILLIAMS, D.J. **1996**, J. Org. Chem. *61*, 9591

ASHIMORI, A.; UCHIDA, T.; OHTAKI, Y.; TANAKA, M.; OHE, K.; FUKAYA, C.; WATANABE, M.; KAGITANI, M.; YOKOYAMA, K. **1991**, Chem. Pharm. Bull. *39*, 108

ASHTON, P.R.; BALLARDINI, R.; BALZANI, V.; BAXTER, I.; CREDI, A.; FYFE, M.C.T.; GANDOLFI, M.T.; GÓMEZ-LÓPEZ, M.; MARTÍNEZ-DÍAZ, M.-V.; PIERSANTI, A.; SPENCER, N.; STODDART, J.F.; VENTURI, M.; WHITE, A.J.P.; WILLIAMS, D.J. **1998**, J. Am. Chem. Soc. *120*, 11932

ASHTON, P.R.; BALZANI, V.; BECHER, J.; CREDI, A.; FYFE, M.C.T.; MATTERSTEIG, G.; MENZER, S.; NIELSEN, M.B.; RAYMO, F.M.; STODDART, J.F.; VENTURI, M.; WILLIAMS, D.J. **1999**, J. Am. Chem. Soc. *121*, 3951

ASINGER, F.; VOGEL, H.H. **1970**, in: HOUBEN-WEYL, *Methoden der Organischen Chemie*, Vol. V/1a, p. 327, Thieme: Stuttgart

ATHERTON, E.; CLIVE, D.L.L.; SHEPPARD, R.C. **1975**, J. Am. Chem. Soc. *97*, 6584

AUGUSTINE, R.L. **1965**, *Catalytic Hydrogenation: Techniques and Applications in Organic Synthesis*, M. Dekker: New York

AUGUSTINE, R.L. **1968**, *Reduction: Techniques and Applications in Organic Synthesis*, M. Dekker: New York

AUGUSTINE, R.L. **1969**, *Oxidation: Techniques and Applications in Organic Synthesis, Vol. 1*, M. Dekker: New York

AUGUSTINE, R.L. **1976**, Catal. Rev. *13*, 285

AUGUSTINE, R.L.; TRECKER, D.J. **1971**, *Oxidation: Techniques and Applications in Organic Synthesis*, Vol. 2, M. Dekker: New York

AURRECOECHEA, J.M.; LÓPEZ, B.; ARRATE, M. **2000**, J. Org. Chem. *65*, 6493

AVENOZA, A.; BUSTO, J.H.; CATIVIELA, C.; CORZANA, F.; PEREGRINA, J.M.; ZURBANO, M.M. **2002**, J. Org. Chem. *67*, 598

AVRAM, M.; NENITZESCU, C.D. **1964**, Chem. Ber. *97*, 372

AYER, W.A.; BOWMAN, W.R.; JOSEPH, T.C.; SMITH, P. **1968**, J. Am. Chem. Soc. *90*, 1648

AYRES, D.C.; RAPHAEL, R.A. **1958**, J. Chem. Soc. *1958*, 1779

BAGGIOLINI, E.G.; IACOBELLI, J.A.; HENNESSY, B.M.; USKOKOVI, M.R. **1982**, J. Am. Chem. Soc. *104*, 2945

BAILEY, E.J.; BARTON, D.H.R.; ELLIS, J.; TEMPLETON, J.F. **1962**, J. Chem. Soc. *1962*, 1578

BAILEY, P.D.; MILLWOOD, P.A.; SMITH, P.D. **1998**, J. Chem. Soc., Chem. Commun. 633

BAILEY, P.S. **1978**, *Ozonation in Organic Chemistry*, Part 1: *Olefinic Compounds*, Academic Press: New York London

BAKER, B.F.; DERVAN, P.B. **1989**, J. Am. Chem. Soc. *111*, 2700

BAKER, B.W.; LINSTEAD, R.P.; WEEDON, B.C.L. **1955**, J. Chem. Soc. *1955*, 2218

BAKER, D.C.; HORTON, D.; TINDALL, C.G., JR. **1976**, in: WHISTLER, R.L.; BEMILLER, J.N. (eds.) *Methods in Carbohydrate Chemistry*, Vol. VII, p. 3, Academic Press: New York London

BAKER, R.; BLACKETT, B.N.; COOKSON, R.C. **1972**, Chem. Commun. 802

BAKER, R.; COOKSON, R.C.; VINSON, J.R. **1974**, Chem. Commun. 515

BALDWIN, J.E. **1976 B**, Chem. Commun. *734*, 738

BALDWIN, J.E.; AU, A.; CHRISTIE, M.A.; HABER, S.B.; HESSON, D. **1975**, J. Am. Chem. Soc. *97*, 5957

BALDWIN, J.E.; CUTTING, J.; DUPONT, W.; KRUSE, L.; SILBERMAN, L.; THOMAS, R.C.

1976 C, J. Chem. Soc. Chem. Commun. *1976*, 736

BALDWIN, J. E.; CHRISTIE, M. A. **1978**, J. Am. Chem. Soc. *100*, 4597

BALDWIN, J. E.; CHRISTIE, M. A.; HABER, S. B.; KRUSE, L. I. **1976 A**, J. Am. Chem. Soc. *98*, 3045

BALKENHOHL, F.; VON DEM BUSSCHE-HUEN-NEFOLD, C.; LANSKY, A.; ZECHEL, C. **1996**, Angew. Chem. *108*, 2436

BALLESTER, P.; CAPÓ, M.; COSTA, A.; DEYÁ, P. M.; GOMILA, R.; DECKEN, A.; DESLONG-CHAMPS, G. **2001**, Org. Lett. *3*, 267

BALZANI, V.; CREDI, A.; LANGFORD, S. J.; RAYMO, F. M.; STODDART, J. F.; VENTURI, M. **2000**, J. Am. Chem. Soc. *122*, 3542

BALZANI, V.; GÓMEZ-LÓPEZ, M.; STODDART, J. F. **1998**, Acc. Chem. Res. *31*, 405

BAN, Y.; SATO, Y.; INOUE, I.; NAGAI, M.; OISHI, T.; TERASHIMA, M.; YONEMITSU, O.; KANAOKA, Y. **1965**, Tetrahedron Lett. 2261

BANWELL, M. G.; McRAE, K. J. **2000**, Org. Lett. *2*, 3583

BARAWKA, D. A.; KWOK, Y.; BRUICE, T. W.; BRUICE, T. C. **2000**, J. Am. Chem. Soc. *122*, 5244

BARAWKAR, D. A.; LINKLETTER, B.; BRUICE, T. C. **1998**, Bioorg. Med. Chem. Lett. *8*, 1517

BARBORAK, J. C.; WATTS, L.; PETTIT, R. **1966**, J. Am. Chem. Soc. *88*, 1328

BARRETT, A. G. M.; BEZUIDENHOUDT, B. C. B.; GASIECKI, A. E.; HOWELL, A. R.; RUSSEL, M. A. **1989**, J. Am. Chem. Soc. *111*, 1392

BARRETT, G. C. **1979**, in *Comprehensive Organic Chemistry, The Synthesis and Reactions of Organic Compounds, Vol. 3*, D. BARTON and W. OLLIS (eds), Pergamon Press

BARRIAULT, L.; DENISSOVE, I. **2002**, Org. Lett. *4*, 1371

BARTH, W. E.; LAWTON, R. G. **1971**, J. Am. Chem. Soc. *93*, 1730

BARTLETT, P. A. **1980**, Tetrahedron *36*, 1

BARTLETT, P. A.; GREEN, F. R. **1978**, J. Am. Chem. Soc. *100*, 4858

BARTLETT, P. D.; STAUFER, C. H. **1935**, J. Am. Chem. Soc. *57*, 2580

BARTON, D. H. R. **1975**, in: STIRLING, C. J. M. (ed.) *Organic Sulphur Chemistry*, p. 181, Butterworths: London

BARTON, D. H. R.; CRICH, D.; KRETZSCHMAR, G. **1986**, J. Chem. Soc., Perkin Trans. *1*, 39

BARTON, D. H. R.; GUZIEC, F. S.; SHAHAK, I. **1974**, J. Chem. Soc. Perkin Trans. *1*, 1794

BARTON, D. H. R.; HESSE, R. H.; MARKWELL, R. E.; PECHET, M. M.; ROZEN, S. **1976**, J. Am. Chem. Soc. *98*, 3034, 3036

BARTON, D. H. R.; HESSE, R. H.; PECHET, M. M.; SMITH, L. C. **1979**, J. Chem. Soc. Perkin Trans. *1*, 1159

BARTON, D. H. R.; LIER, E. E.; McGHIE, J. F. **1968**, J. Chem. Soc. C *1968*, 1031

BARTON, D. H. R.; McCOMBIE, S. W. **1975**, J. Chem. Soc., Perkin Trans. *1*, 1574

BARTON, D. H. R.; SEOANE, E. **1956**, J. Chem. Soc. *1956*, 4150

BARTON, D. H. R.; WILLIS, B. J. **1972**, J. Chem. Soc. Perkin Trans. *1*, 1794

BARTON, J. E. D.; HARLEY-MASON, J. **1965**, Chem. Commun. *1965*, 298

BARTROLI, J.; TURMO, E.; ALGUERÓ, M.; BONCOMPTE, E.; VERICAT, M.; CONTE, L.; RAMIS, J.; MERLOS, M.; GARCÍA-RAFANELL, J.; FORN, J. **1998**, J. Med. Chem. *41*, 1869

BÄSLER, S.; BRUNCK, A.; JAUTELAT, R.; WINTERFELDT, E. **2000**, Helvetica Chimica Acta *83*, 1854

BASTÚS, J. B. **1963**, Tetr. Lett., 955

BASU, S.; WICKSTROM, E. **1997**, Bioconjugate Chem. *8*, 481

BATCHO, A. D.; BERGER, D. A.; USKOKOVI, M. R. **1981**, J. Am. Chem. Soc. *103*, 1293

BATTERSBY, A. R.; BURNETT, A. R.; PARSONS, P. G. **1969**, J. Chem. Soc. C *1969*, 1193

BATTERSBY, A. R.; McDONALD, E. **1979**, Acc. Chem. Res. *12*, 14

BATTERSBY, A. R.; TURNER, J. C. **1960**, J. Chem. Soc. *1960*, 717

BAUM, R. M. **1994**, C&EN February *7*, 20

BAXENDALE, I. R.; LEY, S. V. **2000**, Bioorg. Med. Chem. Lett. *10*, 1983

BEAR, B. R.; SHEA, K. J. **2001**, Org. Lett. *3*, 723

BEAUCAGE, S. L.; CARUTHERS, M. H. **1981**, Tetrahedron Lett. *1981*, 1859

BEEBE, K. D.; WANG, P.; ARABACI, G.; PEI, D. **2000**, Biochemistry *39*, 13251

BÉLANGER, A.; PONPART, J.; DESLONGCHAMPS, P. **1968**, Tetrahedron Lett. *1968*, 2127

BELETSKAYA, I. P.; CHEPRAKOV, A. V. **2000**, Chem. Rev. *100*, 3009

BELLET, P.; NOMINÉ, G.; MATHIEU, J.; VELLUZ, L. **1966**, C. R. Séances Acad. Sci. Sér. C *263*, 88

BERGBREITER, D. E.; WHITESIDE, G. M. **1975**, J. Org. Chem. *40*, 779

BERGELSON, L. D.; SHEMYAKIN, M. M. **1964**, Angew. Chem. *76*, 113 (Int. Ed. Engl. *3*, 250)

BERGERON, R. J.; MCMANIS, J. S. **1987**, J. Org. Chem. *52*, 1700

BERGNER, E. J.; HELMCHEN, G. **2000**, Eur. J. Org. Chem., 419

BERGNER, E. J.; HELMCHEN, G. **2000**, J. Org. Chem. *65*, 5072

BERNARDI, A.; CHECCHIA, A.; BROCCA, P.; SONNINO, S.; ZUCCOTTO, F. **1999**, J. Am. Chem. Soc. *121*, 2032

BERNSTEIN, S.; LENHARD, R. H.; ALLEN, W. S.; HELLER, M.; LITTELL, R.; STOLAR, S. M.; FELDMANN, L. I.; BLANK, R. H. **1956**, J. Am. Chem. Soc. *78*, 5693

BERRISFORD, D. J.; BOLM, C.; SHARPLESS, K. B. **1995**, Angew. Chem. Int. Ed. Engl. *34*, 1059

BESTMANN, H. J. **1979**, Pure Appl. Chem. *51*, 515

BESTMANN, H. J.; STRANSKY, W.; VOSTROWSKY, O. **1976**, Chem. Ber. *109*, 1694

BETZER, J.-F.; DELAGOGE, F.; MULLER, B.; PANCRAZI, A.; PRUNET, J. **1997**, J. Org. Chem. *62*, 7768

BHATT, U.; CHRISTMANN, M.; QUITSCHALLE, M.; CLAUS, E.; KALESSE, M. **2001**, J. Org. Chem. *66*, 1885

BIALER, M.; YAGEN, B.; MECHOULAM, R. **1978**, Tetrahedron *34*, 2389

BICKART, P.; CARSON, F. W.; JACOBUS, J.; MILLER, E. G.; MISLOW, K. **1968**, J. Am. Chem. Soc. *90*, 4869

BIRCH, A. J.; SUBBA RAO, G. **1972**, Adv. Org. Chem. *8*, 1

BIRCH, A. J.; WILLIAMSON, D. H. **1976**, Org. React. (NY) *24*, 1

BLACKBURN, G. M.; OLLIS, W. D.; SMITH, C.; SUTHERLAND, I. O. **1969**, J. Chem. Soc. D *1969*, 99

BLETTNER, C. G.; KÖNIG, W. A.; STENZEL, W.; SCHOTTEN, T. **1999**, J. Org. Chem. *64*, 3885

BLICKENSTAFF, R. T.; GHOSH, A. C.; WOLF, G. C. **1974**, *Total Synthesis of Steroids*, Academic Press: New York

BLOOMFIELD, J. J.; OWSLEY, D. C.; NELKE, J. M. **1976**, Org. React. (NY) *23*, 259

BLUMBERGS, P.; LAMONTAGNE, M. P.; STEVENS, J. I. **1972**, J. Org. Chem. *37*, 1248

BLY, R. S.; DUBOSE, C. M.; KONIZER, G. B. **1968**, J. Org. Chem. *33*, 2188

BODANSZKY, M.; KLAUSNER, Y. S.; ONDETTI, M. A. **1976**, *Peptide Synthesis*, Wiley: New York London

BODEN, C. D. J.; PATTENDEN, G.; YE, T. **1996**, J. Chem. Soc., Perkin Trans. *1*, 2417

BODWELL, G. J.; MILLER, D. O.; VERMEIJ, R. J. **2001**, Org. Lett. *3*, 2093

BOECKMAN, R. K., JR. **1974**, J. Am. Chem. Soc. *96*, 6179

BOECKMAN, R. K., JR.; SILVER, S. M. **1973**, Tetrahedron Lett. *1973*, 3497

BOGER, D. L.; CHAI, W.; JIN, Q. **1998**, J. Am. Chem. Soc. *120*, 7220

BOGER, D. L.; FINK, B. E.; BRUNETTE, S. R.; TSE, W. C.; HEDRICK, M. P. **2001**, J. Am. Chem. Soc. *123*, 5878

BOGER, D. L.; FINK, B. E.; HEDRICK, M. P. **2000**, J. Am. Chem. Soc. *122*, 6382

BOGER, D. L.; ICHIKAWA, S.; ZHONG, W. **2001**, J. Am. Chem. Soc. *123*, 4161

BOGER, D. L.; LEE, J. K.; GOLDBERG, J.; JIN, Q. **2000**, J. Org. Chem. *65*, 1467

BOGER, D. L.; WOLKENBERG, S. E. **2000**, J. Org. Chem. *65*, 9120

BOHLMANN, F.; EICKELER, E. **1979**, Chem. Ber. *112*, 2811

BOLAND, W.; NEY, P.; JAENICKE, L. **1980**, Synthesis *1980*, 1015

BONSE, G.; METZLER, M. **1978**, *Biotransformationen organischer Fremdsubstanzen*, Thieme: Stuttgart

BORCH, R. F. **1968**, Tetr. Lett., 61

BORCH, R. F. **1969**, J. Org. Chem. *34*, 627

BORMAN, S. **2001**, C&EN, 49

BORYSKI, J.; GOLANKIEWICZ, B. **1984**, Nucleosides Nucleotides *3*, 287

BORYSKI, J.; UEDA, T. **1985**, Nucleosides Nucleotides *4*, 595

BOSCHE, H. G. **1975**, in: HOUBEN-WEYL, *Methoden der Organischen Chemie*, Vol. IV/1b: *Oxidation II*, p. 429, Thieme: Stuttgart

BOSSHARD, H. R.; SCHLECHTER, L.; BERGER, A. **1973**, Helv. Chim. Acta *56*, 717

BOUTIGUE, M. H.; JACQUESY, R. **1973**, Bull. Soc. Chim. Fr. (II) *750*, 3062

BOYCE, R.; LI, G.; NESTLER, H. P.; SUENAGA, T.; STILL, W. C. **1994**, J. Am. Chem. Soc. *116*, 7955

BOYSEN, M. M. K.; LINDHORST, T. K. **1999**, Org. Lett. *1*, 1925

BRADY, S. E.; ILTON, M. A.; JOHNSON, W. S. **1968**, J. Am. Chem. Soc. *90*, 2882

Brady, W.T.; Hoff, E.F. **1970**, J. Org. Chem. *35*, 3733

Branda, N.; Grotzfeld, R.M.; Valdés, C.; Rebek, J. Jr. **1995**, J. Am. Chem. Soc. *117*, 85

Bredereck, G.; Gompper, R.; Schuh, H.G.; Theilig, G. **1959**, Angew. Chem. *71*, 753

Bredereck, H.; Effenberger, E.; Simehen, G. **1963, 1965**, Chem. Ber. *96*, 1350; *98*, 1078

Bredereck, H.; Simchen, G.; Rebsdat, S.; Kantlehner, W.; Horn, P.; Wahl, R.; Hoffmann, H.; Grieshaber, P. **1968**, Chem. Ber. *101*, 41

Breslow, R.; Corcoran, R.J.; Snider, B.B.; Doll, R.J.; Khanna, P.L.; Kaleya, R. **1977**, J. Am. Chem. Soc. *99*, 905

Brik, A.; Keinan, E.; Dawson, P.E. **2000**, J. Org. Chem. *65*, 3829

Brockmann, H.; Kluge, E.; Muxfeldt, H. **1957**, Chem. Ber. *90*, 2302

Brown, C.A. **1973**, J. Am. Chem. Soc. *95*, 4100

Brown, H.C. **1972 C**, *Boranes in Organic Chemistry*, Cornell Univ. Press: Ithaca, NY

Brown, H.C. **1975**, *Organic Synthesis via Boranes*, Wiley: New York London

Brown, H.C. **1980**, Science *210*, 485; Angew. Chem. *92*, 675

Brown, H.C.; Chandrasekharan, L.; Rama-chandran, P.V. **1986**, J. Org. Chem. *51*, 3394

Brown, H.C.; Heim, P. **1964**, J. Am. Chem. Soc. *86*, 3566

Brown, H.C.; Kawakami, J.H.; Ikegami, S. **1970**, J. Am. Chem. Soc. *92*, 6914

Brown, H.C.; Krishnamurthy, S. **1972 B**, J. Am. Chem. Soc. *94*, 7159

Brown, H.C.; Negishi, E. **1972 A**, J. Am. Chem. Soc. *94*, 3567

Brown, H.C.; Randad, R.S.; Bhat, K.S.; Zaidlewicz, M.; Racherla, U.S. **1990**, J. Am. Chem. Soc. *112*, 2389

Brown, W.G. **1951**, Org. React. (NY) *6*, 469

Bruggink, A.; Roos, E.C.; de Vroom, E. **1998**, Organic Process Research & Development *2*, 128

Buchanan, J.G.; Clode, D.M.; Vethaviya-sar, N. **1976**, J. Chem. Soc. Perkin Trans. *1*, 1449

Büchi, G.; Carlson, J.A. **1968**, J. Am. Chem. Soc. *90*, 5336

Büchi, G.; Carlson, J.A.; Powell, J.E.; Tietze, L.F. **1973**, J. Am. Chem. Soc. *95*, 540

Büchi, G.; Coffen, D.L.; Kocsis, K.; Sonnet, P.E.; Ziegler, F.E. **1965, 1966**, J. Am. Chem. Soc. *87*, 2073; *88*, 3099

Büchi, G.; Hofheinz, W.; Paukstelis, J.V. **1966**, J. Am. Chem. Soc. *88*, 4113

Bucourt, R.; Pierdet, A.; Costerousse, G.; Toromanoff, E. **1965**, Bull. Soc. Chim. Fr. *1965*, 645

Bull, J.R.; Tuinman, A. **1975**, Tetrahedron *31*, 2151

Bundy, G.L.; Lincoln, F.H.; Nelson, N.A.; Pike, J.E.; Schneider, W.P. **1971**, Ann. N. Y. Acad. Sci. *180*, 76

Bundy, G.L.; Schneider, W.P.; Lincoln, F.H.; Pike, J.E. **1972**, J. Am. Chem. Soc. *94*, 2123

Burgess, K.; Ibarzo, J.; Linthicum, D.S.; Russell, D.H.; Shin, H.; Shitangkoon, A.; Totani, R.; Zhang, A.J. **1997**, J. Am. Chem. Soc. *119*, 1556

Bürgi, H.B.; Dunitz, J.D.; Shefter, E. **1973**, J. Am. Chem. Soc. *95*, 5065

Burke, S.D.; Grieco, P.A. **1979**, Org. React. (NY) *26*, 361

Cadogan, J.I.G. (ed.) **1979**, *Organophosphorus Reagents in Organic Synthesis*, Academic Press: London

Cain, E.N.; Vukov, R.; Masamune, S. **1969**, J. Chem. Soc. D *1969*, 98

Caine, D. **1976**, Org. React. (NY) *23*, 1

Cambie, R.C.; Potter, G.J.; Rutledge, P.S.; Woodgate, P.D. **1977**, J. Chem. Soc. Perkin Trans. *1*, 530

Campbell, C.J.; Laherrère, J.H. **1998**, Sci. Am., 60

Cao, B.; Park, H.; Joullié, M.M. **2002**, J. Am. Chem. Soc. *124*, 520

Carlson, R.M.; Oyler, A.R. **1974**, Tetr. Lett., 2615

Carlson, R.M.; Oyler, A.R.; Peterson, J.R. **1975**, J. Org. Chem. *40*, 1610

Carpino, L.A. **1957**, J. Am. Chem. Soc. *79*, 98

Carpino, L.A. **1973**, Acc. Chem. Res. *6*, 191

Carpino, L.A.; Beyermann, M.; Wenschuh, H.; Bienert, M. **1996**, Acc. Chem. Res. *29*, 268

Carruthers, W. **1973**, Chem. Ind. (London) *1973*, 931

CARRUTHERS, W. **1978**, *Some Modern Methods of Organic Syntheses,* 2nd edn., Cambridge Univ. Press: Cambridge

CARUTHERS, M. H.; BEAUCAGE, S. L.; EFCAVITCH, J. W.; FISCHER, E. F.; MATTEUCCI, M. D.; STABINSKI, Y. **1980**, Nucleic Acids Symp. Ser. 7 [Nucleic Acid Synth.: Appl. Mol. Biol. Genet. Eng.] 215

CASALE, R.; MCLAUGHLIN, L. W. **1990**, J. Am. Chem. Soc. *112*, 5264

CASIMIR, J. R.; DIDIERJEAN, C.; AUBRY, A.; RODRIGUEZ, M.; BRIAND, J.-P.; GUICHARD, G. **2000**, Org. Lett. *2*, 895

CASON, J. A.; ALLEN, C. F. **1949**, J. Org. Chem. *14*, 1036

CASSAR, L.; EATON, P. E.; HALPERN, J. **1970**, J. Am. Chem. Soc. *92*, 6366

CAVA, M. P.; MITCHELL, M. J. **1967**, *Cyclobutadiene and Related Compounds,* Academic Press: New York London

CAVA, M. P.; POHL, R. J. **1960**, J. Am. Chem. Soc. *82*, 5242

CAVILL, G. W. K.; HALL, C. D. **1967**, Tetrahedron *23*, 1119

ČERNÝ, M.; PACÁK, J. **1961**, Coll. Czech. Chem. Commun. *26*, 2084

ČERNÝ, M.; STANĚK, J.; PACÁK, J. **1963**, Monatsh. Chem. *94*, 290

CHAKRABORTY, M.; MCCONVILLE, D. B.; NIU, Y.; TESSIER, C. A.; YOUNGS, W. J. **1998**, J. Org. Chem. *63*, 7563

CHAMBERLIN, A. R.; DEZUBE, M.; REICH, S. R.; SALL, D. J. **1989**, J. Am. Chem. Soc. *111*, 6247

CHAN, T. H.; CHANG, E. **1974**, J. Org. Chem. *39*, 3264

CHAN, T. H.; ONG, B. S. **1976**, Tetr. Lett., 319

CHANDRASEKHARAN, L.; RAMACHANDRAN, P. V.; BROWN, H. C. **1985**, J. Org. Chem. *50*, 5446

CHANG, J.; PAQUETTE, L. A. **2002**, Org. Lett. *4*, 253

CHAPMAN, O. L.; CHANG, C.-C.; KOLC, J.; ROSENQUIST, N. R.; TORMOKA, H. **1975**, J. Am. Chem. Soc. *97*, 6586

CHAPMAN, O. L.; MATTES, K.; MCINTOSH, C. L.; PACANSKY, J. **1973**, J. Am. Chem. Soc. *95*, 6134

CHATTOPADHYAYA, J. B.; REESE, C. B. **1978**, Chem. Commun., 639

CHEESEMAN, J. D.; CORBETT, A. D.; SHU, R.; CROTEAU, J.; GLEASON, J. L.; KAZLAUSKAS, R. J. **2002**, J. Am. Chem. Soc. *124*, 5692

CHEN, S.; JANDA, K. D. **1997**, J. Am. Chem. Soc. *119*, 8724

CHENG, S.; COMER, D. D.; WILLIAMS, J. P.; MYERS, P. L.; BOGER, D. L. **1996**, J. Am. Chem. Soc. *118*, 2567

CHEREST, M.; FELKIN, H.; PRUDENT, N. **1968**, Tetr. Lett., 2199, 2205

CHIKASHITA. H.; HIRAO, K.; ITOH, K. **1993**, Bull. Chem. Soc. Jpn *66*, 1738

CHIU, S.-H.; STODDART, J. F. **2002**, J. Am. Chem. Soc. *124*, 4174

CHOU, T. S.; BURGTORF, J. R.; ELLIS, A. L.; LAMMERT, S. R.; KUKOLJA, S. P. **1974**, J. Am. Chem. Soc. *96*, 1609; see also: CHOU, T. S. **1974**, Tetrahedron Lett. *1974*, 725

CHRISTOPH, G. G.; ENGEL, P.; USHA, R.; BALOGH, D. W.; PAQUETTE, L. A. **1982**, J. Am. Chem. Soc. *104*, 784

CLARK, J. H. **2002**, Acc. Chem. Res. *35*, 791

CLARK, R. D.; KOZAR, L. G.; HEATHCOCK, C. H. **1975**, Synth. Commun. *5*, 1

CLARK, R. D.; HEATHCOCK, C. H. **1976**, J. Org. Chem. *41*, 1396

CLARK-LEWIS, I.; AEBERSOLD, R.; ZILTENER, H.; SCHRADER, J. W.; HOOD, L. E.; KENT, S. B. H. **1986**, Science *231*, 134

CLEZY, P. S.; NICHOL, A. W. **1965**, Aust. J. Chem. *18*, 1835

COATES, G. E.; GREEN, M. L. H.; POWELL, P.; WADE, K. **1977**, *Principles of Organometallic Chemistry,* Chapman & Hall: London

COATES, R. M.; SHAW, J. E. **1968**, Chem. Commun. *1968*, 47

COATES, R. M.; SHAW, J. E. **1970**, J. Org. Chem. *35*, 2597, 2601

COHEN, F.; OVERMAN, L. E. **1997**, J. Org. Chem. *62*, 6379

COHEN, F.; OVERMAN, L. E. **2001**, J. Am. Chem. Soc. *123*, 10782

COLE, J. E.; JOHNSON, W. S.; ROBINS, P. A.; WALKER, J. **1962**, J. Chem. Soc. *1962*, 244

COLLIER, P. N.; CAMPBELL, A. D.; PATEL, I.; RAYNHAM, T. M.; TAYLOR, R. J. K. **2002**, J. Org. Chem. *67*, 1802

COLLIN, J.; NAMY, J.-L.; KAGAN, H. B. **1986**, Nouv. J. Chim. *10*, 229

COLLMAN, J. P. **1975**, Acc. Chem. Res. *8*, 342 [Na₂Fe(CO)₄]

COLLMAN, J. P. **1977**, Acc. Chem. Res. *10*, 265

COLLMAN, J. P.; FINKE, R. G.; CAWSE, J. N.; BRAUMAN, J. I. **1978**, J. Am. Chem. Soc. *100*, 4766

COLLMAN, J.P.; GAGRIE, R.R.; REED, C.A.; HALBERT, T.R.; LANG, G.; ROBINSON, W.T. 1975, J. Am. Chem. Soc. 97, 1427 [picket-fence porphyrin]

COLLMAN, J.P.; WINTER, S.R.; CLARK, D.R. 1972, J. Am. Chem. Soc. 94, 1788

COLUMBUS, I.; HAJ-ZAROUBI, M.; BIALI, S.E. 1998, J. Am. Chem. Soc. 120, 11806

COMMINS, D.L.; DEHGHANI, A. 1992, Tetrahedon Lett. 33, 6299

CONIA, J.M. 1975, Pure Appl. Chem. 43, 317

CONN, M.M.; REBEK, J., JR. 1997, Chem. Rev. 97, 1647

COOK, A.F. 1968, J. Org. Chem. 33, 3589

COOK, C.E.; CORLEY, R.C.; WALL, M.E. 1968, J. Org. Chem. 33, 2789

COOK, G.A. 1969, Enamines, M. Dekker: New York London

COOKE, M.P.; PARLAM, R.M. 1975, J. Am. Chem. Soc. 97, 6863

COOKSON, R.C.; WARIYAR, N.S. 1956, J. Chem. Soc. 1956, 2302

COOPER, A.C.; McALEXANDER, L.H.; LEE, D.-H.; TORRES, M.T.; CRABTREE, R.H. 1998, J. Am. Chem. Soc. 120, 9971

COOPER, S.R. 1992, Crown Compounds. Toward Future Applications, VCH: Weinheim, Ger.

COPE, A.C.; MOORE, P.T.; MOORE, W.R., 1960 A, J. Am. Chem. Soc. 82, 1744

COPE, A.C.; TURNBULL, E.R. 1960 B, Org. React. (NY) 11, 317

COPPOLA, G.M. 1978, J. Heterocycl. Chem. 15, 645

COREY, E.J. 1967 A, Pure Appl. Chem. 14, 19

COREY, E.J. 1971, Quart. Rev. Chem. Soc. 25, 455

COREY, E.J. 1990 C, J. Org. Chem. 55, 1693

COREY, E.J.; GILMAN, N.W.; GANERN, B.E. 1968 A, J. Am. Chem. Soc. 90, 5616

COREY, E.J.; MITRA, R.B.; UDA, H. 1963 A, J. Am. Chem. Soc. 85, 362

COREY, E.J.; BURKE, H.J. 1956 A, J. Am. Chem. Soc. 78, 174

COREY, E.J.; BURKE, H.J.; REMERS, W. 1956 B, J. Am. Chem. Soc. 78, 180

COREY, E.J.; CHAYKOVSKY, M. 1965 A, J. Am. Chem. Soc. 87, 1353

COREY, E.J.; DANHEISER, R.L.; CHANDRASEKARAN, S. 1976, J. Org. Chem. 41, 260

COREY, E.J.; ENDERS, D. 1978, Chem. Ber. 111, 1337 [dimethylhydrazones]

COREY, E.J.; ESTREICHER, H. 1978, J. Am. Chem. Soc. 100, 6294 [1-nitrocycloalkenes]

COREY, E.J.; HAMANAKA, E. 1967 C, J. Am. Chem. Soc. 89, 2757

COREY, E.J.; HOPKINS, P.B.; KIM, S.; YOO, S.; NAMBIAR, K.P.; FALCK, J.R. 1979, J. Am. Chem. Soc. 101, 7131

COREY, E.J.; HORTMANN, A.G. 1965, J. Am. Chem. Soc. 87, 5736

COREY, E.J.; IMWINKELRIED, R.; PIKUL, S.; XIANG, Y.B. 1989 A, J. Am. Chem. Soc. 111, 5493

COREY, E.J.; KATZENELLENBOGEN, J.A.; GILMAN, N.W.; ROMAN, S.A.; ERICKSON, B.W. 1968 D, J. Am. Chem. Soc. 90, 5618

COREY, E.J.; KIM, S.; YOO, S.; NICOLAOU, K.C.; MELVIN, L.S., JR.; BRUNELLE, D.L.; FALCK, J.R.; TRYBULSKI, E.J.; LETT, R.; SHELDRAKE, P.W. 1978, J. Am. Chem. Soc. 100, 4622

COREY, E.J.; KIM, S.; YOO, S.; NICOLAOU, K.C.; MELVIN, L.S.; BRUNELLE, D.J.; FALCK, J.R.; TRYBULDKI, E.J.; LETT, R.; SHELDRAKE, P.W. 1978, J. Am. Chem. Soc. 100, 4620

COREY, E.J.; KIRN, S.S. 1990 B, J. Am. Chem. Soc. 112, 4976

COREY, E.J.; KIRST, H.A. 1972 A, J. Am. Chem. Soc. 94, 667

COREY, E.J.; MITRA, R.B.; UDA, H. 1964 A, J. Am. Chem. Soc. 86, 485 [cyclobutane]

COREY, E.J.; NAEF, R.; HANNON, F.J. 1986, J. Am. Chem. Soc. 108, 7114

COREY, E.J.; NICOLAOU, K.C.; TORU, T. 1975 A, J. Am. Chem. Soc. 97, 2287 [Bu$_2$AlH]

COREY, E.J.; NOYORI, R.; SCHAAF, T.K. 1970, J. Am. Chem. Soc. 92, 2586

COREY, E.J.; NOZOE, S. 1963 B, J. Am. Chem. Soc. 85, 3527

COREY, E.J.; NOZOE, S. 1965 B, J. Am. Chem. Soc. 87, 5728

COREY, E.J.; OHNO, M.; VATAKENCHERRY, P.A.; MITRA, R.B. 1964 B, J. Am. Chem. Soc. 86, 478 [OsO$_4$]

COREY, E.J.; POSNER, G.H. 1968 B, J. Am. Chem. Soc. 90, 5615

COREY, E.J.; SCHAAF, T.K.; HUBER, W.; KOELLIKER, U.; WEINSHENKER, N.M. 1970, J. Am. Chem. Soc. 92, 397

COREY, E.J.; SEMMELHACK, M.F. 1967 B, J. Am. Chem. Soc. 89, 2756

COREY, E.J.; SHULMAN, J.I. 1968 C, Tetrahedron Lett. 1968, 3655

COREY, E. J.; TRYBULSKI, E. J.; MELVIN, L. S.,
JR; NICOLAOU, K. C.; SECRIST, J. A.; LETT, R.;
SHELDRAKE, P. W.; FALCK, J. R.; BRUNELLE,
D. L.; HASLANGER, M. E.; KIM, S.; YOO, S.
1978, J. Am. Chem. Soc. *100*, 4618

COREY, E. J.; TRYBULSKI, E. J.; MELVIN, L. S.;
NICOLAOU, K. C.; SECRIST, J. A.; LETT, R.;
SHELDRAKE, P. W.; FALCK, J. R.; BRUNELLE,
D. J.; HASLANGER, M. F.; KIM, S.; YOO, S.
1978, J. Am. Chem. Soc. *100*, 4618.

COREY, E. J.; ULRICH, P. **1975 B**, Tetrahedron
Lett. *1975*, 3685 [cyclopropanes]

COREY, E. J.; VENKATESWARLU, A. **1972 B**, J.
Am. Chem. Soc. *94*, 6190

COREY, E. J.; WEINSHENKER, N. M.; SCHAAF,
T. K.; HUBER, W. **1969**, J. Am. Chem. Soc.
91, 5675

COREY, E. J.; YU, C.-M.; KIM, S. S. **1989 B**, J.
Am. Chem. Soc. *111*, 5495

COREY, E. L.; YU, C.-M.; LEE, D.-H. **1990 A**, J.
Am. Chem. Soc. *112*, 878

COUSINS, G. R. L.; POULSEN, S.-A.; SANDERS,
J. K. M. **1999**, Chem. Commun., 1575

CRABTREE, R. H. **2001**, *The Organometallic
Chemistry of the Transition Metals*, John Wiley & Sons, Inc.

CRAM, D. J.; ABD El-HAFEZ, F. A. **1952**, J. Am.
Chem. Soc. *74*, 5828

CRAM, D. J.; KARBACH, S.; KIRN, Y. H.; BAC-
ZYNSKYJ, L.; MARTI, K.; SAMPSON, R. M.;
KALLEYMEYN, G. W. **1988**, J. Am. Chem. Soc.
110, 2554

CRAM, D. J.; SAHYUN, M. R. V.; KNOX, G. R.
1962, J. Am. Chem. Soc. *84*, 1734

CREASER, I. I.; HARROWFIELD, J. MacB.;
HERLT, A. J.; SARGESON, A. M.; SPRINGBORG,
J.; GENE, R. J.; SNOW, M. R. **1977**, J. Am.
Chem. Soc. *99*, 3181

CREDI, A.; BALZANI, V.; LANGFORD, S. J.;
STODDART, J. F. **1997**, J. Am. Chem. Soc.
119, 2679

CREGER, P. L. **1972**, J. Org. Chem. *37*, 1907

CREGGE, R. J.; HERRMANN, J. L.; LEE, C. S.;
RICHMAN, J. E.; SCHLESSINGER, R. H. **1973**,
Tetrahedron Lett. *1973*, 2425

CRICH, D.; SMITH, M. **2001**, J. Am. Chem.
Soc. *123*, 9015

CRIEGEE, R.; KROPF, H. **1979**, in: HOUBEN-
WEYL, *Methoden der Organischen Chemie*,
Vol. *VI/1* a-1: *Alkohole I*, p. 592, Thieme:
Stuttgart

CRIMMINS, M. T.; TABET, E. A. **2001**, J. Org.
Chem. *66*, 4012

CRISPINO, G. A.; HO, P. T.; SHARPLESS, K. B.
1993, Science *259*, 64

CROSBY, D. J.; BERTHOLD, R. V. **1962**, J. Org.
Chem. *27*, 3083

CUILLERON, C. Y.; FÉTIZON, M.; GOLFIER, M.
1970, Bull. Soc. Chim. Fr. *1970*, 1193

CZARNIK, A. W.; HOBBS DEWITT, S. **1996**,
Chemistry in Britain, 43

DANIELI, N.; MAZUR, Y.; SONDHEIMER, F.
1966, Tetrahedron *22*, 3189

DANISHEFSKY, S. **1974**, J. Am. Chem. Soc. *96*,
1256

DANISHEFSKY, S.; FUNK, R. L.; KERWIN, J. F.,
JR. **1980**, J. Am. Chem. Soc. *102*, 6889

DANISHEFSKY, S.; HIRAMA, M.; GOMBATZ, K.;
HARAYAMA, T.; BERMAN, E.; SCHUDA, P. F.
1979, J. Am. Chem. Soc. *101*, 7020

DANISHEFSKY, S.; KITAHARA, T.; TSAI, M.; DY-
NAK, J. **1976**, J. Org. Chem. *41*, 1669

DANISHEFSKY, S.; WALKER, F. J. **1979 A**, J.
Am. Chem. Soc. *101*, 7018

DASGUPTA, F.; GAREGG, P. J. **1988**, Carbohydr.
Res. *177*, C 13

DAUBEN, H. J., JR.; LÖKEN, B.; RINGOLD, H. J.
1954, J. Am. Chem. Soc. *76*, 1359

DAUBEN, W. G.; BROOKHART, T. **1981**, J. Am.
Chem. Soc. *103*, 237

DAUBEN, W. G.; LORBER, M.; FULLERTON, D. S.
1969, J. Org. Chem. *34*, 3587

DAUBEN, W. G.; WILLIAMS, R. G.; MCKELVEY,
R. D. **1973**, J. Am. Chem. Soc. *95*, 3932

DAVIES, H. M. L.; STAFFORD, D. G.; HANSEN, T.
1999, Org. Lett. *1*, 233

DAVIS, B. G.; FAIRBANKS, A. J. **2002**, *Carbohydrate Chemistry*, Oxford University Press

DAVIS, F. A.; MOHANTY, P. K. **2002**, J. Org.
Chem. *67*, 1290

DAY, D.; RINGSDORF, H. **1977**, J. Polym. Sci.
Polym. Lett. Ed. *16*, 205

DE LUCA, L.; GIACOMELLI, G.; TADDEI, M.
2001, J. Org. Chem. *66*, 2534

DE LUCCHI, O.; MIOTTI, U.; MODENA, G.
1991, Org. React. (NY) *40*, 157

DEAR, R. E. A.; PATTISON, F. L. M. **1963**, J. Am.
Chem. Soc. *85*, 622

DEARDORFF, D. R.; MATTHEWS, A. J.; MCMEE-
KIN, D. S.; CRANEY, C. L. **1986**, Tetr. Lett.,
1255

DEDIEU, A. **2000**, Chem. Rev. *100*, 543

DEEM, M. L. **1972**, Synthesis *1972*, 675

DEISENHOFER, J.; EPP, O.; MIKI, K.; HUBER,
R.; MICHEL, H. **1984**, J. Mol. Biol. *180*, 385

DEIVE, N.; RODRÍGUEZ, J.; JIMÉNEZ, C. **2001**, J. Med. Chem. *44*, 2612

DE MEIJERE, A. **1979**, Angew. Chem. *91*, 867 (Int. Ed. Engl. *18*, 809)

DENER, J. M.; FANTAUZZI, P. P.; KSHIRSAGAR, T. A.; KELLY, D. E.; WOLFE, A. B. **2001**, Organic Process Research & Development *5*, 445

DENIS, J. M.; CONIA, J. M. **1972**, Tetrahedron Lett. *1972*, 4593

DEPRÉS, J.-P.; GREENE, A. E. **1980**, J. Org. Chem. *45*, 2036

DESCOTES, G. **1993**, *Carbohydrates as Organic Raw Materials II*; VCH: Weinheim

DESS, D. B.; MARTIN, J. C. **1983**, J. Org. Chem. *48*, 4155

DEVILLE, J. P.; BEHAR, V. **2002**, Org. Lett. *4*, 1403

DIAS, L. C.; DE OLIVEIRA, L. G. **2001**, Org. Lett. *3*, 3951

DICKER, I. D.; GRIGG, R.; JOHNSON, A. W.; PINNOCK, H.; RICHARDSON, K.; VAN DEN BROEK, P. **1971**, J. Chem Soc. C *1971*, 536

DIEDERICH, F.; RUBIN, Y.; KNOBLER, C. B.; WHETTEN, R. L.; SCHRIVER, K. E.; HOUK, K. X; LI, Y. **1989**, Science *245*, 1088

DIEDERICH, F.; STAAB, H. A. **1978**, Angew. Chem. *90*, 383 (Int. Ed. Engl. *17*, 372)

DIEDERICH, F.; STANG, P. J., (eds.); *Metal-catalyzed Cross-coupling Reactions*, Wiley-VCH: Weinheim, **1998**

DIEDERICH, F.; WHETTEN, R. L. **1991**, Angew. Chem. *103*, 695 (Int. Ed. Eng). *30*, 678)

DIEDERICH, F.; WHETTEN, R. L. **1992**, Acc. Chem. Res. *25*, 119

DIETER, R. K.; TOKLES, M. **1987**, J. Am. Chem. Soc. *109*, 2040

DIETRICH, B.; LEHN, J.-M.; SAUVAGE, J. P. **1969**, Tetrahedron Lett. *1969*, 2885, 2889; see also: **1973**, Tetrahedron *29*, 1629, 1647

DIMSDALE, M. J.; NEWTON, R. F.; RAINEY, D. K.; WEBB, C. F.; LEE, T. V.; ROBERTS, S. M. **1977**, J. Chem. Soc. Chem. Commun. *1977*, 716

DIXON, D. J.; LEY, S. V.; SHAPPARD, T. **2001**, Org. Lett. *3*, 3749

DIXON, D. J.; LEY, S. V.; RODRÍGUEZ, F. **2001**, Org. Lett. *3*, 3753

DJERASSI, C. **1951**, Org. React. (NY) *6*, 207

DJERASSI, C.; BATRES, E.; ROMO, J.; ROSEN-KRANZ, G. **1952**, J. Am. Chem. Soc. *74*, 3634

DJERASSI, C.; ENGLE, R. R.; BOWERS, A. **1956**, J. Org. Chem. *21*, 1547

DJERASSI, C.; ROSENKRANZ, G.; ROMO, J.; KAUFMANN, S.; PATAKI, J. **1950**, J. Am. Chem. Soc. *72*, 4534 (see also ibid. 4531, 4540; **1950**, J. Org. Chem. *15*, 1289)

DJERASSI, C.; SHAMMA, M.; KHAN, T. Y. **1958**, J. Am. Chem. Soc. *80*, 4723

DOBSON, N. A.; EGLINTON, G.; KRISHNAMURTI, M.; RAPHAEL, R. A.; WILLIS, R. G. **1961**, Tetrahedron *16*, 16

DOBSON, N. A.; RAPHAEL, R. A. **1955**, J. Chem. Soc. *1955*, 3358

DOI, T.; HIJIKURO, I.; TAKAHASHI, T. **1999**, J. Am. Chem. Soc. *121*, 6749

DOMANSKI, T. L.; HE, Y.-A.; KHAN, K. K.; ROUSSEL, F.; WANG, Q.; HALPERT, J. R. **2001**, Biochemistry *40*, 10150

DOMINGUEZ, Z.; DANG, H.; STROUSE, M. J.; GARCIA-GARIBAY **2002**, J. Am. Chem. Soc. *124*, 7719

DONG, V. M.; MACMILLAN, D. W. C. **2001**, J. Am. Chem. Soc. *123*, 2448

DOUGLAS, N. L.; LEY, S. V.; LUCKING, U.; WARRINER, S. L. **1998**, J. Chem. Soc., Perkin Trans. *1*, 51

DOSKOTCH, R. W.; PHILLIPSON, J. D.; RAY, A. B.; BEAL, J. L. **1971**, J. Org. Chem. *36*, 2409

DOYLE, P.; MCLEAN, I. R.; MURRAY, R. D. H.; PARKER, W.; RAPHAEL, R. A. **1965**, J. Chem. Soc. *1965*, 1344

DRAIN, C. M.; GONG, X.; RUTA, V.; SOLL, C. E.; CHICOINEAU, P. F. **1999**, J. Comb. Chem.. *1*, 286

DRÖGEMÜLLER, M.; FLESSNER, T.; JAUTELAT, R.; SCHOLZ, U.; WINTERFELDT, E. **1998**, Eur. J. Org. Chem., 2811

DRÖGEMÜLLER, M.; JAUTELAT, R.; WINTERFELDT, E. **1996**, Angew. Chem. *108*, 1669

DRYDEN, H. L., JR; WEBBER, G. M.; WIECZOREK, J. J. **1964**, J. Am. Chem. Soc. *86*, 742

DUBBER, M.; LINDHORST, T. K. **2000**, J. Org. Chem. *65*, 5275

DURON, S. G.; GIN, D. Y. **2001**, Org. Lett. *3*, 1551

DURST, T. **1979**, *Sulphoxides*, in: BARTON, D.; OLLIS, W. D. (eds.) *Comprehensive Organic Chemistry, Vol. 3*, p. 121 ff., Pergamon Press: Oxford

DYE, J. L.; LOK, M. T.; TEHAN, F. J.; CERASO, J. M.; VORHEES, K. J. **1973**, J. Org. Chem. *38*, 1773

DYKE, S. F. **1972**, Adv. Heterocycl. Chem. *14*, 279

DYKE, S. F. **1973**, *The Chemistry of Enamines*, Cambridge Univ. Press: Cambridge

EATON, E. P. **1979**, Tetrahedron *35*, 2189

EBEL, H. F.; LÜTTRINGHAUS, A. **1970**, in: HOUBEN-WEYL, *Methoden der Organischen Chemie, Vol. XIII/1*, p. 621 ff., Thieme: Stuttgart

ECHAVARREN, A. M.; STILLE, J. K. **1987**, J. Am. Chem. Soc. *109*, 5478

ECKSTEIN, F. **1967**, Chem. Ber. *100*, 2228, 2236

EDWARDS, J. A.; SUNDEEN, J.; SALMOND, W.; IWADARE, T.; FRIED, J. H. **1972**, Tetrahedron Lett. *1972*, 791

EGLINGTON, G.; McCRAE, W. **1963**, Adv. Org. Chem. *4*, 225

EHRHART, G.; RUSCHIG, H. **1972**, *Arzneimittel: Entwicklung, Wirkung, Darstellung*, 2nd edn., Verlag Chemie: Weinheim, Ger.

EICHER, T.; HAUPTMANN, S. **1995**, *The Chemistry of Heterocycles: Structure, Reactions, Syntheses, and Applications*, Thieme: Stuttgart

EISNER, U.; KUTHAN, J. **1972**, Chem. Rev. *72*, 1

ELLISON, R. A. **1973**, Synthesis *1973*, 397

EMERSON, G. F.; WATTS, L.; PETTIT, R. **1965**, J. Am. Chem. Soc. *87*, 131

EMMONS, W. D.; LUCAS, G. B. **1955**, J. Am. Chem. Soc. *77*, 2287

ENDERS, D.; EICHENAUER, H. **1979**, Chem. Ber. *112*, 2933

ENHOLM, E. J.; SATICI, H.; TRIVELLAS, A. **1989 B**, J. Org. Chem. *54*, 5841

ENHOLM, E. L.; TRIVELLAS, A. **1989 A**, Tetrahedron Lett. *1989*, 1063

EPSTEIN, D. M.; CHOUDHARY, S.; CHURCHILL, M. R.; KEIL, K. M.; ELISEEV, A. V.; MORROW, J. R. **2001**, Inorg. Chem. *40*, 1591

ERICKSON, B. W.; MERRIFIELD, R. B. **1976**, in: NEURATH, H.; HILL, R. L. (eds.) *The Proteins*, 3rd edn., Vol. 11, p. 313, Academic Press: New York London

ERNEST, L.; GOSTELI, J.; WOODWARD, R. B. **1979**, J. Am. Chem. Soc. *101*, 6301

ESCHENMOSER, A. **1970**, Quart. Rev. Chem. Soc. *24*, 366; see also: DUBS, P.; GÖTSCHI, E.; ROTH, M.; ESCHENMOSER, A. **1970**, Chimia *24*, 34

ESCHENMOSER, A. **1974**, Naturwissenschaften *61*, 513

ESCHENMOSER, A. **1991**, Nachr. Chem. Tech. Lab. *39*, 795

ESCHENMOSER, A.; DOBLER, M. **1992**, Helv. Chim. Acta *75*, 218

ESCHENMOSER, A.; FELIX, D.; GUT, M.; MEIER, J.; STADLER, P. **1959**, in: WOLSTENHOLME, G. E. W.; O'CONNOR, M. (eds.) *Biosynthesis of Terpenes and Sterols*, Churchill: London

EVANS, D. A.; BAILLARGEON, D. J.; NELSON, J. V. **1978**, J. Am. Chem. Soc. *100*, 2242

EVANS, D. A.; BARTROLI, J.; SHIH, T. L. **1981**, J. Am. Chem. Soc. *103*, 2127

EVANS, D. A.; CLARK, J. S.; METTERNICH, R.; NOVACK, V. J.; SHEPPARD, G. S. **1990**, J. Am. Chem. Soc. *112*, 866

EVANS, D. A.; ENNIS, M. D.; LE, T.; MANDEL, N.; MANDEL, G. **1984**, J. Am. Chem. Soc. *106*, 1154

EVANS, D. A.; FU, G. C.; HOVEYDA, A. H. **1988**, J. Am. Chem. Soc. *110*, 6917

EVANS, D. A.; MacMILLAN, D. W. C.; CAMPOS, K. R. **1997**, J. Am. Chem. Soc. *119*, 10859

EVANS, D. A.; ROVIS, T.; KOZLOWSKI, M. C.; DOWNEY, C. W.; TEDROW, J. S. **2000**, J. Am. Chem. Soc. *122*, 9134

EVANS, D. A.; SCOTT, W. L.; TRUESDALE, L. K. **1972**, Tetrahedron Lett. *1972*, 121

EVANS, D. A.; VOGEL, E.; NELSON, J. V. **1979**, J. Am. Chem. Soc. *101*, 6120

EVANS, D. A.; WILLIS, M. C.; JOHNSTON, J. N. **1999**, Org. Lett. *1*, 865

EVANS, M. E. **1980**, in: WHISTLER, R. L.; BEMILLER, J. N. (eds.) *Methods in Carbohydrate Chemistry, Vol. VIII*, p. 313, Academic Press: New York-London

FAUL, M. M.; RATZ, A. M.; SULLIVAN, K. A.; TRANKLE, W. G.; WINNEROSKI, L. L. **2001**, J. Org. Chem. *66*, 5772

FECHTIG, B.; PETER, H.; BICKEL, H.; FISCHLER, E. **1968**, Helv. Chim. Acta *51*, 1108

FELIX, D.; SCHREIBER, L.; OHLOFF, G.; ESCHENMOSER, A. **1971**, Helv. Chim. Acta *54*, 2896

FELIX, D.; SCHREIBER, L.; PIERS, K.; HORN, U.; ESCHENMOSER, A. **1968**, Helv. Chim. Acta *51*, 1461

FELNER, L.; FISCHLI, A.; WICK, A.; PESARO, M.; BORMANN, D.; WINNACKER, E. L.; ESCHENMOSER, A. **1967**, Angew. Chem. *79*, 863 (Int. Ed. Engl. *6*, 864)

FENSELAU, A. H.; MOFFATT, J. G. **1966**, J. Am. Chem. Soc. *88*, 1762

FERLES, M.; PLIML, J. **1970**, Adv. Heterocycl. Chem. *12*, 43

FERRIER, R. J. **1965**, Adv. Carbohydr. Chem. *20*, 67

FERRIER, R. J.; SANKEY, G. H. **1966**, J. Chem. Soc. C **1966**, 2339

FESSNER, W.-D.; MURTY, B. A. R. C.; WÖRTH, J.; HUNKLER, D.; FRITZ, H.; PRINZBACH, H.; ROTH, W. D.; SCHLEYER, P. V. R.; McEWEN, A. B.; MAIER, W. F. **1987 A**, Angew. Chem. *99*, 484 (Int. Ed. Engl. *26*, 451)

FESSNER, W.-D.; PRINZBACH, H.; RIHS, G. **1983**, Tetrahedron Lett. **1983**, 5857

FESSNER, W.-D.; SEDELMAIER, G.; SPURR, P. R.; RIHS, G.; PRINZBACH, H. **1987 B**, J. Am. Chem. Soc. *109*, 4626

FIESER, L. E.; FIESER, M. **1959**, Steroide, Verlag Chemie: Weinheim, Ger.

FIESER, L. F.; STEVENSON, R. **1954**, J. Am. Chem. Soc. *76*, 1728

FIESER, L. F.; FIESER, M. **1959**, Steroids, p. 507, Reinhold: New York; Chapman & Hall: London

FIESER, L. F.; SMUSZKOVIEZ, J. **1948**, J. Am. Chem. Soc. *70*, 3352

FINN, F. M.; HOFMANN, K. **1976**, in: NEURATH, H.; HILL, R. L. (eds.) *The Proteins*, 3rd edn., Vol. II, chapter 2 (p. 106–237), Academic Press: New York London

FINN, M. G.; SHARPLESS, K. B. **1985**, *On the Mechanism of Asymmetric Epoxidation with Titanium-Tartrate Catalysts*, in: MORRISON, J. D. (ed.) *Asymmetric Synthesis*, Vol. 5: *Chiral Catalysis*, chapter 8, p. 247, Academic Press: New York

FISCHER, E. **1914**, Chem. Ber. *47*, 196

FISCHER, H.; NEBER, M. **1932**, Liebigs Ann. Chem. *496*, 1

FISCHER, H.; STANGLER, G. **1927**, Liebigs Ann. Chem. *459*, 53

FISCHER, H.; STERN, A. **1940**, Die Chemie des Pyrrols, Vol. II, Akademische Verlagsgesellschaft: Leipzig

FLEMING, I. **1973**, *Selected Organic Syntheses*, Wiley: New York London

FLEMING, I.; BARBERO, A.; WALTER, D. **1997**, Chem. Rev. *97*, 2063

FLEMING, I. **1979**, Organic Silicon Chemistry, in: BARTON, D.; OLLIS, W. D. (eds.) *Comprehensive Organic Chemistry*, Vol. 3, p. 539, Pergamon Press: Oxford

FLETCHER, H. G., JR. **1963**, in: WHISTLER, R. L.; WOLFROM, M. L.; BEMILLER, J. N. (eds.) *Methods in Carbohydrate Chemistry*, Vol. II, p. 307, Academic Press: New York London

FLETCHER, M. D.; CAMPBELL, M. M. **1998**, Chem. Rev. *98*, 763

FLYNN, D. L.; CRICH, J. Z.; DEVRAJ, R. V.; HOCKERMAN, S. L.; PARLOW, J. J.; SOUTH, M. S.; WOODWARD, S. **1997**, J. Am. Chem. Soc. *119*, 4874

FONG II, R.; SCHUSTER, D. I.; WILSON, S. R. **1999**, Org. Lett., Vol. 1, No. *5*, 729

FORTIN, S.; BARRIAULT, L.; DORY, Y. L.; DESLONGCHAMPS, P. **2001**, J. Am. Chem. Soc. *123*, 8210

FRANCESCH, A.; ALVAREZ, R.; LÓPEZ, S.; DE LERA, A. R. **1997**, J. Org. Chem. *62*, 310

FRANCIS, M. B.; FINNEY, N. S.; JACOBSEN, E. N. **1996**, J. Am. Chem. Soc. *118*, 8983

FRANCK, B.; BLASCHKE, G. **1966**, Liebigs Ann. Chem. *695*, 144

FRANCK, B.; DUNKELMANN, G.; LUBS, H. J. **1967**, Angew. Chem. *79*, 1066 (Int. Ed. Engl. *6*, 1075)

FRANCK, B.; SCHLINGLOFF, G. **1962**, Liebigs Ann. Chem. *659*, 123

FRANCK, B.; TEETZ, V. **1971**, Angew. Chem. *83*, 509 (Int. Ed. Engl. *10*, 411)

FRANK, S. A.; MERGOTT, D. J.; ROUSH, W. R. **2002**,

FRANKLIN, A. S.; LY, S. K.; MACKIN, G. H.; OVERMAN, L. E.; SHAKA, A. J. **1999**, J. Org. Chem. *64*, 1512

FRANKLIN, A. S.; OVERMAN, L. E. **1996**, Chem. Rev. *96*, 505

FRASER, R. R.; SCHUBER, F. J.; WIGFIELD, Y. Y. **1972**, J. Am. Chem. Soc. *94*, 8795

FRÄTER, G.; MÜLLER, U.; GÜNTHER, W. **1984**, Tetrahedron *40*, 1269

FREIFELDER, M. **1963**, Adv. Catal. *14*, 203

FREIFELDER, M.; ROBINSON, R. M.; STONE, G. R. **1962**, J. Org. Chem. *27*, 284

FRIDKIN, M.; PATCHORNIK, A.; KATCHALSKI, E. **1965**, J. Am. Chem. Soc. *87*, 4646

FRIED, J.; EDWARDS, J. A. **1972**, *Organic Reactions in Steroid Chemistry*, Vols. 1 & 2, Van Nostrand: New York

FRIED, J.; HEIM, S.; ETHEREDGE, S. J.; SUNDER-PLASSMANN, P.; SANTHANAKRISHNAN, T. S.; HIMIZU, J.-I.; LIN, C. H. **1968**, Chem. Commun., 634

FRIED, J.; SZABO, E. F. **1954**, J. Am. Chem. Soc. *76*, 1455

FRIEDRICH, A. **1975**, in: HOUBEN-WEYL, *Methoden der Organischen Chemie, Vol. IV/1b: Oxidation II*, p. 82, Thieme: Stuttgart

FRITSCH, W.; HAEDE, W.; RADSCHEIT, K.; STACHE, U.; RUSCHIG, H. **1974**, Liebigs Ann. Chem. *1974*, 621

FRITSCH, W.; STACHE, U.; HAEDE, W.; RADSCHEIT, K.; RUSCHIG, H. **1969**, Liebigs Ann. Chem. *721*, 168

FROEHLER, B.C. **1986**, Tetr. Lett., 5575; FROEHLER, B.C.; MATTEUCCI, M.; ibid. 469

FUDICKAR, W.; ZIMMERMANN, J.; RUHLMANN, L.; ROEDER, B.; SIGGEL, U.; FUHRHOP, J.-H. **1999**, J. Am. Chem. Soc. *121*, 9539

FUHRER, H.; GANGULY, A.K.; GOPINATH, K.W.; GOVINDACHARI, T.R.; NAGARAJAN, K.; PAI, B.R.; PARTHASARATHY, P.C. **1969**, Tetr. Lett., 133; **1970**, Tetrahedron *26*, 2371

FUHRHOP, J.-H. **1974**, Angew. Chem. *86*, 363 (Int. Ed. Engl. *13*, 321)

FUHRHOP, J.-H.; BACCOUCHE, M. **1976**, Liebigs Ann. Chem. *1976*, 2058

FUHRHOP, J.-H.; BARTSCH, H. **1983**, Liebigs Ann. Chem., 802

FUHRHOP, J.-H.; BARTSCH, H.; FRITSCH, D. **1981**, Angew. Chem. *93*, 797, Internat. Ed. *20*, 804

FUHRHOP, J.-H.; BLUMTRITT, P.; LEHMANN, C.; LUGER, P. **1991**, J. Am. Chem. Soc. *113*, 7437

FUHRHOP, J.-H.; BOETTCHER, C. **1990A**, J. Am. Chem. Soc. *112*, 1768

FUHRHOP, J.-H.; DAVID, H.-H.; MATHIEU, J.; LIMAN, U.; WINTER, H.-J.; BOEKERNA, E. **1986A**, J. Am. Chem. Soc. *108*, 1785

FUHRHOP, J.-H.; DEMOULIN, C.; BOETTCHER, C.; KOENING, J.; SIGGEL, U. **1992**, J. Am. Chem. Soc. *114*, 4159

FUHRHOP, J.-H.; ENDISCH, C. **2000**, *Molecular and Supramolecular Chemistry of Natural Products and Model Compounds*, Marcel Dekker: New York,

FUHRHOP, J.-H.; FRITSCH, D. **1986B**, Acc. Chem. Res. *19*, 130

FUHRHOP, J.-H.; KÖNING, J. **1994**, *Molecular Assemblies and Membranes, Monographs in Supramolecular Chemistry*, Stoddart, J.F. (ed.) Royal Soc. Chem., London

FUHRHOP, J.-H.; MATHIEU, J. **1983**, J. Chem. Soc. Chem. Commun. *1983*, 144

FUHRHOP, J.-H.; SCHNIEDER, P.; BOEKEMA, E.; HELFRICH, W. **1988**, J. Am. Chem. Soc. *110*, 2861

FUHRHOP, J.-H.; SCHNIEDER, P.; ROSENBERG, J.; BOEKEMA, F. **1987**, J. Am. Chem. Soc. *109*, 3387

FUHRHOP, J.-H.; SVENSON, S.; BOETTCHER, C.; RÖSSLER, E.; VIETH, H.-M. **1990B**, J. Am. Chem. Soc. *112*, 4307

FUHRHOP, J.-H.; WITTE, L.; SHELDRICK, W.S. **1976**, Liebigs Ann. Chem. *1976*, 1537

FUJI, K.; NODE, M.; NAGASAWA, H.; NANIWA, Y.; TERADA, S. **1986**, J. Am. Chem. Soc. *108*, 3855

FUJITA, K. **1961**, Bull. Chem. Soc. Jpn *34*, 968

FUJITA, M.; OGURO, D.; MIYAZAWA, M.; OKA, H.; YAMAGUCHI, K.; OGURA, K. **1995**, Nature *378*, 469

FUKUYAMA, T.; LAIRD, A.A.; HOTCHKISS, L.M. **1985**, Tetrahedron Lett. *1985*, 6291

FULTON, D.A.; STODDART, J.F. **2000**, Org. Lett. *2*, 1113

FUNK, R.L.; VOLLHARDT, K.P.C. **1977**, J. Am. Chem. Soc. *99*, 5483

FURLAN, R.L.E.; COUSINS, G.R.L.; SANDERS, J.K.M. **2000**, Chem. Commun., 1761

FURLAN, R.L.E.; NG, Y.-F.; OTTO, S.; SANDERS, J.K.M. **2001**, J. Am. Chem. Soc. *123*, 8876

FÜRSTNER, A. **2000**, Angew. Chem. *112*, 3140

FURUTA, K.; MARUYAMA, T.; YAMAMOTO, H. **1991**, J. Am. Chem. Soc. *113*, 1041; see also Synlett **1991**, *439*, 561

FUSCO, R. **1967**, in: WILEY, R.H. (ed.) *Pyrazoles, Pyrazolines, Pyrazolidines, Indazoles, and Condensed Rings*, p. 3, Interscience: New York

GAIT, M.J.; SHEPPARD, R.C. **1976**, J. Am. Chem. Soc. *98*, 8514

GANAPATHY, S.; SEKHAR, B.B.V.S.; CAIRNS, S.M.; AKUTAGAWA, K.; BENTRUDE, W.G. **1999**, J. Am. Chem. Soc. *121*, 2085

GANEM, B. **1996**, Acc. Chem. Res. *29*, 340

GAO, C.; LAVEY, B.J.; LO, C.-H.L.; DATTA, A.; WENTWORTH, P., JR.; JANDA, K.D. **1998**, J. Am. Chem. Soc. *120*, 2211

GAO, K.; GOROFF, N.S. **2000**, J. Am. Chem. Soc. *122*, 9320

GAO, Y.; HANSON, R.M.; KLUNDER, J.M.; KO, S.Y.; MASAMUNE, H.; SHARPLESS, K.B. **1987**, J. Am. Chem. Soc. *109*, 5765

GARDNER, J.N.; CARLON, F.E.; GNOJ, O. **1968A**, J. Org. Chem. *33*, 3294

GARDNER, J.N.; POPPER, T.L.; CARLON, F.E.; GNOJ, O.; HERZOG, H.L. **1968B**, J. Org. Chem. *33*, 3695

JI, H.; ZHANG, W.; ZHOU, Y.; ZHANG, M.; ZHU, J.; SONG, Y.; LÜ, J.; ZHU, J. **2000**, J. Med. Chem. *43*, 2493

JOHANNSEN, M.; JØRGENSEN, K.A.; HELMCHEN, G. **1998**, J. Am. Chem. Soc. *120*, 7637

JOHANSEN, J.E.; ANGST, C.; KRATKY, C.; ESCHENMOSER, A. **1980**, Angew. Chem. *92*, 141 (Int. Ed. Engl. *19*, 141)

JOHANSSON, R.; SAMUELSSON, B. **1984**, J. Chem. Soc. Chem. Commun. *1984*, 201

JOHNSON, C.R. **1979**, *Sulphur Ylides*, in: BARTON, D.; OLLIS, W.D. (eds.) *Comprehensive Organic Chemistry*, Vol. 3: JONES, D.N. (ed.) p. 247 ff., Pergamon Press: Oxford

JOHNSON, C.R.; HERR, R.W.; WIELAND, P.M. **1973 A**, J. Org. Chem. *38*, 4263

JOHNSON, C.R.; PENNING, T.D. **1988**, J. Am. Chem. Soc. *110*, 4726

JOHNSON, C.R.; SCHROECK, C.W. **1973 B**, J. Am. Chem. Soc. *95*, 7418

JOHNSON, R.A. **1978**, in: TRAHANOVSKY, W.S. (ed.) *Oxidation in Organic Chemistry*, Part C, (*Organic Chemistry*, Vol. 5-C) chapter II, p. 131, Academic Press: New York London

JOHNSON, T.A.; CURTIS, M.D.; BEAK, P. **2001**, J. Am. Chem. Soc. *123*, 1004

JOHNSON, W.S. **1968**, Acc. Chem. Res. *1*, 1

JOHNSON, W.S. **1976**, Angew. Chem. *88*, 33 (Int. Ed. Engl. *15*, 9)

JOHNSON, W.S.; ALLEN, D.S., JR. **1957**, J. Am. Chem. Soc. *79*, 1261, 1995

JOHNSON, W.S.; ALLEN, D.S., JR.; HINDERSINN, R.R.; SAUSEN, G.N.; PAPPO, R. **1962**, J. Am. Chem. Soc. *84*, 2181

JOHNSON, W.S.; COLLINS, J.C., JR.; PAPPO, R.; RUBIN, M.B.; KROPP, P.L.; JOHNS, W.E.; PIKE, J.E.; BARTMANN, W. **1963**, J. Am. Chem. Soc. *85*, 1409

JOHNSON, W.S.; PETERSEN, J.W.; GUTSCHE, C.D. **1945**, J. Am. Chem. Soc. *67*, 2274

JOHNSON, W.S.; PETERSEN, J.W.; GUTSCHE, C.D. **1947**, J. Am. Chem. Soc. *69*, 2942

JOHNSON, W.S.; SZMUSZKOVICZ, S.; ROGIER, E.R.; HADLER, H.L.; WYNBERG, H. **1956**, J. Am. Chem. Soc. *78*, 6285

JONES, G. **1970**, J. Chem. Soc. C, 1230; see also: COLLINGTON, E.W.; JONES, G. **1968**, Chem. Commun., 958, **1969**, J. Chem. Soc. C, 2656

JØRGENSEN, K.A. **2000**, Angew. Chem., Int. Ed. *39*, 3558

JØRGENSEN, K.A.; WHEELER, R.A.; HOFFMANN, R. **1987**, J. Am. Chem. Soc. *109*, 3240

JØRGENSON, M.J. **1970**, Org. React. (NY) *18*, 1

JOSEPH, K.T.; KRISHNA RAO, G.S. **1967**, Tetrahedron *23*, 3215

JULIA, M.; JULIA, S.; TCHEN, S.Y. **1961**, Bull. Soc. Chim. Fr. *1961*, 1849

JULIA, M.; PARIS, J.M. **1974**, Tetr. Lett., 3445

JULIÁN-ORTIZ, J.V. DE; GÁLVEZ, J.; MUOZ-COLLADO, C.; GARCÍA-DOMENECH, R.; GIMENO-CARDONA, C. **1999**, J. Med. Chem. *42*, 3308

JUNG, G., *Combinatorial Chemistry. Synthesis, Analysis, Screening*, **2001**, Wiley-VCH: Weinheim

KAGAN, H.B.; FIAUD, J.C. **1978**, Top. Stereochem. *10*, 175

KAISER, G.; SANDERS, J.K.M. **2000**, Chem. Commun., 1763

KAMETANI, T. **1977 A**, in: APSIMON, J. (ed.) *The Total Synthesis of Natural Products*, Vol. 3, p. 1, Wiley: New York

KAMETANI, T.; IIDA, H.; KIKUCHI, T. **1969**, Chem. Pharm. Bull. *17*, 709

KAMETANI, T.; MATSUMOTO, H.; NEMOTO, H.; FUKOMOTO, K. **1978**, J. Am. Chem. Soc. *100*, 6218

KAMETANI, T.; NEMOTO, H.; ISHIKAWA, H.; SHIROYAMA, K.; MATSUMOTO, H.; FUKUMOTO, K. **1977 B**, J. Am. Chem. Soc. *99*, 3461

KAMETANI, T.; OGASAWARA, K.; TAKAHASHI, T. **1972**, J. Chem. Soc. Chem. Commun. *1972*, 675; and: KAMETANI, T.; OGASAWARA, K.; TAKAHASHI, T. **1973**, Tetrahedron *29*, 72

KAMETANI, T.; TAKAHASHI, T.; OGASAWARA, K.; FUKUMOTO, K. **1974**, Tetrahedron *30*, 1047

KANEKO, C.; SUGIMOTO, A.; TANAKA, S. **1974**, Synthesis *1974*, 876

KARAN, C.; MILLER, B.L. **2001**, J. Am. Chem. Soc. *123*, 7455

KASAI, H.; GOTO, M.; IKEDA, K.; ZAMA, M.; MIZUNO, Y.; TAKEMURA, S.; MATSUURA, S.; SUGIMOTO, T.; GOTO, T. **1976**, Biochemistry *15*, 898

KASATKIN, A.N.; CHECKSFIELD, G.; WHITBY, R.J. **2000**, J. Org. Chem. *65*, 3236

KASATKIN, A.N.; WHITBY, R.J. **1999**, J. Am. Chem. Soc. *121*, 7039

KASHA, M.; RAWLS, H.R.; EL-BAYOUMI, M.A. **1965**, Pure Appl. Chem. *11*, 371

KATAGIRI, N.; ITAKURA, K.; NARANG, S.A. **1975**, J. Am. Chem. Soc. *97*, 7332

KATSUKI, T.; SHARPLESS, K.B. **1980**, J. Am. Chem. Soc. *102*, 5974

KATZ, T.J.; ROTH, R.J. **1972**, J. Am. Chem. Soc. *94*, 4770

KATZ, T.J.; ROTH, R.J.; ACTON, N.; CARNAHAN, E.J. **1999**, J. Org. Chem. *64*, 7663

KATZ, T.J.; WANG, E.J.; ACTON, N. **1971**, J. Am. Chem. Soc. *93*, 3782

KATZENELLENBOGEN, J.A.; BOWLUS, S.B. **1973**, J. Org. Chem. *38*, 627

KAUFMANN, D.; DEMEIJERE, A. **1979**, Tetr. Lett., 779, 783, 787

KAZUTA, Y.; MATSUDA, A.; SHUTO, S. **2002**, J. Org. Chem. *67*, 1669

KEATING, T.A.; ARMSTRONG, R.W. **1996**, J. Am. Chem. Soc. *118*, 2574

KECK, G.E.; ABBOTT, D.E. **1984 C**, Tetr. Lett., 1883

KECK, G.E.; ABBOTT, D.E.; WILEY, M.R. **1987**, Tetr. Lett., 139

KECK, G.E.; BODEN, E.P. **1984 A**, Tetr. Lett., 265; **1984 B**, ibid. 1879

KELLY, J.; PUAR, M.S.; AFONSO, A.; MCPHAIL, A.T. **1998**, J. Org. Chem. *63*, 6039

KEMP, D.S.; PETRAKIS, K.S. **1981**, J. Org. Chem. *46*, 5140

KENNER, G.W.; MCCOMBIE, S.W.; SMITH, K.M. **1973 A**, Liebigs Ann. Chem. *1973*, 1329

KENNER, G.W.; SMITH, K.M.; UNSWORTH, J.F. **1973 B**, J. Chem. Soc. Chem. Commun. *1973*, 43

KENT, S.B.H. **1988**, Annu. Rev. Biochem. *57*, 957

KHAN, M.M.T. **1974**, *Homogeneous Catalysis by Metal Complexes*, Vol. II, *Activation of Alkenes und Alkynes*, Academic Press: New York London

KHORANA, H.G.; JACOB, T.M.; MOON, M.W.; NARANG, S.A.; OHTSUKA, E. **1965**, J. Am. Chem. Soc. *87*, 2954–2995

KIESLICH, K. **1976**, *Microbial Transformations of Non-Steroid Cyclic Compounds*, Thieme: Stuttgart

KIM, B.M.; SHARPLESS, K.B. **1990**, Tetrahedron Lett. *1990*, 3003

KIM, H.-J.; BAMPOS, N.; SANDERS, J.K.M. **1999**, J. Am. Chem. Soc. *121*, 8120

KIM, K.; OKAMOTO, S.; SATO, F. **2001**, Org. Lett. *3*, 67

KIM, N.-S.; CHOI, J.-R.; CHE, J.K. **1993**, J. Org. Chem. *58*, 7096

KIMEL, W.; SAX, N.W.; KAISER, S.; EICHMANN, G.G.; CHASE, G.O.; OFNER, A. **1958**, J. Org. Chem. *23*, 153

KIMEL, W.; SURMATIS, J.D.; WEBER, J.; CHASE, G.O.; SAX, N.W.; OFNER, A. **1957**, J. Org. Chem. *22*, 1611; see also: ROYALS, E.E. **1946**, Ind. Eng. Chem. *38*, 546

KIMURA, T.; VASSILEV, V.P.; SHEN, G.-J.; WONG, C.-H. **1997**, J. Am. Chem. Soc. *119*, 11734

KIRSTEN, C.N.; HERM, M.; SCHRADER, T.H. **1997**, J. Org. Chem. *62*, 6882

KITAHARA, Y.; YOSHIKOSHI, A.; OIDA, S. **1964**, Tetrahedron Lett. *1964*, 1763

KITATANI, K.; HIYAMA, T.; NOZAKI, H. **1976**, J. Am. Chem. Soc. *98*, 2362

KLASS, D.H. **1998**, *Fossil fuel reserves and depletion. Biomass for Renewable Energy, fuels and Chemicals*; Academic Press: San Diego, pp 10–19

KLESCHICK, W.A.; BUSE, C.T.; HEATHCOCK, C.H. **1977**, J. Am. Chem. Soc. *99*, 247

KLIOZE, S.S.; DARMORY, F.P. **1975**, J. Org. Chem. *40*, 1588

KNÖLL, W.; TANIM, C. **1975**, Helv. Chim. Acta *58*, 1162

KNORR, H.; RIED, W. **1978**, Synthesis *1978*, 649

KO, S.Y.; LEE, A.W.M., MASAMUNE, S.; REED, III, L.A.; SHARPLESS, K.B.; WALKER, F.J. **1983**, Science *220*, 949

KOBAYASHI, S.; NAGAYAMA, S. **1996**, J. Am. Chem. Soc. *118*, 8977

KÖBRICH, K.; WERNER, W. **1969**, Tetr. Lett., 2181

KOCIENSKI, P. **1985**, Phosphorus Sulfur *24*, 97 [*Coll. Vol. 23–25*, 477]

KOERT, U.; HARDING, M.M.; LEHN, J.-M. **1990**, Nature (London) *346*, 339

KOIDE, K.; FINKELSTEIN, J.M.; BALL, Z.; VERDINE, G.L. **2001**, J. Am. Chem. Soc. *123*, 398

KOLB, H.C.; VANNIEUWENHZE, M.S.; SHARPLESS, K.B. **1994**, Chem. Rev. *94*, 2483

KOMATSU, T.; HAMAMATSU, K.; TSUCHIDA, E. **1999**, Macromolecules *32*, 8388

KONDO, T.; OKADA, T.; MITSUDO, T. **2002**, J. Am. Chem. Soc. *124*, 186

KÖNIG, B.; PITSCH, W.; KLEIN, M.; VASOLD, R.; PRALL, M.; SCHREINER, P.R. **2001**, J. Org. Chem. *66*, 1742

KÖNIG, W.; GEIGER, R. **1970**, Chem. Ber. *103*, 788

KOSOWER, E.M.; WINSTEIN, S. **1956**, J. Am. Chem. Soc. *78*, 4347, 4354

KÖSTER, H. **1979**, Nachr. Chem. Tech. Lab. *27*, 694

KÖSTER, R.; ARORA, S.; BINGER, P. **1971**, Synthesis 322

KOZIKOWSKI, A.P.; WETTER, H.F. **1976**, Synthesis 561

KRAMER, U.; GUGGISBERG, A.; HESSE, M.; SCHMID, H. **1978**, Angew. Chem. *90*, 210 (Int. Ed. Engl. *17*, 200)

KRAMER, U.; SCHMID, H.; GUGGISBERG, A.; HESSE, M. **1979**, Helv. Chim. Acta *62*, 811

KRÄTSCHMER, W.; FOSTIROPOULOS, K.; HUFFMAN, D.R. **1990 B**, Chem. Phys. Lett. *170*, 167

KRÄTSCHMER, W.; LAMB, L.D.; FOSTIROPOULOS, K; HUFFMAN, D.R. **1990 A**, Nature (London) *347*, 354

KRAUS, G.A.; ANDERSH, B. **1991**, Tetr. Lett., 2189

KRISHNAMURTHY, S.; BROWN, H.C. **1976**, J. Am. Chem. Soc. *98*, 3383

KROPF, H. **1980**, in: HOUBEN-WEYL, *Methoden der Organischen Chemie*, Vol. IV/1c: *Reduktion I*, Thieme: Stuttgart – New York

KUDIS, S.; HELMCHEN, G. **1998**, Angew. Chem. *110*, 3210

KUIVILA, H.G. **1968**, Acc. Chem. Res. *1*, 299

KUMAR, A. **2001**, Chem. Rev. *101*, 1

KUNITAKE, T.; OKAHATA, Y. **1977**, J. Am. Chem. Soc. *99*, 3860

KUO, C.H.; TAUB, D.; WENDLER, N.L. **1968**, J. Org. Chem. *33*, 3126

KUROSU, M.; LORCA, M. **2001**, J. Org. Chem. *66*, 1205

KUTNEY, J.P. **1977**, in: APSIMON, J. (ed.) *The Total Synthesis of Natural Products*, Vol. 3, p. 273, Wiley: New York

KUWAJIMA, L.; NAKAMURA, E. **1975**, J. Am. Chem. Soc. *97*, 3257

KUWAJIMA, L.; SATO, T.; ARAI, M.; MINAMI, N. **1976**, Tetrahedron Lett. *1976*, 1817

KWONG, H.-L.; SORATO, C.; OGINO, Y.; CHEN, H.; SHARPLESS, K.B. **1990**, Tetrahedron Lett. *1990*, 2999

LA, D.S.; SATTELY, E.S.; FORD, J.G.; SCHROCK, R.R.; HOVEYDA, A.H. **2001**, J. Am. Chem. Soc. *123*, 7767

LABADIE, J.W.; STILLE, J.K. **1983**, J. Am. Chem. Soc. *105*, 6129

LACOUR, T.G.; GUO, C.; BHANDARU, S.; BOYD, M.R.; FUCHS, P.L. **1998**, J. Am. Chem. Soc. *120*, 692

LAI, S.; ZERCHER, C.K.; JASINSKI, J.P.; REID, S.N.; STAPLES, R.J. **2001**, Org. Lett. *3*, 4169

LANDO, J.B.; MANN, J.A., JR. **1990**, Langmuir *6*, 293

LANDOR, S.R.; PUNJA, N. **1967**, J. Chem. Soc. C, *1967*, 2495

LANGECKER, H.; SCHEIFFELE, E.; GEIGER, R.; PREZEWOWSKY, K.; STACHE, U.; SCHMITT, K. **1977**, in: *Ullmanns Encyklopädie der technischen Chemie*, 4th edn., Vol. 13, p. 1, Verlag Chemie: Weinheim, Ger.

LARHED, M.; MOBERG, C.; HALLBERG, A. **2002**, Acc. Chem. Res. *35*, 717

LATIMER, W.M. **1952**, *Oxidation Potentials*, Prentice-Hall: Englewood Cliffs, NJ

LAUTENS, M.; ROVIS, T. **1997**, J. Org. Chem. *62*, 5246

LEBRETON, L.; ANNAT, J.; DERREPAS, P.; DUTARTE, P.; RENAUT, P. **1999**, J. Med. Chem. *42*, 277

LEBRETON, L.; JOST, E.; CARBONI, B.; ANNAT, J.; VAULTIER, M.; DUTARTE, P.; RENAUT, P. **1999**, J. Med. Chem. *42*, 4749

LEDNICER, D.; MITSCHER, L.A. **1977**, **1980**, The Organic Chemistry of Drug Synthesis, Vols. 1, Wiley: New York London

LEE, A.W.M.; MARTIN, V.S.; MASAMUNE, S.; SHARPLESS, K.B.; WALKER, F.J. **1982**, J. Am. Chem. Soc. *104*, 3515

LEE, D.G. **1969**, in: AUGUSTINE, R.L. (ed.) *Oxidation, Vol. 1*, chapters 1, 2, M. Dekker: New York

LEE, D.G.; VAN DEN ENGH, M. **1973**, in: TRAHANOVSKY, W.S. (ed.) *Oxidation in Organic Chemistry*, Part B, Academic Press: New York London

LEE, D.; SELLO, J.K.; SCHREIBER, S.L. **2000**, Org. Lett. *2*, 709

LEE, E.; SHIN, I.-J.; KIM, T.S. **1990**, J. Am. Chem. Soc. *112*, 260

LEE, S.Y.; LEE, C.-W.; OH, D.Y. **1999**, J. Org. Chem. *64*, 7017

LEHMANN, H. **1975**, in: HOUBEN-WEYL, *Methoden der Organischen Chemie*, Vol. IV/1 b, *Oxidation II*, p. 905, Thieme: Stuttgart

LEHMANN, J. **1976**, Chemie der Kohlenhydrate, Thieme: Stuttgart

LEHMANN, J. **1996**, *Kohlenhydrate. Chemie und Biologie*, Wiley-VCH: Weinheim

LEHN, J.-M.; RIGAULT, A.; SIEGEL, J.; HAR-ROWFIELD, J.; CHEVRIER, B.; MORAS, D. **1987**, Proc. Natl. Acad. Sci. USA *84*, 2565

LEHN, L.-M. **1988**, Angew. Chem. *100*, 92 (Int. Ed. Engl. *27*, 89)

LEHN, L.-M. **1990**, Angew. Chem. *102*, 1347 (Int. Ed. Engl. *29*, 1304)

LEI, A.; LIU, G.; LU, X. **2002**, J. Org. Chem. *67*, 974

LEMAL, D. M.; LOKENSGARD, J. P. **1966**, J. Am. Chem. Soc. *88*, 5934

LEMICUX, R. U.; NAGABHUSHAN, T. L.; O'NEILL, I. K. **1968**, Canad. L. Chem. *46*, 413

LEMIEUX, R. U.; GUNNER, S. W.; NAGABHU-SHAN, T. L. **1965**, Tetrahedron Lett. *1965*, 2143, 2149

LEPAGE, O.; STONE, C.; DESLONGCHAMPS, P. **2002**, Org. Lett. *4*, 1091

LETSINGER, R. L.; LUNSFORD, W. B. **1976**, J. Am. Chem. Soc. *98*, 3655

LETSINGER, R. L.; SINGMAN, C. N.; HISTAND, G.; SALUNKHE, M. **1988**, J. Am. Chem. Soc. *110*, 4470

LEVIN, M. D.; KASZYNSKI, P.; MICHL, J. **2000**, Chem. Rev. *100*, 169

LEVIN, R. H. **1978**, **Arynes**, in: JONES, M.; MOSS, R. A. (eds.) *Reactive Intermediates*, Vol. 1, Part 1, Wiley: New York

LEVITT, M.; WARSHEL, A. **1978**, J. Am. Chem. Soc. *100*, 2607

LEVY, D. E.; TANG, C. **1995**, *The Chemistry of C-Glycosides*, Vol. 13, 1st edn; Elsevier Science, Oxford

LEWIS, S. N. **1969**, in: AUGUSTINE, R. L. (ed.) *Oxidation*, Vol. 1, chapter 5 (p. 213), M. Dekker: New York

LI, G., SHARPLESS, K. B. **1996**, Angew. Chem. *108*, 449; Angew. Chem. Int. Ed. Engl. *35*, 451

LI, G.; FUDICKAR, W.; SKUPIN, M.; KLYSZCZ, A.; DRAEGER, C.; LAUER, M.; FUHRHOP, J.-H. **2002**, Angew. Chem. *114*, 1906, Int. Ed. *41*, 1828

LICHTENTHALER, F. W. **1991**, *Carbohydrates as Organic Raw Materials*; VCH: Weinheim

LICHTENTHALER, F. W. **2002**, Acc. Chem. Res. *35*, 728

LICHTENTHALER, F. W.; MONDEL, S. **1997**, Pure Appl. Chem. *69*, 1853

LIEBMAN, J. F.; GREENBERG, A. **1976**, Chem. Rev. *76*, 311

LIN, M.; SHAPIRO, M. J.; WAREING, J. R. **1997**, J. Am. Chem. Soc. *119*, 5249

LINDLAR, H.; DUBUIS, R. **1973**, Org. Synth. Coll. Vol. *V*, 880

LINDSTRÖM, U. M. **2002**, Chem. Rev. *102*, 2751

LITTLE, W. F.; REILLEY, C. N.; JOHNSON, J. D.; LYNN, K. N.; SANDERS, A. P. **1964**, J. Am. Chem. Soc. *86*, 1376

LIU, X.; COOK, J. M. **2001**, Org. Lett. *3*, 4023

LIU, Z.; SHEN, Z.; ZHU, T.; HOU, S.; YING, L.; SHI, Z.; GU, Z. **2000**, Langmuir *16*, 3569

LOEWENTHAL, H. J. E. **1959**, Tetrahedron 6, 269

LOGUE, M. W.; MOORE, G. L. **1975**, J. Org. Chem. *40*, 131

LOHMAR, R.; STEGLICH, W. **1980**, Chem. Ber. *113*, 3706

LONG, J.; HU, J.; SHEN, X.; JI, B.; DING, K. **2002**, J. Am. Chem. Soc. *124*, 10

LOSSE, G.; ZEIDLER, D.; GRIESHABER, T. **1968**, Liebigs Ann. Chem. *715*, 196

LOURENS, G. J.; KOEKEMOER, J. M. **1975**, Tetrahedron Lett. *1975*, 3719

LOVELY, C. J.; DU, H.; DIAS, H. V. R. **2001**, Org. Lett. *3*, 1319

LOWN, J. W.; KROWICKI, K. **1985**, J. Org. Chem. *50*, 3774

LÜBKE, K.; SCHRÖDER, E.; KLOSS, G. **1975**, Chemie und Biochemie der Aminosäuren, Peptide und Proteine, Vol. 1, Thieme: Stuttgart

LUCHE, J.-L.; RODRIGUEZ-HAHN, L.; CRABBÉ, P. **1978**, J. Chem. Soc. Chem. Commun. *1978*, 601

LUTZ, G.; HUNKLER, D.; RIHS, G.; PRINZ-BACH, H. **1989**, Angew. Chem. *101*, 307 (Int. Ed. Engl. *28*, 298)

MA, D.; YANG, J. **2001**, J. Am. Chem. Soc. *123*, 9706

MAAHS, G.; HEGENBERG, P. **1966**, Angew. Chem. *78*, 927 (Int. Ed. Engl. *5*, 888)

MACIEJEWSKI, M. **1982**, J. Macromol. Sci., Chem. *17*, 689

MAERCKER, A. **1965**, Org. React. (NY) *14*, 270

MAGNUSSON, G. **1977**, Tetr. Lett., 2713

MAHRWALD, R. **1999**, Chem. Rev. *99*, 1095

MAIER, G.; ALZÉRECA, A. **1973**, Angew. Chem. *85*, 1056 (Int. Ed. Engl. *12*, 1015)

MAIER, G.; PFRIEM, S.; SCHÄFER, U.; MA-TUSCH, R. **1978**, Angew. Chem. *90*, 552 (Int. Ed. Engl. *17*, 520)

MAJERSKI, Z; SCHLEYER, P. V. R. **1968**, Tetr. Lett., 6195

MAK, C. C.; BAMPOS, N.; DARLING, S. L.; MONTALTI, M.; PRODI, L.; SANDERS, J. K. M. **2001**, J. Org. Chem. *66*, 4476

MAKI, Y.; KIKUCHI, K., SUGIYAMA, H.; SETO, S. **1977**, Tetr. Lett., 263

MALECZKA, R. E., JR.; GALLAGHER, W. P. **2001**, Org. Lett. *3*, 4173

MÁLEK, J.; ČERNÝ, M. **1972**, Synthesis 217

MALPASS, J. R.; TWEDDIE, N. J. **1977**, J. Chem. Soc. Perkin Trans. *1*, 874

MANEGOLD, D. **1975**, in: HOUBEN-WEYL, *Methoden der Organischen Chemie*, Vol. IV/1 b, *Oxidation II*, p. 76, Thieme: Stuttgart

MANGONI, L.; ADOLFINI, M.; BARONE, G.; PARILLI, M. **1973**, Tetr. Lett., 4485

MANN, G.; HARTWIG, J. F.; DRIVER, M. S.; FERNÁNDEZ-RIVAS, C. **1998**, J. Am. Chem. Soc. *120*, 827

MANN, L. A.; TJATJOPOULOS, G. J.; AZZAM, M.-O. J.; BOGGS, K. E.; ROBINSON, K. M.; SANDERS, J. N. **1987**, Thin Solid Films *152*, 29

MAREG, E.; ROZEK, J. **1961**, Coll. Czech. Chem. Commun. *26*, 2370

MARINO, J. P.; LANDICK, R. C. **1975**, Tetr. Lett., 4531

MARSHALL, J. A.; BUNDY, G. L. **1966**, J. Am. Chem. Soc. *88*, 4291

MARSHALL, J. A.; BUNDY, G. L.; FANTA, W. I. **1968**, J. Org. Chem. *33*, 3913

MARSHALL, J. A.; FANTA, W. I. **1964**, J. Org. Chem. *29*, 2501

MARSHALL, J. A.; ROEBKE, H. **1969**, J. Org. Chem. *34*, 4188

MARTIN, D. **1971**, in: MARTIN, D.; HAUTHAL, H. G. (eds.) *Dimethylsulfoxide,* chapter 8, p. 306, Akademie Verlag: Berlin

MARTIN, J. C.; ARHART, R. J. **1971**, J. Am. Chem. Soc. *93*, 2339, 2341

MARTIN, J. C.; ARHART, R. J.; FRANZ, J. A.; PEROZZI, E. F.; KAPLAN, L. J. **1977**, Org. Synth. *57*, 22

MARTIN, S. E.; PESAI, S. R.; PHILLIPS, G. W.; MILLER, A. C. **1980**, J. Am. Chem. Soc. *102*, 3294

MARTIN, V. S.; WOODARD, S. S.; KATSUKI, T.; YAMADA, Y.; IKEDA, M.; SHARPLESS, K. B. **1981**, J. Am. Chem. Soc. *103*, 6237

MARUYAMA, K.; NAGATA, T.; ONO, N.; OSUKA, A. **1989**, Bull. Chem. Soc. Jpn. *62*, 3167

MARX, J. X.; ARGYLE, L. C.; NORMAN, L. R. **1974**, J. Am. Chem. Soc. *96*, 2121

MARX, M. A., GRILLOT, A.-L.; LOUER, C. T.; BEAVER, K. A.; BARTLETT, P. A. **1997**, J. Am. Chem. Soc. *119*, 6153

MASAMUNE, S. **1961**, J. Am. Chem. Soc. *83*, 1009

MASAMUNE, S. **1964**, J. Am. Chem. Soc. *86*, 288, 290

MASAMUNE, S.; CHOY, W.; KERDESKY, F. A. L.; IMPERIALI, B. **1981 A**, J. Am. Chem. Soc. *103*, 1566

MASAMUNE, S.; CUTS, H.; HOGBEN, M. G. **1966**, Tetr. Lett., 1017

MASAMUNE, S.; HIRAMA, M.; MORI, S.; ALI, SK. A.; GARVEY, D. S. **1981 B**, J. Am. Chem. Soc. *103*, 1568

MASAMUNE, S.; SATO, T.; KIM, B. M.; WOLLMANN, T. A. **1986**, J. Am. Chem. Soc. *108*, 8279

MASAMUNE, S.; SEIDNER, R. T. **1969**, J. Chem. Soc. D *1969*, 542

MASEREEL, B.; WOUTERS, J.; POCHET, L.; LAMBERT, D. **1998**, J. Med. Chem. *41*, 3239

MATHIAS, L. J. **1979**, Synthesis 561

MATSUMOTO, T.; HOSOYA, T.; SUZUKI, K. **1992**, J. Am. Chem. Soc. *114*, 3568

MATSUYAMA, T.; YAMAOKA, H. **1990**, Langmuir *6*, 291

MATTEUCCI, M. D.; CARUTHERS, M. H. **1980**, Tetr. Lett., 719

MATTEUCCI, M. D.; CARUTHERS, M. H. **1981**, J. Am. Chem. Soc. *103*, 3185

MATTHEWS, R. S.; METEYER, T. E. **1971**, Synthesis *1971*, 1576

MAUZERALL, D. **1960**, J. Am. Chem. Soc. *82*, 2605

MAXAM, A. M.; GILBERT, W. **1980**, Methods Enzymol. *65* [Nucleic Acids, Part 1], 499

MAZUR, Y.; DANIELI, N.; SONDHEIMER, F. **1960**, J. Am. Chem. Soc. *82*, 5889

McCLOSKEY, C. M. **1957**, Adv. Carbohydr. Chem. *12*, 137

McCORMICK, J. P.; BARTON, D. L. **1975**, Chem. Commun., 303

McDONALD, P. D.; HAMILTON, G. A. **1973**, in: TRAHANOVSKY, W. S. (ed.) *Oxidation in Organic Chemistry,* Part B, p. 97, Academic Press: New York London

McELVAIN, S. M. **1948**, Org. React. (NY) *4*, 256

McKILLOP, A.; OLDENZIEL, O. H.; SWANN, B. P.; TAYLOR, E. C.; ROBEY, R. L. **1973**, J. Am. Chem. Soc. *95*, 1296

McMurry, J. E.; Fleming, M. P. **1974**, J. Am. Chem. Soc. *96*, 4708

McMurry, J. E.; Fleming, M. P. **1976 A**, J. Org. Chem. *41*, 896

McMurry, J. E.; Krepski, L. R. **1976 B**, J. Org. Chem. *41*, 3929

McMurry, L. E. **1974**, Acc. Chem. Res. *7*, 281

McOmie, J. F. W. **1973**, *Protective Groups in Organic Chemistry*, Plenum: New York

Meerwein, H.; Bodenbrenner, K.; Borner, P.; Kunert, F. **1960**, Liebigs Ann. Chem. *632*, 38

Meinwald, L.; Frauenglass, E. **1960**, J. Am. Chem. Soc. *82*, 5235

Melder, J.-P.; Fritz, H.; Prinzbach, H. **1989**, Angew. Chem. *101*, 309 (Int. Ed. Engl. *28*, 300)

Melder, J.-P.; Pinkos, R.; Fritz, H.; Prinzbach, H. **1989**, Angew. Chem. *101*, 314 (Int. Ed. Engl. *28*, 305)

Melson, G. A. (ed.) **1979**, *Coordination Chemistry of Macrocyclic Compounds*, Plenum: New York

Melson, G. A. (ed.) Coordination Chemistry of Macrocyclic Compounds, **1979**, Plenum Press, New York

Menapace, L. W. **1964**, J. Am. Chem. Soc. *86*, 3047

Mercier, C.; Soucy, P.; Rosen, W.; Deslongchamps, P. **1973**, Synth. Commun. *3*, 161

Merrifield, R. B. **1969**, in: Nord, F. F. (ed.) *Advances in Enzymology*, Interscience: New York

Messmer, R. P.; Schultz, P. A. **1986**, J. Am. Chem. Soc. *108*, 7407

Meyer zu Reckendorf, W.; Wassiliadou-Micheli, N.; Bischof, E. **1971**, Chem. Ber. *104*, 1

Meyer, W. L.; Biesalski, B. S. **1963**, J. Org. Chem. *28*, 2896

Meyers, A. I.; Seefeld, M. A.; Lefker, B. A.; Blake, J. F.; Williard, P. G. **1998**, J. Am. Chem. Soc. *120*, 7429

Meyers, A. L.; Knauss, G.; Kamata, K.; Ford, M. E. **1976**, J. Am. Chem. Soc. *98*, 567

Michellys, P.-Y.; Maurin, P.; Toupet, L.; Pellissier, H.; Santelle, M. **2001**, J. Org. Chem. *66*, 115

Michelson, A. M. **1963**, *The Chemistry of Nucleosides and Nucleotides*, Academic Press: New York London

Miller, C. E. **1965**, J. Chem. Educ. *42*, 254

Miller, L. L.; Stermitz, F. R.; Falck, J. R. **1971**, **1973**, J. Am. Chem. Soc. *93*, 5941; *95*, 2651

Miller, M. J.; Mattingly, P. G. **1983**, Tetrahedron *39*, 2563

Mills, L. S.; North, P. C. **1983**, Tetr. Lett., 409

Milton, R. C. de L.; Milton, S. C. E.; Adams, P. A. **1990**, J. Am. Chem. Soc. *112*, 6039

Mitchell, R. H.; Boekelheide, V. **1974**, J. Am. Chem. Soc. *96*, 1547, 1558

Mitsunobu, O.; Wada, M.; Sano, T. **1972**, J. Am. Chem. Soc. *94*, 679

Miyake, F. Y.; Ykushijin, K.; Horne, D. A. **2002**, Org. Lett. *4*, 941

Miyashita, A.; Takaya, H.; Souchi, T.; Noyori, R. **1984**, Tetrahedron, *40*, 1245

Miyashita, A.; Yasuda, A.; Takaya, H.; Toriumi, K.; Ito, T.; Souchi, T.; Noyori, R. **1980**, J. Am. Chem. Soc. *102*, 7932

Miyashita, M.; Yanami, T.; Kumazawa, T.; Yoshikoshi, A. **1984**, J. Am. Chem. Soc. *106*, 2149

Miyaura, N.; Itoh, M.; Sasaki, M. **1975**, Synthesis *1975*, 317

Miyaura, N.; Itoh, M.; Suzuki, A. **1976**, Tetrahedron Lett. *1976*, 255

Miyaura, N.; Suginome, H.; Suzuki, A. **1983**, Tetrahedron *39*, 3271

Moffatt, J. G. **1971**, in: Augustine, R. L.; Trecker, D. L. (eds.) *Oxidation*, Vol. 2, chapter 1, p. 1, M. Dekker: New York

Molander, G. A. **1992**, Chem. Rev. *92*, 29

Molander, G. A. **1998**, Acc. Chem. Res. *31*, 603

Molander, G. A.; Harris, C. R. **1996**, Chem. Rev. *96*, 307

Montanari, F. **1975**, in: Stirling, C. J. M. (ed.) *Organic Sulphur Chemistry*, Butterworths: London

Mori, K. **1975**, Tetrahedron *31*, 3011

Mori, K.; Oda, M.; Matsui, M. **1976**, Tetr. Lett., 3173

Mori, T.; Nakahara, T.; Nozaki, H. **1969**, Can. J. Chem. *47*, 3266

Morin, R. B.; Jackson, B. G.; Flynn, E. H.; Roeske, R. W. **1962**, J. Am. Chem. Soc. *84*, 3400

Morrison, J. D.; Mosher, H. S. **1971**, *Asymmetric Organic Reactions*, Prentice-Hall: Englewood Cliffs, NJ

Mousseron, M.; Jacquier, R.; Christol, H. **1957**, Bull. Soc. Chim. Fr. *1957*, 346

MUKAIYAMA, T.; BANNO, K.; NARASAKA, K. **1974**, J. Am. Chem. Soc. *96*, 7503

MUKAIYAMA, T.; MURAI, Y.; SHODA, S. **1981**, Chem. Lett. *1981*, 431

MUKAIYAMA, T.; SATO, T.; HANNA, J. **1973**, Chem. Lett. *1973*, 1041

MULLIGAN, R.C.; BERG, P. **1980**, Science *209*, 1422

MULLIS, K.B. **1990**, Sci. Am. *2624*, 56, 64

MUNDY, B.F. **1972**, L. Chem. Educ. *49*, 91

MURAKATA, C.; OGAWA, T. **1990**, Tetrahedron, Lett. 2439

MUXFELDT, H.; HAAS, G.; HARDTMANN, G.; KATHAWALA, E.; MOOBERRY, L.B.; VEDEJS, E. **1979**, J. Am. Chem. Soc. *101*, 689

NACE, H.R. **1962**, Org. React. (NY) *12*, 57

NADIN, A. **1998**, J. Chem. Soc., Perkin Trans. *1*, 3493

NAGARKATTI, J.P.; ASHLEY, K.R. **1973**, Tetr. Lett., 4599

NAGATA, T.; OSUKA, A.; MARUYAMA, K. **1990**, J. Am. Chem. Soc. *112*, 3054

NAGATA, W.; HIRAI, S.; OKUMURA, T.; KAWATA, K. **1968**, J. Am. Chem. Soc. *90*, 1650

NAGATA, W.; SUGASAWA, T.; NARISADA, M.; WABABAYASHI, T.; HAGASE, Y. **1967**, J. Am. Chem. Soc. *89*, 1483

NAHM, S.; WEINREB, S.M. **1981**, Tetr. Lett. *22*, 3815

NAKAGAWA, H.; NAGANO, T.; HIGUCHI, T. **2001**, Org. Lett. *3*, 1805

NAKAMURA, E.; KUWAJIMA, I. **1977**, J. Am. Chem. Soc. *99*, 7360

NAKATSUBO, F.; KISHI, Y.; GOTO, T. **1970**, Tetr. Lett., 381

NAKATSUKA, M.; RAGAN, L.A.; SAMMAKIA, T.; SMITH, D.B.; UEHLING, D.E.; SCHREIBER, S.L. **1990**, J. Am. Chem. Soc. *112*, 5583

NAKAYA, T.; YAMADA, M.; SHIBATA, K.; IMOTO, M.; TSUCHIYA, H.; OKUNO, M.; NAKAYA, S.; OHNO, S.; NAKAZAKI, M.; NAOMURA, K. **1966**, Tetrahedron Lett. 2615

NARANG, S.A.; WIGHTMAN, R.H. **1973**, in: APSIMON, J. (ed.) *The Total Synthesis of Natural Products*, Vol. 1, p. 279, Wiley: New York London

NARUTA, Y.; USHIDA, S.; MARUYAMA, K. **1979**, Chem. Lett., 919

NAZAROV, I.N.; TORGOV, I.V.; VERKHOLETOVA, G. **1957**, Doki. Akad. Nauk, SSSR *112*, 1067

NEDELEC, L.; GASC, T.; BUCOURT, R. **1974**, Tetrahedron *30*, 3263

NEFKENS, G.H.L. **1960**, Nature (London) *185*, 309

NEFKENS, G.H.L.; TESSER, G.L.; NIVARD, K.J.F. **1960**, Recl. Trav. Chim. Pays-Bas *79*, 688

NEGISHI, E.-I.; TAKAHASHI, T. **1994**, Acc. Chem. Res. *27*, 124

NEGISHI, E.; YOSHIDA, T. **1973**, J. Chem. Soc. Chem. Commun. *1973*, 606

NELSON, O.; CROUCH, R.D. **1996**, Synthesis, 1031

NERDEL, E.; FRANK, D.; LENGERT, H.-J.; WEYERSTAHL, P. **1968**, Chem. Ber. *101*, 1850

NEUMANN, H.; JACOBI VON WANGELIN, A.; GÖRDES, D.; SPANNENBERG, A.; BELLER, M. **2001**, J. Am. Chem. Soc. *123*, 8398

NEUVILLE, L.; BIGOT, A.; DAU, M.E.T.H.; ZHU, J. **1999**, J. Org. Chem. *64*, 7638

NEWHALL, W.F. **1958**, J. Org. Chem. *23*, 1274

NEWKOME, G.R.; YAO, Z.; BAKER, G.R.; GUPTA, V.K.; RUSSE, P.S.; SAUNDERS, M.J. **1986**, J. Am. Chem. Soc. 108, 849

NEWMAN, M.S.; MEKLER, A.B. **1960**, J. Am. Chem. Soc. 82, 4039

NEWTON, R.E.; REYNOLDS, D.P.; FINCH, M.A.W.; KELLY, D.R.; ROBERTS, S.M., **1979**, Tetr. Lett., 3981

NICOLAOU, K.C. **1994**, Angew. Chem., Int. Ed. Engl. *33*, 15 and references cited therein

NICOLAOU, K.C. **1977**, Tetrahedron *33*, 683

NICOLAOU, K.C.; CAULFIELD, T.J.; KATAOKA, H.; STYLIANIDES, N.A. **1990 B**, J. Am. Chem. Soc. *112*, 3693

NICOLAOU, K.C.; LADDUWAHETTY, T.; RANDALL, J.L.; CHUCHOLOWSKI, A. **1986**, J. Am. Chem. Soc. *108*, 2466

NICOLAOU, K.C.; MCGARRY, D.G.; SOMERS, P.K.; KIM, B.H.; OGILVIE, W.W.; YIANNIKOUROS, G.; PRASAD, C.V.C.; VEALE, C.A.; HARK, R.R. **1990 A**, J. Am. Chem. Soc. *112*, 6263

NICOLAOU, K.C.; NAMOTO, K.; RITZÉN, A.; ULVEN, T.; SHOJI, M.; LI, J.; D'AMICO, G.; LIOTTA, D.; FRENCH, C.T.; WARTMANN, M.; ALTMANN, K.-H.; GIANNAKAKOU, P. **2001**, J. Am. Chem. Soc. *123*, 9313

NICOLAOU, K.C.; PASTOR, J.; WINSSINGER, N.; MURPHY, F. **1998**, J. Am. Chem. Soc. *120*, 5132

NICOLAOU, K.C.; PFEFFERKORN, A.; MITCHELL, H.J.; ROECKER, A.J.; BARLUENGA, S.;

Cao, G.-Q.; Affleck, R. L.; Lillig, J. E. **2000**, J. Am. Chem. Soc. *122*, 9954

Nicolaou, K. C., Scott, A. S., Montagnon, T., Vassilikogiannakis, G. E. **2002**, Angew. Chem. *114*, 1742

Nicolaou, K. C.; Sørensen, E. J. **1996**, *Classics in Total Synthesis, Targets, Strategies, Methods*, VCH: Weinheim

Nicolaou, K. C.; Winssinger, N.; Vourloumis, D.; Oshima, T.; Kim, S.; Pfefferkorn, J.; Xu, J.-Y.; Li, T. **1998**, J. Am. Chem. Soc. *120*, 10814

Nielsen, A. T.; Houlihan, W. J. **1968**, Org. React. (NY) *16*, 1

Nigh, W. G. **1973**, in: Trahanovsky, W. S. (ed.) *Oxidation in Organic Chemistry*, Part B, p. *35*, Academic Press: New York London

Nilsson, J. W.; Thorstensson, F.; Kvarnström, I.; Oprea, T.; Samuelsson, B.; Nilsson, I. **2001**, J. Comb. Chem. *3*, 546

Ninomiya, L.; Naito, T. **1973**, J. Chem. Soc. Chem. Commun. 137

Nishimura, J.; Kawabata, N.; Furukawa, J. **1969**, Tetrahedron *25*, 2647

Nitschke, J. R.; Tilley, T. D. **2001**, J. Am. Chem. Soc. *123*, 10183

Noland, W. E. **1955**, Chem. Rev. *55*, 137

Nowick, J. S.; Feng, Q.; Tjivikua, T.; Ballester, P.; Rebek, J., Jr. **1991**, J. Am. Chem. Soc. *113*, 8831

Noyori, R.; Ohkuma, T.; Kitarnura, M.; Takaya, H.; Sayo, N.; Kumobayashi, H.; Akutagawa, S. **1987**, J. Am. Chem. Soc. *109*, 5856

Nutt, R. E.; Veber, D. E.; Saperstein, R. **1980**, J. Am. Chem. Soc. *102*, 6539

O'Brien, P. **1999**, Angew. Chem. *111*, 339

O'Brien, P.; Sliskovic, D. R.; Picard, J. A.; Lee, H. T.; Purchase II, C. F.; Roth, B. D.; White, A. D.; Anderson, M.; Mueller, S. B.; Bocan, T.; Bousley, R.; Hamelehle, K. L.; Homan. R.; Lee, P.; Krause, B. R.; Reindel, J. F.; Stanfield, R. L.; Turluck, D. **1996**, J. Med. Chem. *39*, 2354

Oikawa, Y.; Yoshioka, T.; Yonemitsu, O. **1982**, Tetr. Lett., 885

Olah, G. A.; Brydon, D. L. **1970**, J. Org. Chem. *35*, 313

Oldenziel, O. H.; van Leusen, A. M. **1973**, Tetr. Lett., 1357

Oldenziel, O. H.; van Leusen, A. M. **1974**, Tetr. Lett., 163, 167

Ooi, H.; Urushibara, A.; Esumi, T.; Iwabuchi, Y.; Hatakeyama, S. **2001**, Org. Lett. *3*, 953

Opatz, T.; Liskamp, R. M. J. **2002**, J. Comb. Chem. *4*, 275

Oppolzer, W. **1978 A**, Synthesis 793

Oppolzer, W.; Pfenninger, E.; Keller, K. **1973**, Helv. Chim. Acta *56*, 1807

Oppolzer, W.; Snieckus, V. **1978 B**, Angew. Chem. *90*, 506 (Int. Ed. Engl. *17*, 476)

Oppolzer, W.; Snowden, R. L. **1976**, Tetrahedron Lett., 4187

Oscarson, S.; Sehgelmeble, F. W. **2000**, J. Am. Chem. Soc. *122*, 8869

Ostermeier, M.; Priess, J.; Helmchen, G. **2002**, Angew. Chem. *114*, 625

Otsubo, K.; Inanaga, J.; Yamaguchi, M. **1986**, Tetr. Lett., 5763

Otto, S.; Furlan, R. L. E.; Sanders, J. K. M. **2000**, J. Am. Chem. Soc. *122*, 12063

Otto, S.; Furlan, R. L. E.; Sanders, J. K. M. **2002**, Science *297*, 590

Ouellette, R. J. **1973**, in: Trahanovsky, W. S. (ed.) *Oxidation in Organic Chemistry*, Part B, p. 135, Academic Press: New York London

Overberger, C. G.; Kaye, H. **1967**, J. Am. Chem. Soc. *89*, 5640

Overberger, C. G.; Weise, J. K. **1968**, J. Am. Chem. Soc. *90*, 3525

Page, P. C. B.; McKenzie, M. J.; Gallagher, J. A. **2001**, J. Org. Chem. *66*, 3704

Pappo, R.; Allen, D. S., Jr; Lemieux, R. U.; Johnson, W. S. **1956**, J. Org. Chem. *21*, 478

Paquette, L. A.; Lavrik, P. B.; Summerville, R. H. **1977**, J. Org. Chem. *42*, 2659

Paquette, L. A. **1979**, Top. Curr. Chem. *79*, 41

Paquette, L. A.; Balogh, D. W. **1982**, J. Am. Chem. Soc. *104*, 774

Paquette, L. A.; Balogh, D. W.; Usha, R.; Koumtz, D.; Christoph, G. G. **1981**, Science *211*, 575

Paquette, L. A.; Wang, H.-L.; Su, Z.; Zhao, M. **1998**, J. Am. Chem. Soc. *120*, 5213

Paquette, L. A.; Wyvratt, M. J. **1974**, J. Am. Chem. Soc. *96*, 4671

Paquette, L. A.; Wyvratt, M. J.; Schallner, O.; Schneider, D. E.; Begley, W. J.; Blankenship, R. M. **1976**, J. Am. Chem. Soc. *98*, 6744

Parham, W. E.; Anderson, E. L. **1948**, J. Am. Chem. Soc. *70*, 4187

PASTO, D.I.; TAYLOR, R.T. **1991**, Org. React. (NY) *40*, 91

PATEL, K.M.; SKLAR, L.A.; CURRIE, R.; POWNALL, H.J.; MORISETT, J.D.; SPARROW, J.T. **1979**, Lipids *14*, 816

PATERSON, I.; FLORENCE, G.J.; GERLACH, K.; SCOTT, J.P.; SEREINIG, N. **2001**, J. Am. Chem. Soc. *123*, 9535

PAULSEN, H. **1977**, Pure Appl. Chem. *49*, 1169

PAULSEN, H. **1990**, Angew. Chem. *102*, 851 (Int. Ed. Engl. *29*, 823)

PAULSEN, H.; LOCKHOFF, O. **1981A, B, C**, Chem. Ber. *114*, 3079, 3102, 3115

PAULSEN, H.; SINNWELL, V.; STADLER, P. **1972**, Chem. Ber. *105*, 1978

PAZOS, Y.; IGLESIAS, B.; DE LERA, A.R. **2001**, J. Org. Chem. *66*, 8483

PEARLSTEIN, R.M.; BLACKBURN, B.K.; DAVIS, W.M.; SHARPLESS, K.B. **1990**, Angew. Chem. *102*, 710 (Int. Ed. Engl. *29*, 639)

PEARSON, A.J. **1987**, *Transition Metal-Stabilized Carbocations in Organic Synthesis*, in: HARTLEY, F.R. (ed.) *Chemistry of the Metal-Carbon Bond*, Vol. 4, chapter 10, Wiley: Chichester, UK

PEARSON, A.J.; ZETTLER, M.W. **1989**, J. Am. Chem. Soc. *111*, 3908

PEASE, A.R.; JEPPESEN, J.O.; STODDART, J.F.; LUO, Y.; COLLIER, C.P.; HEATH, J.R. **2001**, Acc. Chem. Res. *34*, 433

PELTER, A.; SMITH, K. **1979**, *Organic Boron Compounds*, in: BARTON, D.; OLLIS, W.D. (eds.) *Comprehensive Organic Chemistry, Vol. 3:* JENES, D.N. (ed.) p. 687 ff., Pergamon Press: Oxford

PENNINGTON, M.W.; DUNN, B.M. (eds.) **1997**, Humana Pr.

PERI, R.; PADMANABHAN, S.; RUTLEDGE, A.; SINGH, S.; TRIGGLE, D.J. **2000**, J. Med. Chem. *43*, 2906

PETASIS, N.A.; ZAVIALOV, I.A. **1997**, J. Am. Chem. Soc. *119*, 445

PETASIS, N.A.; ZAVIALOV, I.A. **1998**, J. Am. Chem. Soc. *120*, 11798

PETERSON, D.J. **1968**, J. Org. Chem. *33*, 781

PETERSON, P.E.; KAMAT, R.J. **1969**, J. Am. Chem. Soc. *91*, 4521

PETTIT, G.R.; PIATAK, D.M. **1962B**, J. Org. Chem. *27*, 2127

PETTIT, G.R.; VAN TAMELEN, E.E. **1962A**, Org. React. (NY) *12*, 356

PFALTZ, A.; BÜHLER, N.; NEIER, R.; HIRAI, K.; ESCHENMOSER, A. **1977**, Helv. Chim. Acta *60*, 2653

PFAU, M.; UGHETTO-MONFRIN, J.; JOULAIN, D. **1979**, Bull. Soc. Chim. Fr. 627

PFEIFFER, F.R.; COHEN, S.R.; WILLIAMS, K.R.; WEISBACH, J.A. **1968**, Tetr. Lett., 3549

PFEIFFER, F.R.; MIAO, C.K.; WEISBACH, J.A. **1970**, J. Org. Chem. *35*, 221

PFITZNER, K.E.; MOFFATT, J.G. **1965**, J. Am. Chem. Soc. *87*, 5661, 5670

PICARD, J.A.; O'BRIEN, P.M.; SLISKOVIC, D.R.; ANDERSON, M.K.; BOUSLEY, R.F.; HAMELEHLE, K.L.; KRAUSE, B.R.; STANFIELD, R.L. **1996**, J. Med. Chem. *39*, 1243

PIERS, E.; CHONG, J.M. **1983**, J. Chem. Soc. Chem. Commun., 934

PINE, S.H.; ZAHLER, R.; EVANS, D.A.; GRUBBS, R.H. **1980**, J. Am. Chem. Soc. *102*, 3270

PINKOS, R.; MELDER, J.-P.; FRITZ, H.; PRINZBACH, H. **1989**, Angew. Chem. *101*, 319 (Int. Ed. Engl.]. *28*, 310)

PINKOS, R.; RIHS, G.; PRINZBACH, H. **1989**, Angew. Chem. *101*, 312 (Int. Ed. Engl. *28*, 303)

PIRRUNG, M.C.; PARK, K.; TUMEY, L.N. **2002**, J. Comb. Chem. *4*, 329

PIWINSKI, J.J.; WONG, J.K.; CHAN, T.-M.; GREEN, M.J.; GANAGULY, A.K. **1990**, J. Org. Chem. *55*, 3341

PIWINSKI, J.J.; WONG, J.K.; GREEN, M.J.; GANGULY, A.K.; BILLAH, M.M.; WEST, R.E.; KREUTNER, W. **1991**, J. Med. Chem. *34*, 457

PLESNIČAR, B. **1978**, in: TRAHANOVSKY, W.S. (ed.) *Oxidation in Organic Chemistry*, Part C, chapter 111, p. 211, Academic Press: New York London

POHMAKOTR, M.; SEEBACH, D. **1977**, Angew. Chem. *89*, 333 (Int. Ed. Engl. *16*, 320)

POLYAKOV, V.A.; NELEN, M.I.; NAZARPACK-KANDLOUSY, N.; RYABOV, A.D.; ELISEEV, A.V. **1999**, J. Phys. Org. Chem. *12*, 357

POMMER, H. **1960**, Angew. Chem. *72*, 811

POMMER, H. **1977**, Angew. Chem. *89*, 437 (Int. Ed. Engl. *16*, 423)

POMMER, H.; NÜRRENBACH, A. **1975**, Pure Appl. Chem. *43*, 527

POP, I.E.; DÉPREZ, B.P.; TARTAR, A.L. **1997**, J. Org. Chem. *62*, 2594

POSNER, G.H. **1972**, Org. React. (NY) *19*, 1; **1975**, Org. React. (NY) *22*, 253

POSTEMA, M. H. D. **1995**, *C-Glycoside Synthesis*, 1st edn.; CRC Press: Boca Raton, FL

POSTEMA, M. H. D.; CALIMENTE, D.; LIU, L.; BEHRMANN, T. L. **2000**, J. Org. Chem. *65*, 6061

PRINZBACH, H.; SEDELMAIER, G.; KRÜGER, C.; GODDARD, R.; MARTIN, H.-D.; GLEITER, R. **1978**, Angew. Chem. *90*, 297 (Int. Ed. Engl. *17*, 271)

RABJOHN, N. **1949**, **1976**, Org. React. *5*, 331; *24*, 261

RAMACHANDRAN, S.; NEWMAN, M. S. **1961**, Org. Synth. *41*, 39

RAMIREZ, E.; EVANGELIDOU-TSOLIS, E.; JANKOWSKI, A.; MARECEK, J. F. **1977**, J. Org. Chem. *42*, 3144

RAMIREZ, E.; MARECEK, J. E.; UGI, I. **1975**, Synthesis, 99

RAMNAUTH, J.; POULIN, O.; BRATOVANOV, S. S.; RAKHIT, S.; MADDAFORD, S. P. **2001**, Org. Lett. *3*, 2571

RAMSTROM, O.; LEHN, J.-M. **2000**, ChemBioChem. *1*, 41

RASMUSSEN, J. K. **1977**, Synthesis 91

RAVID, U.; SILVERSTEIN, R. M. **1977**, Tetrahedron Lett., 423

RAYMO, F. M.; BARTBERGER, M. D.; HOUK, K. N.; STODDART, J. F. **2001**, J. Am. Chem. Soc. *123*, 9264

RAYMO, F. M.; HOUK, K. N.; STODDART, J. F. **1998**, J. Am. Chem. Soc. *120*, 9318

RAYMO, F. M.; STODDART, J. F. **1999**, Chem. Rev. *99*, 1643

REBEK, J., JR.; ASKEW, B.; KILLORAN, M.; NEMETH, D.; LIN, F.-T. **1987**, J. Am. Chem. Soc. *109*, 2426

REDDY, K. L.; SHARPLESS, K. B. **1998**, J. Am. Chem. Soc. *120*, 1207

REESE, C. B.; TRENTHAM, D. R. **1965**, Tetrahedron Lett., 2459

REETZ, M. T. **1989**, S. Afr. J. Chem. *42*, 49

REETZ, M. T.; SEITZ, T. **1987**, Angew. Chem. *99*, 1081 (Int. Ed. Engl. *26*, 1028)

REICH, H. J. **1978**, in: TRAHANOVSKY, W. S. (ed.) *Oxidation in Organic Chemistry*, Part C, p. 1, Academic Press: New York London

REICH, H. J. **1979**, Acc. Chem. Res. *12*, 23

REICH, H. J.; RENGA, J. M.; REICH, I. L. **1975**, J. Am. Chem. Soc. *97*, 5434

REICHERT, B. **1959**, Die Mannich-Reaktion, Springer: Berlin Heidelberg New York

REIF, W.; GRASSNER, H. **1973**, Chem. Ing. Tech. *45*, 646

REINEFELD, E.; HEINCKE, K. D. **1971**, Chem. Ber. *104*, 265

REIST, E. L.; BARTUSKA, V. L.; GOODMAN, L. **1964**, J. Org. Chem. *29*, 3725

RERICK, M. N.; ELIEL, E. L. **1962**, J. Am. Chem. Soc. *84*, 2356

RICCI, A.; ANGELUCCI, F.; BESSETTI, M.; STERZO, C. L. **2002**, J. Am. Chem. Soc. *124*, 1060

RIDLEY, D. D. **2002**, Information Retrieval: Scifinder® and Scifinder® Scholar, Wiley: New York

RIDRÍGUEZ, D.; NAVARRO-VÁZQUEZ, A.; CASTEDO, L.; DIMÍNGUEZ, D.; SAÁ, C. **2001**, J. Am. Chem. Soc. *123*, 9178

RIEDL, R.; TAPPE, R.; BERKESSEL, A. **1998**, J. Am. Chem. Soc. *120*, 8994

RILEY, R. G.; SILVERSTEIN, R. M. **1974**, Tetrahedron *30*, 1171

RISPENS, T.; ENGBERTS, J. B. F. N. **2001**, Org. Lett. *3*, 941

ROBERTS, S. M., *Introduction to Biocatalysis Using Enzymes and Micro-Organisms*, **1994**, Cambridge University Press

ROBERTSON, J., **2000**, *Protecting Group Chemistry* (Chemistry Primers), Oxford University Press

ROBINS, M. J.; WILSON, J. S.; HANSSKE, F. **1983**, J. Am. Chem. Soc. *105*, 4059

ROBINSON, B. **1963**, **1969**, Chem. Rev. *63*, 373; *69*, 227

ROBSON, B.; GARNIER, L. **1986**, *Introduction to Proteins and Protein Engineering*, Elsevier: Amsterdam

ROCZEK, J. **1957**, Coll. Czech. Chem. Commun. *22*, 1509

ROCZEK, J.; MAREG, F. **1959**, Coll. Czech. Chem. Commun. *24*, 2741

ROH, Y.; JANG, H.-Y.; LYNCH, V.; BAULD, N. L.; KRISCHE, M. J. **2002**, Org. Lett. *4*, 611

ROMO, J.; ROSENKRANZ, G.; DJERASSI, C. **1951**, J. Am. Chem. Soc. *73*, 4961

ROSENBERGER, M.; DUGGAN, A. J.; BORER, R.; MÜLLER, R.; SAUCY, G. **1972**, Helv. Chim. Acta *55*, 1663

ROSENBERGER, M.; FRAHER, T. P.; SAUCY, G. **1971**, Helv. Chim. Acta *54*, 2857

ROSENBLUM, M. **1974**, Acc. Chem. Res. *7*, 122

ROSENTHAL, A.; BENZING-NGUYEN, L. **1969**, J. Org. Chem. *34*, 1029

ROSSITER, B. E. **1985**, *Synthetic Aspects and Applications of Asymmetric Epoxidation*, in: MORRISON, J. D. (ed.) *Asymmetric Synthesis*,

Vol. 5: Chiral Catalysis, chapter 7, p. 193, Academic Press: New York

ROSSITER, B. E.; KATSUKI, T.; SHARPLESS, K. B. **1981,** J. Am. Chem. Soc. *103,* 464

ROTELLA, D. P.; SUN, Z.; ZHU, Y.; KRUPINSKI, J.; PONGRAC, R.; SELIGER, L.; NORMANDIN, D.; MACOR, J. E. **2000,** J. Med. Chem. *43,* 1257

ROTERMUND, G. W. **1975,** in: HOUBEN-WEYL, *Methoden der Organischen Chemie,* Vol. IV/1 b: *Oxidation II,* Thieme: Stuttgart

ROWAN, S. J.; CANTRILL, S. J.; STODDART, J. F. **1999,** Org. Lett. *1,* 129

RUBIN, Y.; DIEDERICH, F. **1989,** J. Am. Chem. Soc. *111,* 6870

RUBIN, Y.; KAHR, M.; KNOBLER, C. B.; DIEDERICH, E.; WILKINS, C. L. **1991,** J. Am. Chem. Soc. *113,* 495

RUBIN, Y.; KNOBLER, C. B.; DIEDERICH, F. **1990,** J. Am. Chem. Soc. *112,* 4966

RUDEN, R. A.; BONJOUKLIAN, R. **1974,** Tetr. Lett., 2095

RUFER, C.; KOSMOL, H.; SCHRÖDER, E.; KIESLICH, K.; GIBIAN, H. **1967,** Liebigs Ann. Chem. *702,* 141

RÜHLMANN, K. **1971,** Synthesis 236

RUPPERT, J. E.; WHITE, J. D. **1976,** J. Org. Chem. *41,* 550

RUSSEL, A. E.; GREENBERG, S.; MOFFATT, J. G. **1973,** J. Am. Chem. Soc. *95,* 4025

RUSSO, R.; LAMBERT, Y.; DESLONGCHAMPS, P. **1971,** Can. J. Chem. *49,* 531

RYAN, C. W.; SIMON, R. L.; VAN HEYNINGEN, E. M. **1969,** J. Med. Chem. *12,* 310

RYLANDER, P. N. **1979,** *Catalytic Hydrogenation in Organic Syntheses,* Academic Press: New York London

RYU, I.; SONODA, N. **1996,** Chem. Rev. *96,* 177

SABITHA, G.; BABU, R. S.; RAJKUMAR, M.; SRIVIDYA, R.; YADAV, J. S. **2001,** Org. Lett. *3,* 1149

SAITO, S.; SHIOZAWA, M.; NAGAHARA, T.; NAKADAI, M.; YAMAMOTO, H. **2000,** J. Am. Chem. Soc. *122,* 7847

SAKAN, T.; FUJINO, A.; MURAI, E.; SUZUI, A.; BUTSUGAN, Y. **1960,** Bull. Chem. Soc. Jpn. *33,* 1737

SALAUN, J.; GARNIER, B.; CONIA, J. M. **1974,** Tetrahedron *30,* 1413

SALIMI-MOOSAVI, H.; TANG, T.; HARRISON, D. J. **1997,** J. Am. Chem. Soc. *119,* 8716

SALITURO, G. M.; TOWNSEND, C. A. **1990,** J. Am. Chem. Soc. *112,* 760

SAMBROOK, J.; RUSSEL, D. W. **2001,** *Molecular Cloning: A Laboratory Manual,* Cold Spring Harbor Laboratory Press

SAMMES, P. G. (ed.) **1979,** in: BARTON, D.; OLLIS, W. D. (eds.) *Comprehensive Organic Chemistry, Vol. 4: Heterocyclic Compounds,* Pergamon Press: Oxford

SANDRI, J.; VIALA, J. **1995,** J. Org. Chem. *60,* 6627

SATOH, T.; NANBA, K.; SUZUKI, S. **1971,** Chem. Pharm. Bull. *19,* 817

SATOH, T.; SUZUKI, S.; SUZUKI, Y.; MIYAJI, Y.; IMAI, Z. **1969,** Tetr. Lett., 4555

SAUCY, G.; BORER, R. **1971,** Helv. Chim. Acta *54,* 2121, 2517

SAUERS, R. R.; AHEARN, G. P. **1961,** J. Am. Chem. Soc. *83,* 2759

SAVIN, K. A.; WOO, J. C. G.; DANISHEFSKY, S. J. **1999,** J. Org. Chem. *64,* 4183

SCHAEFER, J. P.; BLOOMFIELD, J. J. **1967,** Org. React. (NY) *15,* 1

SCHÄFER, W.; HELLMANN, H. **1967,** Angew. Chem. *79,* 566 (Int. Ed. Engl. *6,* 518)

SCHERING PROCESS **1957, 1977,** in: *Ullmanns Encyklopädie der technischen Chemie,* 3rd edn., Vol. 8, p. 648, Urban & Schwarzenberg: Munich, Ger; 4th edn., Vol. 13, p. 27, Verlag Chemie: Weinheim, Ger.

SCHICK, H.; LEHMANN, G.; HILGETAG, G. **1969,** Chem. Ber. *102,* 3238

SCHLEICH, S.; HELMCHEN, G. **1999,** Eur. J. Org. Chem., 2515

SCHLUETER, S.; FRAHN, J.; KARAKAYA, B.; SCHLUETER, A. D. **2000,** Macromol. Chem. Phys. *201,* 139

SCHMIDLIN, J.; ARMER, G.; BILLETER, J.-R.; HEUSLER, K.; UEBERWASSER, H.; WIELAND, P.; WETTSTEIN, A. **1957,** Helv. Chim. Acta *50,* 2101

SCHMIDT, A. H.; RIED, W. **1978 A, B,** Synthesis *1,* 869

SCHMIDT, R. R. **1986,** Angew. Chem. *98,* 213 (Int. Ed. Engl. *25,* 212)

SCHMIDT, R. R.; KLÄGER, R. **1985,** Angew. Chem. *97,* 60 (Int. Ed. Engl. *24,* 65)

SCHMIDT, R. R.; KÖHN, A. **1987,** Angew. Chem. *99,* 490 (Int. Ed. Engl. *26,* 482)

SCHMIDT, R. R.; MICHEL, J. **1980,** Angew. Chem. *92,* 763 (Int. Ed. Engl. *19,* 731)

SCHMIDT, R. R.; MICHEL, J.; ROOS, M. **1984,** Liebigs Ann. Chem., 1343

SCHREIBER, J.; LEIMGRUBER, W.; PESARO, M.; SCHUDEL, P.; ESCHENMOSER, A. **1961**, Helv. Chim. Acta *44*, 540

SCHREIBER, L.; FELIX, D.; ESCHENMOSER, A.; WINTER, M.; GAUTSCHI, E.; SCHULTE-ELTE, K.H.; SUNDT, E.; OHLOFF, G.; KALVODA, J.; KAUFMANN, H.; WIELAND, P.; ARMER, G. **1967**, Helv. Chim. Acta *50*, 2101

SCHREIBER, S.L. **2000**, Science *287*, 1964

SCHREIBER, S.L.; FENG, S.; CHEN, J.K.; YU, H.; SIMON, J.A. **1994**, Science *266*, 1241

SCHREIBER, S.L.; SAMMAKIA, T.; UEHLING, D.E. **1989**, J. Org. Chem. *54*, 15

SCHREIBER, S.L.; SCHREIBER, T.S.; SMITH, D.B. **1987**, J. Am. Chem. Soc. *109*, 1525

SCHROCK, R.R. **1976**, J. Am. Chem. Soc. *98*, 5399

SCHRÖDER, E.; LÜBKE, K. **1963**, Experientia *19*, 57

SCHRÖDER, E.; RUFER, C.; SCHMIECHEN, R. **1975**, Arzneimittelchemie, Vol. 1–3, Thieme: Stuttgart

SCHRÖDER, G. **1963**, Angew. Chem. *75*, 722 (Int. Ed. Engl. *2*, 481); **1964**, Chem. Ber. *97*, 3131, 3140

SCHRÖDER, G.; OTH, J.F.M. **1967**, Angew. Chem. *79*, 458 (Int. Ed. Engl. *6*, 414)

SCHWARTZ, J.; LABINGER, J.A. **1976**, Angew. Chem. (Int. Ed. Engl. *15*, 333)

SCHWARZ, J.B.; KUDUK, S.D.; CHEN, X.-T.; SAMES, D.; GLUNZ, P.W.; DANISHEFSKY, S.J. **1999**, J. Am. Chem. Soc. *121*, 2662

SCOTT, A.F.; MERGOTT, D.J.; ROUSH, W.R. **2002**, J. Am. Chem. Soc. *124*, 2404

SCOTT, L.T.; JONES, M., JR. **1972**, Chem. Rev. *72*, 181

SEEBACH, D. **1969**, Synthesis 17

SEEBACH, D. **1979**, Angew. Chem. *91*, 259 (Int. Ed. Engl. *18*, 239)

SEEBACH, D.; AEBI, J.; WASMUTH, D.; MAXWELL, B.; HEATHCOCK, C.H. **1984**, Org. Synth. *63*, 109

SEEBACH, D.; BOES, M.; NAEF, R.; SCHWEIZER, W.B. **1983**, J. Am. Chem. Soc. *105*, 5390

SEEBACH, D.; HOEKSTRA, M.S.; PROTSCHUK, G. **1977**, Angew. Chem. *89*, 334 (Int. Ed. Engl. *16*, 321)

SEEBACH, D.; KOLB, M.; GRÖBEL, B.-T. **1973**, Chem. Ber. *106*, 2277

SEEBACH, D.; PRELOG, V. **1982**, Angew. Chem. *94*, 696 (Int. Ed. Engl. *21*, 654)

SEEBACH, D.; DAUM, H. **1974**, Chem. Ber. *107*, 1748

SEEBERGER, P.H. **2001**, *Solid Support Oligosaccharide Synthesis and Combinatorial Carbohydrate Libraries*, Wiley & Sons

SEEBERGER, P.H.; HAASE, W.-C. **2000**, Chem. Rev. *100*, 4349

SEMMELHACK, M.E.; HEINSOHN, G.E. **1972**, J. Am. Chem. Soc. *94*, 5139

SESSLER, J.L.; HUGDAHL, J.; JOHNSON, M.R. **1986**, J. Org. Chem. *51*, 2838

SESSLER, J.L.; JOHNSON, M.R. **1987**, Angew. Chem. *99*, 679 (Int. Ed. Engl. *26*, 678)

SESSLER, J.L.; JOHNSON, M.R.; LIN, T.-Y.; CREAGER, S.E. **1988A**, J. Am. Chem. Soc. *110*, 3659

SESSLER, J.L.; MURAI, T.; LYNCH, V.; CYR, M. **1988B**, J. Am. Chem. Soc. *110*, 5586

SHAMMA, M.; JENES, C.D. **1970**, J. Am. Chem. Soc. *92*, 4943

SHAPIRO, R.H. **1976**, Org. React. (NY) *23*, 405

SHARMA, M.; KORYTNYK, W. **1977**, Tetrahedron Lett. *573*

SHARPLESS, K.B. **1970**, J. Am. Chem. Soc. *92*, 6999

SHARPLESS, K.B. **1986**, Chem. Britain *22*, 38

SHARPLESS, K.B.; TERANISHI, A.Y. **1973**, J. Org. Chem. *38*, 185

SHARPLESS, K.B.; BEHRENS, C.H.; KATSUKI, T.; LEE, A.W.M.; MARTIN, V.S.; TAKATANI, M.; VITI, S.M.; WALKER, F.J.; WOODARD, S.S. **1983B**, Pure Appl. Chem. *55*, 589

SHARPLESS, K.B.; LAUER, R.F. **1972**, J. Am. Chem. Soc. *94*, 7154

SHARPLESS, K.B.; TERANISHI, A.Y.; BÄCKVALL, J.-E. **1977**, J. Am. Chem. Soc. *99*, 3120

SHARPLESS, K.B.; WOODARD, S.S.; FINN, M.G. **1983A**, Pure Appl. Chem. *55*, 1823

SHEEHAN, J.C.; GUZIEC, F.S., JR. **1972**, J. Am. Chem. Soc. *94*, 6561

SHEEHAN, J.C.; GUZIEC, F.S., JR. **1973**, J. Org. Chem. *38*, 3034

SHELDON, R.A.; KOCHI, J.K. **1972**, Org. React. (NY) *19*, 279

SHELL, A.J. **1996**, Chem. Rev., 195

SHEMYAKIN, M.M.; VINOGRADOVA, E.L.; FEIGINA, M.Yu.; ALDANOVA, N.A.; OLADKINA, V.A.; SHCHUKINA, L.A. **1961**, Dokl. Akad. Nauk. SSSR *140*, 387 (C.A. *56*, 536b)

SHILDNECK, P.R.; WINDUS, W. **1943**, Org. Synth. Coll. Vol. *2*, 411

SHIMIZU, T.; HIRANO, K.; TAKAHASHI, M.; HATANO, M.; FUJII-KURIYAMA, Y. **1988**, Biochemistry *27*, 4138

SHIMIZU, T.; SOGAWA, K.; FUJII-KURIYAMA, Y.; TAKAHASHI, M.; OGOMA, Y.; HATANO, M. **1986**, FEBS Lett. *207*, 217

SHINER, V.J., JR.; TAI, J.J. **1981**, J. Am. Chem. Soc. *103*, 436

SHIPWAY, A.N.; LAHAV, M.; WILLNER, I. **2000**, Adv. Mater. *12*, 993

SHORT, R.P.; MASAMUNE, S. **1989**, J. Am. Chem. Soc. *111*, 1892

SIBI, M.P.; VENKATRAMAN, L.; LIU, M.; JASPERSE, C.P. **2001**, J. Am. Chem. Soc. *123*, 8444

SIDDALL, J.B.; MARSHALL, J.P.; BOWERS, A.; CROSS, A.D.; EDWARDS, J.A.; FRIED, J.H. **1966**, J. Am. Chem. Soc. *88*, 379

SIEGEL, C.; THORNTON, E.R. **1989**, J. Am. Chem. Soc. *111*, 5722

SIGMAN, M.S.; JACOBSEN, E.N. **1998**, J. Am. Chem. Soc. *120*, 4901

SIMMONS, H.E.; CAIRNS, T.L.; VLADUCHIK, S.A.; HOINESS, C.M. **1973**, Org. React. (NY) *20*, 1

SKUPIN, M.; LI, G.; FUDICKAR, W.; ZIMMERMANN, J.; ROEDER, B.; FUHRHOP, J.-H. **2001**, J. Am. Chem. Soc. *123*, 3454

SLUKA, J.P.; GRIFFIN, J.H.; MACK, D.P.; DERVAN, P.B. **1990**, J. Am. Chem. Soc. *112*, 6369

SMALLEY, R.E. **1992**, Acc. Chem. Res. *25*, 98

SMEETS, S.; ASOKAN, C.V.; MOTMANS, F.; DEHAEN, W. **2000**, J. Org. Chem. *65*, 5882

SMITH, III, A.B. III; FRIESTAD, G.K.; DUAN, J.J.-W.; BARBOSA, J.; HULL, K.G.; IWASHIMA, M.; QIU, Y.; SPOORS, P.G.; BERTOUNESQUE, E.; SALVATORE, B.A. **1998**, J. Org. Chem. *63*, 7596

SMITH, H.; HUGHES, G.A.; DOUGLAS, G.H.; WENDT, G.R.; BUZBY, G.C., JR.; EDGREN, R.A.; FISHER, J.; FOELL, T.; GADSBY, B.; HARTLEY, D.; HERBST, D.; JANSEN, A.B.A.; LEDIG, K.; McLOUGHLIN, B.J.; McMENAMIN, J.; PATTISON, T.W.; PHILLIPS, P.C.; REES, R.; SIDDALL, J.; SIUDA, J.; SMITH, L.L.; TOKOLICS, J.; WATSON, D.H.P. **1964**, J. Chem. Soc., 4472

SMITH III, A.B.; FRIESTAD, G.K.; BARBOSA, J.; BERTOUNESQUE, E.; HULL, K.G.; IWASHIMA, M.; QIU, Y.; SALVATORE, B.A.; SPOORS, P.G.; DUAN, J.J.-W. **1999**, J. Am. Chem. Soc. *121*, 10468

SMITH, III, A.B.; VERHOEST, P.R.; MINBIOLE, K.P.; SCHELHAAS, M. **2001**, J. Am. Chem. Soc. *123*, 4834

SMITH, K.M. (ed.) **1975**, *Porphyrins and Metalloporphyrins,* Elsevier: Amsterdam New York

SMITH, M.; KHORANA, H.G. **1959**, J. Am. Chem. Soc. *81*, 2911

SODERQUIST, J.A. **1991**, Aldrichim. Acta *24*, 15

SONDHEIMER, E.; McQUILKIN, R.M.; GARRATT, P.J. **1970**, J. Am. Chem. Soc. *92*, 6682

SONDHEIMER, E.; WOLOVSKY, R.; AMIEL, Y. **1962**, J. Am. Chem. Soc. *84*, 274

SONDHEIMER, F. **1950**, J. Chem. Soc. *1950*, 877

SONDHEIMER, F. **1963**, Pure Appl. Chem. *7*, 363

SONG, Y.; OKAMOTO, S.; SATO, F. **2001**, Org. Lett. *3*, 3543

SØRENSEN, D.; KVÆRNØ, L.; BRYLD, R.; HÅKANSSON, A.E.; VERBEURE, B.; GAUBERT, G.; HERDEWIJN, P.; WENGEL, J. **2002**, J. Am. Chem. Soc. *124*, 2164

SOWADA, R. **1971**, *Dimethylsulfoniummethylid und Dimethyloxosulfoniummethylid,* in: MARTIN, D.; HAUTHAL, H.G. (eds.) *Dimethylsulfoxid,* p. 367 ff., Akademie Verlag: Berlin

SPANGLER, R.J.; BECKMANN, B.G. **1976**, Tetr. Lett., 2517

SPANGLER, R.J.; BECKMANN, B.G.; KIRN, J.H. **1977**, J. Org. Chem. *42*, 2989

SPANGLER, R.J.; KIM, J.H. **1973**, Synthesis 107

SPENCER, J.L.; FLYNN, E.H.; ROESKE, R.W.; SIU, F.Y.; CHAUVETTE, R.R. **1966**, J. Med. Chem. *9*, 746

SPERO, D.M.; KAPADIA, S.R. **1997**, J. Org. Chem. *62*, 5537

SPINO, C.; BEAULIEU, C. **1998**, J. Am. Chem. Soc. *120*, 11832

ST. HILAIRE, P.M.; LOWARY, T.L.; MELDAL, M.; BOCK, K. **1998**, J. Am. Chem. Soc. *120*, 13312

STAAB, H.A.; SCHWENDEMANN, M. **1979**, Liebigs Ann. Chem., 1258

STADLER, P.A.; NECHVATAL, A.; FREY, A.J.; ESCHENMOSER, A. **1957**, Helv. Chim. Acta *40*, 1373

STALLBERG, G.; STALLBERG, S.; STENHAGEN, E. **1956**, Acta Chem. Scand. *6*, 313

STÄTZ, A. **1987**, Angew. Chem. *99*, 323 (Int. Ed. Engl. *26*, 320)

STECHL, H. H. **1975**, in: HOUBEN-WEYL, *Methoden der Organischen Chemie, Vol. IV/1 b: Oxidation II*, p. 873, Thieme: Stuttgart

STEIN, C.; DE JESO, B.; POMMIER, J. C. **1982**, Synth. Commun. *12*, 495

STEITZ, A., JR. **1968**, J. Org. Chem. *33*, 2978

STEPHENS, C. R.; BEEREBOORN, J. J.; RENNHARD, H. R.; GORDON, P. X; MURAI, K.; BLACKWOOD, R. K.; SCHACH VON WITTENAU, M. **1963**, J. Am. Chem. Soc. *85*, 2643; see also: **1962**, ibid. *84*, 2645, and **1961**, ibid. *83*, 2773

STEVENS, J. D. **1972**, in: WHISTLER, R. L.; BEMILLER, J. N. (eds.) *Methods in Carbohydrate Chemistry*, Vol. VI, p. 124, Academic Press: New York London

STEVENS, R. V.; ELLIS, M. C. **1967**, Tetrahedron Lett., 5185

STEVENS, R. V.; FITZPATRICK, J. M.; KAPLAN, M.; ZIMMERMAN, R. L. **1971**, J. Chem. Soc. D, 857

STEVENS, R. V.; WENTLAND, M. P. **1968**, J. Am. Chem. Soc. *90*, 5580

STEWART, R. **1965**, in: WIBERG, K. B. (ed.) *Oxidation in Organic Chemistry*, Part A, chapter 1, p. 2, Academic Press: New York London

STILL, W. C. **1996**, Acc. Chem. Res. *29*, 155

STILL, W. C.; GENNARI, C. **1983**, Tetr. Lett. *24*, 4405

STILL, W. C.; MacDONALD, T. L. **1976**, Tetr. Lett., 2659

STILLE, J. K. **1985**, Pure Appl. Chem. *57*, 1771

STILLE, J. K. **1986**, Angew. Chem. *98*, 504 (Int. Ed. Engl. *25*, 508)

STINSON, S. C. **1998**, C&EN, 83

STOKES, J. C.; ESSLINGER, W. G. **1975**, J. Chem. Educ. *52*, 784

STOREY, H. T.; BEACHAM, J.; CERNOSEK, S. E.; FINN, F. M.; YANAIHARA, C.; HOFMANN, K. **1972**, J. Am. Chem. Soc. *94*, 6170

STORK, G. **1968A**, Pure Appl. Chem. *17*, 383

STORK, G.; BURGSTAHLER, A. W. **1955**, J. Am. Chem. Soc. *77*, 5068

STORK, G.; DARLING, S. D. **1964**, J. Am. Chem. Soc. *86*, 1761 [Li-NH3 reduction]

STORK, G.; DARLING, S. D.; HARRISON, I. T.; WHARTON, P. S. **1962A**, J. Am. Chem. Soc. *84*, 2018

STORK, G.; DOLFINI, J. E. **1963**, J. Am. Chem. Soc. *85*, 2872

STORK, G.; GANEM, B. **1973**, J. Am. Chem. Soc. *95*, 6152

STORK, G.; HUDRLIK, P. F. **1968 C**, J. Am. Chem. Soc. *90*, 4462, 4464

STORK, G.; KRAUS, G. A.; GARCIA, G. A. **1974A**, J. Org. Chem. *39*, 3459

STORK, G.; KRETCHMER, R. A.; SCHLESSINGER, R. H. **1968B**, J. Am. Chem. Soc. *90*, 1647

STORK, G.; SCHULENBERG, J. W. **1962**, J. Am. Chem. Soc. *84*, 284

STORK, G.; SINGH, J. **1974B**, J. Am. Chem. Soc. *96*, 6181

STORK, G.; TERRELL, R.; SZMUSZKOVICZ, J. **1954**, J. Am. Chem. Soc. *76*, 2029

STORK, G.; TOMASZ, M. **1962B**, J. Am. Chem. Soc. *84*, 310

STORK, G.; TOMASZ, M. **1964**, J. Am. Chem. Soc. *86*, 471 [spiroanellation]

STORK, G.; VAN TAMELEN, E. E.; FRIEDMANN, L. J.; BURGSTAHLER, A. W. **1953**, J. Am. Chem. Soc. *75*, 384

STÜTZ, A.; PETRANYI, G. **1984**, J. Med. Chem. *27*, 1539

SUBRAMANIAN, L. R.; HANACK, M.; CHANG, L. W. K.; IMHOFF, M. A.; SCHLEYER, P. R.; EFFENBERGER, F.; KURTZ, W.; STANG, P. J.; DUEBER, T. E. **1976**, J. Org. Chem. *41*, 4099

SUDAU, A.; MUENCH, W.; NUBBEMEYER, U. **2000**, J. Org. Chem. *65*, 1710

SUGINOME, M.; ITO, Y. **2000**, Chem. Rev. *100*, 3221

SUHARA, Y.; NIHEI, K.-I.; KURIHARA, M.; KITTAKA, A.; YAMAGUCHI, K.; FUJISHIMA, T.; KONNO, K.; MIYATA, N.; TAKAYAMA, H. **2001**, J. Org. Chem. *66*, 8760

SUM, F. W.; WEILER, L. **1979**, J. Am. Chem. Soc. *101*, 4401

SUN, P.; SUN, C.; WEINREB, S. M. **2001**, Org. Lett. *3*, 3507

SUNDBERG, R. J. **1970**, *The Chemistry of Indoles*, Academic Press: New York London

SURMAN, M. D.; MULVIHILL, M. J.; MILLER, M. J. **2002**, Org. Lett. *4*, 139

SUZUKI, A. **1986**, Pure & Appl. Chem. *58*, 629

SVOBODA, M.; ZÁVADA, J.; SICHER, J. **1965**, Coll. Czech. Chem. Commun. *30*, 413

SWANN, P. G.; CASANOVA, R. A.; DESAI, A.; FRAUENHOFF, M. M.; URBANCIC, M.; SLOMCZYNSKA, U.; HOPFINGER, A. J.; LE BRETON, G. C.; VENTON, D. L. **1996**, Biopolymers *40*, 617

SWERN, D. **1953**, Org. React. (NY) *7*, 378

SWERN, D. **1970**, *Organic Peroxides*, Vol. 1, chapter 6, p. 313, Wiley: New York

SZANTAY, C.; TÖKE, L.; HONTI, K. **1965**, Tetrahedron Lett. 1665

SZURDOKI, F.; REN, D.; WALT, D. R. **2000**, Anal. Chem. *72*, 5250

TABER, D. F.; NAKAJIMA, K. **2001**, J. Org. Chem. *66*, 2515

TABER, D. F.; TENG, D. **2002**, J. Org. Chem. *67*, 1607

TAGUCHI, H.; TANAKA, S.; YAMAMOTO, H.; NOZAKI, H. **1973**, Tetrahedron Lett. 2465

TAKAI, K.; NITTA, K.; UTIMOTO, K. **1986**, J. Am. Chem. Soc. *108*, 7408

TAKANO, S.; IWABUCHI, Y.; OGASAWARA, K. **1991**, J. Am. Chem. Soc. *113*, 2786

TAKASU, K.; UENO, M.; IHARA, M. **2001**, J. Org. Chem. *66*, 4667

TAKAYA, H.; OHTA, T.; SAYO, N.; KUMOBAYASHI, H.; AKUTAGAWA, S.; INOUE, S.; KASAHARA, I.; NOYORI, R. **1987**, J. Am. Chem. Soc. *109*, 1596.

TANABE, M.; CROWE, D. F.; DEHN, R. L. **1967**, Tetrahedron Lett., 3739, 3943

TANAKA, K.; MORI, H.; YAMAMOTO, M.; KATSUMURA, S. **2001**, J. Org. Chem. *66*, 3099

TANAKA, T.; OBA, M.; TAMAI, K.; SUEMUNE, H. **2001**, J. Org. Chem. *66*, 2667

TANIS, S. P.; NAKANISHI, K. **1979**, J. Am. Chem. Soc. *101*, 4398

TANTILLO, D. J.; HOUK, K. N.; JUNG, M. E. **2001**, J. Org. Chem. *66*, 1938

TARBELL, D. S.; WILLIAMS, K. I. H.; SEHM, E. J. **1959**, J. Am. Chem. Soc. *81*, 3443

TAYLOR, E. C.; MCKILLOP, A. **1970**, Acc. Chem. Res. *3*, 338

TAYLOR, E. C.; ROBEY, R. L.; LIU, K.-T.; FAVRE, B.; BOZIMO, H. T.; CONLEY, R. A.; CHIANG, C.-S. **1976**, J. Am. Chem. Soc. *98*, 3037

TAYLOR, E. L.; DJERASSI, C. **1976**, J. Am. Chem. Soc. *98*, 2275

TAYLOR, J. W.; OTT, J.; ECKSTEIN, F. **1985 B**, Nucleic Acids Res. *13*, 8764

TAYLOR, J. W.; SCHMIDT, W.; COSSTICK, R.; OKRUSZEK, A.; ECKSTEIN, F. **1985 A**, Nucleic Acids Res. *13*, 8749

TEBBE, F. N.; PARSHALL, G. W.; REDDY, G. S. **1978**, J. Am. Chem. Soc. *100*, 3611

TEJIMA, S.; FLETCHER, H. G., JR. **1963**, J. Org. Chem. *28*, 2999

TEMAL-LAIB, T.; CHASTANET, J.; ZHU, J. **2002**, J. Am. Chem. Soc. *124*, 583

TERRETT, N. K.; BELL, A. S.; BROWN, D.; ELLIS, P. **1996**, Bioorg. & Med. Chem. Lett. *6*, 15

THAPER, R. K.; KUMAR, Y.; KUMAR, S. M. D.; MISRA, S.; KHANNA, J. M. **1999**, Organic Process Research & Development *3*, 476

THOMPSON, S. E.; SMITHRUD, D. B. **2002**, J. Am. Chem. Soc. *124*, 442

THUMMEL, R. P. **1974**, J. Chem. Soc. Chem. Commun. *1974*, 899

THUMMEL, R. P. **1980**, Acc. Chem. Res. *13*, 70

TI, G. S.; GAFFNEY, B. L.; JONES, R. A. **1982**, J. Am. Chem. Soc. *104*, 1316

TIETZE, L. F. **1996**, Chem. Rev. *96*, 115

TJIVIKUA, T.; BALLESTER, P.; REBEK, J., JR. **1990 A**, J. Am. Chem. Soc. *112*, 1249

TJIVIKUA, T.; DESLONGCHAMPS, G.; REBEK, J., JR. **1990 B**, J. Am. Chem. Soc. *112*, 8408

TODD, D. **1948**, Org. React. (NY) *4*, 378

TOMALIA, D. A.; BAKER, H.; DEWALD, J.; HALL, M.; KALLOS, G.; MARTIN, S.; ROECK, J.; RYDER, J.; SMITH, P. **1985**, Polym. J. (Tokyo) *17*, 117; **1986**, Macromolecules *19*, 2466

TOMALIA, D. A.; HALL, M.; HEDSTRAND, D. M. **1987**, J. Am. Chem. Soc. *109*, 1601

TOMALIA, D. A.; NAYLOR, A. M.; GODDARD, W. A., **1990**, Angew. Chem. *102*, 119 (Int. Ed. Engl. *29*, 113)

TOMOHIRO, Y.; SATAKE, A.; KOBUKE, Y. **2001**, J. Org. Chem. *66*, 8442

TORII, S.; OKUMOTO, H.; AKAHOSHI, F.; KOTANI, T. **1989**, J. Am. Chem. Soc. *111*, 8932

TOSHIMA, K.; TATSUTA, K. **1993**, Chem. Rev. *93*, 1503

TRACHTENBERG, E. N. **1969**, in: AUGUSTINE, R. L. (ed.) *Oxidation*, Vol. 1, chapter 3, p. 125, M. Dekker: New York

TRAHANOVSKY, W. S. (ed.) **1973**, **1978**, *Oxidation in Organic Chemistry*, Parts B & C, Academic Press: New York London

TRAN, Y.; AUROY, P. **2001**, J. Am. Chem. Soc. *123*, 3644

TRAYNELIS, V. J.; HERGENROTHER, W. L.; HANSON, H. T.; VALICENTI, J. A. **1964**, J. Org. Chem. *29*, 123

TRAYNELIS, V. J.; HERGENROTHER, W. L.; LIVINGSTON, J. R.; VALICENTI, J. A. **1962**, J. Org. Chem. *27*, 2377

TREIBS, A. **1971**, *Das Leben und Wirken von Hans Fischer*, Hans-Fischer-Gesellschaft: Munich

TREMBLAY, M. R.; POIRIER, D. **2000**, J. Comb. Chem. *2*, 48

TROST, B. M. **1986**, Angew. Chem. *98*, 1 (Int. Ed. Eng. *25*, 1)

TROST, B.M.; BOGDANOWICZ, M.J. **1973**, J. Am. Chem. Soc. *95*, 289, 5311

TROST, B.M.; EDSTROM, E.D.; CARTER-PETIL-LO, M.B. **1989B**, L. Org. Chem. *54*, 4489

TROST, B.M.; KEINAN, E. **1978**, J. Am. Chem. Soc. *100*, 7779

TROST, B.M.; LEE, D.C. **1989A**, J. Org. Chem. *54*, 2271

TROST, B.M.; MELVIN, L.S., JR. **1975A**, *Sulfur Ylides*, Academic Press: New York

TROST, B.M.; PRECKEL, M.; LEICHTER, L.M. **1975B**, J. Am. Chem. Soc. *97*, 2224

TROST, B.M.; STREGE, P.E. **1974**, Tetr. Lett., 2603

TROST, B.M.; VERHOEVEN, T.R. **1976**, J. Am. Chem. Soc. *98*, 630

TRUEX, T.J.; HOHN, R.H. **1972**, J. Am. Chem. Soc. *94*, 4529

TRUPP, B.; FRITZ, H.; PRINZBACH, H. **1989**, Angew. Chem. *101*, 1381 (Int. Ed. Engl. *28*, 1345)

TSUNODA, T.; YAMAMIYA, Y.; KAWAMURE, Y.; ITO, S. **1995**, Tetr. Lett. *36*, 2529

TURRO, N.L.; HAMMOND, W.B. **1966**, J. Am. Chem. Soc. *88*, 3672

TYRLIK, S.; WOLOCHOWICZ, I. **1973**, Bull. Soc. Chim. Fr., 2147

ULRICH, H. (ed.) **1967**, *Cycloaddition Reactions of Heterocumulenes*, Academic Press: New York London

UMEMOTO, K.; YAMAGUCHI, K.; FUJITA, M. **2000**, J. Am. Chem. Soc. *122*, 7150

UMINO, N.; IWAKUMA, T.; ITOH, N. **1976**, Tetr. Lett., 763

UPASANI, R.B.; YANG, K.C.; ACOSTA-BURRUEL, M.; KONKOY, C.S.; MCLELLAN, J.A.; WOODWARD, R.M.; LAN, N.C.; CARTER, R.B.; HAWKONSON, J.E. **1997**, J. Med. Chem. *40*, 73

UTIMOTO, K.; TANAKA, T.; FURUBAYASHI, T.; NOZAKI, H. **1973**, Tetr. Lett., 787

VALENTINE, D., JR.; SCOTT, J.W. **1978**, Synthesis, 329

VAN BEKKUM, H.; RÖPER, H.; VORAGEN, A. **1996**, *Carbohydrates as Organic Raw Materials III*; VCH: Weinheim

VAN DAEHNE, W.; FREDERIKSEN, E.; GUNDERSEN, E.; LUND, F.; MORCH, P.; PETERSEN, H.J.; ROHOLT, K.; TYBRING, L.; GODTFREDSEN, W.O. **1970**, J. Med. Chem. *13*, 607

VAN TAMELEN, E.E. **1968**, Acc. Chem. Res. *1*, 111

VAN TAMELEN, E.E.; JAMES, D.R. **1977**, J. Am. Chem. Soc. *99*, 950

VAN TAMELEN, E.E.; PLACEWAY, C.; SCHIEMENZ, G.P.; WRIGHT, I.G. **1969**, J. Am. Chem. Soc. *91*, 7359

VAN TAMELEN, E.E.; SPENCER, T.A.; ALLEN, D.S.; ORVIS, R.L. **1961**, Tetrahedron *14*, 8

VEDEJS, E. **1975**, Org. React. (NY) *22*, 401

VEDEJS, E.; ENGLER, D.A.; TEISCHOW, J.E. **1978**, J. Org. Chem. *43*, 188

VEDEJS, E.; MARTH, C.F. **1988B**, J. Am. Chem. Soc. *110*, 3948

VEDEJS, E.; MARTH, C.F. **1990**, J. Am. Chem. Soc. *112*, 3905

VEDEJS, E.; MARTH, C.F.; RUGGERI, R. **1988A**, J. Am. Chem. Soc. *110*, 3940

VERNON, L.P.; SEELY, G.R. (eds.) **1966**, *The Chlorophylls*, Academic Press: New York London

VILLACORTA, G.M.; RAO, C.P.; LIPPARD, S.J. **1988**, J. Am. Chem. Soc. *110*, 3175

VIPPAGUNTA, S.R.; DORN, A.; MATILE, H.; BHATTACHARJEE, A.K.; KARLE, J.M.; ELLIS, W.Y.; RIDLEY, R.G.; VENNERSTROM, J.L. **1999**, J. Med. Chem. *42*, 4630

VOEGTLE, F. **1991**, *Supramolecular Chemistry*, Wiley: New York London

VOGEL, E.; BISKUP, M.; PRETZER, W.; BÖLL, W.A. **1964**, Angew. Chem. *76*, 785 (Int. Ed. Engl. *3*, 642)

VOGEL, E.; FELDMANN, R.; DÜWEL, H. **1970**, Tetrahedron Lett. 1941

VOGEL, E.; SOMBROEK, J.; WAGEMANN, W. **1975**, Angew. Chem. *87*, 591 (Int. Ed. Engl. *14*, 564)

VREEKAMP, R.H.; VAN DUYNHOVEN, J.P.M.; HUBERT, M.; VERBOOM, W.; REINHOUDT, D.N. **1996**, Angew. Chem., Int. Ed. Engl. *35*, 1215

WACKETT, L.P., HERSHBERGER, C.D. **2001**, *Biocatalysis and Biodegradation: Microbial Transformation of Organic Compounds*, Amer. Society for Microbiology

WADSWORTH, W.S. **1977**, Org. React. (NY) *25*, 73

WAGEMANN, W.; LYODA, M.; DEGER, H.M.; SOMBROEK, J.; VOGEL, E. **1978**, Angew. Chem. *90*, 988 (Int. Ed. Engl. *17*, 956)

WALDMANN, H. **1992**, Nachr. Chem. Tech. Lab. *40*, 702

WALKER, B.J. **1972**, *Organophosphorus Chemistry*, Penguin: London

WALKER, D.; HIEBERT, J.D. **1967**, Chem. Rev. *67*, 153

WANG, J.; RAMNARAYAN, K. **1999**, J. Comb. Chem. *1*, 524

WANG, Y.-F.; CHEN, C.-S.; GIRDAUKAS, G.; SIH, C. J. **1984**, J. Am. Chem. Soc. *106*, 3695

WARREN, S. **1978**, *Designing Organic Synthesis – A Programmed Introduction to the Synthon Approach*, Wiley: New York London

WASERMAN, H. H.; GLAZER, E. **1975**, J. Org. Chem. *40*, 1505

WATSON, C. **1999**, Angew. Chem. *111*, 2025

WEAVER, J. H.; CHAI, Y.; KROLL, G. H.; JIN, C.; OHNO, T. R.; HAUFLER, R. E.; GUO, T.; ALFORD, J. M.; CONCEICAO, J.; CHIBANTE, L. P. F.; JAIN, A.; PALMER, G.; SMALLEY, R. E. **1992**, Chem. Phys. Lett. *190*, 460

WEGMANN, B.; SCHMIDT, R. R. **1988**, Carbohydr. Res. *184*, 254

WEISSERMEL, K.; ARPE, H.-J. **1978**, *Industrielle organische Chemie*, 2nd edn., Verlag Chemie: Weinheim, Ger.

WELCH, S. C.; CHAYABUNJONGLERD, S. **1979**, J. Am. Chem. Soc. *101*, 6768

WENDEBORN, S.; DE MESMAEKER, A.; BRILL, W. K.-D.; BERTEINA, S. **2000**, Acc. Chem. Res. *33*, 215

WENDISCH, D. **1971**, in: HOUBEN-WEYL, *Methoden der Organischen Chemie*, Vol. IV/3, p. 126 ff., Thieme: Stuttgart

WENGEL, J. **1999**, Acc. Chem. Res. *32*, 301

WENKERT, E.; MUELLER, R. A.; REARDON, E. J.; SATHE, S. S.; SCHARF, D. J.; TOSI, G. **1970**, J. Am. Chem. Soc. *92*, 7428

WENKERT, E.; YODER, J. E. **1970 B**, J. Org. Chem. *35*, 2985

WEST, R.; GLAZE, W. H. **1961**, J. Org. Chem. *26*, 2096

WHALEY, W. M. **1951 A, B**, Org. React. (NY) *6*, 74, 151

WHARTON, P. S.; HIEGEL, G. A. **1965**, J. Org. Chem. *30*, 3254

WHITESIDES, G. M.; FISCHER, W. F.; SAN FILIPPO, J.; BASHE, R. W.; HOUSE, H. O. **1969**, J. Am. Chem. Soc. *91*, 4871

WIBERG, K. B. (ed.) **1965**, *Oxidation in Organic Chemistry*, Part A, Academic Press: New York London

WIBERG, K. B.; HAMMER, J. D.; CASTEJON, H.; BAILEY, W. F.; DeLEON, E. L.; JARRAT, R. M. **1999**, J. Org. Chem. *64*, 2085

WIESNER, K.; MUSIL, V.; WIESNER, K. J. **1968**, Tetrahedron Lett., 5643

WIESNER, K.; VALENTA, Z.; AYER, W. A.; FOWLER, L. R.; FRANCIS, J. E. **1958**, Tetrahedron *4*, 87

WIGFIELD, D. C.; TAYMAZ, K. **1973**, Tetrahedron Lett. 4841

WILLIAMS, I. D.; PEDERSEN, S. F.; SHARPLESS, K. B.; LIPPARD, S. J. **1984**, J. Am. Chem. Soc. *106*, 6430

WILLIAMS, J. M.; RICHARDSON, A. C. **1967**, Tetrahedron *23*, 1369

WILLIAMS, N. R. **1970**, Adv. Carbohydr. Chem. Biochem. *25*, 109

WILLIAMS, R. M.; GLINKA, T.; KWAST, E.; COFFMAN, H.; STILLE, J. K. **1990**, J. Am. Chem. Soc. *112*, 808

WILLIAMS, T. M.; BERGMAN, J. M.; BRASHEAR, K.; BRESLIN, M. J.; DINSMORE, C. J.; HUTCHINSON, J. H.; MacTOUGH, S. C.; STUMP, C. A.; WEI, D. D.; ZARTMAN, C. B.; BOGUSKY, M. J.; CULBERSON, J. C.; BUSER-DOEPNER, C.; DAVIDE, J.; GREENBERG, I. B.; HAMILTON, K. A.; KOBLAN, K. S.; KOHL, N. E.; LIU, D.; LOBELL, R. B.; MOSSER, S. D.; O'NEILL, T. J.; RANDS, E.; SCHABER, M. D.; WILSON, F.; SENDERAK, E.; MOTZEL, S. L.; GOBBS, J. B.; GRAHAM, S. L.; HEIMBROOK, D. C.; HARTMAN, G. D.; OLIFF, A. I.; HUFF, J. R. **1999**, J. Med. Chem. *42*, 3779

WILLIAMSON, R. **1979–1981**, *Genetic Engineering*, Vol. 1–3, Academic Press: New York

WILSON, C. V. **1957**, Org. React. (NY) *9*, 332, 350, 380

WILSON, G. E., JR.; HUANG, M. G.; SCHLOMANN, W. W., JR. **1968**, J. Org. Chem. *33*, 2133

WILSON, M. D.; FERGUSON, G. S.; WHITESIDES, G. M. **1990**, J. Am. Chem. Soc. *112*, 1244

WINDRIDGE, G. C.; JORGENSEN, E. C. **1971**, J. Am. Chem. Soc. *93*, 6318

WINKLER, J. D. **1996**, Chem. Rev. *96*, 167

WINTERFELDT, E. **1975**, Synthesis 617

WIRTH, D. D.; MILLER, M. S.; BOINI, S. K.; KOENIG, T. M. **2000**, Organic Process Research & Development *4*, 513

WITTIG, G. **1980**, Science *210*, 600; Angew. Chem. *92*, 671

WITTIG, G.; DAVIS, P.; KOENIG, G. **1951**, Chem. Ber. *84*, 627

WITTIG, G.; POHMER, L. **1956**, Chem. Ber. *89*, 1334

WITTIG, G.; REIFF, H. **1968**, Angew. Chem. *80*, 8 (Int. Ed. Engl. *7*, 7)

WOLF, F. J.; WEIJLARD, J. **1963**, Org. Synth. Coll. Vol. *IV*, 124

WOLFF, M. E.; KERWIN, J. F.; OWINGS, F. F.; LEWIS, B. B.; BLANK, B. **1963**, J. Org. Chem. *28*, 2729

WOLFROM, M. L.; BHAT, H. B. **1967**, J. Org. Chem. *32*, 1821

WOLFROM, M. L.; THOMPSON, A. **1963**, in: WHISTLER, R. L.; WOLFROM, M. L., BEMILLER, J. N. (eds.) *Methods in Carbohydrate Chemistry*, Vol. 2, p. 211, Academic Press: New York London

WOLINSKY, J.; CHAN, D. **1965**, J. Org. Chem. *30*, 41

WOLKENBERG, S. E.; BOGER, D. L. **2002**, Chem. Rev. *102*, 2477

WONG, C.-H.; HENDRIX, M.; MANNING, D.; ROSENBOHM, C. K.; GREENBERG, W. A. **1998**, J. Am. Chem. Soc. *120*, 8319

WOODWARD, R. B. **1960**, Angew. Chem. *72*, 651

WOODWARD, R. B. **1961**, Pure Appl. Chem. *2*, 383; see also: WOODWARD, R. B.; AYER, W. A.; BEATON, J. M.; BICKELHAUPT, E.; BONNETT, R.; BUCHSCHACHER, P.; CLOSS, G. L.; DUTLER, H.; HANNAH, L.; HAUCK, F. P.; ITÖ, S.; LANGEMANN, A.; LEGOFF, E.; LEIMGRUBER, W.; LWOWSKI, W.; SAUER, J.; VALENTA, Z.; VOLZ, H. **1960**, J. Am. Chem. Soc. *82*, 3800

WOODWARD, R. B. **1967**, Spec. Publ. Chem. Soc. *21*, 217

WOODWARD, R. B. **1977**, Spec. Publ. Chem. Soc. *28*, 167

WOODWARD, R. B. et al. **1990**, Tetrahedron *46*, 7599

WOODWARD, R. B., LOGUSCH, E.; NAMBIAR, K. P.; SAKAN, K.; WARD, D. E.; AU-YEUNG, B.-W.; BALARAM, P.; BROWNE, L. J.; CARD, P. J.; CHEN, C. H.; CHENEVERT, R. B.; FLIRI, A.; FROBEL, K.; GAIS, H. J.; GARRATT, D. G.; HAYAKAWA, K.; HEGGIE, W.; HESSON, D. P.; HOPPE, D.; HOPPE, L.; HYATT, J. A.; IKEDA, D.; JACOBI, P. A.; KIM, K. S.; KOBUKE, Y.; KOJIMA, K.; KROWICKI, K.; LEE, V. J.; LEUTERT, T.; MALCHENKO, S.; MARTENS, J.; MATTHEWS, R. S.; ONG, B. S.; PRESS, J. B.; RAJAN BABU, T. V.; ROUSSEAU, G.; SAUTER, H. M.; SUZUKI, M.; TATSMA, K.; TOLBERT, L. M.; TRUESDALE, E. A.; UCHIDA, L.; UEDA, Y.; UYEHARA, T.; VASELLA, A. T.; VLADUCHICK, W. C.; WADE, P. A.; WILLIAMS, R. M.; WONG,

H. N.-C. **1981**, J. Am. Chem. Soc. *103*, 3210, 3213, 3215

WOODWARD, R. B.; BADER, F. E.; BICKEL, H.; FREY, A. J.; KIERSTEAD, R. W. **1958**, Tetrahedron *2*, 1

WOODWARD, R. B.; CAVA, M. P.; OLLIS, W. D.; HUNGER, A.; DAENIKER, H. U.; SCHENKER, K. **1963**, Tetrahedron *19*, 247

WOODWARD, R. B.; PACHTER, I. J.; SCHEINBAUM, M. L. **1971**, J. Org. Chem. *36*, 1137

WOODWARD, R. B.; PATCHETT, A. A.; BARTON, D. H. R.; IVES, D. A. J.; KELLY, R. B. **1957**, J. Chem. Soc., 1131

WOODWARD, R. B.; SONDHEIMER, F.; TAUB, D.; HEUSLER, K.; MCLAMORE, W. M. **1952**, J. Am. Chem. Soc. 74, 4223

WOODWORTH, C. W.; BUSS, V.; SCHLEYER, P. V. R. **1968**, Chem. Commun. 569

XU, R.; GREIVELDINGER, G.; MARENUS, L. E.; COOPER, A.; ELLMAN, J. A. **1999**, J. Am. Chem. Soc. *121*, 4898

YAKELIS, N. A.; ROUSH, W. R. **2001**, Org. Lett. *3*, 957

YAMADA, M.; YURA, T.; MORIMOTO, M.; HARADA, T.; YAMADA, K.; HONMA, Y.; KINOSHITA, M.; SUGIURA, M. **1996**, J. Med. Chem. *39*, 596

YAMAGUCHI, M.; HIRAO, L. **1983**, Tetrahedron Lett. *1983*, 391

YAMANOI, Y.; SAKAMOTO, Y.; KUSUKAWA, T.; FUJITA, M.; SAKAMOTO, S.; YAMAGUCHI, K. **2001**, J. Am. Chem. Soc. *123*, 980

YAMAURA, Y.; HYAKUTAKE, M.; MORI, M. **1997**, J. Am. Chem. Soc. *119*, 7615

YANG, H.; FOSTER, K.; STEPHENSON, C. R. J.; BROWN, W.; ROBERTS, E. **2000**, Org. Lett. *2*, 2177

YANG, M.; WANG, X.; LI, H.; LIVANT, P. **2001**, J. Org. Chem. *66*, 6729

YANG, N. C.; SHANI, A.; LENZ, G. R. **1966**, J. Am. Chem. Soc. *88*, 5369

YOHOYAMA, M.; TOYOSHIMA, A.; AKIBA, T.; TOGO, H. **1994**, Chem. Lett., 265

YOUNGBLOOD, W. J.; GRYKE, D. T.; LAMMI, R. K.; BOCIAN, D. F.; HOLTEN, LD.; LINDSEY, J. S. YU, W.; JIN, Z. **2001**, J. Am. Chem. Soc. *123*, 3369

ZAHN, H.; SCHMIDT, G. **1970**, Liebigs Ann. Chem. *731*, 91, 101

ZARAGOZA, F.; STEPHENSEN, H. **2001**, J. Org. Chem. *66*, 2518

ZHANG, Y.; GONG, X.; ZHANG, H.; LAROCK, R.C.; YEUNG, E.S. **2000**, J. Comb. Chem. *2*, 450

ZHANG, Z.; OLLMANN, I.R.; YE, X.-S.; WISCH-NAT, R.; BAASOV, T.; WONG, C.-H. **1999**, J. Am. Chem. Soc. *121*, 734

ZHAO, D.; MOORE, J.S. **2002**, J. Org. Chem. *67*, 3548

ZHAO, M.; LI, J.; MANO, E.; SONG, Z.; TSCHAEN, D.M.; GRABOWSKI, E.J.J.; REIDER P.J. **1999**, J. Org. Chem. *64*, 2564

ZIEGLER, K.; KRUPP, E.; ZOSEL, K. **1960**, Liebigs Ann. Chem. *629*, 241

ZIMMERMAN, H.E.; GRUNEWALD, G.L.; PAUFLER, R.M.; SHERWIN, M.A. **1969A**, J. Am. Chem. Soc. *91*, 2330

ZIMMERMAN, H.E.; BINKLEY, R.W.; GIVENS, R.S.; GRUNEWALD, G.L.; SHERWIN, M.A. **1969B**, J. Am. Chem. Soc. *91*, 3316

ZIMMERMAN, H.E.; IWAMURA, H. **1970**, J. Am. Chem. Soc. *92*, 2015

ZIMMERMAN, H.E.; IWAMURA, H. **1968**, J. Am. Chem. Soc. *90*, 4763

ZOLLER, M.L.; SMITH, M. **1983**, Methods Enzymol. *100* [Recomb. DNA, Pt. B], 468

ZORBACH, W.W.; TIO, C.O. **1961**, L. Org. Chem. *26*, 3543

ZORETIC, P.A.; FANG, H. **1998**, J. Org. Chem. *63*, 7213

ZURFLÜH, R.; WALL, E.N.; SIDDALL, J.B.; EDWARDS, J.A. **1968**, J. Am. Chem. Soc. *90*, 6224

ZWEIFEL, G.; ARZOUMANIAN, H.; WHITNEY, C.C. **1967**, J. Am. Chem. Soc. *89*, 3652, 5086

ZWEIFEL, G.; BROWN, H.C. **1963**, Org. React. (NY) *13*, 1

ZWEIFEL, G.; POLSTON, N.L.; WHITNEY, C.C. **1968**, J. Am. Chem. *90*, 6243

Subject Index